ATLANTIS STUDIES IN MATHEMATICS FOR ENGINEERING AND SCIENCE

VOLUME 8

SERIES EDITOR: **C.K. CHUI**

Atlantis Studies in Mathematics for Engineering and Science

Series Editor:

C. K. Chui, Stanford University, USA

(ISSN: 1875-7642)

Aims and scope of the series

The series 'Atlantis Studies in Mathematics for Engineering and Science' (AMES) publishes high quality monographs in applied mathematics, computational mathematics, and statistics that have the potential to make a significant impact on the advancement of engineering and science on the one hand, and economics and commerce on the other. We welcome submission of book proposals and manuscripts from mathematical scientists worldwide who share our vision of mathematics as the engine of progress in the disciplines mentioned above.

All books in this series are co-published with World Scientific.

For more information on this series and our other book series, please visit our website at:

www.atlantis-press.com/publications/books

ATLANTIS PRESS

AMSTERDAM – PARIS

Nonlinear Hybrid Continuous/Discrete-Time Models

Marat Akhmet

Middle East Technical University, Ankara, Turkey

ATLANTIS PRESS

AMSTERDAM – PARIS

Atlantis Press

8, square des Bouleaux
75019 Paris, France

For information on all Atlantis Press publications, visit our website at: *www.atlantis-press.com*

Atlantis Studies in Mathematics for Engineering and Science
Volume 1: Continued Fractions: Volume 1: Convergence Theory – L. Lorentzen, H. Waadeland
Volume 2: Mean Field Theories and Dual Variation – T. Suzuki
Volume 3: The Hybrid Grand Unified Theory – V. Lakshmikantham, E. Escultura, S. Leela
Volume 4: The Wavelet Transform – R.S. Pathak
Volume 5: Theory of Causal Differential Equations – V. Lakshmikantham, S. Leela, Z. Drici
Volume 6: The Omega Problem of all Members of the United Nations – E.N. Chukwu
Volume 7: Boundary Element Methods with Applications to Nonlinear Problems – G. Chen, J. Zhou

ISBN 978-94-91216-02-2
ISBN 978-94-91216-03-9 (eBook)
ISSN: 1875-7642

To my family.

Preface

The dynamical systems theory was shaped by the investigation of stellar and planetary motions. The applications were later extended to isolated mechanisms, electronics and other processes. Analysis was dominated by assumptions of continuity and extension to infinity. At present we face new challenges, such as events with a finite life and models of large dimensions. The processes are interconnected, rather than isolated, and the motions are alternately continuous and discrete. To meet these challenges we need appropriate tools: differential equations with a discontinuous right-hand side, differential equations with impulses and hybrid systems.

Hybrid systems are a recent concept in dynamical systems theory and have important applications. Hybrid dynamical systems have not been defined in a precise way yet. Roughly speaking, we call a model as hybrid if it is a combination of continuous and discrete dynamics. One can say that a system is hybrid if some of the dependent variables satisfy differential, while others - discrete equations. In addition, one may have a variable that satisfies a differential and a discrete equation alternately. More formal and general definitions of hybrid systems are given in [279, 280]. Further information can be found in [55, 80, 96, 97, 101, 128, 145, 194–196, 201, 209, 210, 213, 227, 231, 279, 280, 282, 338].

In the early 80's, K. Cook, S. Busenberg, J. Wiener and S. Shah started developing a new type of differential equations. They called these differential equations with piecewise constant argument. Many interesting results and many applications of this theory have been produced in the last three decades. The existence and uniqueness of solutions, oscillations and stability, integral manifolds and periodic solutions, and numerous other issues have been intensively discussed. Besides the theoretical analysis, various models in biology, mechanics and electronics were introduced through these systems.

The original method of investigation of these equations was based on the reduction to discrete systems. That is, only the values of solutions at moments that are integers or multiples of integers were discussed. Moreover, systems must be *linear* with respect to the values of solutions, if the argument is not deviated. These requirements significantly limit the theoretical depth of investigation, as well as the scope of real world problems that can be modeled using these equations. From the analysis of *MathSciNet*, we found that the number of papers on differential equations with piecewise constant argument published over the last ten and five years is 138 and 69, respectively. This number is 12 in 2009, and 11 in 2010. Thus, it is clear that the rate of publication is quite low and has not changed over the last ten years, and that a new methodological approach is necessary to develop the theory further.

In the Conference on Differential and Difference Equations at the Florida Institute of Technology, 2005, it was proposed [8] to consider non-linear differential equations with a more general type of piecewise constant argument, and equivalent integral equations, as the basis of investigation. In our opinion, this line of thought should provide new direction for the theory. In this book, we will show in detail how this can be done. The reader will see that not only can the scope of problems studied be expanded through this approach, but one can also observe entirely new phenomena and deepen the parallelism with the theory of ordinary differential equations, despite the fact that the systems under discussion are functional differential equations.

Next, a compartmental model of blood pressure distribution is considered. The systemic arterial pressure is assumed to be dominating. The heart contraction moments are supposed to be prescribed, and the cases where they behave periodically and almost periodically are examined.

We also consider the case where the moments are defined recursively, so that chaotic phenomena appear. Additionally, the case where the moments are not prescribed and are variable is investigated.

Finally, a biological model of integrate-and-fire oscillators is examined. The main result is the solution of the synchronization problem. That is, we find conditions that ensure that identical or not quite identical units of the system fire in unison. This problem is of extreme importance for the cardiac pacemaker research. Examples with numerical simulations are provided to validate the theoretical results, and prospects for further investigation are discussed.

The first chapter serves as the introductory part of the book. We present the description of the systems and the main definitions, and outline the literature.

In the following chapter, linear and quasilinear systems with argument functions that are both advanced and delayed, are considered. It is shown that the set of all solutions of the linear homogeneous equation is a finite-dimensional linear space under certain conditions. The fundamental matrix of solutions is built. For quasilinear systems, the integral representation formula is defined. Moreover, basic concepts of stability theory are provided for these equations.

In the third chapter, we prove the reduction principle for differential equations with piecewise constant arguments. Theorems on the existence of integral manifolds, their stability, and the stability of the zero solution are proved.

The fourth chapter is devoted to periodic solutions of perturbed linear systems. The method of the small parameter is used as an instrument in this chapter. Both non-critical and critical cases are investigated. Continuous and differentiable dependence of solutions on initial data is examined to prove the main results. Simulations are provided to illustrate the theoretical analysis.

The stability of nonlinear systems is analyzed in the fifth chapter. We use the Lyapunov-Razumikhin technique, as well as the method of Lyapunov functions, to investigate stability. We apply the obtained theorems to prove the stability of the zero solution of the logistic equation.

Differential equations with state-dependent piecewise constant argument are described in the sixth chapter. We analyze the conditions of existence and uniqueness of the solutions of these systems in a general case, and for quasilinear equations. Then periodic solutions and the stability of the zero solution are discussed. The theory of differential equations with discontinuous right-hand-sides and with variable moments of impulses shows the way to developing the theory of systems with state-dependent piecewise constant argument.

The existence of almost periodic solutions is the main subject of the seventh chapter. Using Bohner type discontinuous almost periodic functions and the technique developed in [117], we prove that the exponential dichotomy is a sufficient condition for the existence of these solutions.

The following chapter deals with the stability of neural networks. The biological explanations of advance and delay arguments are discussed. The methods of Lyapunov functionals and functions are applied to obtain the main results. Several examples are provided to illustrate the theorems.

Blood pressure distribution is investigated in the ninth chapter. A model where the systemic arterial pressure dominates is developed. The existence of periodic, almost periodic motions, stability, and chaotic behavior are examined. We consider both cases where the moments of jumps of the pressure are prescribed, and where they are variable. In the last case the threshold concept is formalized.

The last chapter of the book is devoted to the integrate-and-fire model of biological oscillators. This model is appropriate for cardiac pacemaker analysis, as well as for neural networks research. Mathematical models of integrate-and-fire biological oscillators, as far as we know, were initiated in [260,262]. In this book, a method of analysis of integrate-and-fire models that consist of pulse-coupled biological oscillators is developed. The method is based on a thoroughly constructed map and the technique of investigation of differential equations with discontinuities at non-fixed moments.

Synchronization and existence of periodic motions of identical and not quite identical oscillators are investigated. The second Peskin's conjecture, [262], has been solved. The synchronized regime of identical oscillators admits moments of fire that are equally-distanced. Their discrete dynamics can be represented as a solution of simple differential equations, if needed. Consequently, the model belongs to the class of systems considered in our book.

This book contains very recent results, and any comments and suggestions would be greatly appreciated.

The author would like to thank Duygu Aruğaslan, Cemil Büyükadalı, Mehmet Onur Fen, Mehmet Turan and Enes Yılmaz for discussions of the problems considered in the book and for collaboration on several joint papers.

Contents

Chapter 1

Introduction

It is observed, [280], that there are at least three scientific communities, which contribute to theory of hybrid systems: computer science, modeling and simulation, systems and control communities. The results of our book are about hybrid systems, which are obtained through modeling activity, and they are initiated in analysis of systems which operate in different modes, continuous and discrete. That is, we consider a narrow class of hybrid systems. Let us describe it more precisely. Denote by $\mathbb{N}$, $\mathbb{Z}$ and $\mathbb{R}$ the sets of all natural numbers, integers and real numbers, respectively. The state variable in our book is finite dimensional, $x \in \mathbb{R}^n$, $n \in \mathbb{N}$. Denote by $x(t)$ the position of the state variable, where $t \in \mathbb{R}$ is the time argument. Suppose that there is a discrete set of moments θ_k, $k \in \mathbb{Z}$, which one calls, *switching moments* [279]. We shall call also θ_k, an *event time* [280]. If t is in the continuous part, then $x(t)$ satisfies a differential equation, otherwise the value of jumps of x is evaluated by *jumps equation* [20]. With each event time we associate a *switch* and a *jump*. That is, at each event time an *event* occurs, such that variables x and t jump, and the right-hand-side of the differential equation and the jumps equation switch at the event time. If one specify a hybrid system, then events and switching moments have to be determined. If events are externally induced, then the switching and jumps are *controlled*, otherwise they are autonomous [280]. In chapters 2–8 we analyze hybrid systems without jumps equations. In the last two chapters jumps equations are considered to model blood pressure distribution and biological integrate-and-fire oscillators.

Differential equations which we consider in the next seven chapters belong rather to a new class of hybrid systems. Definitely, they can be classified as differential equations with deviating argument or functional differential equations. Nevertheless, methods of investigations, which we will see in this book are different than those for functional differential equations. In pioneer papers they were reduced to discrete equations. We have found out a subclass of the systems, which is very close to ordinary differential equations in their properties. Thus, one can say that these equations are in a focus of several theories of differential and difference equations. In this part of the book, it can be considered as a comprehensive introduction to the theory of differential equations with piecewise constant argument. We have given a rigorous theoretical background useful for the future analysis and indicated new problems. The results are not only development of a particular theory of differential equations. A bridge is built to the future strong connection between functional differential equations and differential equations with piecewise constant arguments. From another point of view, we find a subclass of differential equations with piecewise constant arguments, which in their properties is very close to ordinary differential equations. We try to extend the class of systems with piecewise constant argument, and simultaneously to restrict it such that the two classes of differential equation: ordinary and functional differential equations can be closer to each other. So, possibly the theory of the triad: 1) ordinary differential equations; 2) differential equations with piecewise constant arguments; 3) functional differential equations, can be developed in more strong connections. This is our idea, and we believe that our proposals may stimulate new investigations advancing the theory and adding to the previous significant achievements.

Systematic study of theoretical and practical problems involving piecewise constant arguments was initiated in the early 80's. Since then, differential equations with piecewise constant arguments have attracted great attention from the researchers in mathematics, biology, engineering and other fields. A mathematical model including piecewise constant argument was first considered by Busenberg and Cooke [65] in 1982. They constructed a first-order linear equation to investigate vertically transmitted diseases. Following this work, using the method of reduction to discrete equations, many authors have analyzed various types of differential equations with piecewise constant argument.

The deep study of such equations has been initiated by Cooke and Wiener [85], Wiener [319], and Shah and Wiener [286], and in each of the areas: existence and uniqueness of solutions, asymptotic behavior, periodic and oscillating solutions, approximation, application to control theory, biomedical models and problems of mathematical physics, there appears to be ample opportunity for extending the known results. For specific references see [2–7], [43–45], [66, 67, 69, 70], [79–85], [93, 94], [110–112], [131–137], [139–144], [169–174], [176, 186], [204–206], [208, 211, 212, 224, 232, 235, 236, 238, 239, 245], [249–256], [258,263,283,284,286,287,291], [302–313], [317–330], [333–336], [340–345].

We should mention that the interest of differential equations with piecewise constant argument was emphasized in the paper "On Certain Problems in the Theory of Differential Equations with Deviating Argument," [238], where A. Myshkis shortly noted that "bordering on difference equations are the impulsive differential equations with deviating argument with impacts and switching, loaded equations (that is, those including values of the unknown solution for given constant values of the argument), equations $x'(t) = f(t,x(t),x(h(t)))$ with lagging arguments of the form $h(t) = [t]$ (that is, having intervals of constancy), etc." In [81], Cooke and Wiener point out that their attention was directed to these equations by the paper. However, the foundation of the theory was lied by K. Cooke, J. Wiener and their coauthors.

Typical differential equations with piecewise constant arguments considered by K. Cooke and his coauthors are of the form

$$\frac{dx(t)}{dt} = f(t,x(t),x([t])), \tag{1.1}$$

or

$$\frac{dx(t)}{dt} = f(t,x(t),x(2[(t+1)/2])), \tag{1.2}$$

where $t \in \mathbb{R}$, $x \in \mathbb{R}^n$, and $[\cdot]$ denotes the greatest integer function. Research of these equations is motivated by the fact that they represent a hybrid of continuous and discrete systems and thus combine the properties of both differential and difference equations.

Equations (1.1) and (1.2) are hybrid continuous/discrete-time systems. Indeed, let us consider (1.1), since equation (1.2) can be described very similarly. It is easily seen that the sequence of integers is the sequence of switching moments, $\theta_k = k$, $k \in \mathbb{Z}$, for the system. If one considers the function $f(t,x,w)$, then the choice of $w = x([t])$ determines the switch of the right-hand-side. The dynamics of the time is continuous as well as discrete. The discrete dynamics is defined by the equation $\theta_{k+1} = \theta_k + 1$, $\theta_0 = 0$. Since the dynamics is provided by the greatest integer function inserted in (1.1), one can say that the switching is autonomous.

Let us observe more carefully the role of differential equations with piecewise constant argument in the theory of differential equations and for applications.

First of all, it should be mentioned that the large amount of results of the early stage is collected in the book [317]. This is the unique book, which contains the theory of differential equations with piecewise constant argument in its main part. One can also find interesting results of the theory in books [133] of K. Gopalsamy and [143] of I. Györi and G. Ladas, where stability and oscillatory solutions are discussed as well as the manuscript [93], where applications are widely presented.

V. Lakshmikhantham and J. Wiener [328] prove existence and uniqueness theorems for the initial value problem

$$x'(t) = f(x(t), x(g(t))), \quad x(0) = x_0, \tag{1.3}$$

where f is a continuous function, and $g : [0, \infty) \to [0, \infty)$, $g(t) \leqslant t$, is a step function, that is, it is constant and equal to $g(t_n)$ on each interval $[t_n, t_{n+1})$, where $\{t_n\}$ is a strictly increasing sequence of real numbers with $\lim_{n\to\infty} t_n = \infty$.

In paper [327] the asymptotic behavior of a second-order differential equation with piecewise constant argument of the form

$$x'' + \omega^2 x(t) = -bx'([t-1]), \tag{1.4}$$

where b and ω are positive constants, was studied. It was found that the last equation may generate periodic or even unbounded solutions whereas all solutions of the corresponding ordinary differential equation $x'' + bx'(t) + \omega^2 x(t) = 0$ tend to zero as $t \to \infty$.

The numerical approximation of differential equations is one of the benefits of differential equations with piecewise constant argument. The Euler scheme for a differential equation $x'(t) = f(x(t))$ has the form $x_{n+1} - x_n = hf(x_n)$, where $x_n = x(nh)$ and h is the step size. One can consider it as the differential equation with piecewise constant argument $x'(t) = f(x([t/h]h))$. In [139] it was realized that equations with piecewise constant arguments can be used to approximate delay differential equations that contain discrete delays. The results were used to compute numerical solutions of ordinary and delay differential equations. Later, in papers [140, 141] the results were extended. See, also [79]. Thus, limit relations between the solutions of differential equations with delay and the solutions of some differential equations with piecewise constant argument have been proved.

Due to the the exceptional role of manifolds in the reduction of the dimensions of equations, it is natural that the exploration of manifolds as well as their properties and neighborhoods, is one of the most interesting problems. In [249–251] the subject was discussed for differential equations with piecewise constant argument.

From the current literature, one can see that the interest on investigation of differential equations with piecewise constant arguments is continuously growing. Let us mention the published papers by grouping them in subjects below.

Exceptionally many papers are about oscillatory solutions of these systems [2, 3], [5–7], [81–83, 85, 115, 121, 135, 136, 144, 172–174, 189, 190, 287, 305, 309, 322].

Advanced results of the theory of stability can be found in numerous articles [81, 131, 133, 211, 254, 256, 258, 291].

As it is well known, the problem of existence of periodic solutions and almost periodic solutions has been one of the most attractive topics in the qualitative theory of ordinary

and functional differential equations since significance for the physical sciences. Authors of [44,161,170,171,283,333,336], [340–345] investigated the existence of almost periodic solutions. Examples of research articles on the existence of quasi-periodic solutions [186], and of periodic solutions [306–308,310,315,318,325] can be found.
Green's function and comparison principles were discussed in [67,245].
One can find that the theoretical potential of the results mentioned above is determined by the method of investigation of differential equations with piecewise constant argument [65,85,286,320]: the reduction of equations to discrete equations. Thus, extension of the theory have been very restricted and the tremendous number of results, known for ordinary differential equations, continuous functional differential equations were left out of analysis. To demonstrate the method of the reduction, we give the following simple example.

Example 1.1. Consider the scalar equation

$$\frac{dx}{dt} = ax([t]), \tag{1.5}$$

where a is a real constant. Solve the equation with the initial condition $x(0) = x_0$, where x_0 is a fixed real number. If $t \in [0,1]$, then the equation has a form

$$\frac{dx}{dt} = ax_0,$$

and, consequently, $x(t) = x_0(1+at), x(1) = x_0(1+a)$. Continue in this way to obtain that

$$x(k+1) = x_0(1+a)^k, k \geqslant 0. \tag{1.6}$$

One must say the analysis of equation (1.5) on the basis of (1.6) is, in fact, the essence of the method of the reduction to discrete equations.
Considering the model (1.6) we can find, for example, that if $a = -2$, then the solution of (1.5) is 2-periodic. Moreover, the zero solution is asymptotically stable if $-2 < a < 0$, and $t_0 = 0$. One must emphasize that existence of solutions of (1.5), which starts at non-integer moments of time, and their stability need additional analysis. Simultaneously, the example shows that introduction of piecewise constant argument is similar to perturbation of equations and it produces oscillations for the Malthus model.

The theoretical importance of the piecewise constant argument was emphasized in [238], but the first differential equation was formed in the paper [65]. It is very important that the system is a response to the application request. Consequently, we may underline that piecewise constant argument problems emerged first of all from the analysis of real world problems. Other than mathematicians, this class of differential equations has attracted the attention of many scientists due to their wide range of applications in biology, mechanics, electronics, control theory, neural networks, biomedical models of disease [25–27, 38, 65, 66, 93, 94, 133, 204, 206, 216, 217, 224, 234–236, 284, 304, 306, 327].

The equations with piecewise constant arguments play an important role in mathematical modeling of biological problems. Busenberg and Cooke [65] constructed a first-order linear differential equation with piecewise constant argument to investigate vertically transmitted diseases. In several papers [45, 131, 133, 134, 204, 206, 235, 236, 283, 284, 304] authors investigated different types of population models based on logistic equations with piecewise constant arguments and obtained mathematical results on oscillations or chaotic behavior. In the paper [94], a direct analytical and numerical method independent of the existing classical methods for solving linear and nonlinear vibration problems was given with the introduction of a piecewise constant argument $[Nt]/N$. A new numerical method which produces sufficiently accurate results with good convergence was introduced. Development of the formula for numerical calculations was based on the original governing differential equations. For the details we refer to the interesting book of L. Dai [93].

Another application of differential equations with piecewise constant argument is the stabilization of hybrid control systems with feedback delay. Some of these systems have been described in [80]. Moreover, Magni and Scattolini [216] considered recently a new model predictive control algorithm for nonlinear systems based on differential equations with piecewise constant argument. The plant under control, the state and control constraints, and the performance index to be minimized are described in continuous time, while the manipulated variables are allowed to change at fixed and uniformly distributed sampling times. In so doing, the optimization is performed with respect to sequences, as in a discrete-time nonlinear model predictive control, but the continuous-time evolution of the system is considered as in a continuous-time nonlinear model predictive control. More precisely, the authors consider the following control system

$$\frac{dx(t)}{dt} = f(t, x(t), u(t)), \quad t \geqslant 0, \quad x(0) = x_0, \tag{1.7}$$

where x and u are n and m dimensional vectors respectively, x is the state, and u is the input. The function f is continuously differentiable in its both arguments $(x, u) \in X \times U$, where X and U are compact sets, both containing the origin as an interior point. If one denotes by T_s a suitable sampling period and $t_k = kT_s, k$ nonnegative integers, the sampling instants, then the investigation goal is to find a "sampled" feedback control law: $u(t) \equiv \kappa(t, x(t_k))$, $\kappa(t, 0) = 0, t \in [t_k, t_{k+1})$, which asymptotically stabilizes the origin of the associated closed-loop system. For more details see [76, 216, 217, 269].

The significance of the equations for practice can be seen from the following result. The authors of [234] investigated the damped loading system subjected to a piecewise constant voltage described by the equation of charge:

$$Lq''(t)+Rq'(t)+C^{-1}q(t)=Aq\left(\frac{[Nt]}{N}\right), \tag{1.8}$$

which was compared with a similar linear loading system governed by the following equation of charge

$$Lq''(t)+Rq'(t)+C^{-1}q(t)=Aq(t). \tag{1.9}$$

They considered, through numerical simulation, the phenomena of sensitivity on the initial data, stability and existence of oscillating solutions.

The ideology of anticipation [38,87,105,274] became very fruitful in applications of many mathematical theories in the last decades. The theory of differential equations is one of the most useful branches of mathematics in this sense. We can state that an anticipatory approach can also stimulate proper mathematical investigations [38,87]. Anticipatory systems have been started to be considered in the context of our research to obtain criteria of existence of oscillatory solutions and stability criteria for these type of models. We suppose that involving an anticipatory assumption in a model means that we take into account not only objective external conditions, but a will, a wish, an anticipation of internal as well of external individuals, too. Anticipatory capability of species, which can be represented by an advance term might be utilized in maintaining the future populations at dynamical states with minimum risk of vanishing. As an evolutionary fact species which might have attained anticipatory capabilities have a higher chance of survival. Thus, we assume that anticipation is a highly realistic assumption for successful species. In our modeling of population dynamics, anticipation means a qualitative kind of prediction enriched by the active moment of decisions made at present real time. Moreover, it seems that complex of factors which are not necessary subjective can be considered as a reason of an anticipation. In various kinds of biological systems, we observe similar phenomena of a preadjusting or thresholds prediction. There are several possibilities to introduce anticipation in models. Those of the most adequate to the real world problems, and effective from the mathematical point of view are models with piecewise constant argument of advanced type.

It is important to note that the equations provide the simplest examples of differential equations capable of displaying chaotic behavior. Let us see the following example.

Example 1.2. Consider the initial value problem

$$\begin{aligned} x'(t) &= (\mu-1)x([t])-\mu x^2([t]),\\ x(0) &= x_0. \end{aligned} \tag{1.10}$$

One can see that for $t \in [n, n+1)$, the corresponding ordinary differential equation is of the form

$$x'(t) = (\mu - 1)x(n) - \mu x^2(n).$$

Then, by integrating the last equation from n to $n+1$, we obtain the relation

$$x(n+1) = \mu x(n)(1 - x(n)), \quad n = 0, 1, \ldots.$$

which is the famous logistic map. Therefore, we conclude that, for example, if $\mu = 3.83$, independent of choice of x_0, the unique solution of Eq. (1.10) exhibits chaos [98].

More complex examples, including application of the theorem 'period three implies chaos' [207] for differential equations with piecewise constant arguments can be found in [133, 189, 212].

In this book, the concept of differential equations with piecewise constant argument has been generalized by considering arbitrary piecewise constant argument. In addition, an integral representation formula has been developed as another approach of investigation, to meet the challenges discussed above. This is significant novelty in the theory, which is beneficial: we do not need additional assumptions on the reduced discrete equations; the new method requires more easily verifiable conditions similar to those for ordinary differential equations. So, investigation of the problems may become less cumbersome if this is applied. Detailed comparison of values of a solution at a point and at the neighbor moment, where the argument function has discontinuity, helps to extend the discussion. Our new approach based on the construction of an equivalent integral equation embraces the existence and uniqueness of solutions, dependence on initial data, and, exceptionally important is those of stability results. We consider the initial value problem in the general form, that is when the initial moment is an arbitrary real number, not necessarily one of the switching moments of the piecewise constant argument, as it was considered by our predecessors.

Let us give a general description of the systems with piecewise constant argument. Since the main peculiarity is the involvement of piecewise constant functions as an arguments, it is reasonable to give at first the description on these functions.

Fix an interval $J \subseteq \mathbb{R}$. Denote by $\theta = \{\theta_i\}$, $\theta \subset J$, a strictly ordered sequence of real numbers such that the set $\mathscr{A}$ of indices i is an interval of $\mathbb{Z}$. Let, also, $\zeta = \{\zeta_i\}$, $i \in \mathscr{A}$, be another sequence of elements of J. We do not impose any restriction on ζ.

We say that a function, which is defined on J, is of the η-type, and denote it $\eta(t)$, if it is equal to ζ_i if $\theta_i \leqslant t < \theta_{i+1}$, $i \in \mathscr{A}$. This is the most general type of argument-functions considered in our book. Specifically, we shall use the following η-type functions.

We consider in our book the following subclasses of η-type functions.

We say that a function is of the β-type, and denote it $\beta(t)$, if $\zeta_i = \theta_i$, $i \in \mathscr{A}$. One can see that the greatest integer function $[t]$, which is equal to the maximal among all integers less than t, is a $\beta(t)$ function with $\theta_i = i$, $i \in \mathbb{Z}$. Similarly, $\beta(t) = 2[t/2]$ if $\theta_i = 2i$, $i \in \mathbb{Z}$. One can see a sketch of the graph of a $\beta(t)$ function in Figure 1.1.

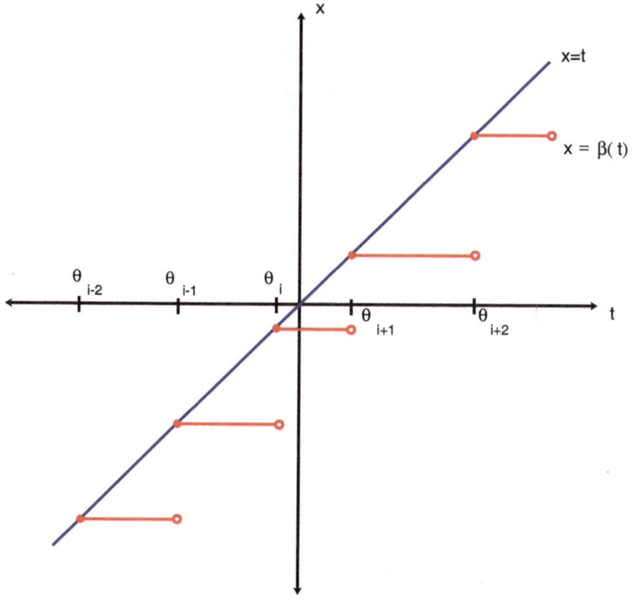

Fig. 1.1 The graph of the argument $\beta(t)$.

We say that a function is of the γ-type, and denote it $\gamma(t)$, if $\theta_i \leqslant \zeta_i \leqslant \theta_{i+1}$ for all $i \in \mathscr{A}$. One can easily find, for example, that $2[\frac{t+1}{2}]$ is $\gamma(t)$ function with $\theta_i = 2i-1$, $\zeta_i = 2i$. In Figure 1.2 the typical graph of $\gamma(t)$ function is seen.

Fix a subset $X \times J$ of $\mathbb{R}^{n+1}$, where X is an open and connected set, J is an interval of $\mathbb{R}$. It is a big interest to investigate systems of the following type

$$x' = f(t, x(t), x(\eta_1(t)), x(\eta_1(t)), \ldots, x(\eta_k(t))), \tag{1.11}$$

where $\eta_i, i = 1, \ldots, k$, are η-type functions. The function $f(t, x, y_1, y_2, \ldots, y_k)$ is assumed to be continuous in all of its arguments. We shall call systems of the form (1.11) as differential equations with piecewise arguments.

Nevertheless, we restrict our interest, in this book, mainly with systems, where the argu-

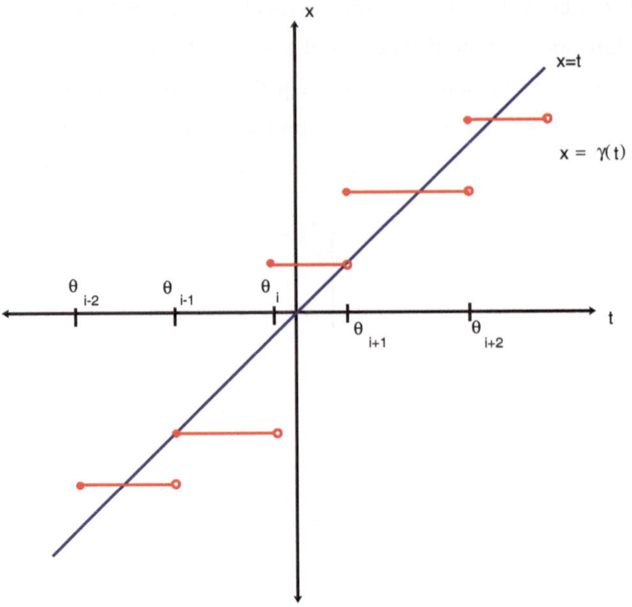

Fig. 1.2 The graph of the argument $\gamma(t)$.

ments are of β or γ type. Thus, the following two systems,

$$x' = f(t, x(t), x(\beta(t))), \tag{1.12}$$

and

$$x' = f(t, x(t), x(\gamma(t))), \tag{1.13}$$

are of the main interest in the book. They are hybrid continuous/discrete-time dynamics systems. Indeed, consider (1.13) as more general one. The sequence of switching moments is θ_k, $k \in \mathbb{Z}$. If we set $f(t,x,w)$, then the switching of the right-hand-side at the moment $t = \theta_k$ is defined by $w = x(\zeta_k)$. Since the sequence of switching moments is prescribed, we have the event time controlled.

Let us clarify that the argument function $\gamma(t)$ is of mixed type. Fix $k \in \mathbb{N}$ and consider the system on the interval $[\theta_k, \theta_{k+1})$. Then, the identification function $\gamma(t)$ is equal to ζ_k. If the argument t satisfies $\theta_k \leqslant t < \zeta_k$, then $\gamma(t) > t$ and (1.13) is an equation with advanced argument. Similarly, if $\zeta_k < t < \theta_{k+1}$, then $\gamma(t) < t$ and, hence, (1.13) is an equation with delayed argument. Consequently, it is worth pointing out that the equation (1.13) is of mixed type.

The subject of an additional interest is the system with piecewise constant arguments of the following form

$$\frac{dx(t)}{dt} = A(t)x(t) + f(t, x(\theta_{\upsilon(t)-p_1}), x(\theta_{\upsilon(t)-p_2}), \ldots, x(\theta_{\upsilon(t)-p_k})), \tag{1.14}$$

where $x \in \mathbb{R}^n$, $t \in \mathbb{R}$, $\upsilon(t) = i$ if $\theta_i \leqslant t < \theta_{i+1}$, $i = \ldots -2, -1, 0, 1, 2, \ldots$, is an identification function, θ_i is a strictly ordered sequence of real numbers, unbounded on the left and on the right, p_j, $j = 1, 2, \ldots, m$, are fixed integers. One can easily see that these equations belong to the more general class of systems (1.11), such that $\eta_j(t) = \theta_{\upsilon(t)-p_j}$, $j = 1, 2, \ldots, k$.
In this book we assume that the solutions of equations (1.12), (1.13) and (1.14) are *continuous functions*. But the argument-functions $\beta(t)$ and $\gamma(t)$ are discontinuous. Hence, being in the process of solution, the right-hand sides of the systems have discontinuities at moments θ_i, $i \in \mathbb{Z}$. Summarizing, we consider the solutions of the equations as continuous functions with discontinuous derivatives, and continuously differentiable within intervals $[\theta_i, \theta_{i+1})$, $i \in \mathscr{A}$.

Definition 1.1. A function $x(t)$ is a solution of a differential equation with piecewise constant arguments on J, if

(i) $x(t)$ is continuous on J;
(ii) the derivative $x'(t)$ exists for $t \in J$ with the possible exception of the points θ_i, $i \in \mathscr{A}$, where one-sided derivatives exist;
(iii) the equation (1.12) ((1.13)) is satisfied by $x(t)$ on each interval (θ_i, θ_{i+1}), $i \in \mathscr{A}$, and it holds for the right derivative of $x(t)$ at the points θ_i, $i \in \mathscr{A}$.

The following example shows that even for simple differential equations with piecewise constant argument solutions not always exist to the left, and moreover, difficulties with uniqueness may appear.

Example 1.3. Consider the following equation

$$x' = 2x - x^2([t]), \tag{1.15}$$

where $x \in \mathbb{R}$, $t \in \mathbb{R}$. Fix $z \in \mathbb{R}$ and let $x(t) = x(t, 1, z)$, $x(1) = z$, be a solution of (1.15). Let us try to define the solution for $t < 1$. It is known that if $t \in [0, 1]$ then $x(t)$ satisfies the equation

$$x' = 2x - x^2(0). \tag{1.16}$$

Denote $x(0) = x_0$, introduce an operator $T : \mathbb{R} \to \mathbb{R}$ such that

$$Tx_0 = \exp(2)x_0 + \int_0^1 \exp(2(1-s))x_0^2 ds.$$

If $x(t)$ exists on $[0,1]$ then the equation $Tx_0 = z$ must be solvable with respect to x_0. But the last equation has a form $[\exp(2)-1]x_0^2 + 2\exp(2)x_0 - 2z = 0$. Since the latter can be solved not for all $z \in \mathbb{R}$, not all solutions $x(t,1,z)$, $z \in \mathbb{R}$, can be proceeded to $t = 0$.

Further we shall consider the uniqueness of solutions continued to the left. Fix numbers $x_0, x_1 \in \mathbb{R}$, $x_1 \neq x_2$, such that $(x_0 + x_1)(1-\exp(2)) = 2\exp(2)$. Denote $x_0(t) = x(t,0,x_0)$ and $x_1(t) = x(t,0,x_1)$, solutions of (1.15). For $t \in [0,1]$, they are solutions of the equations $x' = 2x - x_0^2$ and $x' = 2x - x_1^2$, respectively. One can easily check that $Tx_0 = Tx_1$, that is $x_0(1) = x_1(1)$ and the solution $x(t,1,x_1(1))$ of (1.15) can not be continued to the $t = 0$ uniquely.

In Chapter 9 we investigate a model of the blood pressure distribution. The idea to consider the blood circulation as a discontinuous process is very popular in literature [53, 166, 177]. Denote $P(t)$– the value of systemic arterial pressure at time $t \geqslant 0$. After some simplifications, one can show [166], Chapter 5, that the pressure P satisfies the following two equations

$$\frac{dP(t)}{dt} = -\alpha P(t), \quad \text{if} \quad t \neq t_i, \tag{1.17}$$

where $t_{i+1} - t_i = T,\ \ i \in Z,\ \ T$-positive fixed number, and

$$P(t_i+) = P(t_i-) + I, \tag{1.18}$$

where $I > 0$ is fixed, $P(t_i+)$ and $P(t_i-)$ are values of systemic arterial pressure right after and before the heart contraction at time t_i.

One an rewrite the last two equations in the form of the following system

$$\begin{aligned} &\frac{dP(t)}{dt} = -\alpha P(t), \quad t \neq t_i, \\ &\Delta P|_{t=t_i} = I_i, \end{aligned} \tag{1.19}$$

where $\Delta P|_{t=t_i} = P(t_i+) - P(t_i-)$, if we assume an particular case $I_{i+1} = I_i$, $i \in \mathbb{Z}$. System (1.19) which involves the differential equation and the jumps equation is a system of differential equations with impulses at fixed moments of time.

Beside the system with with fixed moments of the heart contraction, we consider, in the chapter, a model with variable moments of jumps such that discontinuities of the systemic arterial pressure are committed when it achieves a special level $\overline{P} \in \mathbb{R}$, $\overline{P} > 0$. In this case one can suppose that an impulsive differential equation which describes the systemic arterial pressure has the form

$$\begin{aligned} &\frac{dP(t)}{dt} = -\alpha P(t), \quad P \neq \overline{P}, \\ &\Delta P|_{P=\overline{P}} = I, \end{aligned} \tag{1.20}$$

where $P \in \mathbb{R}, P \geqslant \overline{P}$, and $\alpha > 0, I > 0$, are fixed real numbers.

To investigate the blood pressure distribution globally, we propose a prototype compartmental model. Compartmental models have been effectively applied in chemistry [244], medicine [47, 139, 175, 191], epidemiology [237], ecology [223], pharmacokinetics [47,57,271,292].

One can suppose that the total volume of the blood is constant in the time, and consequently all parts of the cardiovascular system should be involved if the blood flow problem is investigated. The pressure could be discussed also locally, as, for example, in [166, 202], where blood pressure for only compartment is considered. In our book we focus on the interactions of several parts of the system, but not all of them. It is assumed that there exists a mutual interaction of blood pressures in different parts of the system. The systemic arterial pressure, should be singled out, as it is disturbed impulsively with large simultaneous inflow of blood in the time of heart contraction. Other parts of the system do not directly undergo the impulsive action, so their change is continuous.

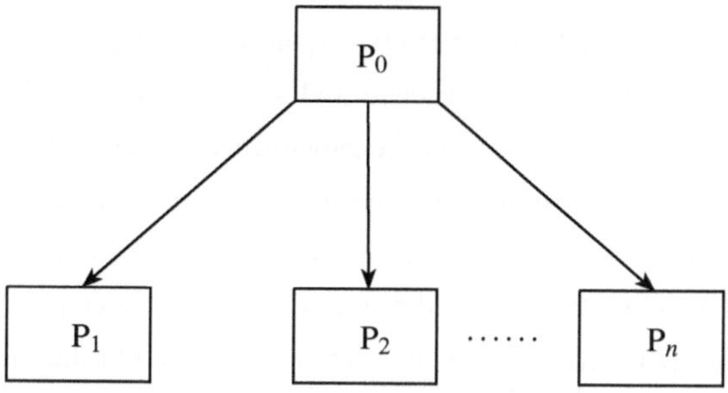

Fig. 1.3 The compartmental model of the blood pressure.

Let C_0 be the aorta and arteries, and P_0 blood pressure in C_0. Moreover, denote C_i, $i = \overline{1,n}$, other vessels and organs, which adjoin to C_0, and have an essential influence on fluctuations of P_0, and P_i, $i = \overline{1,n}$, the blood pressure values in compartments C_i, $i = \overline{1,n}$ (see Figure 1.3). We investigate that the pressure variables satisfy the following system of differential equations with impulses

$$\begin{aligned}
\frac{dP_0(t)}{dt} &= -k_0P_0 - g_0(P_0-P_1,P_0-P_2,\ldots,P_0-P_m),\\
\frac{dP_1(t)}{dt} &= -k_1P_1 + g_1(P_0-P_1),\\
\frac{dP_2(t)}{dt} &= -k_2P_2 + g_2(P_0-P_2),\\
&\ldots\ldots\\
\frac{dP_n(t)}{dt} &= -k_mP_m + g_m(P_0-P_m),\\
\Delta P_0|_{t=\theta_i} &= I_0 + J_0(P_0),
\end{aligned} \tag{1.21}$$

where $\Delta P_0|_{t=\theta_i} \equiv P_0(\theta_i+) - P_0(\theta_i), P_0(\theta_i+) = \lim_{t\to\theta_i+} P_0(t)$.

The theory of impulsive differential equations have been developed very intensively over the last decades (see [8, 19, 20, 36, 41, 42, 113, 119, 192, 193, 197, 278, 334] and papers cited there). There have been many studies that consider application of the theory to the fields of biology, mechanics and medicine; some examples of these works include [20, 192, 278]. Exponential decay of systemic arterial pressure in time starting from the systolic pressure of one heart beat towards the diastolic pressure, and a sudden jump from the diastolic pressure to the systolic pressure of the next cycle makes systemic arterial pressure a perfect candidate application for impulsive differential equations theory.

From a mathematical perspective, existence and stability of periodic and almost periodic solutions in our manuscript can be considered as special cases of general results from [41, 42, 278], however, periodicity of solutions for equations with non fixed impulses, positiveness of solutions, ε-oscillatory behavior of systemic arterial pressure in connection with periodic as well with almost periodic solutions, and its asymptotic properties, have been studied for the first time in this book (see, also [32]). On the other hand, application of the theory of impulsive differential equations to the modeling of blood pressure has been investigated for the first time in all cases. The continuous/discrete dynamics of time presents in this chapter either through controlled switching moments explicitly or as settled dynamics, first of all periodic, implicitly. It is very important that the discrete dynamics of time can be chaotic with properly chosen discrete equations. Another model of heart beat using circle maps has been considered in [50, 124].

Last chapter of the book is devoted to analysis of integrate-and-fire biological models. The basic model under discussion is the model of the cardiac pacemaker. The cells that create rhythmical impulses for contraction of cardiac muscle, and control the heart rate, are called pacemaker cells. C. Peskin developed the model of a cell interacted with a stimulus [181] to a population of identical pulse-coupled oscillators [262]. Synchronization of the system, assumed as firing in unison was proved for two [262] and more than two [216] identical oscillators. The last paper has been the most stimulating intensive analysis of the problem [60, 109, 122, 188, 222, 285, 296, 298, 299]. In [24], we propose a method of investigation for pulse-coupled integrate-and-fire biological oscillators of the general type. It is efficient if they are not quite identical. The second Peskin's conjecture, [262], has been solved. Sufficient conditions for synchronization are found. The present research utilizes results and proposals from [62, 95, 106, 107, 129, 181, 262, 331, 332].

The main object of the chapter is an integrate-and-fire model of n oscillators, x_i, $i = 1,2,\ldots,n$. They satisfy the following equations

$$x_i' = f(x_i) + \phi_i(x). \tag{1.22}$$

The domain of the model consists of all points $x = (x_1, x_2, \ldots, x_n)$ such that $0 \leqslant x_i \leqslant 1 + \zeta_i(x)$ for all $i = 1,2,\ldots,n$. If the oscillator x_j increases from zero, and meets the surface such that $x_j(t) = 1 + \zeta_j(x(t))$, then it fires, $x_j(t+) = 0$. This firing changes the values of all oscillators with $i \neq j$,

$$x_i(t+) = \begin{cases} 0 & \text{if } x_i(t) + \varepsilon + \varepsilon_i \geqslant 1 + \zeta_i(x), \\ x_i(t) + \varepsilon + \varepsilon_i & \text{otherwise.} \end{cases} \tag{1.23}$$

The coupling in the model is all-to-all. That is, each firing elicits jumps in all non-firing oscillators. If several oscillators fire simultaneously, then other oscillators react as if just one oscillator fires. In other words, the intensity of the signal is not important, and pulse strengths are not additive. The discrete dynamics of time is defined for synchronous regimes of the models. Exceptionally it is valid for identical oscillators.

Chapter 2

Linear and quasi-linear systems with piecewise constant argument

In this chapter we start investigation with the most simple linear systems of differential equations with piecewise constant argument. Then the analysis will be extended to quasi-linear systems. Existence-uniqueness of solutions, the linear space of solutions, fundamental matrix, stability problems are under discussion.

We consider the following two equations

$$z'(t) = A_0(t)z(t) + A_1(t)z(\gamma(t)), \tag{2.1}$$

and

$$z'(t) = A_0(t)z(t) + A_1(t)z(\gamma(t)) + f(t, z(t), z(\gamma(t))), \tag{2.2}$$

where $z \in \mathbb{R}^n$, $t \in \mathbb{R}$. The argument-function γ was introduced in the last chapter. However, in this chapter, it is considered with $J = \mathbb{R}$ and $\mathscr{A} = \mathbb{Z}$.

The equation (2.1) is a linear homogeneous system with argument-function $\gamma(t)$, and equation (2.2) is a quasilinear system.

The following assumptions will be needed throughout this chapter:

(C1) $A_0, A_1 \in C(\mathbb{R})$ are $n \times n$ real valued matrices;

(C2) $f(t,x,y) \in C(\mathbb{R} \times \mathbb{R}^n \times \mathbb{R}^n)$ is an $n \times 1$ real valued function;

(C3) $f(t,x,y)$ satisfies the condition

$$\|f(t,x_1,y_1) - f(t,x_2,y_2)\| \leqslant L(\|x_1 - x_2\| + \|y_1 - y_2\|), \tag{2.3}$$

for some positive constant L, and satisfies the condition

$$f(t,0,0) = 0, \quad t \in \mathbb{R}; \tag{2.4}$$

(C4) matrices A_0, A_1 are uniformly bounded on $\mathbb{R}$;

(C5) $\inf_{\mathbb{R}} \|A_1(t)\| > 0$;

(C6) there exists a number $\overline{\theta} > 0$ such that $\theta_{i+1} - \theta_i \leqslant \overline{\theta}, i \in \mathbb{Z}$;

(C7) there exists a number $\theta > 0$ such that $\theta_{i+1} - \theta_i \geqslant \theta, i \in \mathbb{Z}$;

(C8) there exists a positive real number p such that

$$\lim_{t\to\infty} \frac{i(t_0,t)}{t-t_0} = p$$

uniformly with respect to $t_0 \in \mathbb{R}$, where $i(t_0,t)$ denotes the number of points θ_i in the interval (t_0,t).

Condition (C7) implies immediately that $|\theta_i| \to \infty$ as $|i| \to \infty$. One can easily see that equations (2.1) and (2.2) have the form of functional differential equations

$$z'(t) = A_0(t)z(t) + A_1(t)z(\zeta_i), \tag{2.5}$$

and

$$z'(t) = A_0(t)z(t) + A_1(t)z(\zeta_i) + f(t,z(t),z(\zeta_i)), \tag{2.6}$$

respectively, if $t \in [\theta_i, \theta_{i+1}), i \in \mathbb{Z}$.
That is, these systems have the structure of a continuous dynamical system within the intervals $[\theta_i, \theta_{i+1}), i \in \mathbb{Z}$.
It is useful to specify the Definition 1.1 in the following way.

Definition 2.1. A continuous function $z(t)$ is a solution of (2.1), (2.2) on $\mathbb{R}$ if:

(i) the derivative $z'(t)$ exists at each point $t \in \mathbb{R}$ with the possible exception of the points θ_i, $i \in \mathbb{Z}$, where the one-sided derivatives exist;
(ii) the equation is satisfied for $z(t)$ on each interval (θ_i, θ_{i+1}), $i \in \mathbb{Z}$, and it holds for the right derivative of $z(t)$ at the points θ_i, $i \in \mathbb{Z}$.

Remark 2.1. One can easily see that the last definition is appropriate for differential equations with β type arguments, as they also are γ functions.

2.1 Linear homogeneous systems

Let $\mathscr{I}$ be an $n \times n$ identity matrix. Denote by $X(t,s), X(s,s) = \mathscr{I}, t, s \in \mathbb{R}$, the fundamental matrix of solutions of the system

$$x'(t) = A_0(t)x(t) \tag{2.7}$$

which is associated with systems (2.1) and (2.2). We introduce the following matrix-function

Notation 2.1.

$$M_i(t) = X(t,\zeta_i) + \int_{\zeta_i}^{t} X(t,s)A_1(s)ds, \;\; i \in \mathbb{Z}.$$

This matrix is very useful in what follows. From now on we make the assumption:

(C9) For every fixed $i \in \mathbb{Z}$, $\det[M_i(t)] \neq 0$, $\forall t \in [\theta_i, \theta_{i+1}]$.

Remark 2.2. One can easily see that the last condition is equivalent to the following one: $\det[\mathscr{I} + \int_{\zeta_i}^{t} X(\zeta_i,s)A_1(s)ds] \neq 0$, for all $t \in [\theta_i, \theta_{i+1}]$, $i \in \mathbb{Z}$.

Theorem 2.1. *If condition* (C1) *is fulfilled, then for every* $(t_0,z_0) \in \mathbb{R} \times \mathbb{R}^n$ *there exists a unique solution* $z(t) = z(t,t_0,z_0), z(t_0) = z_0$, *of* (2.5) *in the sense of Definition* 2.1 *if and only if condition* (C9) *is valid.*

Proof. *Sufficiency.* Fix a $(t_0,z_0) \in \mathbb{R} \times \mathbb{R}^n$. Without loss of generality assume that $\theta_i \leqslant \zeta_i < t_0 \leqslant \theta_{i+1}$, for a fixed $i \in \mathbb{Z}$. We consider only the construction of the solution for decreasing t, since forward continuation can be investigated in a similar manner.
Condition (C9) implies that the equation

$$v_i = X(\zeta_i,t_0)z(t_0) + \int_{t_0}^{\zeta_i} X(\zeta_i,s)A_1(s)v_i ds$$

can be solved for v_i uniquely. Indeed, since we have

$$\left[\mathscr{I} + \int_{\zeta_i}^{t_0} X(\zeta_i,s)A_1(s)ds\right] v_i = X(\zeta_i,t_0)z(t_0),$$

multiplying both sides of the last expression by $X(t_0,\zeta_i)$, we obtain

$$v_i = M_i^{-1}(t_0)z(t_0).$$

Define $z(t) : [\theta_i,t_0] \to \mathbb{R}^n$ as the unique solution of the system

$$z'(t) = A_0(t)z(t) + A_1(t)v_i \tag{2.8}$$

with the initial condition $z(t_0) = z_0$. One can easily see that $z(\zeta_i) = v_i$.
Consider, now, the interval $[\theta_{i-1}, \theta_i]$. Again, by condition (C9), the equation

$$v_{i-1} = X(\zeta_{i-1},\theta_i)\psi(\theta_i) + \int_{\theta_i}^{\zeta_{i-1}} X(\zeta_{i-1},s)A_1(s)v_{i-1}ds$$

is uniquely solvable with respect to v_{i-1}. Let $z(t)$ be equal to the solution of the equation

$$z'(t) = A_0(t)z(t) + A_1(t)v_{i-1}, \tag{2.9}$$

on $[\theta_{i-1}, \theta_i]$ with the initial data $(\theta_i, \psi(\theta_i))$. Obviously, the solution exists and is unique, and $z(\zeta_{i-1}) = v_{i-1}$.

Assume that we have defined the solution $z(t)$ on the interval $[\theta_j, t_0], j < i-1$. Then, the equation

$$v_{j-1} = X(\zeta_{j-1}, \theta_j)\psi(\theta_j) + \int_{\theta_j}^{\zeta_{j-1}} X(\zeta_{j-1}, s)A_1(s)v_{j-1}ds$$

is uniquely solvable with respect to v_{j-1}. We assume that $z(t)$ is a solution of the equation

$$z'(t) = A_0(t)z(t) + A_1(t)v_{j-1}, \tag{2.10}$$

on $[\theta_{j-1}, \theta_j]$, with the initial data $(\theta_j, z(\theta_j))$. Consequently, the function $z(t)$ could be continued up to $-\infty$ by induction. One can easily see that $z(t)$ is the unique solution of (2.5) on $(-\infty, t_0]$ by construction.

Necessity. Assume that condition (C9) is not true for some fixed $i \in \mathbb{Z}$ and $\xi \in [\theta_i, \theta_{i+1}]$. That is, $\det M_i(\xi) = 0$. Definitely, $\xi \neq \zeta_i$. If $z(t) = z(t, \xi, z(\xi))$ is a solution, then $z(\xi) = M_i(\xi)z(\zeta_i)$, and $z(\zeta_i)$ could not be defined uniquely. This fact proves the necessity of (C9). The theorem is proved. □

The last theorem is of major importance for this chapter. It arranges the correspondence between the points $(t_0, z_0) \in \mathbb{R} \times \mathbb{R}^n$ and the solutions of (2.1) in the sense of Definition 2.1, and there exists no solution of the equation out of the correspondence. Using this assertion, we can say that the definition of the initial value problem for differential equations with piecewise constant arguments is similar to the problem for ordinary differential equations. Particularly, the dimension of the space of all solutions is n. Hence, the investigation of problems considered in this chapter does not need to be supported by the results of the theory of functional differential equations [133, 150].

The following useful, in some particular cases, assertion is implied by the proof of the last theorem.

Theorem 2.2. *Assume that condition* (C1) *is fulfilled, and a number* $t_0 \in \mathbb{R}$, $\theta_i \leqslant t_0 < \theta_{i+1}$, *is fixed. For every* $z_0 \in \mathbb{R}^n$ *there exists a unique solution* $z(t) = z(t, t_0, z_0)$ *of* (2.1) *in the sense of Definition* 2.1 *such that* $z(t_0) = z_0$ *if and only if* $\det[M_i(t_0)] \neq 0$ *and* $\det[M_j(t)] \neq 0$ *for* $t = \theta_j, \theta_{j+1}, j \in \mathbb{Z}$.

System (2.1) is a differential equation with a deviated argument. That is why it is reasonable to suppose that the initial "interval" must consist of more than one point. The following arguments show that in our case we need only one initial moment. Indeed, assume that (t_0, z_0) is fixed, and $\theta_i \leqslant t_0 < \theta_{i+1}$ for a fixed $i \in \mathbb{Z}$. We suppose that $t_0 \neq \zeta_i$. The solution satisfies, on the interval $[\theta_i, \theta_{i+1}]$, the functional differential equation

$$z'(t) = A_0(t)z + A_1(t)z(\zeta_i). \tag{2.11}$$

Formally, we need the pair of initial points (t_0, z_0) and $(\zeta_i, z(\zeta_i))$ to proceed with the solution. Indeed, since $z_0 = M_i(t_0)z(\zeta_i)$ and the matrix $M_i(t_0)$ is nonsingular, we can say that the initial condition $z(t_0) = z_0$ is sufficient to define the solution.

Theorem 2.1 implies that the set of all solutions of (2.1) is an n-dimensional linear space. Hence, for a fixed $t_0 \in \mathbb{R}$ there exists a fundamental matrix of solutions of (2.1), $Z(t) = Z(t,t_0), Z(t_0,t_0) = I$ such that

$$\frac{dZ}{dt} = A_0(t)Z(t) + A_1(t)Z(\gamma(t)).$$

Let us construct $Z(t)$. Without loss of generality assume that $\theta_i < t_0 < \zeta_i$ for a fixed $i \in \mathbb{Z}$, and define the matrix only for increasing t, as the construction is similar for decreasing t. We first note that $Z(\zeta_i) = M_i^{-1}(t_0)I = M_i^{-1}(t_0)$. Hence, on the interval $[\theta_i, \theta_{i+1}]$, we have $Z(t,t_0) = M_i(t)M_i^{-1}(t_0)$. Therefore

$$Z(\zeta_{i+1}) = M_{i+1}^{-1}(\theta_{i+1})Z(\theta_{i+1}) = M_{i+1}^{-1}(\theta_{i+1})M_i(\theta_{i+1})M_i^{-1}(t_0),$$

and hence for $t \in [\theta_{i+1}, \theta_{i+2}]$, we have

$$Z(t,t_0) = M_{i+1}(t)Z(\zeta_{i+1}) = M_{i+1}(t)M_{i+1}^{-1}(\theta_{i+1})M_i(\theta_{i+1})M_i^{-1}(t_0).$$

One can continue by induction to obtain

$$Z(t) = M_l(t)\left[\prod_{k=l}^{i+1} M_k^{-1}(\theta_k)M_{k-1}(\theta_k)\right]M_i^{-1}(t_0), \tag{2.12}$$

if $t \in [\theta_l, \theta_{l+1}]$, for arbitrary $l > i$.

Similarly, if $\theta_j \leqslant t \leqslant \theta_{j+1} < \cdots < \theta_i \leqslant t_0 \leqslant \theta_{i+1}$, then

$$Z(t) = M_j(t)\left[\prod_{k=j}^{i-1} M_k^{-1}(\theta_{k+1})M_{k+1}(\theta_{k+1})\right]M_i^{-1}(t_0). \tag{2.13}$$

One can easily see that

$$Z(t,s) = Z(t)Z^{-1}(s), \quad t,s \in \mathbb{R}, \tag{2.14}$$

and a solution $z(t)$ of (2.5) with $z(t_0) = z_0$ for $(t_0, z_0) \in \mathbb{R} \times \mathbb{R}^n$, is given by

$$z(t) = Z(t,t_0)z_0, \quad t \in \mathbb{R}. \tag{2.15}$$

Differential inequalities

Let $I \subseteq \mathbb{R}$ be an open interval, a scalar function $f(t,x,y)$ is defined on $\Omega = [t_0, t_0+T] \times I \times I$, $0 < T$. Function f is continuous in t except possibly moments θ_k, $k \in \mathbb{Z}$, where it admits discontinuities of the first kind, and is locally Lipschitzian in the second argument.

Lemma 2.1. *Assume that a continuous function $u(t)$ satisfies*

$$u'(t) \leqslant f(t, u(t), u(\gamma(t))), \tag{2.16}$$

for $t_0 \leqslant t \leqslant t_0 + T$, and a continuous function $v(t)$ is a solution of the equation

$$v'(t) = f(t, v(t), u(\gamma(t))), \quad t_0 \leqslant t \leqslant t_0 + T, \tag{2.17}$$

except possibly moments θ_k, $k \in \mathbb{Z}$. Then $u(t) \leqslant v(t)$, if $u(t_0) \leqslant v(t_0)$.

Proof. Assume that $\theta_i \leqslant t_0 \leqslant \theta_{i+1}$. One can easily verify the assertion, by using Theorem 4.1, [153] on the differential inequality for ordinary differential equations, on intervals $[t_0, \theta_{i+1}]$ and $[\theta_k, \theta_{k+1}]$, $k > 1$, consecutively. □

Example 2.1. Consider a non-negative function $u(t) : \mathbb{R} \to \mathbb{R}$, which is continuous, continuously differentiable, except, possibly, points $\theta_k, k \in \mathbb{Z}$, where the right-side derivative exists. Assume that $u(t)$ satisfies the differential inequality

$$u'(t) \leqslant a(t)u(t) + b(t)u(\gamma(t)), \quad t_0 \leqslant t, \tag{2.18}$$

where a, b are non-negative piecewise continuous functions with discontinuities of the first kind at θ_k, $k \in \mathbb{Z}$. The moment $t_0 \in [\theta_i, \theta_{i+1}]$, $t_0 \leqslant \zeta_i$, is fixed. We assume that

$$\begin{aligned} &\int_{t_0}^{\zeta_i} e^{\int_s^{\zeta_i} a(r)dr} b(s)ds < 1, \\ &\int_{\theta_k}^{\zeta_k} e^{\int_s^{\zeta_k} a(r)dr} b(s)ds < 1, \quad k > i. \end{aligned} \tag{2.19}$$

One can easily verify that these last inequalities relate to conditions of Theorem 2.2. Moreover, consider the linear equation

$$v'(t) = a(t)v(t) + b(t)u(\gamma(t)), \quad t_0 \leqslant t, \tag{2.20}$$

and its solution $v(t), v(t_0) = v_0$.

By the last lemma $u(t)$ satisfies $u(t) \leqslant v(t)$, $t \geqslant t_0$, if $u(t_0) \leqslant v_0$.

If $t \in [t_0, \theta_{i+1}]$, we have that u is not greater than the solution

$$v(t) = e^{\int_{t_0}^t a(s)ds} v_0 + \int_{t_0}^t e^{\int_s^t a(r)dr} b(s)u(\zeta_i)ds,$$

of the equation

$$v'(t) = a(t)v(t) + b(t)u(\zeta_i). \tag{2.21}$$

One can also verify that

$$u(\zeta_i) \leqslant \frac{e^{\int_{t_0}^{\zeta_i} a(s)ds} v_0}{1 - \int_{t_0}^{\zeta_i} e^{\int_s^{\zeta_i} a(r)dr} b(s)ds},$$

and

$$u(t) \leqslant (e^{\int_{t_0}^{t} a(s)ds} + \frac{e^{\int_{t_0}^{\zeta_i} a(s)ds}}{1 - \int_{t_0}^{\zeta_i} e^{\int_s^{\zeta_i} a(r)dr} b(s)ds} \int_{t_0}^{t} e^{\int_s^t a(r)dr} b(s)ds) v_0.$$

Consider the initial moment $t = \theta_{i+1}$, and interval $[\theta_{i+1}, \theta_{i+2}]$. We have that $u(\theta_{i+1}) \leqslant v(\theta_{i+1})$. Consequently, $u(t) \leqslant v(t)$ on the interval. Recursively we can obtain that the following inequality is correct

$$u(t) \leqslant (e^{\int_{\theta_l}^{t} a(s)ds} + \frac{e^{\int_{\theta_l}^{\zeta_l} a(s)ds}}{1 - \int_{\theta_l}^{\zeta_l} e^{\int_s^{\zeta_l} a(r)dr} b(s)ds} \int_{\theta_l}^{t} e^{\int_s^t a(r)dr} b(s)ds) \times$$

$$\prod_{k=l}^{i+1} (e^{\int_{\theta_k}^{\theta_{k+1}} a(s)ds} + \frac{e^{\int_{\theta_k}^{\zeta_k} a(s)ds}}{1 - \int_{\theta_k}^{\zeta_k} e^{\int_s^{\zeta_k} a(r)dr} b(s)ds} \int_{\theta_k}^{\theta_{k+1}} e^{\int_s^{\theta_{k+1}} a(r)dr} b(s)ds) \times$$

$$(e^{\int_{t_0}^{\theta_{i+1}} a(s)ds} + \frac{e^{\int_{t_0}^{\zeta_i} a(s)ds}}{1 - \int_{t_0}^{\zeta_i} e^{\int_s^{\zeta_i} a(r)dr} b(s)ds} \int_{t_0}^{\theta_{i+1}} e^{\int_s^{\theta_{i+1}} a(r)dr} b(s)ds) v_0, \tag{2.22}$$

if $t \in [\theta_l, \theta_{l+1}]$.

The results of this example will be useful in Chapter 4.

2.2 Quasi-linear systems

Let us consider system (2.2). One can easily see that (C4)–(C7) imply the existence of positive constants m, M and $\overline{M}$ such that $m \leqslant \|Z(t,s)\| \leqslant M$, $\|X(t,s)\| \leqslant \overline{M}$ for $t, s \in [\theta_i, \theta_{i+1}]$, $i \in \mathbb{Z}$.

From now on we make the assumption

(C10) $2\overline{M}L(1+M)\overline{\theta} < 1$.

Then, we can see that $\overline{M}(1+M)L\overline{\theta}e^{\overline{M}L(1+M)\overline{\theta}} < 1$, and the expression

Notation 2.2. $\kappa(L) = \frac{Me^{\overline{M}L(1+M)\overline{\theta}}}{1-\overline{M}(1+M)L\overline{\theta}e^{\overline{M}L(1+M)\overline{\theta}}}$

can be introduced. The following assumption is also needed:

(C11) $2\overline{M}L\overline{\theta}\kappa(L)(1+M) < m$.

The following Lemma is the most important auxiliary result of this chapter.

Lemma 2.2. *Suppose that conditions* (C1)–(C7), (C9)–(C11) *hold, and fix* $i \in \mathbb{Z}$. *Then, for every* $(\xi, z_0) \in [\theta_i, \theta_{i+1}] \times \mathbb{R}^n$, *there exists a unique solution* $z(t) = z(t, \xi, z_0)$ *of* (2.6) *on* $[\theta_i, \theta_{i+1}]$.

Proof. *Existence.* Fix $i \in \mathbb{Z}$. We assume without loss of generality that $\theta_i \leqslant \zeta_i < \xi \leqslant \theta_{i+1}$. Set $\|z(t)\|_0 = \max_{[\theta_i, \theta_{i+1}]} \|z(t)\|$, take $z_0(t) = Z(t,\xi)z_0$ and define a sequence $\{z_k(t)\}$ by

$$z_{k+1}(t) = Z(t,\xi)\left[z_0 + \int_{\xi}^{\zeta_i} X(\zeta_i,s)f(s,z_k(s),z_k(\zeta_i))ds\right] + \int_{\zeta_i}^{t} X(t,s)f(s,z_k(s),z_k(\zeta_i))ds, \quad k \geqslant 0.$$

The last expression implies that

$$\|z_{k+1}(t) - z_k(t)\|_0 \leqslant [2\overline{M}L(1+M)\overline{\theta}]^{k+1}M\|z_0\|.$$

Thus, there exists a unique solution $z(t) = z(t,\xi,z_0)$ of the equation

$$z(t) = Z(t,\xi)\left[z_0 + \int_{\xi}^{\zeta_i} X(\zeta_i,s)f(s,z(s),z(\zeta_i))ds\right] + \int_{\zeta_i}^{t} X(t,s)f(s,z(s),z(\zeta_i))ds, \quad (2.23)$$

which is a solution of (2.6) on $[\theta_i, \theta_{i+1}]$ as well. This proves the existence.

Uniqueness. Denote by $z_j(t) = z(t,\xi,z_0^j)$, $z_j(\xi) = z_0^j$, $j = 1,2$, the solutions of (2.6), where $\theta_i \leqslant \xi \leqslant \theta_{i+1}$. Without loss of generality, we assume that $\xi \leqslant \zeta_i$. It is sufficient to check that for every $t \in [\theta_i, \theta_{i+1}]$, $z_0^1 \neq z_0^2$ implies $z_1(t) \neq z_2(t)$. We have that

$$z_1(t) - z_2(t) = Z(t,\xi)(z_0^1 - z_0^2) + \int_{\xi}^{\zeta_i} X(\zeta_i,s)[f(s,z_1(s),z_1(\zeta_i)) - f(s,z_2(s),z_2(\zeta_i))]ds$$
$$+ \int_{\zeta_i}^{t} X(t,s)[f(s,z_1(s),z_1(\zeta_i)) - f(s,z_2(s),z_2(\zeta_i))]ds.$$

Hence,

$$\|z_1(t) - z_2(t)\| \leqslant M\|z_0^1 - z_0^2\| + \overline{M}L\overline{\theta}(1+M)\|z_1(\zeta_i) - z_2(\zeta_i)\| + \overline{M}L(1+M)\left|\int_{\xi}^{t} \|z_1(s) - z_2(s)\|ds\right|.$$

From the Gronwall-Bellman Lemma, it follows that

$$\|z_1(t) - z_2(t)\| \leqslant [M\|z_0^1 - z_0^2\| + \overline{M}L\overline{\theta}(1+M)\|z_1(\zeta_i) - z_2(\zeta_i)\|]e^{\overline{M}L(1+M)\overline{\theta}}.$$

In particular,

$$\|z_1(\zeta_i) - z_2(\zeta_i)\| \leqslant [M\|z_0^1 - z_0^2\| + \overline{M}L\overline{\theta}(1+M)\|z_1(\zeta_i) - z_2(\zeta_i)\|]e^{\overline{M}L(1+M)\overline{\theta}},$$

which gives us

$$\|z_1(\zeta_i) - z_2(\zeta_i)\| \leqslant \kappa(L)\|z_0^1 - z_0^2\|,$$

and hence

$$\|z_1(t) - z_2(t)\| \leqslant \kappa(L)\|z_0^1 - z_0^2\|. \quad (2.24)$$

Assume on the contrary that there exists $t \in [\theta_i, \theta_{i+1}]$ such that $z_1(t) = z_2(t)$. Then

$$Z(t,\xi)(z_0^1 - z_0^2) = Z(t,\xi)\int_{\xi}^{\zeta_i} X(\zeta_i,s)[f(s,z_1(s),z_1(\zeta_i)) - f(s,z_2(s),z_2(\zeta_i))]ds$$
$$+\int_{\zeta_i}^{t} X(t,s)[f(s,z_1(s),z_1(\zeta_i)) - f(s,z_2(s),z_2(\zeta_i))]ds. \quad (2.25)$$

We have that

$$\|Z(t,\xi)(z_0^2 - z_0^1)\| \geqslant m\|z_0^2 - z_0^1\|, \quad (2.26)$$

and (2.24) implies that

$$\|Z(t,\xi)\int_{\xi}^{\zeta_i} X(\zeta_i,s)[f(s,z_1(s),z_1(\zeta_i)) - f(s,z_2(s),z_2(\zeta_i))]ds$$
$$+\int_{\zeta_i}^{t} X(t,s)[f(s,z_1(s),z_1(\zeta_i)) - f(s,z_2(s),z_2(\zeta_i))]ds\|$$
$$\leqslant 2\overline{M}L\overline{\theta}\kappa(L)(1+M)\|z_0^2 - z_0^1\|. \quad (2.27)$$

Finally, we can see that (C11), (2.26) and (2.27) contradict (2.25). The lemma is proved. □

Remark 2.3. Inequality (2.24) implies the continuous dependence of solutions of (2.2) on initial value.

Theorem 2.3. *Suppose that conditions* (C1)–(C7), (C9)–(C11) *are fulfilled. Then for every* $(t_0, z_0) \in \mathbb{R} \times \mathbb{R}^n$ *there exists a unique solution* $z(t) = z(t,t_0,z_0)$ *of* (2.2) *in the sense of Definition* 2.1 *such that* $z(t_0) = z_0$.

Proof. We prove the theorem only for decreasing t, but one can easily see that the proof is similar for increasing t.
Let us assume without loss of generality that $\theta_i \leqslant \zeta_i < t_0 \leqslant \theta_{i+1}$ for some $i \in \mathbb{Z}$.
Using Lemma 2.2 for $\xi = t_0$ one can check that solution $z(t) = z(t,t_0,z_0)$ of (2.2) exists on $[\zeta_i, t_0]$ as a solution of (2.6) and is unique. Then conditions (C1)–(C3) imply that $z(t)$ can be continued to $t = \theta_i$, as it is a solution of the system of ordinary differential equations $z' = A_0(t)z(t) + A_1(t)z(\zeta_i) + f(t,z(t),z(\zeta_i))$ on $[\theta_i, \theta_{i+1})$.
Next, using the lemma again, we can continue $z(t)$ from $t = \theta_i$ to $t = \zeta_{i-1}$, and then to $t = \theta_{i-1}$. Since $\theta_i \to -\infty$ as $i \to -\infty$, the induction completes the proof. □

Lemma 2.3. *Suppose that conditions* (C1)–(C7), (C9)–(C11) *hold. Then, the solution*

$z(t) = z(t, t_0, z_0)$, $(t_0, z_0) \in \mathbb{R} \times \mathbb{R}^n$, of (2.2) is a solution, on $\mathbb{R}$, of the integral equation

$$\begin{aligned} z(t) = Z(t,t_0) &\left[z_0 + \int_{t_0}^{\zeta_i} X(t_0,s) f(s,z(s),z(\gamma(s)))\, ds \right] \\ &+ \sum_{k=i}^{k=j-1} Z(t,\theta_{k+1}) \int_{\zeta_k}^{\zeta_{k+1}} X(\theta_{k+1},s) f(s,z(s),z(\gamma(s)))\, ds \\ &+ \int_{\zeta_j}^{t} X(t,s) f(s,z(s),z(\gamma(s)))\, ds, \end{aligned} \tag{2.28}$$

where $\theta_i \leqslant t_0 \leqslant \theta_{i+1}$ and $\theta_j \leqslant t \leqslant \theta_{j+1}$, $i < j$.

Proof. We shall prove the Lemma only for $\theta_i < t_0 < \theta_{i+1} < t \leqslant \theta_{i+2}$. All other cases can be proved analogously. Consider at first $t \in [\theta_i, \theta_{i+1}]$. The solution uniquely satisfies the integral equation

$$z(t) = X(t,\zeta_i) z(\zeta_i) + \int_{\zeta_i}^{t} X(t,s) A_1(s) z(\zeta_i) ds + \int_{\zeta_i}^{t} X(t,s) f(s,z(s),z(\gamma(s))) ds.$$

Using the last expression one can easily see that

$$z(\zeta_i) = M_i^{-1}(t_0) \left[z_0 - \int_{\zeta_i}^{t_0} X(t_0,s) f(s,z(s),z(\gamma(s))) ds \right].$$

Hence,

$$z(t) = M_i(t) M_i^{-1}(t_0) \left[z_0 - \int_{\zeta_i}^{t_0} X(t_0,s) f(s,z(s),z(\gamma(s))) ds \right] + \int_{\zeta_i}^{t} X(t,s) f(s,z(s),z(\gamma(s))) ds,$$

and

$$\begin{aligned} z(\theta_{i+1}) = M_i(\theta_{i+1}) M_i^{-1}(t_0) &\left[z_0 - \int_{\zeta_i}^{t_0} X(t_0,s) f(s,z(s),z(\gamma(s))) ds \right] \\ &+ \int_{\zeta_i}^{\theta_{i+1}} X(\theta_{i+1},s) f(s,z(s),z(\gamma(s))) ds. \end{aligned}$$

Then, for $t \in [\theta_{i+1}, \theta_{i+2}]$,

$$z(t) = X(t,\zeta_{i+1}) z(\zeta_{i+1}) + \int_{\zeta_{i+1}}^{t} X(t,s) A_1(s) z(\gamma(s)) ds + \int_{\zeta_{i+1}}^{t} X(t,s) f(s,z(s),z(\gamma(s))) ds,$$

and

$$z(\zeta_{i+1}) = M_{i+1}^{-1}(\theta_{i+1}) [z(\theta_{i+1}) - \int_{\zeta_{i+1}}^{\theta_{i+1}} X(\theta_{i+1},s) f(s,z(s),z(\gamma(s))) ds].$$

Hence,

$$
\begin{aligned}
z(t) &= M_{i+1}(t)M_{i+1}^{-1}(\theta_{i+1})[z(\theta_{i+1}) - \int_{\zeta_{i+1}}^{\theta_{i+1}} X(\theta_{i+1},s)f(s,z(s),z(\gamma(s)))ds] \\
&\quad + \int_{\zeta_{i+1}}^{t} X(t,s)f(s,z(s),z(\gamma(s)))ds \\
&= M_{i+1}(t)M_{i+1}^{-1}(\theta_{i+1})\{M_i^{-1}(\theta_{i+1})M_i^{-1}(t_0)[z_0 \\
&\quad - \int_{\zeta_i}^{t_0} X(t_0,s)f(s,z(s),z(\gamma(s)))ds] \\
&\quad + \int_{\zeta_i}^{\theta_{i+1}} X(\theta_{i+1},s)f(s,z(s),z(\gamma(s)))ds \\
&\quad - \int_{\zeta_{i+1}}^{\theta_{i+1}} X(\theta_{i+1},s)f(s,z(s),z(\gamma(s)))ds\} \\
&\quad + \int_{\zeta_{i+1}}^{t} X(t,s)f(s,z(s),z(\gamma(s)))ds \\
&= M_{i+1}(t)M_{i+1}^{-1}(\theta_{i+1})M_i^{-1}(\theta_{i+1})M_i^{-1}(t_0)z_0 \\
&\quad + M_{i+1}(t)M_{i+1}^{-1}(\theta_{i+1})M_i^{-1}(\theta_{i+1})M_i^{-1}(t_0)\int_{t_0}^{\zeta_i} X(t,s)f(s,z(s),z(\gamma(s)))ds \\
&\quad + M_{i+1}(t)M_{i+1}^{-1}(\theta_{i+1})\int_{\zeta_i}^{\zeta_{i+1}} X(\theta_{i+1},s)f(s,z(s),z(\gamma(s)))ds \\
&\quad + \int_{\zeta_{i+1}}^{t} X(t,s)f(s,z(s),z(\gamma(s)))ds \\
&= Z(t,t_0)z_0 + Z(t,t_0)\int_{t_0}^{\zeta_i} X(t_0,s)f(s,z(s),z(\gamma(s)))ds \\
&\quad + Z(t,\theta_{i+1})\int_{\zeta_i}^{\zeta_{i+1}} X(\theta_{i+1},s)f(s,z(s),z(\gamma(s)))ds \\
&\quad + \int_{\zeta_{i+1}}^{t} X(t,s)f(s,z(s),z(\gamma(s)))ds.
\end{aligned}
$$

The lemma is proved. □

2.3 Stability

In this section, we assume that conditions (C1)–(C11) are fulfilled, and hence, all solutions of the considered systems are defined on the whole real axis and their integral curves do not intersect each other. For the stability investigation, we consider the systems on $\mathbb{R}_+ = [0, \infty)$. Definitions of Lyapunov stability for the solutions of both discussed systems can be done in the same way as for ordinary differential equations. Let us formulate them.

Definition 2.2. The zero solution of (2.1), (2.2) is stable if to any positive ε and t_0, there corresponds a number $\delta = \delta(t_0, \varepsilon) > 0$ such that $\|x(t, t_0, x_0)\| < \varepsilon$, $t \geqslant t_0 \geqslant 0$, whenever $\|x_0\| < \delta$.

Definition 2.3. The zero solution of (2.1), (2.2) is uniformly stable, if δ in the previous definition is independent of t_0.

Definition 2.4. The zero solution of (2.1), (2.2) is asymptotically stable if it is stable in the sense of Definition 2.2, and there exists a positive number $\kappa(t_0)$ such that if $\psi(t)$ is any solution of (2.1), (2.2) with $\|\psi(t_0)\| < \kappa(t_0)$, then $\|\psi(t)\| \to 0$ as $t \to \infty$.

Definition 2.5. The zero solution of (2.1), (2.2) is uniformly asymptotically stable if it is uniformly stable in the sense of Definition 2.3, and for any $\varepsilon > 0$, we can find $T(\varepsilon) > 0$ such that any solution $\psi(t)$ of (2.1), (2.2) with $\|\psi(t_0)\| < \kappa$, where κ is independent of t_0, satisfies $\|\psi(t)\| < \varepsilon$ for $t \geqslant t_0 + T(\varepsilon)$.

Definition 2.6. The zero solution of (2.1), (2.2) is unstable if there exist numbers $\varepsilon_0 > 0$ and $t_0 \in I$ such that for any $\delta > 0$ there exists a solution $y_\delta(t)$, $\|y_\delta(t_0)\| < \delta$, of the system such that either it is not continuable to ∞ or there exists a moment t_1, $t_1 > t_0$ such that $\|y_\delta(t_1)\| \geqslant \varepsilon_0$.

Let $Z(t)$ be a fundamental matrix of (2.1). We can prove the following assertions, using representations (2.14) and (2.15) in exactly the same way as theorems for ordinary differential equations [87, 153].

Theorem 2.4. *The zero solution of* (2.1) *is stable if and only if* $Z(t)$ *is bounded on* $t \geqslant 0$.

Theorem 2.5. *The zero solution of* (2.1) *is asymptotically stable if and only if* $Z(t) \to 0$ *as* $t \to \infty$.

Theorem 2.6. *The zero solution of* (2.1) *is uniformly stable if and only if there exists a number* $M > 0$ *such that* $\|Z(t)Z^{-1}(s)\| \leqslant M$, $t \geqslant s \geqslant 0$.

Theorem 2.7. *The zero solution of* (2.1) *is uniformly asymptotically stable if and only if there exist two positive numbers* N *and* ω *such that* $\|Z(t)Z^{-1}(s)\| \leqslant N e^{-\omega(t-s)}$, $t \geqslant s \geqslant 0$.

On the basis of the last theorems we can formulate the following theorems which provide sufficient conditions for the stability of linear systems.

Theorem 2.8. *Suppose* (C1)–(C6) *hold and* $\|M_k^{-1}(\theta_k)M_{k-1}(\theta_k)\| \leqslant 1$, $k \in \mathbb{N}$. *Then the zero solution of* (2.1) *is stable.*

Theorem 2.9. *Suppose* (C1)–(C6) *hold and there exists a nonnegative number* $\kappa < 1$ *such that* $\|M_k^{-1}(\theta_k)M_{k-1}(\theta_k)\| \leqslant \kappa$, $k \in \mathbb{N}$. *Then the zero solution of* (2.1) *is asymptotically stable.*

Theorem 2.10. *Suppose* (C1)–(C7) *hold. The zero solution of* (2.1) *is uniformly stable if* $\|M_k^{-1}(\theta_k)M_{k-1}(\theta_k)\| \leqslant 1, k \in \mathbb{N}$.

Theorem 2.11. *Suppose* (C1)–(C5), (C8) *hold. The zero solution of* (2.1) *is uniformly stable if* $\|M_k^{-1}(\theta_k)M_{k-1}(\theta_k)\| \leqslant 1, k \in \mathbb{N}$.

Theorem 2.12. *Suppose* (C1)–(C7) *hold. The zero solution of* (2.1) *is uniformly asymptotically stable if there exists a nonnegative number* $\kappa < 1$ *such that* $\|M_k^{-1}(\theta_k)M_{k-1}(\theta_k)\| \leqslant \kappa$, $k \in \mathbb{N}$.

Theorem 2.13. *Suppose* (C1)–(C5), (C8) *hold. The zero solution of* (2.1) *is uniformly asymptotically stable if there exists a nonnegative number* $\kappa < 1$ *such that* $\|M_k^{-1}(\theta_k)M_{k-1}(\theta_k)\| \leqslant \kappa, k \in \mathbb{N}$.

Interesting results comparable to the last theorems can be found in [317].

Example 2.2. Consider the following differential equation with piecewise constant function as argument

$$x'(t) = \alpha x(t) + \beta x(\gamma(t)), \tag{2.29}$$

in which α, β are fixed real constants, the identification function $\gamma(t)$ is defined by the sequences $\theta_i = \kappa i$, $\zeta_i = \theta_i + \kappa_1$, $i \in \mathbb{Z}$, where $\kappa > 0$, $\kappa > \kappa_1 > 0$, are fixed numbers. We will find conditions on the coefficients and the sequences to provide uniform asymptotic stability of the zero solution. One can evaluate that

$$M_i(t) = \mathrm{e}^{\alpha(t-\zeta_i)} + \int_{\zeta_i}^{t} \mathrm{e}^{\alpha(t-s)}\beta ds = \mathrm{e}^{\alpha(t-\zeta_i)} + \frac{\beta}{\alpha}\left(\mathrm{e}^{\alpha(t-\zeta_i)} - 1\right).$$

Then

$$M_i(\theta_i) = \mathrm{e}^{-\alpha\kappa_1} + \frac{\beta}{\alpha}(\mathrm{e}^{-\alpha\kappa_1} - 1).$$

Moreover, if we denote $\kappa_2 = \kappa - \kappa_1$, then

$$M_{i-1}(\theta_i) = \mathrm{e}^{\alpha\kappa_2} + \frac{\beta}{\alpha}(\mathrm{e}^{-\alpha\kappa_1} - 1),$$

and

$$M_i^{-1}(\theta_i)M_{i-1}(\theta_i) = \frac{\mathrm{e}^{\alpha\kappa_2} + \frac{\beta}{\alpha}(\mathrm{e}^{\alpha\kappa_2} - 1)}{\mathrm{e}^{-\alpha\kappa_1} + \frac{\beta}{\alpha}(\mathrm{e}^{-\alpha\kappa_1} - 1)}.$$

On the basis of Theorem 2.13 the inequality

$$\left|\frac{e^{\alpha\kappa_2}+\frac{\beta}{\alpha}(e^{\alpha\kappa_2}-1)}{e^{-\alpha\kappa_1}+\frac{\beta}{\alpha}(e^{-\alpha\kappa_1}-1)}\right|<1$$

is necessary and sufficient for the zero solution to be uniformly asymptotically stable.

It is of particular interest to consider the case when $\kappa_1 = 0$ and $\kappa_1 = \kappa$. Consider the first one. Then conditions from the last inequality imply that the zero solution is uniformly asymptotically stable if and only if $\alpha < 0$ and $|\beta| < -\alpha$.

Similarly, if $\kappa_1 = \kappa$, then for the zero solution to be uniformly asymptotically stable it is necessary and sufficient that one of the following two conditions is true: $\beta < -\alpha$ or $\beta > \alpha\frac{1+e^{-\alpha\kappa}}{1-e^{-\alpha\kappa}}$.

Let us obtain an evaluation of the fundamental solution $Z(t,t_0)$ of the equation using (2.12). Set

$$\xi = \left|\frac{e^{\alpha\kappa_2}+\frac{\beta}{\alpha}(e^{\alpha\kappa_2}-1)}{e^{-\alpha\kappa_1}+\frac{\beta}{\alpha}(e^{-\alpha\kappa_1}-1)}\right|,$$

and assume that condition of the uniform asymptotic stability is valid. Then $\xi < 1$ and

$$\min_{\theta_i\leqslant t\leqslant\theta_{i+1}} M_i(t) = e^{-\alpha\kappa_1}+\frac{\beta}{\alpha}(e^{-\alpha\kappa_1}-1), \quad \max_{\theta_i\leqslant t\leqslant\theta_{i+1}} M_i(t) = e^{\alpha\kappa_2}+\frac{\beta}{\alpha}(e^{\alpha\kappa_2}-1).$$

Hence, for

$$N = \max\left\{e^{\alpha\kappa_2}+\frac{\beta}{\alpha}(e^{\alpha\kappa_2}-1), \left(e^{-\alpha\kappa_1}+\frac{\beta}{\alpha}(e^{-\alpha\kappa_1}-1)\right)^{-1}\right\}e^{-\ln\xi},$$

we have

$$|Z(t,t_0)| \leqslant N e^{\frac{\ln\xi}{\kappa}(t-t_0)}, \quad t\geqslant t_0.$$

If the zero solution of (2.1) is uniformly asymptotically stable, then by Theorem 2.7 there exist two numbers $K \geqslant 1$ and $\omega > 0$ such that $\|Z(t,s)\| \leqslant Ke^{-\omega(t-s)}$, $t \geqslant s \geqslant 0$.

Let $\gamma(L) = \frac{M}{1-2\bar{M}L(1+M)\bar{\theta}}$. We may make the following assumption

(C12) $2K e^{2\omega\bar{\theta}}\bar{M}L\max(1,\gamma(L)) < \omega$.

Theorem 2.14. *Suppose* (C1)–(C7), (C9)–(C12) *are fulfilled and the zero solution of* (2.1) *is uniformly asymptotically stable. Then the zero solution of* (2.2) *is uniformly asymptotically stable.*

Proof. If $z(t) = z(t,t_0,z_0)$ is a solution of (2.2), then by (2.23), (2.28) and assuming, without loss of generality, that $t_0 < \zeta_i \leqslant \cdots \leqslant \zeta_j < t$, we have that

$$\begin{aligned}
\|z(t)\| &\leqslant K\mathrm{e}^{-\omega(t-t_0)}\|z_0\| + K\mathrm{e}^{-\omega(t-t_0)}\int_{t_0}^{\zeta_i} 2\overline{M}L\max(1,\gamma(L))\|z(s)\|\,ds \\
&\quad + \sum_{k=i}^{j} K\mathrm{e}^{-\omega(t-\theta_{k+1})}\int_{\zeta_k}^{\zeta_{k+1}} 2\overline{M}L\max(1,\gamma(L))\|z(s)\|\,ds \\
&\quad + \int_{\zeta_j}^{t} 2\overline{M}L\max(1,\gamma(L))\|z(s)\|\,ds \\
&\leqslant K\mathrm{e}^{-\omega(t-t_0)}\|z_0\| + \int_{t_0}^{t} 2K\mathrm{e}^{-\omega(t-s-2\overline{\theta})}\overline{M}L\max(1,\gamma(L))\|z(s)\|\,ds.
\end{aligned}$$

If we set $u(t) = \|z(t)\|K\mathrm{e}^{\omega t}$, then the last inequality implies that

$$u(t) \leqslant Ku(t_0) + \int_{t_0}^{t} 2K\mathrm{e}^{2\omega\overline{\theta}}\overline{M}L\max(1,\gamma(L))u(s)ds.$$

Now, by virtue of the Gronwall-Bellmann Lemma, we obtain

$$\|z(t)\| \leqslant K\mathrm{e}^{(-\omega+2K\mathrm{e}^{2\omega\overline{\theta}}\overline{M}L\max(1,\gamma(L)))(t-t_0)}\|z_0\|.$$

The last inequality, in conjunction with (C11), proves that the zero solution is uniformly asymptotically stable. The theorem is proved. □

Notes

Since the method of reduction to discrete systems, presumes that an equation has to be solved explicitly or implicitly with respect to the values of the solution at switching moments, the most appropriate equations considered by the method are either linear systems or equations, where solutions present linearly if their argument is non-deviated. These equations have been investigated mostly in literature [2–7], [81–85], [121, 133, 143, 317]. Consequently, basic problems of the theory have been left out of the discussion. Some of these problems are: Existence of solutions with initial moments, which are not moments of switching of constancy; stability of solutions with an arbitrary initial moment; uniform stability, etc.

The method of equivalent integral equations proposed in the papers [8–15] combined with the comparison of values of solutions at switching moments help to construct the fundamental matrix of solutions (2.13) and the Cauchy's formula (2.28). These formulas will play decisive role for the theory, similar they have for ordinary differential equations. We attract the reader's attention to the result that a linear differential equation with piecewise constant arguments has a finite dimensional space of solutions. It may be used as a basis for a new deep analysis of functional differential equations. The main results of this chapter were published in [11] for the first time. Lemma 2.1 and results of Example 2.1 are newly obtained in this chapter. Based on these achievements, the theory of linear systems for differential equations with piecewise constant arguments can be effectively developed further. Using this method, in papers [13,263], asymptotic behavior of solutions of quasilinear systems with piecewise constant argument is analyzed.

Chapter 3

The reduction principle for systems with piecewise constant argument

3.1 Introduction

The theory of integral manifolds founded by H. Poincaré and A. M. Lyapunov [215, 267] became a very powerful instrument for investigating problems of the qualitative theory of differential equations. Over the past several decades, many researchers have been studying the methods of reducing high dimensional problems to low dimensional ones. If we discuss this problem for long-time dynamics of differential equations, we should consider the Reduction Principle [264, 265]. For a brief history of the principle, the reader is referred to the papers [199, 219, 264]. As it is well known that the principle was utilized in the center manifold theory, as well as in the theory of inertial manifolds [68, 118, 155]. On the other hand, it is natural that the exploration of the properties and neighborhoods of manifolds is one of the most interesting problems of the theory of differential equations [59, 68, 74, 153, 180, 248, 268]. One should not be surprised that integral manifolds and the reduction principle are among the major subjects of investigation for specific types of differential and difference equations [39, 51, 73, 75, 100, 118, 149, 150, 200, 249–251, 268, 294]. The main novelty of this chapter is to extend the principle to differential equations with piecewise constant arguments.

Fix two real-valued sequences θ_i, ζ_i, $i \in \mathbb{Z}$, such that $\theta_i < \theta_{i+1}$, $\theta_i \leqslant \zeta_i \leqslant \theta_{i+1}$ for all $i \in \mathbb{Z}$, $|\theta_i| \to \infty$ as $|i| \to \infty$, and there exists a number $\theta > 0$ such that $\theta_{i+1} - \theta_i \leqslant \theta$, $i \in \mathbb{Z}$.

In this chapter we deal with the quasilinear system

$$z' = Az + f(t, z(t), z(\gamma(t))), \tag{3.1}$$

where $z \in \mathbb{R}^n$, $t \in \mathbb{R}$, $\gamma(t) = \zeta_i$, if $t \in [\theta_i, \theta_{i+1})$, $i \in \mathbb{Z}$.

We shall need the following assumptions throughout the chapter:

(C1) A is a constant $n \times n$ real valued matrix;

(C2) $f(t,x,z)$ is continuous in the first argument, $f(t,0,0)=0, t\in\mathbb{R}$, and f is Lipschitzian in the second and third arguments with a positive Lipschitz constant l such that

$$\|f(t,z_1,w_1)-f(t,z_2,w_2)\|\leqslant l(\|z_1-z_2\|+\|w_1-w_2\|)$$

for all $t\in\mathbb{R}$ and $z_1,\ z_2,\ w_1,\ w_2\in\mathbb{R}^n$.

(C3) If we denote by $\lambda_j,\ j=\overline{1,n}$, the eigenvalues of the matrix A, then there exists a positive integer k such that $\mu=\max_{j=\overline{1,k}}\Re\lambda_j<0$, and $\min_{j=\overline{k+1,n}}\Re\lambda_j=0$, where $\Re\lambda_j$ denotes the real part of the eigenvalue λ_j of the matrix A.

By using the last condition, we can suppose, without loss of generality

(C4)

$$A=\begin{pmatrix} B_- & 0\\ 0 & B_+\end{pmatrix},$$

where square matrices B_- and B_+ are of dimension k and $n-k$, respectively, $\lambda_j, j=\overline{1,k}$, are the eigenvalues of the matrix B_+ and $\lambda_j,\ j=\overline{k+1,n}$, are the eigenvalues of the matrix B_-.

Equation (3.1) can be written as

$$z'=Az+f(t,z(t),\overline{z}), \tag{3.2}$$

if $t\in[\theta_i,\theta_{i+1}),\ \overline{z}=z(\zeta_i), i\in\mathbb{Z}$.

In other words, system (3.1) on $[\theta_i,\ \theta_{i+1}),\ i\in\mathbb{Z}$, has the form of a special functional-differential equation

$$z'=Az+f(t,z(t),z(\zeta_i)). \tag{3.3}$$

That is, (3.1) has the structure of a continuous dynamical system within intervals $[\theta_i,\theta_{i+1})$, $i\in\mathbb{Z}$. In this chapter, we assume that the solutions of the equation are continuous functions. But the deviating function $\gamma(t)$ is discontinuous. Thus, in general, the right-hand side of (3.1) has discontinuities at moments θ_i. As a result, we consider the solutions of the equation as functions, which are continuous and continuously differentiable within intervals $[\theta_i,\theta_{i+1}), i\in\mathbb{Z}$.

The theory of differential equations with piecewise constant arguments necessitates a more careful discussion of the initial value problem.

In the sequel, we use the uniform norm $\|T\|=\sup\{\|Tz\|:\|z\|\leqslant 1\}$ for matrices.

It is known that there exists a constant $\Omega>0$ such that $\|e^{A(t-s)}\|\leqslant e^{\Omega|t-s|},\ t,s\in\mathbb{R}$. Hence, one can show that $\|e^{A(t-s)}\|\geqslant e^{-\Omega|t-s|}, t,s\in\mathbb{R}$.

The last two inequalities imply the following very simple but useful estimates: $\|e^{A(t-s)}\|\leqslant M,\ \|e^{A(t-s)}\|\geqslant m$, if $|t-s|\leqslant\theta$, where $M=e^{\Omega\theta}, m=e^{-\Omega\theta}$.

From now on we make the assumption:

(C5) $M\ell\theta\, e^{M\ell\theta} < 1,\ 2M\ell\theta < 1,\ \dfrac{2M^2\ell\theta e^{M\ell\theta}}{1 - M\ell\theta\, e^{M\ell\theta}} < m.$

Lemma 3.1. *Assume that conditions* (C1)–(C3), *and* (C5) *are fulfilled, and fix* $i \in \mathbb{Z}$. *Then for every* $(\xi, z_0) \in [\theta_i, \theta_{i+1}] \times \mathbb{R}^n$ *there exists a unique vector* $v \in \mathbb{R}^n$ *such that there exists a unique solution* $z(t) = z(t, \xi, z_0)$ *of* (3.3) *on* $[\theta_i, \theta_{i+1}]$ *with* $z(\zeta_i) = v$.

Theorem 3.1. *Assume that conditions* (C1)–(C3), *and* (C5) *are fulfilled. Then for every* $(t_0, z_0) \in \mathbb{R} \times \mathbb{R}^n$ *there exists a unique solution* $z(t) = z(t, t_0, z_0)$ *of* (3.1) *in the sense of Definition* 2.1 *such that* $z(t_0) = z_0$.

The proof of the last assertions can be verified exactly using the same procedure as Lemma 2.2 and Theorem 2.3. Analogously to the inequality 2.24 the following relation can be proved

$$\|z_1(t) - z_2(t)\| \leqslant M e^{M\ell\theta} \left[1 + \frac{M\ell\theta\, e^{M\ell\theta}}{1 - M\ell\theta\, e^{M\ell\theta}}\right] \|z_0^2 - z_0^1\|, \tag{3.4}$$

which yields the continuous dependence on the initial value immediately.
The last theorem is of major importance as well as Theorem 2.1. The correspondence between points $(t_0, z_0) \in \mathbb{R} \times \mathbb{R}^n$ and solutions of (3.1) helps us to prove main theorems of this chapter.

3.2 Existence of integral surfaces

Definition 3.1. A set Σ in the (t,z)-space is said to be an integral set of system (3.1) if any solution $z(t) = z(t, t_0, z_0)$, $z(t_0) = z_0$, with $(t_0, z_0) \in \Sigma$, has the property that $(t, z(t)) \in \Sigma$, $t \in \mathbb{R}$. In other words, for every $(t_0, z_0) \in \Sigma$ the solution $z(t) = z(t, t_0, z_0)$, $z(t_0) = z_0$, is continuable on $\mathbb{R}$ and $(t, z(t)) \in \Sigma$, $t \in \mathbb{R}$.

Definition 3.2. A set Σ in the (t,z)-space is said to be a local integral set of system (3.1) if for every $(t_0, z_0) \in \Sigma$ there exists $\varepsilon > 0$, $\varepsilon = \varepsilon(t_0, z_0)$, such that if $z(t) = z(t, t_0, z_0)$ is a solution of (3.1) and $|t - t_0| < \varepsilon$, then $(t, z(t)) \in \Sigma$.

Using condition (C4) we can write equation (3.1) in the following form

$$\begin{aligned} \frac{du}{dt} &= B_- u + f_-(t, z(t), z(\gamma(t))), \\ \frac{dv}{dt} &= B_+ v + f_+(t, z(t), z(\gamma(t))), \end{aligned} \tag{3.5}$$

where $z = (u, v)$, $u \in \mathbb{R}^k$, $v \in \mathbb{R}^{n-k}$, $(f_-, f_+) = f(t, z(t), z(\gamma(t)))$.

Throughout the chapter we use fixed positive numbers σ and α such that $\alpha < \sigma < -\mu$, and denote

$$\beta = \int_0^\infty (1+t^m)\mathrm{e}^{-\alpha t}dt.$$

It is obvious that there exist constants $K \geqslant 1$ and $m \in \mathbb{N}$, $m < n-k$, such that

$$\|\mathrm{e}^{B_- t}\| \leqslant K\mathrm{e}^{-\sigma t} \text{ and } \|\mathrm{e}^{-B_+ t}\| \leqslant K(1+t^m),$$

for all $t \in R_+ = [0,\infty)$.

We shall establish the validity of the lemma below.

Lemma 3.2. *Fix* $N \in \mathbb{R}$, $N > 0$, *and assume that conditions* (C1)–(C3) *are valid. A continuous function* $z(t) = (u,v)$, $\|z(t)\| \leqslant N\mathrm{e}^{-\alpha(t-t_0)}$, $t \geqslant t_0$, *is a solution of* (3.1) *on* $\mathbb{R}$ *if and only if* $z(t)$ *is a solution on* $\mathbb{R}$ *of the following system of integral equations*

$$\begin{aligned} u(t) &= \mathrm{e}^{B_-(t-t_0)}u(t_0) + \int_{t_0}^t \mathrm{e}^{B_-(t-s)}f_-(s,z(s),z(\gamma(s)))ds, \\ v(t) &= -\int_t^\infty \mathrm{e}^{B_+(t-s)}f_+(s,z(s),z(\gamma(s)))ds. \end{aligned} \tag{3.6}$$

Proof. *Necessity.* Assume that $z(t) = (u,v)$, $\|z(t)\| \leqslant N\mathrm{e}^{-\alpha(t-t_0)}$, $t \in [t_0,\infty)$, is a solution of (3.1). Denote

$$\begin{aligned} \phi(t) &= \mathrm{e}^{B_-(t-t_0)}u(t_0) + \int_{t_0}^t \mathrm{e}^{B_-(t-s)}f_-(s,z(s),z(\gamma(s)))ds, \\ \psi(t) &= -\int_t^\infty \mathrm{e}^{B_+(t-s)}f_+(s,z(s),z(\gamma(s)))ds. \end{aligned} \tag{3.7}$$

By straightforward evaluation we can see that the integrals converge, are bounded on $[t_0,\infty)$, and, moreover,

$$\begin{aligned} \|\phi(t)\| &\leqslant K\mathrm{e}^{-\sigma(t-t_0)}\|u(t_0)\| + Kl\left[\frac{N(1+\mathrm{e}^{\alpha\theta})}{\sigma-\alpha} + 2\frac{1}{\sigma}\max_{[\theta_i,\theta_{i+1}]}\|z(s)\|\mathrm{e}^{\sigma(|t_0|+\theta)}\right]\mathrm{e}^{-\alpha(t-t_0)}, \\ \|\psi(t)\| &\leqslant Kl\beta N(1+\mathrm{e}^{\alpha\theta})\mathrm{e}^{-\alpha(t-t_0)}. \end{aligned} \tag{3.8}$$

We can show that $(\phi(t)-u(t),\ \psi(t)-v(t))$ is a continuously differentiable on $\mathbb{R}$ satisfying $u'(t) = B_-u(t)$, $v'(t) = B_+v(t)$ with the initial condition $\phi(t_0)-u(t_0) = 0$. Assume on the contrary that $\psi(t_0)-v(t_0) \neq 0$. Then $\psi(t)-v(t)$ is not a decay solution, which contradicts (3.8). Thus, $\phi(t)-u(t) = 0$, $\psi(t)-v(t) = 0$ on $\mathbb{R}$.

Sufficiency. Suppose that $z(t)$ is a solution of (3.6). Differentiating $z(t)$ in $t \in (\theta_i,\theta_{i+1})$, $i \in \mathbb{Z}$, one can see that the function satisfies (3.1). Moreover, letting $t \to \theta_i+$, and remembering that $z(\gamma(t))$ is a right-continuous function, we find that $z(t)$ satisfies (3.1) on $[\theta_i,\theta_{i+1})$. The Lemma is proved. □

For convenience, we adopt the following notation below

$$p = K(1+\mathrm{e}^{\alpha\theta})\left[\frac{1}{\sigma-\alpha}+\beta\right].$$

In what follows we mainly use the same technique in [264]. See also [68,248].
We assume that the following condition

$$2p\ell < 1 \tag{3.9}$$

is fulfilled in the sequel.

Theorem 3.2. *Suppose that conditions* (C1)–(C5) *are fulfilled. Then, there exists a function* $F(\zeta_i,u)$, $i\in\mathbb{Z}$, *satisfying*

$$F(\zeta_i,0)=0, \tag{3.10}$$

$$\|F(\zeta_i,u_1)-F(\zeta_i,u_2)\| \leqslant pK\ell\|u_1-u_2\|, \tag{3.11}$$

for all i, u_1, u_2, *such that a solution* $z(t)$ *of* (3.1) *with* $z(\zeta_i)=(c,F(\zeta_i,c))$, $c\in\mathbb{R}^k$, *is defined on* $\mathbb{R}$ *and satisfies*

$$\|z(t)\| \leqslant 2K\|c\|\mathrm{e}^{-\alpha(t-\zeta_i)}, \quad t\geqslant\zeta_i. \tag{3.12}$$

Proof. Let us apply the method of successive approximations to system (3.6). Denote $z_0(t)=(0,0)^T$, $z_m=(u_m,v_m)^T$, $m\geqslant 0$,

$$u_{m+1}(t)=\mathrm{e}^{B_-(t-\zeta_i)}c+\int_{\zeta_i}^{t}\mathrm{e}^{B_-(t-s)}f_-(s,z_m(s),z_m(\gamma(s)))ds,$$

$$v_{m+1}(t)=-\int_{t}^{\infty}\mathrm{e}^{B_+(t-s)}f_+(s,z_m(s),z_m(\gamma(s)))ds.$$

First we need to show that

$$\|z_m(t)\| \leqslant 2K\|c\|\mathrm{e}^{-\alpha(t-\zeta_i)}, \quad t\geqslant\zeta_i, \quad m\geqslant 0. \tag{3.13}$$

Indeed, z_0 satisfies the relation. Assume that z_{m-1} satisfies (3.13). Then

$$\|u_m(t)\| \leqslant K\mathrm{e}^{-\sigma(t-\zeta_i)}\|c\|+\frac{2K^2\ell(1+\mathrm{e}^{\alpha\theta})}{\sigma-\alpha}\mathrm{e}^{-\alpha(t-\zeta_i)}\|c\|,$$

$$\|v_m(t)\| \leqslant 2\beta K^2\ell(1+\mathrm{e}^{\alpha\theta})\mathrm{e}^{-\alpha(t-\zeta_i)}\|c\|, \tag{3.14}$$

and (3.13) are valid provided that (3.9) is given. Similarly, one can establish the following inequality

$$\|z_{m+1}(t)-z_m(t)\| \leqslant K\|c\|(2p\ell)^m\mathrm{e}^{-\alpha(t-\zeta_i)}. \tag{3.15}$$

The last inequality and assumption (3.9) imply that the sequence z_m converges uniformly for all c and $t \geqslant \zeta_i$. Let $z(t,\zeta_i,c) = (u(t,\zeta_i,c), v(t,\zeta_i,c))$ be the limit function. It is obvious that the function is a solution of (3.6) with $t_0 = \zeta_i$. By Lemma 3.2, $z(t,\zeta_i,c)$ is a solution of (3.1), too. We have that

$$u(\zeta_i,\zeta_i,c) = c,$$
$$v(\zeta_i,\zeta_i,c) = -\int_{\zeta_i}^{\infty} e^{B_+(t-s)} f_+(s, z(s,\zeta_i,c), z(\gamma(s),\zeta_i,c))ds.$$

Denote $F(\zeta_i,c) = v(\zeta_i,\zeta_i,c)$. Since

$$\|v_m(t,\zeta_i,c_1) - v_m(t,\zeta_i,c_2)\| \leqslant pK\ell\|c_1 - c_2\|, \quad m \geqslant 1, \tag{3.16}$$

inequality (3.11) is valid, and (3.10) is obvious. The theorem is proved. □

For every $i \in \mathbb{Z}$, consider a set Ψ_i of continuous functions on $\mathbb{R}$ such that if $\psi \in \Psi_i$ then there exists a positive constant K_ψ, satisfying $\|\psi(t)\| \leqslant K_\psi e^{-\alpha(t-\zeta_i)}$, if $\zeta_i \leqslant t$.

Lemma 3.3. *For every ζ_i, $i \in \mathbb{Z}$, $c \in \mathbb{R}^k$, the system*

$$u(t) = e^{B_-(t-t_0)}c + \int_{\zeta_i}^{t} e^{B_-(t-s)} f_-(s, z(s), z(\gamma(s)))ds,$$
$$v(t) = -\int_{t}^{\infty} e^{B_+(t-s)} f_+(s, z(s), z(\gamma(s)))ds.$$

has only one solution from Ψ_i.

Proof. If $z_1(t)$ and $z_2(t)$ are two bounded solutions of the system on $[\zeta_i,\infty)$, then by straightforward evaluation one can show that

$$\sup_{[\zeta_i,\infty)} \|z_1(t) - z_2(t)\| \leqslant 2p\ell \sup_{[\zeta_i,\infty)} \|z_1 - z_2\|.$$

Hence, in the light of (3.9) the lemma is proved. □

Let us denote by S_i^+ the set of all points from the (t,z)-space $(z = (u,v))$ such that $t = \zeta_i$, $v = F(\zeta_i,u)$.

Lemma 3.4. *If $z(t,\zeta_i,z_0)$ is a solution of (3.1) from Ψ_i, then $(\zeta_i,z_0) \in S_i^+$.*

Proof. Assume that $z(t) \in \Psi_i, z(t) = z(t,\zeta_i,z_0) = (u,v)$, is a solution of (3.1). It is obvious that

$$u(t) = e^{B_-(t-\zeta_i)}u(\zeta_i) + \int_{\zeta_i}^{t} e^{B_-(t-s)} f_-(s, z(s), z(\gamma(s)))ds,$$
$$v(t) = e^{B_+(t-\zeta_i)}\widehat{\kappa} - \int_{t}^{\infty} e^{B_+(t-s)} f_+(s, z(s), z(\gamma(s)))ds, \tag{3.17}$$

where

$$\widehat{\kappa} = v(\zeta_i) + \int_{\zeta_i}^{\infty} e^{B_+(\zeta_i-s)} g_-(s, z(s), z(\gamma(s)))ds,$$

and the improper integral converges and is bounded on $[\zeta_i,\infty)$. But conditions (C3), (C4) on eigenvalues of matrix B_+ imply that $\|e^{B_+(t-\zeta_i)}\widehat{\kappa}\| \to 0$ as $t \to \infty$ only if $\widehat{\kappa} = 0$. In other words, $z(t)$ satisfies (3.6) with $t_0 = \zeta_i$. Hence, by Theorem 3.2 and Lemma 3.3 $(\zeta_i, z_0) \in S_i^+$. The Lemma is proved. □

Let us introduce the following set of points in the (t,z)-space

$$S^+ = \{(t,z) : \zeta_i \leqslant t < \zeta_{i+1}, z = z(t,\zeta_i,z_0), \ \ (\zeta_i,z_0) \in S_i^+, \ \text{ for some } \ i \in \mathbb{Z}\}.$$

Theorem 3.3. *S^+ is an invariant set in the sense of Definition* 3.1.

Proof. Consider the following two cases in turn.

a) Assume that $(t_0,z_0) \in S_i^+, z_0 = (u_0,v_0)$ for a fixed $i \in \mathbb{Z}$. That is, $t_0 = \zeta_i$ and $v_0 = F(\zeta_i,u_0)$. Consider the solution $z(t), z(\zeta_i) = z_0$. We shall need to show that $(t,z(t)) \in S^+$ for all $t \in \mathbb{R}$. We know that $\|z(t)\| \leqslant 2K\|c\|e^{-\alpha(t-\zeta_i)}, t \geqslant \zeta_i$.

If $j > i$, then from the last expression it follows that

$$\|z(t)\| \leqslant 2K\|u_0\|e^{-\alpha(\zeta_j-\zeta_i)}e^{-\alpha(t-\zeta_j)}, \ \ t \geqslant \zeta_j.$$

That is $z(t) \in \Psi_j$, and by Lemma 3.4 $(\zeta_j, z(\zeta_j)) \in S_j^+$.

If $j < i$ then

$$\|z(t)\| \leqslant \max \max_{[\zeta_j,\zeta_i]} \|z(t)\|, \ \ 2K\|u_0\|e^{\alpha(\zeta_i-\zeta_j)}e^{-\alpha(t-\zeta_j)}, \ \ t \geqslant \zeta_j,$$

and again $z(t) \in \Psi_j$. We have obtained that $(\zeta_j, z(\zeta_j)) \in S_j^+$ for all $j \in \mathbb{Z}$.

Consider now point $(t,z(t))$, $\zeta_j < t < \zeta_{j+1}$, for some $j \in \mathbb{Z}$. We have already shown that $z(t) \in \Psi_j$. That is, $(\zeta_j, z(\zeta_j)) \in S_j^+, z(t) = z(t,\zeta_j,z(\zeta_j))$ and $(t,z(t)) \in S^+$ by the definition of S^+. In the light of above discussion one can conclude that $(t,z(t)) \in S^+$ for all $t \in \mathbb{R}$.

b) Assume that $(t_0,z_0) \in S^+ \setminus \bigcup_{i\in\mathbb{Z}} S_i^+$. Then $\zeta_j < t_0 < \zeta_{j+1}$ for some $j \in \mathbb{Z}$. By definition of S^+ there exists a solution $z(t) = (u,v) = z(t,\zeta_j,z_1)$, $z(\zeta_j) = z_1$, $(\zeta_j,z_1) \in S_j^+$, such that $z_0 = z(t_0)$. Particularly,

$$\|z(t)\| \leqslant 2K\|u(\zeta_j)\|e^{-\alpha(t_0-\zeta_j)}e^{-\alpha(t-t_0)}, \ \ t \geqslant t_0.$$

By repeating the discussion in part (a) for the solution and applying the existence and uniqueness Theorem 2.1, one can verify that $(t,z(t)) \in S^+$ for all $t \in \mathbb{R}$.

The theorem is proved. □

On the basis of Theorem 3.3, Lemmas 3.2 and 3.4 we can conclude that there exists an invariant surface S^+ of equation (3.1), such that every solution starting at S^+ tends to zero as $t \to \infty$.

Let $z(t) = (u,v)$ be a solution on S^+. From the proof of the previous theorem it follows that for every $t_0 \in \mathbb{R}$, $\|z(t)\| \leqslant K(t_0,z(t_0))e^{-\alpha(t-t_0)}$, $t \geqslant t_0$, where $K(t_0,z(t_0))$ is a positive constant. Next, It can be shown in the same way as for Lemma 3.4 that the solution satisfies the following equation

$$u(t) = e^{B_-(t-t_0)}u(t_0) + \int_{t_0}^{t} e^{B_-(t-s)} f_-(s,z(s),z(\gamma(s)))ds,$$
$$v(t) = -\int_{t}^{\infty} e^{B_+(t-s)} f_+(s,z(s),z(\gamma(s)))ds.$$

If $t = t_0$ in the last equation, then

$$u(t_0) = u(t_0),$$
$$v(t_0) = -\int_{t_0}^{\infty} e^{B_+(t-s)} f_+(s,z(s,t,u(t_0)),z(\gamma(s),t,u(t_0)))ds.$$

That is, if $(u,v) \in S^+$ then $v = F(t,u)$, where

$$F(t,u) = -\int_{t}^{\infty} e^{B_+(t-s)} f_+(s,z(s,t,u),z(\gamma(s),t,u))ds.$$

It is obvious that $F(t,u)$ is a continuous function in both arguments.

Theorem 3.4. *Assume that conditions* (C1)–(C5) *are fulfilled. Then for an arbitrarily small positive* $\tilde{\alpha}$ *and a sufficiently small Lipschitz constant* l *there exists a function* $G(\zeta_i,v)$, $i \in \mathbb{Z}$, *from* $\mathbb{R}^{n-k}$ *to* $\mathbb{R}^k$, *satisfying*

$$G(\zeta_i,0) = 0, \tag{3.18}$$
$$\|G(\zeta_i,d_1) - G(\zeta_i,d_2)\| \leqslant P\ell\|d_1 - d_2\| \tag{3.19}$$

for all d_1, d_2, *such that a solution* $z(t)$ *of* (3.1) *with* $z(\zeta_i) = (G(\zeta_i,v_0),v_0)$, $v_0 \in \mathbb{R}^{n-k}$, *is defined on* $\mathbb{R}$ *and satisfies*

$$\|z(t)\| \leqslant D\|v_0\|e^{-\tilde{\alpha}(t-\zeta_i)}, \quad t \leqslant \zeta_i, \tag{3.20}$$

where P, $D > 0$ *are constants.*

Proof. Let us denote $\kappa = \frac{\sigma}{2}$ and $\eta(t) = z(t)e^{\kappa t}$. Then system (3.1) is transformed into the following form

$$\frac{d\xi}{dt} = (B_- + \kappa I)\xi + g_-(t,\eta(t),\eta(\gamma(t))),$$
$$\frac{d\omega}{dt} = (B_+ + \kappa I)\zeta + g_+(t,\eta(t),\eta(\gamma(t))), \tag{3.21}$$

where $\eta = (\xi,\omega)$,I is an identity matrix, $\eta(\gamma(t)) = z(\gamma(t))e^{-\kappa\gamma(t)}$, and $g(t,z,y) = (g_-,g_+) = e^{\kappa t}f(t,ze^{-\kappa t},ye^{-\kappa\gamma(t)})$. It is easy to see that the function $g(t,z,y)$ satisfies the

Lipschitz condition in z, y with a constant $\ell e^{\kappa\theta}$, and the eigenvalues of the matrices $B_- + \kappa I$ and $B_+ + \kappa I$ have negative and positive real parts, respectively, such that

$$\mu + \kappa = \max_{j=\overline{1,k}} \Re\lambda_j(B_- + \kappa I) < -\sigma + \kappa < 0,$$

and

$$\min_{j=\overline{k+1,n}} \Re\lambda_j(B_+ + \kappa I) = \kappa > 0.$$

Fix a positive number $\overline{\kappa} < \min\{\sigma - \kappa, \kappa\} = \kappa$. There exists a positive number $\overline{K}$ such that

$$\|e^{(B_- + \kappa I)(t-s)}\| \leqslant \overline{K} e^{-\overline{\kappa}(t-s)}, \ t \geqslant s,$$

and

$$\|e^{(B_+ + \kappa I)(t-s)}\| \leqslant \overline{K} e^{\overline{\kappa}(t-s)}, \quad t \leqslant s.$$

In order to finish the proof we need the following two assertions which can be proved similarly to Lemma 3.2 and Theorem 3.2.

Lemma 3.5. *Fix $N \in \mathbb{R}$, $N > 0$, and assume that conditions* (C1)–(C3) *are valid. A continuous function $\eta(t) = (\xi, \omega)$, $\|\eta(t)\| \leqslant N e^{\widetilde{\alpha}(t-t_0)}$, $0 < \widetilde{\alpha} < \overline{\kappa}$, $t \leqslant t_0$, $\theta_j < t_0 \leqslant \theta_{j+1}$, is a solution of* (3.21) *on $(-\infty, t_0]$ if and only if $\eta(t)$ is a solution of the following system of integral equations*

$$\xi(t) = \int_{-\infty}^{t} e^{(B_- + \kappa I)(t-s)} g_-(s, \eta(s), \eta(\gamma(s)))ds,$$

$$\omega(t) = e^{(B_+ + \kappa I)(t-t_0)} \omega(t_0) + \int_{t_0}^{t} e^{(B_+ + \kappa I)(t-s)} g_+(s, \eta(s), \eta(\gamma(s)))ds. \tag{3.22}$$

Lemma 3.6. *Assume that conditions* (C1)–(C5) *are fulfilled. Then for an arbitrary $\alpha_1 \in (0, \overline{\kappa})$ and a sufficiently small Lipschitz constant l there exists a function $\overline{G}(\zeta_i, u)$, $i \in \mathbb{Z}$, satisfying*

$$\overline{G}(\zeta_i, 0) = 0, \tag{3.23}$$

$$\|\overline{G}(\zeta_i, d_1) - \overline{G}(\zeta_i, d_2)\| \leqslant Pl\|d_1 - d_2\|, \tag{3.24}$$

where P is a positive constant, and such that $\xi_0 = \overline{G}(\zeta_i, \omega_0)$ defines a solution $\eta(t)$ of (3.1) *with $\eta(\zeta_i) = (\overline{G}(\zeta_i, \omega_0), \omega_0)$ and*

$$\|\eta(t)\| \leqslant 2\overline{K}\|\omega_0\| e^{\alpha_1(t-\zeta_i)}, \ t \leqslant \zeta_i. \tag{3.25}$$

One can show that if $\eta(t,r,c) = (\xi,\omega)$ is a solution of (3.1) such that $\omega(r) = c$, then

$$\overline{G}(t,\omega) = \int_{-\infty}^{t} e^{(B_- + \kappa I)(t-s)} g_-(s,\eta(s,t,\omega),\eta(\gamma(s),t,\omega))ds. \tag{3.26}$$

Next, applying the inverse transformation $z(t) = \eta(t)e^{-\kappa t}$ we can define a new function $G(\zeta_i,v) = e^{-\kappa t}\overline{G}(\zeta_i, ve^{\kappa t})$ and check that

$$\|G(\zeta_i,v_1) - G(\zeta_i,v_2)\| \leqslant Pl\|v_1 - v_2\|, \tag{3.27}$$

and $\|z(t)\| \leqslant 2\overline{K}\|v_0\|e^{(\alpha_1-\kappa)(t-\zeta_i)}$, $t \leqslant \zeta_i$, if $u_0 = G(\zeta_i,v_0)$. If we denote now $D = 2\overline{K}$ and choose $\overline{\kappa}$ sufficiently close to κ then we can take $\alpha_1 = \kappa - \widetilde{\alpha} > 0$ such that the last inequality implies (3.20). The theorem is proved. □

Using the equation $\xi = \overline{G}(t,\omega)$ we can define, similarly to S^+, an integral surface $\widetilde{S}_0$ such that every solution of (3.21) starting on $\widetilde{S}_0$ tends to the origin as $t \to -\infty$. Let us denote by S_0 the integral set defined by the equation $u = G(t,v)$.

3.3 Stability of the zero solution

The stability is investigated for $t_0 = 0$ in the theory of differential equations with piecewise constant argument where the argument is delayed [85, 317]. Nevertheless, Theorem 3.1 and continuous dependence on the initial value allow us to investigate stability, assuming that the initial moment t_0 can be an arbitrary real number, and there is not any necessity to involve the concept of initial interval for an initial value problem despite the fact that (3.1) is an equation with a deviating argument. Thus we arrive that definitions of stability for differential equations with piecewise constant arguments under conditions of our book coincide with the definitions for ordinary differential equations [153]. Additionally to the definitions of the Section 2.3 we consider the following one.

Definition 3.3. The zero solution of (3.1) is exponentially stable if there exists an $\alpha > 0$, and for every $\varepsilon > 0$ and t_0 there exists a $\delta(\varepsilon,t_0) > 0$, such that

$$\|z(t,t_0,z_0)\| \leqslant \varepsilon e^{-\alpha(t-t_0)}$$

for all $t \geqslant t_0$, whenever $\|z_0\| < \delta$. If the δ above is independent of t_0, then the zero solution is uniformly exponentially stable.

Together with Theorem 3.2 with $k = n$ and inequality (2.24), the following assertion is valid.

Theorem 3.5. *Assume that conditions* (C1), (C2) *and* (C5) *are fulfilled, and all eigenvalues of the matrix A have negative real parts. Then the zero solution of* (3.1) *is uniformly exponentially stable if the Lipschitz constant l is sufficiently small.*

3.4 Stability of the integral surface

We shall introduce a notion of stability for an integral set [68, 264, 336]. Let $M \subset \mathbb{R} \times \mathbb{R}^n$ and $M(\sigma)$ be an integral surface of (3.1) and the intersection of M with the hyperplane $t = \sigma$, respectively. Moreover, we denote by $d(z, M(t))$ the distance between a point $z \in \mathbb{R}^n$ and the set $M(t)$, $t \in \mathbb{R}$.

Definition 3.4. M is a stable integral surface of (3.1), if for any $\varepsilon > 0$ there exists a number $\delta > 0$, $\delta = \delta(\varepsilon, t_0)$, such that if $d(z_0, M(t_0)) < \delta$, then $d(z(t, t_0, z_0), M(t)) < \varepsilon$ for all $t \geqslant t_0$.

Definition 3.5. M is a uniformly stable integral surface of (3.1) if the δ above is independent of t_0.

Definition 3.6. A stable integral surface M is stable in large, if every solution of (3.1) approaches M as $t \to \infty$.

Lemma 3.7. *If the Lipschitz constant l is sufficiently small, then for every solution $z(t) = (u, v)$ of* (3.1) *there exists a solution $\mu(t) = (\phi, \psi)$ on S_0 such that*

$$\|u(t) - \phi(t)\| \leqslant 2K\|u(\zeta_i) - \phi(\zeta_i)\| \mathrm{e}^{-\alpha(t-\zeta_i)},$$
$$\|v(t) - \psi(t)\| \leqslant K\|u(\zeta_i) - \phi(\zeta_i)\| \mathrm{e}^{-\alpha(t-\zeta_i)}, \quad \zeta_i \leqslant t, \tag{3.28}$$

where α is the coefficient defined for Theorem 3.2.

Proof. Fix a solution $z(t, \zeta_i, z_0)$ of (3.1). Denote by $\mu(t, \zeta_i, d) = (\phi, \psi)$ a solution of (3.1) such that $\psi(\zeta_i, \zeta_i, d) = d$, $\phi(\zeta_i, \zeta_i, d) = G(\zeta_i, d)$. Let us carry out the transformation

$$X(t) = u - \phi(t), \quad Y(t) = v - \psi(t), \tag{3.29}$$

in system (3.1) and denote $Z = (X, Y)$. The transformed equation has the form

$$\frac{dX}{dt} = B_- X + Q_-(t, Z(t), Z(\gamma(t)), d),$$
$$\frac{dY}{dt} = B_+ Y + Q_+(t, Z(t), Z(\gamma(t)), d), \tag{3.30}$$

where $Q(t, X, Y, d) = (Q_-, Q_+) = f(t, z(t), z(\gamma(t))) - f(t, \mu(t), \mu(\gamma(t)))$. One can see that Q satisfies the Lipschitz condition with respect to Z and d, with the same constant l. In view of the Theorem 3.2 and according to (3.10), (3.11) there exists a function $\widetilde{F}$ such that the equation $Y = \widetilde{F}(\zeta_i, X, d)$ defines a set for (3.30) which satisfies the following properties

$$\widetilde{F}(\zeta_i, 0, d) = 0,$$
$$\|\widetilde{F}(\zeta_i, X_1, d) - \widetilde{F}(\zeta_i, X_2, d)\| \leqslant pK\ell\|X_1 - X_2\|. \tag{3.31}$$

Using formulas similar to (3.14), one can see that every solution $Z(t)$, $Z(\zeta_i) = (X_0, \widetilde{F}(\zeta_i, X_0, d))$, satisfies

$$\|X(t)\| \leqslant K\mathrm{e}^{-\sigma(t-\zeta_i)}\|X_0\| + \frac{2K^2\ell(1+\mathrm{e}^{\alpha\theta})}{\sigma-\alpha}\mathrm{e}^{-\alpha(t-\zeta_i)}\|X_0\|,$$
$$\|Y(t)\| \leqslant 2\beta K^2\ell(1+\mathrm{e}^{\alpha\theta})\mathrm{e}^{-\alpha(t-\zeta_i)}\|X_0\|. \tag{3.32}$$

Let us show that there exist X_0 and d such that for solutions $z(t)$ and $(X(t), Y(t))$ of systems (3.1) and (3.30), respectively, we have

$$X(t) = u(t, \zeta_i, z_0) - \phi(t), \quad Y(t) = v(t, \zeta_i, z_0) - \psi(t).$$

The last equalities for $t = \zeta_i$ have the form

$$X_0 = u_0 - G(\zeta_i, d), \quad \widetilde{F}(\zeta_i, X_0, d) = v_0 - d. \tag{3.33}$$

Now, we consider the system as an equation with respect to X_0 and d and we need to show that it has a solution for every pair (u_0, v_0). Equation (3.33) implies that

$$d = v_0 - \widetilde{F}(\zeta_i, u_0 - G(\zeta_i, d), d). \tag{3.34}$$

Applying properties (3.31) of the function $\widetilde{F}$ and equality (3.34) we can write that $\|d - v_0\| \leqslant pK\ell\|u_0 - G(\zeta_i, d)\|$. Using the fact that the function G satisfies the Lipschitz condition and the inequality $\|d - v_0\| \leqslant pK\ell\|u_0 - G(\zeta_i, v_0)\| + pK\ell\|G(\zeta_i, d) - G(\zeta_i, v_0)\|$, one can show that

$$\|d - v_0\| \leqslant \frac{pK\ell}{1 - pPK\ell^2}\|u_0 - G(\zeta_i, v_0)\|. \tag{3.35}$$

We assume that $1 - pPK\ell^2 > 0$, $pK\ell(1 + P\ell) \leqslant 1$, and will consider the ball $\widehat{B} = \{d : \|d - v_0\| \leqslant \|u_0 - G(\zeta_i, v_0)\|\}$. Inequality (3.35) implies that (3.34) transforms $\widehat{B}$ into itself, and by Brauer's theorem there exists a fixed point of the transformation. Denote the point by $\overline{d}$. Substituting $\overline{d}$ into the first equation of (3.33) we obtain the value $\overline{X}_0$. The pair $(\overline{X}_0, \overline{d})$ satisfies system (3.33). Now, applying (3.32), (3.9) and the theorem of existence and uniqueness we complete the proof of the Lemma. □

Theorem 3.6. *If the Lipschitz constant ℓ is sufficiently small, then the surface S_0 is stable in large.*

Proof. First, let us consider stability. We shall show that the surface is in fact uniformly stable. Due to the continuous dependence of the solutions on the initial values, without loss of generality, one can consider $t_0 = \zeta_i$, $i \in \mathbb{Z}$.

Fix $t_0 = \zeta_i$ for some $i \in \mathbb{Z}$ and $\varepsilon > 0$. Consider a solution $z(t) = z(t, t_0, z_0)$, $z_0 \in \mathbb{R}^n$, $z(t) = (u, v)$. By Lemma 3.7 for a given solution $z(t) = (u, v)$ there exists a solution $\mu(t) = \mu(t, \zeta_i, d) = (\phi, \psi)$, $\psi(\zeta_i) = d$, on S_0 such that (3.28) is valid and, hence,

$$\|z(t) - \mu(t)\| \leqslant 2K\|z_0 - \mu(t_0)\|e^{-\alpha(t-t_0)}, \quad t \geqslant t_0. \tag{3.36}$$

Denote by S_0^i a set of all points $z = (u, v)$ such that $u = G(\zeta_i, v)$ and $\rho_0 = d(z_0, S_0^i)$. There exists a point $z_1 \in S_0^i$, $z_1 = (u_1, v_1)$, such that $\rho_0 = \|z_0 - z_1\|$. Using (3.19) we can obtain that

$$\begin{aligned}\|u_0 - G(\zeta_i, v_0)\| &\leqslant \|u_0 - u_1\| + \|u_1 - G(\zeta_i, v_1)\| + \|G(\zeta_i, v_1) - G(\zeta_i, v_0)\| \\ &\leqslant \|u_0 - u_1\| + P\ell\|v_1 - v_0\| \\ &\leqslant (1 + P\ell)\rho_0. \end{aligned} \tag{3.37}$$

Next, applying (3.35) and the last inequality we have that

$$\|d - v_0\| \leqslant (1 + P\ell)\frac{pK\ell}{1 - pPK\ell^2}\rho_0. \tag{3.38}$$

Then

$$\begin{aligned}\|u_0 - G(\zeta_i, d)\| &\leqslant \|u_0 - G(\zeta_i, v_0)\| + \|G(\zeta_i, v_0) - G(\zeta_i, d)\| \\ &\leqslant P\ell\|d - v_0\| + (1 + P\ell)\rho_0 \\ &\leqslant P\ell(1 + P\ell)\frac{pK\ell}{1 - pPK\ell^2}\rho_0 + (1 + P\ell)\rho_0. \end{aligned} \tag{3.39}$$

Finally, using (3.38) and (3.39) we conclude that

$$\begin{aligned}\|z_0 - \mu(t_0)\| &\leqslant \|u_0 - \phi(t_0)\| + \|v_0 - \psi(t_0)\| \\ &= \|u_0 - G(\zeta_i, d)\| + \|v_0 - d\| \\ &\leqslant \frac{(1 + P\ell)(1 + pK\ell)}{1 - pPK\ell^2}\rho_0. \end{aligned} \tag{3.40}$$

Now, inequalities (3.36) and (3.40) imply uniform stability with

$$\delta(\varepsilon) = \varepsilon\frac{1 - pPK\ell^2}{2K(1 + P\ell)(1 + pK\ell)}.$$

Stability in large follows immediately from (3.36). The theorem is proved. □

3.5 The reduction principle

In this section we need the following conditions.

(C6) The function $f(t, z, w)$ is uniformly continuously differentiable in z, w for all t, z, w, and

$$\frac{\partial f(t, 0, 0)}{\partial z} = 0, \quad \frac{\partial f(t, 0, 0)}{\partial w} = 0.$$

(C7) If we denote by $\lambda_j, j = \overline{1,n}$, the eigenvalues of the matrix A, then there exists a positive integer k such that $\mu = \max_{j=\overline{1,k}} \Re\lambda_j < 0$, and $\Re\lambda_j = 0,\ j = \overline{k+1,n}$, where $\Re\lambda_j$ denotes the real part of the eigenvalue λ_j of the matrix A.

Denote $T(h) = \{(t,z) \in \mathbb{R} \times \mathbb{R}^n : \|z\| < h\}$ for a fixed number $h > 0$. Assume that $\varepsilon_0 > 0$ is sufficiently small for the Lipschitz constant ℓ, provided by (C6), to satisfy all conditions of Theorem 3.4 in $T(\varepsilon_0)$.
Denote $\varepsilon_1 = \frac{\varepsilon_0}{2\overline{K}}$, where $\overline{K}$ is the constant from (3.25).
By Lemma 3.6 there exists a local integral manifold of (3.21) in $T(\varepsilon_1)$, such that a solution starting on the manifold is continuable to $-\infty$, and is exponentially decaying.
By using the inverse transformation $z = \eta e^{-\kappa t}$, we obtain a local integral manifold of (3.1) in $T(\varepsilon_1)$ given by the equation $u = G(t,v)$. Solutions of (3.1) on the manifold are not necessarily continuable to $-\infty$ in $T(\varepsilon_1)$. For the function G, condition (3.27) is true, and $G(t,0) = 0,\ t \in \mathbb{R}$. On the local manifold, solutions of (3.1) satisfy the following system

$$\frac{dv}{dt} = B_+ v + f_+(t, G(t,v(t)), v(t)), (G(t,v(\gamma(t))), v(\gamma(t)))). \tag{3.41}$$

We can see that the function $f_+(t,(G(t,v),v),(G(t,\overline{v}),\overline{v}))$ satisfies the Lipschitz condition in v, $\overline{v}$ with the constant $\ell(1+P\ell)$.

Theorem 3.7. *Assume that conditions* (C1)–(C2), (C4)–(C7) *are fulfilled. The trivial solution of* (3.1) *is stable, asymptotically stable or unstable in Lyapunov sense, if the trivial solution of* (3.41) *is stable, asymptotically stable or unstable, respectively.*

Proof. Consider system (3.1) in $T(\varepsilon_1)$. We assume, additionally, that ε_0 is sufficiently small, so that the conditions of Theorem 3.7 on smallness of the Lipschitz constant are valid in $T(\varepsilon_0)$, and, we also have

$$1 + P\ell \leqslant 2. \tag{3.42}$$

Suppose that the zero solution of (3.41) is stable in the sense of Definition 2.2. Fix an $\varepsilon > 0$ and assume $\varepsilon < \varepsilon_1$ without loss of generality.
In view of the equation (3.4), we can assume that $t_0 = \zeta_i$ for some fixed $i \in \mathbb{Z}$. Fix a positive number ν such that the inequality $2\nu(1+P\ell) < 1$ is true. Stability implies the existence of $\delta > 0,\ 0 < 2\delta < \varepsilon$, such that if $d \in \mathbb{R}^{n-k}$, $\|d\| < 2\delta$, then the solution $v = \psi(t,\zeta_i,d)$ of (3.41) satisfies the inequality

$$\|\psi(t,\zeta_i,d)\| < \nu\varepsilon, \quad \zeta_i \leqslant t. \tag{3.43}$$

Let u_0 and v_0 be arbitrary vectors satisfying $\|u_0\| + \|v_0\| < \delta$. Denote by $z(t) = z(t,\zeta_i,z_0)$, $z(\zeta_i) = z_0$, $z_0 = (u_0, v_0)$, a solution of (3.1). From now on we shall follow the proof of

Theorem 3.7, modifying it for the local case. Let $\mu(t) = \mu(t, \zeta_i, d) = (\phi, \psi)$ be a solution of (3.1) such that $\psi(\zeta_i, \zeta_i, d) = d$, $\phi(\zeta_i, \zeta_i, d) = G(\zeta_i, d)$ and $\psi(\zeta_i, \zeta_i, d)$ satisfies (3.43). Applying (3.43) and the Lipschitz condition on G, we have that $\|\phi(t, \zeta_i, d)\| \leqslant P\ell v\varepsilon$, $\zeta_i \leqslant t$. Then $\|\mu(t, \zeta_i, d)\| \leqslant (1 + P\ell)v\varepsilon$, $\zeta_i \geqslant t$. Finally, using the choice of v we can write that

$$\|\mu(t)\| < \frac{1}{2}\varepsilon. \tag{3.44}$$

Applying the transformation (3.29) equation (3.30) can be obtained. It follows from (3.44) that (3.29) transforms neighborhood $T(\frac{\varepsilon}{2})$ for (3.30) into neighborhood $T(\varepsilon)$ for (3.1). So, the conditions set by Theorem 3.7 for the coefficient ℓ are valid if (3.30) is considered in the $\frac{\varepsilon}{2}$-neighborhood of $X = 0, Y = 0, t \in \mathbb{R}$.

Now, if we assume that

$$\|X(\zeta_i)\| < \frac{\varepsilon}{2K(1+2p\ell)}, \tag{3.45}$$

then similarly to the sequence (u_m, v_m) in Theorem 3.2 we can construct a sequence $Z_m = (X_m, Y_m)$, $m \geqslant 0$, such that $(X_0, Y_0) = (0,0)^T$,

$$X_{m+1}(t) = e^{B_-(t-\zeta_i)}X(\zeta_i) + \int_{\zeta_i}^{t} e^{B_-(t-s)}Q_+(s, Z_m(s), Z_m(\gamma(s)))ds,$$

$$Y_{m+1}(t) = -\int_t^{\infty} e^{B_+(t-s)}Q_-(s, Z_m(s), Z_m(\gamma(s)))ds,$$

$$\|X_m(t)\| \leqslant Ke^{-\sigma(t-\zeta_i)}\|X(\zeta_i)\| + \frac{2K^2\ell(1+e^{\alpha\theta})}{\sigma-\alpha}e^{-\alpha(t-\zeta_i)}\|X(\zeta_i)\|,$$

$$\|Y_m(t)\| \leqslant 2\beta K^2\ell(1+e^{\alpha\theta})e^{-\alpha(t-\zeta_i)}\|X(\zeta_i)\|, \tag{3.46}$$

and, therefore,

$$\|Z_m(t)\| \leqslant K(1+2p\ell)\|X(\zeta_i)\|e^{-\alpha(t-\zeta_i)} < \frac{\varepsilon}{2}, \quad \zeta_i \leqslant t.$$

The limit function $Z(t) = (X(t), Y(t))$ of the sequence is a solution of (3.30) and satisfies

$$\|Z(t)\| \leqslant K(1+2p\ell)\|X(\zeta_i)\|e^{-\alpha(t-\zeta_i)} < \frac{\varepsilon}{2}, \ \zeta_i \leqslant t. \tag{3.47}$$

Thus, define a function $\widetilde{F}(\zeta_i, X, d)$ such that $Y(\zeta_i) = \widetilde{F}(\zeta_i, X(\zeta_i), d)$, which satisfies (3.31). Next, by using (3.42) and (3.46) the existence of a pair $(\overline{X}_0, \overline{d})$ can be proved such that

$$\overline{X}_0 = u_0 - G(\zeta_i, \overline{d}), \ v_0 - \overline{d} = \widetilde{F}(\zeta_i, \overline{X}_0, \overline{d}), \ \|\overline{X}_0\| < \frac{\varepsilon}{2K(1+2p\ell)}, \ \|\overline{d}\| < 2\delta.$$

Now, the transformation (3.29) and the inequality (3.47) imply that

$$\|z(t, \zeta_i, z_0) - \mu(t, \zeta_i, \overline{d})\| \leqslant K(1+2p\ell)\|\overline{X}_0\|e^{-\alpha(t-\zeta_i)}, \ \zeta_i \leqslant t. \tag{3.48}$$

From (3.44) and (3.48) it follows that $\|z(t, \zeta_i, z_0)\| < \varepsilon$, $\zeta_i \leqslant t$.

In view of the previous inequality, we can conclude that the zero solution of (3.1) is stable. Moreover, if the zero solution of (3.41) is asymptotically stable, then (3.48) implies that the zero solution of (3.1) is also asymptotically stable. Finally, it is obvious that if the zero solution of (3.41) is unstable, then the trivial solution of (3.1) is unstable as well. The theorem is proved. □

Notes

The method of integral manifolds, is a very powerful instrument of any theory of differential equations. It has been considered for differential equations with piecewise constant argument in a series of papers by Papaschinopoulos, G. [249–251]. The author considers systems which are reducible to discrete equations. In articles [8–10] by applying the method of equivalent integral equations, and analyzing thoroughly existence and uniqueness of solutions the investigation is in the general case with solutions at the non-deviated time are present non-linearly. Papers [8,9] consider existence of integral manifolds of perturbed hyperbolic linear system. The main results of this chapter were published in [10]. In the present chapter we extend the method of the center manifold [264, 265] for non-autonomous systems. This result is of strong importance for reduction of dimension, particularly in bifurcation theory [68].

Chapter 4

The small parameter and differential equations with piecewise constant argument

The problem of the existence of periodic solutions is one of the most interesting topics for applications. The method of small parameter is introduced by Poincaré [267] to investigate the problem, and it has been developed by many authors (see, for example, [218, 273], and the references cited therein). This method remains as one of the most effective methods for this problem and it is important that the results obtained in this field can be extended to the bifurcation theory [48, 233]. In this chapter, we investigate the existence and stability of periodic solutions of quasilinear system with piecewise constant an argument and a small parameter in noncritical and critical cases. Theorems on continuous dependence of solutions with respect to initial conditions and parameters, and an analogue of the Gronwall-Bellman lemma are also proved. Examples illustrating the obtained results are constructed as well. In the first section of this chapter we consider a simpler problem of a non-critical case with the delay argument-function $\beta(t)$. In the last section critical case is considered as well as the argument-function $\gamma(t)$ of the mixed, advanced-delayed, type is used for the system.

4.1 Periodic solutions: the non-critical case

In this section, the quasilinear system with piecewise constant an argument and a small parameter is considered. The noncritical case, when the corresponding linear ordinary differential equations have no nontrivial periodic solution is investigated, and the existence and stability of periodic solutions of the system are discussed.

The main goal of this section is the application of the method of small parameter to the following quasilinear system

$$x'(t) = A(t)x(t) + f(t) + \mu g(t, x(t), x(\beta(t)), \mu), \tag{4.1}$$

where $t \in \mathbb{R}, x \in \mathbb{R}^n$, and μ is a small parameter which belongs to an interval $(-\mu_0, \mu_0) \subset \mathbb{R}$.

The functions $f(t)$, $g(t,x,y,\mu)$ are n-dimensional vectors, $A(t)$ is an $n\times n$ matrix for $n\in\mathbb{N}$. The argument-function $\beta(t)=\theta_j$ if $\theta_j\leqslant t<\theta_{j+1}$, $j\in\mathbb{Z}$, is defined on $\mathbb{R}$. Here, θ_j, $j\in\mathbb{Z}$, is a strictly ordered sequence of real numbers, $|\theta_i|\to\infty$ as $|i|\to\infty$, and there exist two positive real numbers $\theta,\overline{\theta}$ such that $\theta\leqslant\theta_{j+1}-\theta_j\leqslant\overline{\theta}$, $j\in\mathbb{Z}$.
In the non-critical case, when the corresponding linear homogeneous system does not have any nontrivial periodic solution, we apply the method of small parameter [218,233,267] to investigate the problem of the existence of a periodic solution of (4.1).
Throughout this chapter, the following assumptions for the system (4.1) will be needed :

(H1) Functions $A(t)$, $f(t)$ and $g(t,x,y,\mu)$ are continuous in all of their arguments.

(H2) The function $g(t,x,y,\mu)$ satisfies Lipschitz continuity with a constant L such that

$$\|g(t,\widetilde{x},\widetilde{y},\mu)-g(t,x,y,\mu)\|\leqslant L[\|\widetilde{x}-x\|+\|\widetilde{y}-y\|],$$

for all $t\in\mathbb{R}$, $\widetilde{x}, x, \widetilde{y}, y\in\mathbb{R}^n$, $\mu\in J$.

The following definitions and theorems on the existence, uniqueness and continuation of solutions are very similar to those from the Chapter 3. Let $X(t)$ be a fundamental matrix solution of the homogeneous system, corresponding to the system (4.1),

$$x'(t)=A(t)x(t), \tag{4.2}$$

such that $X(0)=\mathscr{I}$, where $\mathscr{I}$ is an $n\times n$ identity matrix. Denote by $X(t,s)=X(t)X^{-1}(s)$, $t, s\in\mathbb{R}$, the transition matrix of (4.2). Let $\kappa=\sup_{t\in\mathbb{R}}\|A(t)\|<\infty$.

Lemma 4.1 ([148]). *Assume* (H1) *is satisfied. Then, the inequality*

$$\|X(t,s)\|\leqslant\exp(\kappa|t-s|),\ \ t, s\in\mathbb{R}, \tag{4.3}$$

holds.

Lemma 4.2 ([148]). *Assume* (H1) *is satisfied. Then, the inequality*

$$m\leqslant\|X(t,s)\|\leqslant M,$$

where $m=\exp(-\kappa\overline{\theta})$, $M=\exp(\kappa\overline{\theta})$, *holds for* $|t-s|\leqslant\overline{\theta}$.

In the following two lemmas we introduce the integral equations corresponding to equation (4.1). The proof is very similar to that of Lemma 3.2.

Lemma 4.3. *Suppose* (H1) *is satisfied. A function* $x(t)=x(t,t_0,x_0,\mu)$, *where* t_0 *is a real fixed number, is a solution of* (4.1) *on* $\mathbb{R}$ *if and only if it is a solution on* $\mathbb{R}$ *of the following integral equation*

$$x(t)=X(t,t_0)x_0+\int_{t_0}^{t}X(t,s)[f(s)+\mu g(s,x(s),x(\beta(s)),\mu)]\,ds. \tag{4.4}$$

Lemma 4.4. *Suppose* (H1) *is satisfied. A function* $x(t) = x(t,t_0,x_0,\mu)$, *where* t_0 *is a real fixed number, is a solution of* (4.1) *on* $\mathbb{R}$ *if and only if it is a solution on* $\mathbb{R}$ *of the following integral equation*

$$x(t) = x_0 + \int_{t_0}^{t} [A(s)x(s) + f(s) + \mu g(s,x(s),x(\beta(s)),\mu)]\,ds. \tag{4.5}$$

The following example emphasizes that a solution of an equation with piecewise constant argument may exist on a half-axis but not on the whole real axis, unless some conditions hold.

Example 4.1. Consider the differential equation

$$x'(t) = \alpha x(t) - \mu x^2(\beta(t)), \tag{4.6}$$

where $x \in \mathbb{R}$, $t \in \mathbb{R}$, α is a real positive constant, and $\beta(t) = \theta_j$ if $\theta_j \leqslant t < \theta_{j+1}$, $j \in \mathbb{Z}$, $\theta_{2i-1} = 4i-1$, $\theta_{2i} = 4i$, $i \in \mathbb{Z}$. The distance $\theta_{j+1} - \theta_j$, $j \in \mathbb{Z}$, is either equal to $\theta = 1$ or to $\overline{\theta} = 3$. Let us fix $x_0 \in \mathbb{R}$. We shall look for conditions on α and μ such that a solution $x(t) = x(t,\theta_0,x_0,\mu)$, $x(\theta_0) = x_0$, $x_0 > 0$, of (4.6) exists.

If $\mu = 0$, it is easy to see that the solution $x(t)$ of (4.6) exists uniquely, and it is positive and not bounded on $\mathbb{R}$.

Suppose $\mu > 0$. Using the transformation $x(t) = y(t)/\mu$, from (4.6) we obtain

$$y'(t) = \alpha y(t) - y^2(\beta(t)). \tag{4.7}$$

Let $y(t) = y(t,\theta_0,y_0)$ be a solution of (4.7) with $y(\theta_0) = y_0$, $y_0 > 0$. Denote $y_k = y(\theta_k)$, $k \in \mathbb{Z}$. We first consider the existence and uniqueness of the solution $y(t)$. Let us start with $t \in [\theta_0,\infty)$, that is, when the time is increasing.

If $t \in [\theta_0,\theta_1]$, then $y(t)$ is a solution of the equation

$$y'(t) = \alpha y(t) - y_0^2,$$

which is a linear non-homogeneous differential equation with a constant coefficient, and that is why the solution $y(t)$ is uniquely defined on $[\theta_0,\theta_1]$. The rest can be deduced from the arguments of mathematical induction. That is, the solution $y(t)$ and the corresponding solution $x(t)$ exist uniquely on $[\theta_0,\infty)$.

Next, let us consider the solution for decreasing time. We will show that if

$$y_0 \leqslant \frac{\alpha \exp(2\alpha)}{4[\exp(\alpha)-1]}, \tag{4.8}$$

$$\frac{\alpha \exp(\alpha)}{\exp(\alpha)-1} \leqslant \frac{\alpha \exp(6\alpha)}{4[\exp(3\alpha)-1]}, \tag{4.9}$$

$$\frac{\alpha \exp(3\alpha)}{\exp(3\alpha)-1} \leqslant \frac{\alpha \exp(2\alpha)}{4[\exp(\alpha)-1]} \tag{4.10}$$

are satisfied, then the solution $y(t) = y(t, \theta_0, y_0)$ exists on $(-\infty, \theta_0]$.

If $t \in [\theta_{-1}, \theta_0]$, then $y(t)$ coincides with the solution of the following ordinary differential equation

$$y'(t) = \alpha y(t) - y_{-1}^2. \tag{4.11}$$

Using the equivalent integral equation of (4.11), it can be written that

$$y(t) = \exp(\alpha(t-\theta_{-1}))y_{-1} + \frac{1}{\alpha}[1-\exp(\alpha(t-\theta_{-1}))]y_{-1}^2. \tag{4.12}$$

Denote $z = y_{-1}$. It is easy to see that the solution $y(t)$ exists on $[\theta_{-1}, \theta_0]$, if the quadratic equation for z, obtained from (4.12) with $t = \theta_0$,

$$z^2 - \frac{\alpha \exp(\alpha)}{\exp(\alpha)-1} z + \frac{\alpha}{\exp(\alpha)-1} y_0 = 0 \tag{4.13}$$

has a real root. The last equation has a real root, if the inequality (4.8) is valid. Hence, if (4.8) holds, then the solution $y(t)$ exists on $[\theta_{-1}, \theta_0]$, but is not necessarily unique.

Suppose that the inequality (4.8) is satisfied. In this case, it can be easily checked that the roots $z_{1,2}$ of the equation (4.13) satisfy the inequality

$$0 \leqslant z_{1,2} \leqslant \frac{\alpha \exp(\alpha)}{\exp(\alpha)-1}. \tag{4.14}$$

Denote $z = y_{-2}$. If $t \in [\theta_{-2}, \theta_{-1}]$, one can similarly obtain that the solution $y(t)$ exists on $[\theta_{-2}, \theta_{-1}]$, if the quadratic equation

$$z^2 - \frac{\alpha \exp(3\alpha)}{\exp(3\alpha)-1} z + \frac{\alpha}{\exp(3\alpha)-1} y_{-1} = 0 \tag{4.15}$$

has a real root. The last equation has a real root, if the inequality

$$y_{-1} \leqslant \frac{\alpha \exp(6\alpha)}{4[\exp(3\alpha)-1]} \tag{4.16}$$

is satisfied. By means of (4.14) and (4.16), it is clear that if the inequality (4.9) holds, then the solution $y(t)$ exists on $[\theta_{-2}, \theta_{-1}]$.

It is easy to see that the roots $z_{3,4}$ of equation (4.15) satisfy the inequality

$$0 \leqslant z_{3,4} \leqslant \frac{\alpha \exp(3\alpha)}{\exp(3\alpha)-1} \tag{4.17}$$

provided that the inequality (4.9) is valid.

If $t \in [\theta_{-3}, \theta_{-2}]$, we then have a quadratic equation similar to (4.13), and

$$y_{-2} \leqslant \frac{\alpha \exp(2\alpha)}{4[\exp(\alpha)-1]} \tag{4.18}$$

holds. Therefore, the solution $y(t)$ exists on $[\theta_{-3}, \theta_{-2}]$. Finally, using the inequalities (4.10), (4.17) and (4.18) one can see that the solution $y(t)$ exists on $[\theta_{-4}, \theta_{-3}]$.

By using the arguments of mathematical induction, we can conclude that if the inequalities (4.8)–(4.10) are satisfied, then the solution $y(t,\theta_0,y_0)$ exists on $(-\infty,\theta_0]$, but is not necessarily unique.
Consequently, if the inequalities (4.9), (4.10) and

$$0<\mu \leqslant \frac{\alpha \exp(2\alpha)}{4x_0[\exp(\alpha)-1]}, \tag{4.19}$$

obtained from (4.8), are satisfied for $x_0>0$, then $x(t)=x(t,\theta_0,x_0,\mu)$ exists on $\mathbb{R}$. Moreover, if one of the inequalities (4.9), (4.10) or (4.19) is violated, then the solution $x(t)$ exists on the half-axis, but it does not exist on $\mathbb{R}$.

From now on, we need the following assumptions:

(H3) $|\mu|<1/(2ML\overline{\theta})$;
(H4) $|\mu|ML\overline{\theta}[1+M(1+L|\mu|\overline{\theta})\exp(ML|\mu|\overline{\theta})]<m$.

The following theorem is valid, which can be verified in the same way as Theorem 2.3.

Theorem 4.1. *Suppose that* (H1)–(H4) *hold. Then, for each* $(t_0,x_0)\in\mathbb{R}\times\mathbb{R}^n$, *there exists a unique solution* $x(t)$ *of* (4.1) *on* $\mathbb{R}$ *with* $x(t_0)=x_0$.

4.2 Dependence of solutions on parameters

Consider a continuous non-negative function $u(t):\mathbb{R}\to\mathbb{R}$, which is continuously differentiable, except, possibly points θ_k, $k\in\mathbb{Z}$, where the right-side derivative exists.
Assume that $u(t)$ satisfies the differential inequality

$$u'(t)\leqslant a(t)u(t)+b(t)u(\beta(t)),\ \ t\geqslant\theta_i, \tag{4.20}$$

where i is fixed and a, b are non-negative piecewise continuous functions with discontinuities of the first kind at θ_k, $k\in\mathbb{Z}$. Consider also the linear equation

$$v'(t)=a(t)v(t)+b(t)u(\beta(t)),\ \ t\geqslant\theta_i. \tag{4.21}$$

By applying the differential inequality result for ordinary differential equations [87, 153] one can prove that $u(t)\leqslant v(t)$, where $v(t)$ is a solution of (4.21) with $u(\theta_i)\leqslant v(\theta_i)$. The discussion is similar to that in Section 2.1, Example 2.1.
Now, set

$$K_k(t)=\mathrm{e}^{\int_{\theta_k}^t a(s)ds}+\int_{\theta_k}^t \mathrm{e}^{\int_s^t a(r)dr}b(s)ds.$$

One can easily find that the solution $v(t)$, $v(t_0) = v_0$, of (4.21) is equal to

$$v(t) = K_l(t)\prod_{k=i}^{l-1} K_k(\theta_{k+1})v_0, \tag{4.22}$$

if $t \in [\theta_l, \theta_{l+1}]$. Consequently,

$$u(t) \leqslant K_l(t)\prod_{k=i}^{l-1} K_k(\theta_{k+1})v_0. \tag{4.23}$$

The following assertion is of importance for our next analysis.

Lemma 4.5. *Suppose that $u(t)$, nonnegative continuous scalar function, and nonnegative real constant, α, satisfy the inequality*

$$u(t) \leqslant \alpha + \int_{\theta_i}^{t} [a(s)u(s) + b(s)u(\beta(s))]\,ds, \tag{4.24}$$

for $t \geqslant t_0$, functions a, b are non-negative.
Then

$$u(t) \leqslant K_l(t)\prod_{k=i}^{l-1} K_k(\theta_{k+1})\alpha, \tag{4.25}$$

where $t \in [\theta_l, \theta_{l+1}]$.

Proof. Let $V(t) = \alpha + \int_{\theta_i}^{t} [a(s)u(s) + b(s)u(\beta(s))]\,ds$. It is easily seen that $V' \leqslant a(t)V(t) + b(t)V(\gamma(t))$. Now, apply the results obtained for inequality (4.20). □

Let us fix a positive real number T.

Theorem 4.2. *Suppose* (H1)–(H4) *are valid. If $x(t) = x(t,\theta_i,x_0,\mu)$ and $\tilde{x}(t) = x(t,\theta_i,x_0 + \Delta x,\mu)$ are solutions of* (4.1), *where Δx is an n-dimensional real vector, then*

$$\|\tilde{x}(t) - x(t)\| = O(\|\Delta x\|), \tag{4.26}$$

where $t \in [\theta_i, \theta_i + T]$, if $|\mu|$ is sufficiently small.

Proof. If $t \in [\theta_i, \theta_i + T]$, then

$$\begin{aligned}\|\tilde{x}(t) - x(t)\| \leqslant{}& X(t,\theta_i)\|\Delta x\| + |\mu| \int_{\theta_i}^{t} X(t,s)\|g(s,\tilde{x}(s),\tilde{x}(\beta(s)),\mu) \\ &- g(s,x(s),x(\beta(s)),\mu)\|ds.\end{aligned}$$

Hence,

$$\|\tilde{x}(t) - x(t)\| \leqslant M\|\Delta x\| + |\mu|ML\int_{\theta_i}^{t} [\|\tilde{x}(s) - x(s)\| + \|\tilde{x}(\beta(s)) - x(\beta(s))\|]ds.$$

The theorem is proved by applying Lemma 4.5 to the last inequality. □

In the next theorem, we establish the differential dependence of a solution of (4.1) on an initial value. The following assumption is required:

(H5) The function $g(t,x,y,\mu)$ has continuous first order partial derivatives in all its arguments $t \in \mathbb{R}$, x, $y \in \mathbb{R}^n$, $\mu \in J$.

Let us introduce the following equations:

$$U'(t) = A(t)U(t) + \mu[A_1(t)U(t) + A_2(t)U(\beta(t))], \tag{4.27}$$

$$U(\theta_i) = \mathscr{I}, \tag{4.28}$$

where $U \in \mathbb{R}^{n\times n}$, and the functions

$$A_1(t) = \frac{\partial g}{\partial x}(t,x(t),x(\beta(t)),\mu), A_2(t) = \frac{\partial g}{\partial y}(t,x(t),x(\beta(t)),\mu)$$

are $n \times n$ matrices.

Theorem 4.3. *Suppose that conditions* (H1)–(H5) *are valid, and let* $e_i = (0,\ldots,0,\ 1,\ 0,\ldots,0)^T$ *be the n-tuple whose i-th component is* 1 *and all others are* 0 *for* $k = 1,\ldots,n$, *and* δ *a real positive constant. If* $U(t)$ *is the solution of* (4.27) *and* (4.28) *and* $x(t) = x(t,\theta_i,x_0,\mu)$ *and* $\widetilde{x}_k(t) = x(t,\theta_i,x_0+\Delta x_k,\mu)$ *are the solutions of* (4.1), *where* $\Delta x_k = \delta e_k$ *is an n-dimensional vector, then*

$$\widetilde{x}_k(t) - x(t) - U(t)\Delta x_k = o(\delta) \tag{4.29}$$

is satisfied for $t \in [\theta_i,\ \theta_i + T]$, *if* $|\mu|$ *is sufficiently small.*

Proof. By the equivalence Lemma 4.4, it can be verified that the integral equations

$$\widetilde{x}_k(t) = X(t,\ \theta_i)(x_0 + \Delta x_k) + \int_{\theta_i}^{t} X(t,\ s)[f(s) + \mu g(s,\widetilde{x}_k(s),\widetilde{x}_k(\beta(s)),\mu)]ds,$$

$$x(t) = X(t,\ \theta_i)x_0 + \int_{\theta_i}^{t} X(t,\ s)[f(s) + \mu g(s,x(s),x(\beta(s)),\mu)]ds,$$

$$U(t) = X(t,\ \theta_i) + \mu\int_{\theta_i}^{t} X(t,\ s)[A_1(s)U(s) + A_2(s)U(\beta(s))]\,ds,$$

are satisfied by $\widetilde{x}_k(t), x(t)$ and $U(t)$ respectively. For $t \in [\theta_i, \theta_i + T]$, the equation

$$\widetilde{x}_k(t) - x(t) - U(t)\Delta x_k = \mu\int_{\theta_i}^{t} X(t,\ s)[g(s,\widetilde{x}_k(s),\widetilde{x}_k(\beta(s)),\mu)$$
$$-g(s,x(s),x(\beta(s)),\mu) - A_1(s)U(s)\Delta x_k - A_2(s)U(\beta(s))\Delta x_k]ds$$

can be easily verified.

By expanding $g(s,\widetilde{x}_k(s),\widetilde{x}_k(\beta(s)),\mu)$ about $(s,x(s),x(\beta(s)),\mu)$, we have

$$g(s,\widetilde{x}_k(s),\widetilde{x}_k(\beta(s)),\mu) = g(s,x(s),x(\beta(s)),\mu) + A_1(s)[\widetilde{x}_k(s) - x(s)]$$
$$+A_2(s)[\widetilde{x}_k(\beta(s)) - x(\beta(s))] + \xi(s),$$

where $\xi(t) = o(\Delta x_i)$. Hence,

$$\|\widetilde{x}_k(t) - x(t) - U(t)\Delta x_k\| \leqslant \zeta + |\mu| M \int_{\theta_j}^{t} [\|A_1(s)\| \, \|\widetilde{x}_k(s) - x(s) - U(s)\Delta x_k\|$$

$$+ \|A_2(s)\| \, \|\widetilde{x}_k(\beta(s)) - x(\beta(s)) - U(\beta(s))\Delta x_k\|] \, ds,$$

where $\zeta = |\mu| M \int_{\theta_i}^{\theta_i+T} \|\xi(s)\| ds$. Consequently, by applying Lemma 4.5 to the last inequality, the theorem is proved. □

As a result of the last theorem, we have shown that the initial value problem (4.27) and (4.28) is a variation of (4.1).

Next, we prove the main theorem of this section. We need the following assumptions:

(H6) The functions $A(t)$, $f(t)$ and $g(t,x,y,\mu)$ are ω-periodic in t, for some positive real number ω.

(H7) The sequence $\{\theta_i\}$ satisfies an (ω, p)-property, that is, $\theta_{i+p} = \theta_i + \omega$, $i \in \mathbb{Z}$, for some positive integer p.

Let us consider the following version of the Poincaré criterion.

Lemma 4.6. *Suppose that* (H1)–(H4) *and* (H6), (H7) *hold. Then, the solution* $x(t) = x(t, \theta_i, x_0, \mu)$ *of* (4.1), *with* $x(t_0) = x_0$, *is* ω*-periodic if and only if*

$$x(\omega) = x(0). \tag{4.30}$$

Proof. If $x(t)$ is ω-periodic, then (4.30) is obviously satisfied. Conversely, suppose that (4.30) holds. Define $y(t) = x(t+\omega)$ on $\mathbb{R}$. Then, equation (4.30) is equivalent to $y(0) = x(0)$. One can show that $\beta(t+\omega) = \beta(t) + \omega$. Hence,

$$\begin{aligned} y'(t) &= x'(t+\omega) \\ &= A(t+\omega)x(t+\omega) + f(t+\omega) + \mu g(t+\omega, x(t+\omega), x(\beta(t+\omega)), \mu) \\ &= A(t)y(t) + f(t) + \mu g(t, y(t), y(\beta(t)), \mu). \end{aligned}$$

That is, $y(t)$ is a solution of (4.1). By uniqueness of the solution, $x(t) = y(t)$ on $\mathbb{R}$. The lemma is proved. □

The following theorem is a generalization of a classical theorem originally due to Poincaré [267] for differential equations with piecewise constant arguments.

Theorem 4.4. *Assume that* (H1)–(H7) *hold, and*

$$x'(t) = A(t)x(t) \tag{4.31}$$

has no nontrivial periodic solution with period ω. Then, for sufficiently small $|\mu|$, equation (4.1) *has a unique ω-periodic solution, which tends to the unique periodic solution with period ω of*

$$x'(t) = A(t)x(t) + f(t), \tag{4.32}$$

as $\mu \to 0$.

Proof. Assume, without loss of generality, that $0 = \theta_i$. Let $x(t,\zeta,\mu)$ be a solution of equation (4.1), satisfying the initial condition $x(0,\zeta,\mu) = \zeta$, and let $x_0(t) = x(t,\zeta_0,0)$ be a unique periodic solution of period ω of equation (4.32). To show, using Lemma 4.6, that for a sufficiently small μ the ω-periodic solution $x(t,\zeta,\mu)$ exists, it is necessary and sufficient that the equation

$$x(\omega,\zeta,\mu) - \zeta = 0 \tag{4.33}$$

be solvable with respect to ζ.
Let $P(\zeta,\mu) = x(\omega,\zeta,\mu) - \zeta$. In order to apply the implicit function theorem, we will show that the determinant of $P'_\zeta(\zeta_0,0)$ exists and is different from zero.
Let $Z(t,\zeta,\mu) = (\partial x_i/\partial \zeta_k)$, $i, k = 1,\ldots,n$. Differentiating equation (4.1) with respect to ζ, (use Theorem 4.3) we can see that $Z(t,\zeta_0,0)$ is the fundamental matrix of equation (4.31). On the other hand, $P_\zeta{}'(\zeta_0,0) = \det(Z(\omega,\zeta_0,0) - I)$ and, since the eigenvalues of the matrix $Z(\omega,\zeta_0,0)$ are different from unity, it follows that $P_\zeta{}'(\zeta_0,0) \neq 0$. Therefore, in a sufficiently small neighborhood of the point $(0,\zeta_0)$, equation (4.33) is solvable with respect to ζ. The existence and uniqueness of an ω-periodic solution are proved. The fact that the solution $x(t,\zeta,\mu)$ tends to $x_0(t)$, when $\mu \to 0$, follows from Theorem 4.2. The theorem is proved. □

The following example demonstrates the last theorem.

Example 4.2. Consider the following system

$$x'(t) = \begin{pmatrix} \alpha & \gamma \\ -\gamma & \alpha \end{pmatrix} x(t) + \begin{pmatrix} \sin(\pi t) \\ \cos(\pi t) \end{pmatrix} + \mu g(t,x(t),x(\beta(t)),\mu), \tag{4.34}$$

where $x \in \mathbb{R}^2$, $\alpha \neq 0$, γ, $\mu \geqslant 0$, $\beta(t) = \theta_i$ if $\theta_i \leqslant t < \theta_{i+1}, i \in \mathbb{Z}$, with $\theta_i = i + (-1)^i/3, i \in \mathbb{Z}$; $g(t,x,y,\mu)$ is 2-periodic in t, continuous function, having continuous first order partials in all of its arguments, and satisfying Lipschitz continuity with a constant L, that is,

$$\|g(t,x_1,y_1,\mu) - g(t,x_2,y_2,\mu)\| \leqslant L\,[\|x_1 - x_2\| + \|y_1 - y_2\|],$$

where $x_1,\ y_1,\ x_2,\ y_2 \in \mathbb{R}^2$. One can see that the sequence $\{\theta_i\}$ fulfills $\theta_{i+2} = \theta_i + 2$ for all $i \in \mathbb{Z}$. By fixing a sufficiently small $|\mu|$ satisfying the inequalities

$$|\mu| < 1/(2ML\overline{\theta}),$$
$$|\mu| ML\overline{\theta}[1 + M(1 + L|\mu|\overline{\theta})\exp(ML|\mu|\overline{\theta})] < m,$$

where $\overline{\theta} = 5/3$, $\kappa = \sqrt{\alpha^2 + \gamma^2}$, $m = \mathrm{e}^{-5\kappa/3}$, and $M = \mathrm{e}^{5\kappa/3}$, we conclude that assumptions (H1)–(H7) are fulfilled. Therefore, through every point (t_0, ζ) of $\mathbb{R}^3$, there passes exactly one solution $x(t,\mu) = x(t,t_0,\zeta,\mu),\ x(t_0,\mu) = \zeta$ of (4.34).

The monodromy matrix of (4.34) is

$$X(2) = \begin{bmatrix} \mathrm{e}^{2\alpha}\cos(2\gamma) & \mathrm{e}^{2\alpha}\sin(2\gamma) \\ -\mathrm{e}^{2\alpha}\sin(2\gamma) & \mathrm{e}^{2\alpha}\cos(2\gamma) \end{bmatrix},$$

and it has no unit multiplier for $\alpha \neq 0$. Therefore, there is a unique 2-periodic solution $x_0(t)$ of the system

$$x'(t) = \begin{pmatrix} \alpha & \gamma \\ -\gamma & \alpha \end{pmatrix} x(t) + \begin{pmatrix} \sin(\pi t) \\ \cos(\pi t) \end{pmatrix},$$

with the initial value

$$x_0(\theta_0) = (I - X(2))^{-1} \int_{\theta_0}^{\theta_2} X(\theta_2 - s) \begin{pmatrix} \sin(\pi s) \\ \cos(\pi s) \end{pmatrix} ds.$$

Hence, by Theorem 4.4, there is a unique 2-periodic solution $x(t,\mu)$ of (4.34), satisfying $x(t,\mu) \to x_0(t)$ as $\mu \to 0$.

Definition 4.1. The solution $x(t,x_0) = x(t,t_0,x_0,\mu), x(t_0,x_0) = x_0$ of (4.1) is said to be uniformly stable if for every $\varepsilon > 0$, there exists a real number $\delta = \delta(\varepsilon) > 0$ such that $\|\overline{x_0} - x_0\| < \delta$ implies $\|x(t,\overline{x_0}) - x(t,x_0)\| < \varepsilon$ for every $t \geqslant t_0$.

Definition 4.2. The solution $x(t,x_0) = x(t,t_0,x_0,\mu), x(t_0,x_0) = x_0$ of (4.1) is said to be uniformly asymptotically stable if it is uniformly stable and there is a real number $b > 0$ such that for every $\zeta > 0$ there exists $T(\zeta) > 0$ such that $\|\overline{x_0} - x_0\| < b$ implies that $\|x(t,\overline{x_0}) - x(t,x_0)\| < \zeta$ if $t > t_0 + T(\zeta)$.

Theorem 4.5. *Suppose that* (H1)–(H7) *hold. Let $x(t) = x(t,t_0,x_0,\mu)$ be a solution of* (4.1). *If all the characteristic multipliers of the equation*

$$x'(t) = A(t)x(t) \tag{4.35}$$

are less than unity in modulus, then for sufficiently small $|\mu|$, the solution $x(t)$ is uniformly asymptotically stable.

Proof. Let $u(t)$ be a solution of (4.1) satisfying the initial condition $u(t_0) = x_0 + \eta$. Let us define $z(t) = u(t) - x(t)$. Since all the multipliers are less than unity in modulus, we have

$$\|X(t,s)\| \leqslant K\exp(-\alpha(t-s)), \quad s \leqslant t,$$

where K and α are positive constants. By Lemma 4.3, it is easy to see that

$$\|z(t)\| \leqslant \|X(t,t_0)\|\|\eta\| + \int_{t_0}^{t} \|X(t,s)\||\mu|[\|g(s,x(s)+z(s),x(\beta(s))+z(\beta(s)),\mu)$$

$$-g(s,x(s),x(\beta(s)),\mu)\|]ds$$

and

$$\|z(t)\| \leqslant K\exp(-\alpha(t-t_0))\|\eta\| + \int_{t_0}^{t} \exp(-\alpha(t-s))|\mu|KL[\|z(s)\| + \|z(\beta(s))\|]ds.$$

Then,

$$\exp(\alpha t)\|z(t)\| \leqslant K\exp(\alpha t_0)\|\eta\| + \int_{\theta_j}^{t} \exp(\alpha s)\,|\mu|KL[\|z(s)\| + \|z(\beta(s))\|]ds.$$

Applying Lemma 4.5 to the last inequality, we have

$$\|z(t)\| \leqslant K\exp[(-\alpha + 2|\mu|KL)(t-\theta_j)]\|\eta\|$$

Therefore, for $|\mu| < \alpha/(2KL)$, the solution $x(t)$ is uniformly asymptotically stable. The theorem is proved. □

4.3 Periodic solutions: the critical case

In this section, the conditions for the existence of periodic solutions for forced weakly nonlinear ordinary differential equations with alternately retarded – advanced piecewise constant argument of generalized type are given. The resonant case is studied, that is, when the unperturbed linear ordinary differential equation has a nontrivial periodic solution. The dependence of solutions on initial values and parameters also takes place.

Consider the argument-function $\gamma(t)$, which is defined on the whole real line with two real-valued sequences θ_i, ζ_i, $i \in \mathbb{Z}$, such that $\theta_i < \theta_{i+1}$, $\theta_i \leqslant \zeta_i \leqslant \theta_{i+1}$ for all $i \in \mathbb{Z}$, $|\theta_i| \to \infty$ as $|i| \to \infty$.

In this section, we shall consider the equation

$$z'(t) = A(t)z(t) + f(t) + \mu g(t,z(t),z(\gamma(t)),\mu), \tag{4.36}$$

where $z \in \mathbb{R}^n$, $t \in \mathbb{R}$, $\mu \in (-\mu,\mu)$, and $\gamma(t) = \zeta_i$, if $t \in [\theta_i, \theta_{i+1})$, $i \in \mathbb{Z}$.

The following assumptions will be needed throughout the section:

(C1) $A : \mathbb{R} \to \mathbb{R}^{n\times n}$, $f : \mathbb{R} \to \mathbb{R}^n$ and $g : \mathbb{R} \times \mathbb{R}^n \times \mathbb{R}^n \times J \to \mathbb{R}^n$ are continuous functions.

(C2) The function $g(t,x,y,\mu)$ satisfies Lipschitz continuity in the second and third arguments with a positive Lipschitz constant L such that

$$\|g(t,x_1,y_1,\mu) - g(t,x_2,y_2,\mu)\| \leqslant L(\|x_1 - x_2\| + \|y_1 - y_2\|)$$

for all $t \in \mathbb{R}$, $\mu \in J$ and $x_1, x_2, y_1, y_2 \in \mathbb{R}^n$.

(C3) The matrix A is uniformly bounded on $\mathbb{R}$.

(C4) There exists a number $\overline{\theta} > 0$ such that $\theta_{i+1} - \theta_i \leqslant \overline{\theta}$, $i \in \mathbb{Z}$.

(C5) There exists a number $\theta > 0$ such that $\theta_{i+1} - \theta_i \geqslant \theta$, $i \in \mathbb{Z}$.

We combine that method with the method of small parameter [218,233,267] to investigate the problem of the existence of periodic solutions of (4.36) in the so called critical case, when the corresponding linear homogeneous system admits nontrivial periodic solutions. Let $X(t)$ be the fundamental matrix solution of the homogeneous system, corresponding to (4.36),

$$x'(t) = A(t)x(t), \ \ t \in \mathbb{R}, \tag{4.37}$$

such that $X(0) = \mathscr{I}$, where $\mathscr{I}$ is the $n \times n$ identity matrix. Denote by $X(t,s) = X(t)X^{-1}(s)$, $t, s \in \mathbb{R}$ the transition matrix.

Lemma 4.7. *Suppose that* (C1) *is satisfied. A function $z(t) = z(t,t_0,z_0,\mu)$, where t_0 is a fixed real number, is a solution of* (4.36) *on $\mathbb{R}$ if and only if it is a solution, on $\mathbb{R}$, of the integral equation*

$$z(t) = X(t,t_0)z_0 + \int_{t_0}^{t} X(t,s)[f(s) + \mu g(s,z(s),z(\gamma(s)),\mu)]ds. \tag{4.38}$$

In the sequel, we shall use the following evaluations, which are obtained in the last section, for the transition matrix, $X(t,s)$,

$$m \leqslant \|X(t,s)\| \leqslant M. \tag{4.39}$$

From now on, we make the following assumption:

(C6) $2|\mu|ML\overline{\theta} < 1$,

$$|\mu|M^2L\overline{\theta}\left\{\frac{1+|\mu|ML\overline{\theta}\exp\left(|\mu|ML\overline{\theta}\right)}{1-|\mu|ML\overline{\theta}\exp\left(|\mu|ML\overline{\theta}\right)} + \exp\left(|\mu|ML\overline{\theta}\right)\right\} < m.$$

The following assertion can be proved exactly in the way it is done for its counterpart in the last section.

Lemma 4.8. *Assume that the conditions* (C1)–(C6) *are fulfilled. Then for every $(t_0,z_0) \in \mathbb{R} \times \mathbb{R}^n$ there exists a unique solution $z(t) = z(t,t_0,z_0,\mu)$ of the equation* (4.36) *on $\mathbb{R}$ such that $z(t_0) = z_0$.*

Lemma 4.9. *Suppose that $u(t)$, nonnegative continuous scalar function, and nonnegative real constant, α, satisfy the inequality*

$$u(t) \leqslant \alpha + \int_{t_0}^{t} [a(s)u(s) + b(s)u(\gamma(s))]\,ds, \tag{4.40}$$

for $t \geqslant t_0$, functions a, b are non-negative, and all conditions of Example 2.1 *are satisfied. Then*

$$\begin{aligned} u(t) \leqslant & \left(e^{\int_{\theta_l}^{t} a(s)ds} + \frac{e^{\int_{\theta_l}^{\zeta_l} a(s)ds}}{1 - \int_{\theta_l}^{\zeta_l} e^{\int_s^{\zeta_l} a(r)dr} b(s)ds} \int_{\theta_l}^{t} e^{\int_s^t a(r)dr} b(s)ds \right) \times \\ & \prod_{k=l-1}^{i+1} \left(e^{\int_{\theta_k}^{\theta_{k+1}} a(s)ds} + \frac{e^{\int_{\theta_k}^{\zeta_k} a(s)ds}}{1 - \int_{\theta_k}^{\zeta_k} e^{\int_s^{\zeta_k} a(r)dr} b(s)ds} \int_{\theta_k}^{\theta_{k+1}} e^{\int_s^{\theta_{k+1}} a(r)dr} b(s)ds \right) \times \\ & \left(e^{\int_{t_0}^{\theta_{i+1}} a(s)ds} + \frac{e^{\int_{t_0}^{\zeta_i} a(s)ds}}{1 - \int_{t_0}^{\zeta_i} e^{\int_s^{\zeta_i} a(r)dr} b(s)ds} \int_{t_0}^{\theta_{i+1}} e^{\int_s^{\theta_{i+1}} a(r)dr} b(s)ds \right) \alpha, \end{aligned} \tag{4.41}$$

where $t \in [\theta_l, \theta_{l+1}]$, $\theta_i \leqslant t_0 \leqslant \zeta_i \leqslant \theta_{i+1}$.

Proof. Let $V(t) = \alpha + \int_{t_0}^{t} [a(s)u(s) + b(s)u(\gamma(s))]\,ds$. It is easily seen that $V' \leqslant a(t)V(t) + b(t)V(\gamma(t))$. Now, apply the results of Example 2.1. □

Let us fix a positive real number T. By the following theorem, we set continuous dependence of solutions of (4.36) on the initial value z_0.

Theorem 4.6. *Suppose that* (C1)–(C6) *are valid. If $z(t) = z(t, t_0, y_0, \mu)$ and $\tilde{z}(t) = z(t, t_0, z_0 + \Delta z, \mu)$, $\theta_i \leqslant t_0 \leqslant \zeta_i \leqslant \theta_{i+1}$, are the solutions of* (4.36)*, where Δz is an n-dimensional vector, then*

$$\|\tilde{z}(t) - z(t)\| = O(\|\Delta z\|) \tag{4.42}$$

for $t \in [t_0, t_0 + T]$, if $|\mu|$ is sufficiently small.

Proof. Set $u(t) = \|\tilde{z}(t) - z(t)\|$. Find that $u(t) \leqslant Mu(t_0) + \int_{t_0}^{t} \mu ML[u(s) + u(\gamma(s))]ds$. It is easily seen that all conditions of the last Lemma are valid, if $|\mu|$ is small. Now, apply (4.41) to complete the proof. □

The differential dependence of a solution of the equation (4.36) on an initial value is established by our next theorem, which requires the following assumption:

(C7) $g(t, x, y, \mu)$ has continuous first order partial derivatives in all of its arguments $t \in \mathbb{R}$, $x, y \in \mathbb{R}^n$, $\mu \in J$.

Let us introduce the following equations

$$U'(t) = A(t)U(t) + \mu[A_1(t)U(t) + A_2(t)U(\gamma(t))], \tag{4.43}$$

$$U(t_0) = \mathscr{I}, \tag{4.44}$$

where $U \in \mathbb{R}^{n\times n}$ and the functions

$$A_1(t) = \frac{\partial g}{\partial x}(t, z(t), z(\gamma(t)), \mu), \qquad A_2(t) = \frac{\partial g}{\partial y}(t, z(t), z(\gamma(t)), \mu)$$

are $n \times n$ matrices.

One can see that condition (C9), Chapter 2 is valid on the finite interval $[t_0, t_0 + T]$, if $|\mu|$ is sufficiently small. That is, the solution $U(t)$ exists on that interval.

Theorem 4.7. *Suppose that conditions* (C1)–(C7) *are valid, and let* $e_i = (0,\ldots,0,1,0,\ldots,0)^T$ *be the n-tuple whose i-th component is* 1 *and all others are* 0 *for* $i = 1,\ldots,n$, *and* δ *a real positive constant. If* $U(t)$ *is the solution of* (4.43) *and* (4.44), *and* $z(t) = z(t, t_0, z_0, \mu)$, $\widetilde{z}_i(t) = z(t, t_0, z_0 + \Delta z_i, \mu)$, $\theta_i \leqslant t_0 \leqslant \zeta_i \leqslant \theta_{i+1}$, *are solutions of the equation* (4.36), *where* $\Delta z_i = \delta e_i$ *is an n-dimensional vector, then*

$$\widetilde{z}_i(t) - z(t) - U(t)\Delta z_i = o(\delta) \tag{4.45}$$

is satisfied on a section $[t_0, t_0 + T]$, *if* $|\mu|$ *is sufficiently small.*

Proof. According to Lemma 4.7, the functions $\widetilde{z}_i(t)$, $z(t)$ and $U(t)$ satisfy the integral equations

$$\widetilde{z}_i(t) = X(t,t_0)(z_0 + \Delta z_i) + \int_{t_0}^{t} X(t,s)[f(s) + \mu\, g(s, \widetilde{z}_i(s), \widetilde{z}_i(\gamma(s)), \mu)]\, ds,$$

$$z(t) = X(t,t_0)z_0 + \int_{t_0}^{t} X(t,s)[f(s) + \mu\, g(s, z(s), z(\gamma(s)), \mu)]\, ds,$$

$$U(t) = X(t, t_0) + \mu \int_{t_0}^{t} X(t,s)[A_1(s)U(s) + A_2(s)U(\gamma(s))]\, ds,$$

respectively. If $t \in [t_0, t_0 + T]$, by an easy computation, one can see that

$$\widetilde{z}_i(t) - z(t) - U(t)\Delta z_i = \mu \int_{t_0}^{t} X(t,s)[g(s, \widetilde{z}_i(s), \widetilde{z}_i(\gamma(s)), \mu)$$
$$- g(s, z(s), z(\gamma(s)), \mu) - A_1(s)U(s)\Delta z_i - A_2(s)U(\gamma(s))\Delta z_i]\, ds.$$

Use Theorem 4.6 to obtain that

$$g(s, \widetilde{z}_i(s), \widetilde{z}_i(\gamma(s)), \mu) = g(s, z(s), z(\gamma(s)), \mu) + A_1(s)[\widetilde{z}_i(s) - z(s)]$$
$$+ A_2(s)[\widetilde{z}_i(\gamma(s)) - z(\gamma(s))] + \xi(s),$$

where $\xi(s) = o(\delta)$, for $s \in [t_0, t_0 + T]$. Hence, the inequality

$$\|\widetilde{z}_i(t) - z(t) - U(t)\Delta z_i\| \leqslant \zeta + |\mu| M \int_{t_0}^{t} [\|A_1(s)\| \, \|\widetilde{z}_i(s) - z(s) - U(s)\Delta z_i\| \\ + \|A_2(s)\| \, \|\widetilde{z}_i(\gamma(s)) - z(\gamma(s)) - U(\gamma(s))\Delta z_i\|] \, ds,$$

where $\zeta = |\mu| M \int_{t_0}^{t_0+T} \|\xi(s)\| \, ds$, is valid. Consequently, by an application of Lemma 4.9 to the last inequality, accuracy of (4.45) is shown. □

As a consequence of the last theorem, it is remarkable that the initial value problem (4.43) and (4.44) is a variation of equation (4.36).

Next, the main result of this section will be demonstrated. Let us introduce the following assumptions:

(C8) The functions $A(t)$, $f(t)$ and $g(t, x, y, \mu)$ are periodic in t with a fixed positive real period ω.

(C9) The sequences θ_i and ζ_i, $i \in \mathbb{Z}$, satisfy an (ω, p)-property, that is there is a positive integer p such that the equations $\theta_{i+p} = \theta_i + \omega$ and $\zeta_{i+p} = \zeta_i + \omega$ hold for all $i \in \mathbb{Z}$.

Let us prove the following version of the Poincaré criterion.

Lemma 4.10. *Suppose that* (C1)–(C6), (C8) *and* (C9) *hold. Then, the solution* $z(t) = z(t, t_0, x_0, \mu)$ *of the equation* (4.36), *is* ω*-periodic if and only if*

$$z(\omega) = z(0). \tag{4.46}$$

Proof. If $z(t)$ is ω-periodic, then (4.46) is obviously satisfied. Suppose that the equation (4.46) holds. Let $y(t) = z(t + \omega)$ on $\mathbb{R}$. Then, (4.46) means that $y(0) = z(0)$.

It can be verified that $\gamma(t + \omega) = \gamma(t) + \omega$ for all $t \in \mathbb{R}$. Hence,

$$\begin{aligned} y'(t) &= z'(t + \omega) \\ &= A(t + \omega) z(t + \omega) + f(t + \omega) + \mu g(t + \omega, z(t + \omega), z(\gamma(t + \omega)), \mu) \\ &= A(t) y(t) + f(t) + \mu g(t, y(t), y(\gamma(t)), \mu). \end{aligned}$$

That is, $y(t)$ is a solution of (4.36). By the uniqueness of the solution, we have $z(t) = y(t)$ on $\mathbb{R}$. The lemma is proved. □

In the previous section, the noncritical case is taken into consideration. Now, we suppose that the homogeneous equation, corresponding to (4.36), has a nontrivial ω-periodic solution.

Let ϕ_j, $j = 1, \ldots, k$, $k \leqslant n$, be the solutions of (4.37), which form a maximal set of linearly independent ω-periodic solutions. Then, the corresponding adjoint system of (4.37),

$$x'(t) = -A^T(t) x(t), \tag{4.47}$$

has a maximal set of linearly independent ω-periodic solutions, ψ_j, $j = 1,\ldots,k$. We constitute an $n \times k$ matrix $K_1(t)$, whose columns are solutions ψ_j, $j = 1,\ldots,k$. Let us introduce the following condition:

(C10) $\int_0^{\omega} K_1^T(s) f(s) ds = 0.$

Theorem 4.8 ([218,273]). *Suppose that conditions* (C1)–(C3), (C8) *and* (C10) *hold. Then, if* (4.37) *admits* $k \leqslant n$ *linearly independent ω-periodic solutions, then there exists a family of k linearly independent ω-periodic solutions of the equation*

$$z'(t) = A(t)z(t) + f(t), \tag{4.48}$$

of the form $z(t,\alpha) = \alpha_1\phi_1(t) + \cdots + \alpha_k\phi_k(t) + \tilde{z}(t)$, *where* $\alpha = (\alpha_1,\ldots,\alpha_k)$ *is a real constant vector and* $\tilde{z}(t)$ *is a particular ω-periodic solution of* (4.48).

Now, our purpose is to investigate the existence of periodic solutions of (4.36). The next theorem is a generalization of a classical theorem due to Malkin [218] for differential equations with piecewise constant arguments.

Theorem 4.9. *Suppose that* (C1)–(C10) *hold and* (4.48) *admits a family of ω-periodic solutions* $z(t,\alpha)$. *Let* α^0 *be a solution of the equation* $h(\alpha) = 0$, *where the function h is given by*

$$h(\alpha) = \int_0^{\omega} K_1^T(s) g(s, z(s,\alpha), z(\gamma(s),\alpha), 0) ds, \tag{4.49}$$

such that

$$\det\left(\left.\frac{\partial h}{\partial \alpha}\right|_{\alpha=\alpha^0} \right) \neq 0.$$

Then for sufficiently small $|\mu,|$ (4.36) *has an ω-periodic solution that converges to* $z(t,\alpha^0)$ *when* $\mu \to 0$.

Proof. Let $z(t)$ be a solution of (4.36) and let us extend the matrix $K_1(t)$ to a fundamental matrix of solutions $K(t)$ by supplementing the columns ψ_j, $j = k+1,\ldots,n$, which are solutions of (4.47). Performing the substitution $y(t) = K^T(0)z(t)$ in (4.36), the equation

$$y'(t) = P(t)y(t) + r(t) + \mu F(t, y(t), y(\gamma(t)), \mu), \tag{4.50}$$

is achieved where

$$P(t) = K^T(0)A(t)K^T(0)^{-1}, \quad r(t) = K^T(0)f(t),$$
$$F(t, y(t), y(\gamma(t)), \mu) = K^T(0)g(t, K^T(0)^{-1}z(t), K^T(0)^{-1}z(\gamma(t)), \mu).$$

Denote $y(t,\alpha) = K^T(0)z(t,\alpha)$, $\beta = (\beta_{k+1},\ldots,\beta_n)$ and let $v(t) = y(t,\alpha,\beta)$ be a solution of (4.50) with the initial condition $v(0) = y(0,\alpha) + (0,\beta)^T$. Further, let $L(t) = K^{-1}(0)K(t)$, $L_1(t) = K^{-1}(0)K_1(t)$, $L_2(t)$ be the matrix composed of the entries of the last $n-k$ columns and $n-k$ rows of the matrix $L(t)$, and $L_3(t)$ be the matrix composed of the last $n-k$ rows of $L^T(t)$. Denote

$$U(\alpha,\beta,\mu) = \int_0^{\omega} L_1^T(s)F(s,v(s),v(\gamma(s)),\mu)ds,$$

$$V(\alpha,\beta,\mu) = (L_2^T(\omega)-I)\beta - \mu\int_0^{\omega} L_3(s)F(s,v(s),v(\gamma(s)),\mu)ds.$$

Then the ω-periodicity condition for the solution $v(t)$ takes the form of the equations

$$U(\alpha,\beta,\mu) = 0, \tag{4.51}$$

$$V(\alpha,\beta,\mu) = 0. \tag{4.52}$$

Taking $\mu = 0$ in (4.52), $\beta = 0$ is obtained and therefore the equation (4.51) turns out to

$$U(\alpha,0,0) = \int_0^{\omega} L_1^T(s)F(s,y(s,\alpha),y(\gamma(s),\alpha),0)ds = 0. \tag{4.53}$$

Let $\alpha^0 = (\alpha_1^0,\ldots,\alpha_k^0)$ be a solution of (4.53). Since the function U has continuous partial derivatives with respect to α_j, $j = 1,\ldots,k$, in a sufficiently small neighborhood of the point $(\alpha_0,0,0)$, it follows that under the assumption

$$\det\left(\left.\frac{\partial U}{\partial \alpha}\right|_{\alpha=\alpha^0}\right) \neq 0,$$

the system of equations (4.51) and (4.52) is solvable with respect to α and β so that the functions $\alpha_j(\mu)$ and $\beta_s(\mu)$, $j = 1,\ldots,k$, $s = k+1,\ldots,n$, are continuous and $\alpha_j(\mu) \to \alpha_j^0$, $\beta_s(\mu) \to 0$ as $\mu \to 0$.

Thus, for sufficiently small $|\mu|$, the system (4.36) admits an ω-periodic solution, which converges to the solution $z(t,\alpha^0)$ of (4.48) as $\mu \to 0$. This argumentation completes the proof of the theorem. □

The next examples will show the feasibility of our theory. The equations of Duffing type are widely investigated in the field of nonlinear dynamics, and used to model many processes in mechanics and electronics [138, 241]. These type of equations are used to construct the following examples.

Example 4.3. In this example, we deal with the equation

$$q''(t) = -q(t) + 3\sin^2(t) + \mu\left(q(t) + q'\left(2\pi\left[\frac{t+\pi}{2\pi}\right]\right)\cos t\right), \tag{4.54}$$

which has an argument of γ type. The form of the perturbation of this equation is chosen to be linear since the simulation of the solutions for the equation with retarded and advanced argument is difficult in nonlinear case.

The equation (4.54) can be transformed into a system as

$$z'(t) = \begin{pmatrix} 0 & 1 \\ -1 & 0 \end{pmatrix} z(t) + \begin{pmatrix} 0 \\ 3\sin^2 t \end{pmatrix} + \mu \begin{pmatrix} 0 \\ z_1(t) + z_2\left(2\pi\left[\frac{t+\pi}{2\pi}\right]\right)\cos t \end{pmatrix}. \quad (4.55)$$

Let us slightly generalize it to the system

$$z'(t) = \begin{pmatrix} 0 & 1 \\ -1 & 0 \end{pmatrix} z(t) + \begin{pmatrix} 0 \\ 3\sin^2 t \end{pmatrix} + \mu \begin{pmatrix} a z_1\left(2\pi\left[\frac{t+\pi}{2\pi}\right]\right)\sin t + b z_2(t) \\ c z_1(t) + d z_2\left(2\pi\left[\frac{t+\pi}{2\pi}\right]\right)\cos t \end{pmatrix}, \quad (4.56)$$

where a, b, c and d are real constants.

Obviously, (4.55) is a specific case of (4.56) when $a = 0$, $b = 0$, $c = 1$, and $d = 1$.

If $\mu = 0$, (4.56) takes the form

$$z'(t) = \begin{pmatrix} 0 & 1 \\ -1 & 0 \end{pmatrix} z(t) + \begin{pmatrix} 0 \\ 3\sin^2 t \end{pmatrix}. \quad (4.57)$$

It can be easily verified that

$$\psi_1 = \begin{pmatrix} \cos t \\ -\sin t \end{pmatrix} \text{ and } \psi_2 = \begin{pmatrix} \sin t \\ \cos t \end{pmatrix}$$

are two 2π-periodic solutions of the adjoint system corresponding to the last equation. Then,

$$\int_0^{2\pi} K_1^T(s) f(s) ds = \int_0^{2\pi} \begin{pmatrix} \cos s & -\sin s \\ \sin s & \cos s \end{pmatrix} \begin{pmatrix} 0 \\ 3\sin^2 s \end{pmatrix} ds = 0,$$

that is, the condition (C10) is achieved. Hence, the family of 2π-periodic solutions of (4.57) is given by

$$z(t,\alpha) = \begin{pmatrix} \alpha_1 \cos t + \alpha_2 \sin t + \frac{3}{2} + \frac{\cos 2t}{2} \\ -\alpha_1 \sin t + \alpha_2 \cos t - \sin 2t \end{pmatrix}, \quad (4.58)$$

where $\alpha_1, \alpha_2 \in \mathbb{R}$ are the parameters.

Next, the existence of a 2π-periodic solution of the equation (4.56) will be clarified.

The function $h(\alpha)$ in Theorem 4.9 can be evaluated as

$$\begin{aligned} h(\alpha) &= \int_0^{2\pi} K_1^T(s) g(s, z(s,\alpha), z(\gamma(s),\alpha), 0) ds \\ &= \int_0^{2\pi} \begin{pmatrix} \cos s & -\sin s \\ \sin s & \cos s \end{pmatrix} \begin{pmatrix} a z_1\left(2\pi\left[\frac{s+\pi}{2\pi}\right], \alpha\right)\sin s + b z_2(s,\alpha) \\ c z_1(s,\alpha) + d z_2\left(2\pi\left[\frac{s+\pi}{2\pi}\right], \alpha\right)\cos s \end{pmatrix} ds \\ &= \int_0^{2\pi} \begin{pmatrix} -c\alpha_2 \sin^2 s + b\alpha_2 \cos^2 s \\ (a(\alpha_1 + 2) - b\alpha_1)\sin^2 s + (c\alpha_1 + d\alpha_2)\cos^2 s \end{pmatrix} ds \\ &= \begin{pmatrix} \pi(b-c)\alpha_2 \\ \pi((a-b+c)\alpha_1 + d\alpha_2 + 2a) \end{pmatrix}. \end{aligned}$$

Suppose that $b \neq c$ and $a \neq b-c$. By a straightforward calculation one can point out that the zero of the equation $h(\alpha) = 0$ is $\alpha^0 = (\frac{-2a}{a-b+c}, 0)$, and the determinant evaluated at this point is

$$\det\left(\left.\frac{\partial h}{\partial \alpha}\right|_{\alpha=\alpha^0}\right) = \det\begin{pmatrix} 0 & \pi(b-c) \\ \pi(a-b+c) & d \end{pmatrix}$$
$$= -\pi^2(b-c)(a-b+c) \neq 0.$$

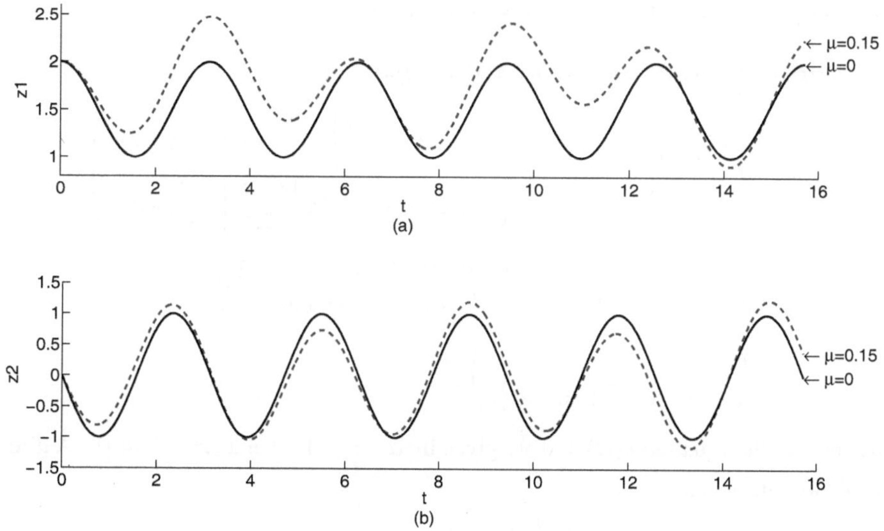

Fig. 4.1 Simulation of the periodic solution of (4.57) (solid) and the solution of (4.56) (dashed) which is near the periodic solution of the perturbed system if $a = 0$, $b = 0$, $c = 1$, $d = 1$, with identical initial data, $z(0) = (2,0)^T$. In (a) the first coordinates are shown, and second coordinates of the solutions are given in (b).

Hence, by means of Theorem 4.9, we can conclude that for sufficiently small $|\mu|$, the equation (4.56) has a 2π-periodic solution and this solution tends to $z(t,\alpha^0)$ as $\mu \to 0$. Since the initial value of the solution is close to the initial value of the periodic solution of the equation (4.57), and since there is continuous dependence on the parameter μ, the following simulations with identical initial data, $z(0) = (2,0)^T$, can be constructed. These can be observed from the Figure 4.1, where the solid lines are graphs of the periodic solution of the equation (4.57), and graphs of two coordinates of the periodic solution of the equation (4.56) are near the dashed lines.

Example 4.4. Distinctively from the previous example, this time the situation is considered

with nonlinear perturbation. In this case, a numerical simulation cannot be provided, but by the result of this chapter, the existence of periodic solutions can be indicated.
In this example, the equation

$$z'(t) = \begin{pmatrix} 0 & 1 \\ -1 & 0 \end{pmatrix} z(t) + \begin{pmatrix} 0 \\ 3\sin^2 t \end{pmatrix} + \mu \begin{pmatrix} z_1\left(2\pi\left[\frac{t+\pi}{2\pi}\right]\right)^2 \sin t + z_2(t) \\ 2z_1(t) + z_2\left(2\pi\left[\frac{t+\pi}{2\pi}\right]\right)^2 \cos t \end{pmatrix} \tag{4.59}$$

is investigated, and similar to the previous one, it can be verified that the conditions (C1)–(C10) hold and the function $h(\alpha)$ can be evaluated as

$$\begin{aligned} h(\alpha) &= \int_0^{2\pi} K_1^T(s) g(s, z(s,\alpha), z(\gamma(s),\alpha), 0) ds, \\ &= \int_0^{2\pi} \begin{pmatrix} \cos s & -\sin s \\ \sin s & \cos s \end{pmatrix} \begin{pmatrix} z_1\left(2\pi\left[\frac{s+\pi}{2\pi}\right], \alpha\right)^2 \sin s + z_2(s,\alpha) \\ 2z_1(s,\alpha) + z_2\left(2\pi\left[\frac{s+\pi}{2\pi}\right], \alpha\right)^2 \cos s \end{pmatrix} ds \\ &= \int_0^{2\pi} \begin{pmatrix} -2\alpha_2 \sin^2 s + \alpha_2 \cos^2 s \\ ((\alpha_1+2)^2 - \alpha_1)\sin^2 s + (2\alpha_1 + \alpha_2^2)\cos^2 s \end{pmatrix} ds \\ &= \begin{pmatrix} -\pi\alpha_2 \\ \pi((\alpha_1+2)^2 + \alpha_1 + \alpha_2^2) \end{pmatrix}. \end{aligned}$$

The zeros of the equation $h(\alpha) = 0$ are given by $\alpha^1 = (-1,0)$ and $\alpha^2 = (-4,0)$. A direct calculation shows that

$$\det\left(\left.\frac{\partial h}{\partial \alpha}\right|_{\alpha=\alpha^i} \right) \neq 0, \quad i = 1,2.$$

By the help of Theorem 4.9, for sufficiently small $|\mu|$, (4.59) has two 2π-periodic solutions and these solutions tend to $z(t,\alpha^1)$ and $z(t,\alpha^2)$, respectively, as $\mu \to 0$.

Notes

The Poincaré method of small parameter [267], with Lyapunov's stability [215] development was applied in [218, 219] for intensive investigation of the existence of periodic solutions. As far as we aware the first papers about the method for differential equations with piecewise constant arguments are [34, 35]. The results for a quasilinear system, considering both critical and non-critical cases, have been obtained. In view of this approach, higher order approximations can be achieved, if the dependence on the parameters will be developed. That is, one needs to extend results of higher order differentiability. Another opportunity is the extension of the results for equations with state-dependent piecewise argument of Chapter 7. We believe that theorems on the smoothness of solutions of the differential equations with impulses at non-fixed moments [20] will be very useful in this extension. The method of small parameter can be applied also for synchronization of the systems [56, 187, 240, 266].

Chapter 5

Stability

5.1 The Lyapunov-Razumikhin method

In this section we develop the Lyapunov's second method for stability of differential equations with piecewise constant argument of generalized type by employing the Razumikhin technique [150, 270]. Sufficient conditions are established for stability, uniform stability and uniform asymptotic stability of the trivial solution of such equations. We also provide appropriate examples to illustrate our results.

Detailed comparison of values of a solution at a point and at the neighbor moment, where the argument function has discontinuity, helps us to extend the discussion. It embraces the existence and uniqueness of solutions, continuous dependence of solutions on initial data, and, exceptionally important is those of stability results, which we intend to consider in the present chapter. To give more sense to the last words, in Example 5.3 at the end of the chapter, we will do additional stability analysis of the results obtained by Gopalsamy and Liu [131] for the logistic equation

$$N'(t) = rN(t)(1 - aN(t) - bN([t])), \quad t > 0, \tag{5.1}$$

where $[t]$ denotes the maximal integer not greater than t.

5.1.1 *Theoretical results*

Let $\mathbb{N}$ and $\mathbb{R}^+$ be the set of natural numbers and nonnegative real numbers, respectively, i.e., $\mathbb{N} = \{0, 1, 2, \ldots\}$ and $\mathbb{R}^+ = [0, \infty)$. Denote the n-dimensional real space by $\mathbb{R}^n$ for $n \in \mathbb{N}$, and the Euclidean norm in $\mathbb{R}^n$ by $\|\cdot\|$. Fix a real-valued sequence θ_i such that $0 = \theta_0 < \theta_1 < \cdots < \theta_i < \cdots$ with $\theta_i \to \infty$ as $i \to \infty$.

In this section, we shall consider the following equation

$$x'(t) = f(t, x(t), x(\beta(t))), \tag{5.2}$$

where $x \in S(\rho) = \{x \in \mathbb{R}^n : \|x\| < \rho\}, t \in \mathbb{R}^+, \beta(t) = \theta_i$ if $t \in [\theta_i, \theta_{i+1}), i \in \mathbb{N}$.
The following assumptions will be needed throughout the section:

(C1) $f(t,y,z) \in C(\mathbb{R}^+ \times S(\rho) \times S(\rho))$ is an $n \times 1$ real valued function;
(C2) $f(t,0,0) = 0$ for all $t \geqslant 0$;
(C3) $f(t,y,z)$ satisfies the condition

$$\|f(t,y_1,z_1) - f(t,y_2,z_2)\| \leqslant \ell(\|y_1 - y_2\| + \|z_1 - z_2\|) \tag{5.3}$$

for all $t \in \mathbb{R}^+$ and $y_1, y_2, z_1, z_2 \in S(\rho)$, where $\ell > 0$ is a Lipschitz constant;
(C4) there exists a positive number θ such that $\theta_{i+1} - \theta_i \leqslant \theta, i \in \mathbb{N}$;
(C5) $\ell\theta[1 + (1 + \ell\theta)e^{\ell\theta}] < 1$;
(C6) $3\ell\theta e^{\ell\theta} < 1$.

Let us use the following sets of functions:

$\mathscr{K} = \{a \in C(\mathbb{R}^+, \mathbb{R}^+) : a$ is strictly increasing and $a(0) = 0\}$,

$\Omega = \{b \in C(\mathbb{R}^+, \mathbb{R}^+) : b(0) = 0, \ b(s) > 0$ for $s > 0\}$.

Definition 5.1. A function $x(t)$ is a solution of (5.2) on $\mathbb{R}^+$ if

(i) $x(t)$ is continuous on $\mathbb{R}^+$;
(ii) the derivative $x'(t)$ exists for $t \in \mathbb{R}^+$ with the possible exception of the points θ_i, $i \in \mathbb{N}$, where one-sided derivatives exist;
(iii) equation (5.2) is satisfied by $x(t)$ on each interval (θ_i, θ_{i+1}), $i \in \mathbb{N}$, and it holds for the right derivative of $x(t)$ at the points θ_i, $i \in \mathbb{N}$.

The following lemma is an important auxiliary result.

Notation 5.1. $K(\ell) = \dfrac{1}{1 - \ell\theta[1 + (1 + \ell\theta)e^{\ell\theta}]}$.

Lemma 5.1. *Let* (C1)–(C5) *be fulfilled. Then the following inequality*

$$\|x(\beta(t))\| \leqslant K(\ell)\|x(t)\| \tag{5.4}$$

holds for all $t \geqslant 0$.

Proof. Let us fix $t \in \mathbb{R}^+$. Then there exists $k \in \mathbb{N}$ such that $t \in [\theta_k, \theta_{k+1})$. We have

$$x(t) = x(\theta_k) + \int_{\theta_k}^{t} f(s, x(s), x(\theta_k))ds, \ \ t \in [\theta_k, \theta_{k+1}).$$

Hence,

$$\begin{aligned}\|x(t)\| &\leqslant \|x(\theta_k)\| + \ell\int_{\theta_k}^{t}(\|x(s)\| + \|x(\theta_k)\|)\,ds\\ &\leqslant (1+\ell\theta)\,\|x(\theta_k)\| + \ell\int_{\theta_k}^{t}\|x(s)\|ds.\end{aligned}$$

The Gronwall-Bellman Lemma yields that $\|x(t)\| \leqslant (1+\ell\theta)e^{\ell\theta}\,\|x(\theta_k)\|$. Moreover, for $t \in [\theta_k, \theta_{k+1})$ we have

$$x(\theta_k) = x(t) - \int_{\theta_k}^{t} f(s, x(s), x(\theta_k))ds.$$

Thus,

$$\begin{aligned}\|x(\theta_k)\| &\leqslant \|x(t)\| + \ell\int_{\theta_k}^{t}(\|x(s)\| + \|x(\theta_k)\|)\,ds\\ &\leqslant \|x(t)\| + \ell\int_{\theta_k}^{t}\left[(1+\ell\theta)e^{\ell\theta}+1\right]\|x(\theta_k)\|\,ds\\ &\leqslant \|x(t)\| + \ell\theta\left[(1+\ell\theta)e^{\ell\theta}+1\right]\|x(\theta_k)\|.\end{aligned}$$

It follows from condition (C5) that $\|x(\theta_k)\| \leqslant K(\ell)\|x(t)\|$ for $t \in [\theta_k, \theta_{k+1})$. Hence (5.4) holds for all $t \geqslant 0$. □

Theorem 5.1. *Assume that conditions* (C1) *and* (C3)–(C6) *are satisfied. Then for every* $(t_0, x_0) \in \mathbb{R}^+ \times S(\rho)$ *there exists a unique solution* $x(t) = x(t, t_0, x_0)$ *of* (5.2) *on* $\mathbb{R}^+$ *in the sense of Definition* 5.1 *such that* $x(t_0) = x_0$.

Definition 5.2. Let $V : \mathbb{R}^+ \times S(\rho) \to \mathbb{R}^+$. Then V is said to belong to the class ϑ if

(i) V is continuous on $\mathbb{R}^+ \times S(\rho)$ and $V(t, 0) \equiv 0$ for all $t \in \mathbb{R}^+$;

(ii) $V(t, x)$ is continuously differentiable on $(\theta_i, \theta_{i+1}) \times S(\rho)$ and for each $x \in S(\rho)$, right derivative exists at $t = \theta_i$, $i \in \mathbb{N}$.

Definition 5.3. Given a function $V \in \vartheta$, the derivative of V with respect to system (5.2) is defined by

$$V'(t,x,y) = \frac{\partial V(t,x)}{\partial t} + grad_x^T V(t,x) f(t,x,y) \tag{5.5}$$

for all $t \neq \theta_i$ in $\mathbb{R}^+$ and $x, y \in S(\rho)$.

Next, we assume that conditions (C1)–(C6) are satisfied and we will obtain the stability of the zero solution of (5.2) based on the Lyapunov-Razumikhin method. We can formulate the definitions of Lyapunov stability in the same way as for ordinary differential equations.

Definition 5.4. The zero solution of (5.2) is said to be

(i) stable if for any $\varepsilon > 0$ and $t_0 \in \mathbb{R}^+$, there exists a $\delta = \delta(t_0, \varepsilon) > 0$ such that $\|x_0\| < \delta$ implies $\|x(t, t_0, x_0)\| < \varepsilon$ for all $t \geqslant t_0$;

(ii) uniformly stable if δ is independent of t_0.

Definition 5.5. The zero solution of (5.2) is said to be uniformly asymptotically stable if it is uniformly stable and there is a $\delta_0 > 0$ such that for every $\varepsilon > 0$ and $t_0 \in \mathbb{R}^+$, there exists a $T = T(\varepsilon) > 0$ such that $\|x(t, t_0, x_0)\| < \varepsilon$ for all $t > t_0 + T$ whenever $\|x_0\| < \delta_0$.

Theorem 5.2. *Assume that there exists a function $V \in \vartheta$ such that*

(i) $u(\|x\|) \leqslant V(t, x)$ *on* $\mathbb{R}^+ \times S(\rho)$, *where* $u \in \mathscr{K}$;

(ii) $V'(t, x, y) \leqslant 0$ *for all* $t \neq \theta_i$ *in* $\mathbb{R}^+$ *and* $x, y \in S(\rho)$ *such that*

$$V(\beta(t), y) \leqslant V(t, x).$$

Then the zero solution of (5.2) *is stable.*

Proof. First of all, we show stability for $t_0 = \theta_j$ for some $j \in \mathbb{N}$. Then it will allow us to prove stability for an arbitrary $t_0 \in \mathbb{R}^+$ due to Lemma 5.1.

Let $\rho_1 \in (0, \rho)$. Given $\varepsilon \in (0, \rho_1)$ and $t_0 = \theta_j$, choose $\delta_1 > 0$ sufficiently small so that $V(\theta_j, x(\theta_j)) < u(\varepsilon)$ if $\|x(\theta_j)\| < \delta_1$. Let $\delta = \delta_1 / K(\ell)$. We note $\delta < \delta_1$ as $K(\ell) > 1$ and show that this δ is the needed one.

Let us fix $k \in \mathbb{N}$ and consider the interval $[\theta_k, \theta_{k+1})$. Using the condition (ii), we shall show that

$$V(t, x(t)) \leqslant V(\theta_k, x(\theta_k)) \quad \text{for} \quad t \in [\theta_k, \theta_{k+1}). \tag{5.6}$$

Set $V(t) = V(t, x(t))$. If (5.6) is not true, then there exist points κ and τ with $\theta_k \leqslant \kappa < \tau < \theta_{k+1}$ such that

$$V(\kappa) = V(\theta_k) \quad \text{and} \quad V(t) > V(\theta_k) \quad \text{for} \quad t \in (\kappa, \tau].$$

By applying the Mean-Value Theorem to the function V, we get

$$\frac{V(\tau) - V(\kappa)}{\tau - \kappa} = V'(\zeta) > 0 \tag{5.7}$$

for some $\zeta \in (\kappa, \tau)$. Indeed, being $V(\zeta) > V(\theta_k)$, it follows from the condition (ii) that $V'(\zeta) \leqslant 0$, which contradicts to (5.7). Hence, (5.6) is true. Using the continuity of V and $x(t)$, we can obtain by induction that

$$V(t, x(t)) \leqslant V(\theta_j, x(\theta_j)) \quad \text{for all} \quad t \geqslant \theta_j. \tag{5.8}$$

If $\|x(\theta_j)\| < \delta$, we have $V(\theta_j, x(\theta_j)) < u(\varepsilon)$ since $\delta < \delta_1$. This together with (5.8) leads us to the inequality $V(t, x(t)) < u(\varepsilon)$ which implies immediately that $\|x(t)\| < \varepsilon$ for all $t \geqslant \theta_j$. Hence, stability for the case $t_0 = \theta_j$, $i \in \mathbb{N}$ is proved.
Now, let us consider the case $t_0 \in \mathbb{R}^+$, $t_0 \neq \theta_i$ for all $i \in \mathbb{N}$. Then there is $j \in \mathbb{N}$ such that $\theta_j < t_0 < \theta_{j+1}$. Given $\varepsilon > 0$ $(\varepsilon < \rho_1)$, we choose $\delta_1 > 0$ such that $V(\theta_j, x(\theta_j)) < u(\varepsilon)$ if $\|x(\theta_j)\| < \delta_1$. Take a solution $x(t)$ of (5.2) such that $\|x(t_0)\| < \delta$, where $\delta = \delta_1/K(\ell)$. By Lemma 5.1, $\|x(t_0)\| < \delta$ results in $\|x(\theta_j)\| < \delta_1$. Then by the discussion used for $t_0 = \theta_j$, we obtain that $\|x(t)\| < \varepsilon$ for all $t \geqslant \theta_j$ and hence for all $t \geqslant t_0$, proving the stability of the zero solution. □

Theorem 5.3. *Assume that there exists a function $V \in \vartheta$ such that*

(i) $u(\|x\|) \leqslant V(t,x) \leqslant v(\|x\|)$ *on* $\mathbb{R}^+ \times S(\rho)$, *where* $u, v \in \mathscr{K}$;
(ii) $V'(t,x,y) \leqslant 0$ *for all* $t \neq \theta_i$ *in* $\mathbb{R}^+$ *and* $x, y \in S(\rho)$ *such that*
$$V(\beta(t), y) \leqslant V(t,x).$$

Then the zero solution of (5.2) *is uniformly stable.*

Proof. Fix $\varepsilon > 0$ and choose $\delta_1 > 0$ such that $v(\delta_1) \leqslant u(\varepsilon)$. Define $\delta = \delta_1/K(\ell)$. Similar to the previous discussion, we consider two cases: when $t_0 = \theta_j$ for some $j \in \mathbb{N}$ and, when $t_0 \neq \theta_i$ for all $i \in \mathbb{N}$, to show that this δ is the needed one. If $t_0 = \theta_j$, where j is a fixed nonnegative integer and $\|x(\theta_j)\| < \delta$, then as a consequence of the condition (i) we have $V(\theta_j, x(\theta_j)) \leqslant v(\delta) < v(\delta_1) \leqslant u(\varepsilon)$. Using the same argument used in the proof of Theorem 5.2, we get the inequality $V(t, x(t)) \leqslant V(\theta_j, x(\theta_j))$ for all $t \geqslant \theta_j$ and see that $V(t, x(t)) < u(\varepsilon)$ for all $t \geqslant \theta_j$. Hence $\|x(t)\| < \varepsilon$ for all $t \geqslant \theta_j$. We note that evaluation of δ does not depend on the choice of $j \in \mathbb{N}$.
Now, take $t_0 \in \mathbb{R}^+$ with $t_0 \neq \theta_i$ for all $i \in \mathbb{N}$. Then there exists $j \in \mathbb{N}$ such that $\theta_j < t_0 < \theta_{j+1}$. Take a solution $x(t)$ of (5.2) such that $\|x(t_0)\| < \delta$. It follows by Lemma 5.1 that $\|x(\theta_j)\| < \delta_1$. From a similar idea used for the case $t_0 = \theta_j$, we conclude that $\|x(t)\| < \varepsilon$ for $t \geqslant \theta_j$ and indeed for all $t \geqslant t_0$. Finally, one can see that the evaluation is independent of $j \in \mathbb{N}$ and correspondingly for all $t_0 \in \mathbb{R}^+$. □

Theorem 5.4. *Assume that all of the conditions in Theorem* 5.3 *are valid and there exist a continuous nondecreasing function ψ such that $\psi(s) > s$ for $s > 0$ and a function $w \in \Omega$. If condition* (ii) *is replaced by*

(iii) $V'(t,x,y) \leqslant -w(\|x\|)$ *for all* $t \neq \theta_i$ *in* $\mathbb{R}^+$ *and* $x, y \in S(\rho)$ *such that*
$$V(\beta(t), y) < \psi(V(t,x)),$$

then the zero solution of (5.2) *is uniformly asymptotically stable.*

Proof. When $V(\beta(t),y) \leqslant V(t,x)$, we have $V(\beta(t),y) < \psi(V(t,x))$. Then by the condition (iii), we have $V'(t,x,y) \leqslant 0$. From Theorem 5.3, it follows that the zero solution of (5.2) is uniformly stable.

First, we show "uniform" asymptotic stability with respect to all elements of the sequence θ_i, $i \in \mathbb{N}$.

Fix $j \in \mathbb{N}$ and $\rho_1 \in (0,\rho)$. If $t_0 = \theta_j$ and $\delta > 0$ are such that $v(K(\ell)\delta) = u(\rho_1)$, $K(\ell) > 1$, arguments of Theorem 5.3 show that $V(t,x(t)) \leqslant v(\delta) < v(K(\ell)\delta)$ for all $t \geqslant \theta_j$ and hence $\|x(t)\| < \rho_1$ if $\|x(\theta_j)\| < \delta$. In what follows, we shall show that this δ can be taken as δ_0 in the Definition 5.5 of uniform asymptotic stability. That is, for arbitrary ε, $0 < \varepsilon < \rho_1$, we need to show that there exists a $T = T(\varepsilon) > 0$ such that $\|x(t)\| < \varepsilon$ for $t \geqslant \theta_j + T$ if $\|x(\theta_j)\| < \delta$.

Set $\gamma = \inf\{w(s) : v^{-1}(u(\varepsilon)) \leqslant s \leqslant \rho_1\}$. We note that this set is not empty since $\varepsilon < \rho_1$ and $u, v \in \mathscr{K}$ implies that $u(\varepsilon) < v(\rho_1)$, which, in turn, leads us to the inequality $v^{-1}(u(\varepsilon)) < \rho_1$.

Denote $\delta_1 = K(\ell)\delta$. From the properties of the function $\psi(s)$, there is a number $a > 0$ such that $\psi(s) - s > a$ for $u(\varepsilon) \leqslant s \leqslant v(\delta_1)$.

Let N be the smallest positive integer such that $u(\varepsilon) + Na \geqslant v(\delta_1)$.

Choose $t_k = k(\dfrac{v(\delta_1)}{\gamma} + \theta) + \theta_j$, $k = 1,2,\ldots,N$. We will prove that

$$V(t,x(t)) \leqslant u(\varepsilon) + (N-k)a \ \text{ for } \ t \geqslant t_k, \quad k = 0,1,2,\ldots,N. \tag{5.9}$$

We have $V(t,x(t)) \leqslant v(\delta_1) \leqslant u(\varepsilon) + Na$ for $t \geqslant t_0 = \theta_j$. Hence, (5.9) holds for $k = 0$. Now, we suppose that (5.9) is true for some $0 \leqslant k < N$. Let us show that

$$V(t,x(t)) \leqslant u(\varepsilon) + (N-k-1)a \ \text{ for } \ t \geqslant t_{k+1}. \tag{5.10}$$

Let $I_k = [\beta(t_k) + \theta, t_{k+1}]$. To prove (5.10), we first claim that there exists a $t^* \in I_k$ such that

$$V(t^*,x(t^*)) \leqslant u(\varepsilon) + (N-k-1)a. \tag{5.11}$$

Otherwise, $V(t,x(t)) > u(\varepsilon) + (N-k-1)a$ for all $t \in I_k$.

On the other side, we have

$$V(t,x(t)) \leqslant u(\varepsilon) + (N-k)a \ \text{ for } \ t \geqslant t_k, \tag{5.12}$$

which implies that $V(\beta(t),x(\beta(t))) \leqslant u(\varepsilon) + (N-k)a$ for $t \geqslant \beta(t_k) + \theta$.

Hence, for $t \in I_k$,

$$\psi(V(t,x(t))) > V(t,x(t)) + a > u(\varepsilon) + (N-k)a \geqslant V(\beta(t),x(\beta(t))).$$

Since $v^{-1}(u(\varepsilon)) \leqslant \|x(t)\| \leqslant \rho_1$ for $t \in I_k$, it follows from the hypothesis (iii) that

$$V'(t,x(t),x(\beta(t))) \leqslant -w(\|x(t)\|) \leqslant -\gamma \text{ for all } t \neq \theta_m \text{ in } I_k.$$

Using the continuity of the function V and the solution $x(t)$, we get

$$\begin{aligned} V(t_{k+1},x(t_{k+1})) &\leqslant V(\beta(t_k)+\theta,x(\beta(t_k)+\theta)) - \gamma(t_{k+1}-\beta(t_k)-\theta) \\ &< v(\delta_1) - \gamma(t_{k+1}-t_k-\theta) = 0, \end{aligned}$$

which is a contradiction. Thus (5.11) holds, that is, there is a $t^* \in I_k$ such that $V(t^*,x(t^*)) \leqslant u(\varepsilon)+(N-k-1)a$.

Next, we show that

$$V(t,x(t)) \leqslant u(\varepsilon)+(N-k-1)a \text{ for all } t \in [t^*,\infty). \tag{5.13}$$

If (5.13) does not hold, then there exists a $\bar{t} \in (t^*,\infty)$ such that

$$V(\bar{t},x(\bar{t})) > u(\varepsilon)+(N-k-1)a \geqslant V(t^*,x(t^*)).$$

Thus, we can find a $\tilde{t} \in (t^*,\bar{t})$ such that $\tilde{t} \neq \theta_m$, $V'(\tilde{t},x(\tilde{t}),x(\beta(\tilde{t}))) > 0$ and $V(\tilde{t},x(\tilde{t})) > u(\varepsilon)+(N-k-1)a$. If there is no such $\tilde{t}$, then for all $t \in (t^*,\bar{t})$, $t \neq \theta_m$, we have $V'(t,x(t),x(\beta(t))) \leqslant 0$ or $V(t,x(t)) \leqslant u(\varepsilon)+(N-k-1)a$. But, $V'(t,x(t),x(\beta(t))) \leqslant 0$ leads to $V(\bar{t},x(\bar{t})) \leqslant V(t^*,x(t^*))$, a contradiction. If $V(t,x(t)) \leqslant u(\varepsilon)+(N-k-1)a$, then $V(t,x(t)) < V(\bar{t},x(\bar{t}))$ for $t \in (t^*,\bar{t})$, $t \neq \theta_m$, is also a contradiction. Hence, $\tilde{t}$ exists.

However,

$$\psi(V(\tilde{t},x(\tilde{t}))) > V(\tilde{t},x(\tilde{t}))+a > u(\varepsilon)+(N-k)a \geqslant V(\beta(\tilde{t}),x(\beta(\tilde{t})))$$

implies that $V'(\tilde{t},x(\tilde{t}),x(\beta(\tilde{t}))) \leqslant -\gamma < 0$, a contradiction. Then, we conclude that $V(t,x(t)) \leqslant u(\varepsilon)+(N-k-1)a$ for all $t \geqslant t^*$ and thus for all $t \geqslant t_{k+1}$. This completes the induction and we know that (5.9) holds. For $k = N$, we have

$$V(t,x(t)) \leqslant u(\varepsilon), \ t \geqslant t_N = N\left(\frac{v(\delta_1)}{\gamma}+\theta\right)+t_0.$$

Hence, $\|x(t)\| < \varepsilon$ for $t > \theta_j + T$ where $T = N(\dfrac{v(\delta_1)}{\gamma}+\theta)$, proving uniform asymptotic stability for $t_0 = \theta_j$, $j \in \mathbb{N}$.

Consider the case $t_0 \neq \theta_i$ for all $i \in \mathbb{N}$. Then $\theta_j < t_0 < \theta_{j+1}$ for some $j \in \mathbb{N}$. $\|x(t_0)\| < \delta$ implies by Lemma 5.1 that $\|x(\theta_j)\| < \delta_1$. Hence, the argument used above for the case $t_0 = \theta_j$ yields that $\|x(t)\| < \varepsilon$ for $t > \theta_j + T$ and in turn for all $t > t_0 + T$. □

5.1.2 *Examples*

In the following examples, we assume that the sequence θ_i, which is in the basis of the definition of the function $\beta(t)$, satisfies the condition (C4).

Example 5.1. Consider the following linear equation

$$x'(t) = -a(t)x(t) - b(t)x(\beta(t)) \tag{5.14}$$

where a and b are bounded continuous functions on $\mathbb{R}^+$ such that $|b(t)| \leqslant a(t)$ for all $t \geqslant 0$. Clearly, one can check that conditions (C1)-(C2) and (C3) with a Lipschitz constant $\ell = \sup_{t\in I} a(t)$ are fulfilled. Moreover, we assume that the sequence θ_i and ℓ satisfy (C5) and (C6). Let $V(x) = x^2/2$. Then for $t \neq \theta_i$, $i \in \mathbb{N}$, we have

$$\begin{aligned} V'(x(t)) &= -a(t)x^2(t) - b(t)x(t)x(\beta(t)) \\ &\leqslant -a(t)x^2(t) + |b(t)|\,|x(t)|\,|x(\beta(t))| \\ &\leqslant -[a(t) - |b(t)|]x^2(t) \leqslant 0 \end{aligned}$$

whenever $|x(\beta(t))| \leqslant |x(t)|$. Since $V = x^2/2$, $V(x(\beta(t))) \leqslant V(x(t))$ implies that $V'(x(t)) \leqslant 0$. Thus by Theorem 5.3, the trivial solution of (5.14) is uniformly stable.
Next, let us investigate uniform asymptotic stability. If there are constants $\lambda > 0$, $\omega \in [0,1)$ and $q > 1$ with $\lambda \leqslant a(t)$, $|b(t)| \leqslant \omega\lambda$ and $1 - q\omega > 0$, then for $\psi(s) = q^2 s$, $w(s) = (1 - q\omega)\lambda s^2$ and $V(x) = x^2/2$ as above, we obtain that

$$V'(x(t)) \leqslant -w(|x(t)|), \quad t \neq \theta_i,$$

whenever $V(x(\beta(t))) < \psi(V(x(t)))$. Theorem 5.4 implies that $x = 0$ is uniformly asymptotically stable.

The following illustration is a development of an example from [270].

Example 5.2. Let us now consider a nonlinear scalar equation

$$x'(t) = f(x(t), \mu x(\beta(t))) \tag{5.15}$$

where $f(x,y)$ is a continuous function with $f(0,0) = 0$, $f(x,0)/x = -\sigma$ for some $\sigma > 0$ satisfying $\sigma \geqslant \ell|\mu|$ and $|f(x_1,y_1) - f(x_2,y_2)| \leqslant \ell(|x_1 - x_2| + |y_1 - y_2|)$. Then conditions (C1)–(C3) are valid. We take a sequence θ_i so that (C5)-(C6) hold true together with the Lipschitz constant ℓ.
Choosing $V(x) = x^2$, we get for $t \neq \theta_i$

$$\begin{aligned} V'(x(t)) &= 2x(t)f(x(t), \mu x(\beta(t))) \\ &= 2\left[\frac{f(x(t), \mu x(\beta(t))) - f(x(t),0)}{x(t)} + \frac{f(x(t),0)}{x(t)}\right]x^2(t) \\ &\leqslant 2\left[\frac{\ell|\mu||x(\beta(t))|}{|x(t)|} - \sigma\right]x^2(t) \leqslant 2(\ell|\mu| - \sigma)x^2(t) \leqslant 0 \end{aligned}$$

whenever $V(x(\beta(t))) \leqslant V(x(t))$. It follows from Theorem 5.3 that the solution $x = 0$ of (5.15) is uniformly stable.

Example 5.3 (A logistic equation with harvesting). In [131], stability of the positive equilibrium $N^* = 1/(a+b)$ of equation (5.1) has been studied. Equation (5.1) models the dynamics of a logistically growing population subjected to a density-dependent harvesting. There, $N(t)$ denotes the population density of a single species and the model parameters r, a and b are assumed to be positive.

Gopalsamy and Liu showed that N^* is globally asymptotically stable if $\alpha \geqslant 1$ where $\alpha = a/b$. Particularly, it was shown that the equilibrium is stable for integer initial moments. The restriction is caused by the method of investigation: the reduction to difference equation. Our results are for all initial moments from $\mathbb{R}^+$, not only integers. Moreover, we consider uniform stability for the general case $\beta(t)$. Consequently, we may say that the approach of the book allows to study stability of the class of equations in complete form.

Let us discuss the following equation

$$N'(t) = rN(t)(1 - aN(t) - bN(\beta(t))), \ t > 0, \tag{5.16}$$

which is a generalization of (5.1). One can see that (5.1) is of type (5.16) when $\beta(t) = [t]$. For our needs, we translate the equilibrium point N^* to origin by the transformation $x = b(N - N^*)$, which takes (5.16) into the following form

$$x'(t) = -r\left[x(t) + \frac{1}{1+\alpha}\right][\alpha x(t) + x(\beta(t))]. \tag{5.17}$$

Note that $f(x,y) := -r\left(x + \frac{1}{1+\alpha}\right)(\alpha x + y)$ is a continuous function and has continuous partial derivatives for $x, y \in S(\rho)$. If we evaluate the first order partial derivatives of the function $f(x,y)$, we see that

$$\left|\frac{\partial f}{\partial x}\right| \leqslant r\left(2\alpha\rho + \rho + \frac{\alpha}{1+\alpha}\right), \quad \left|\frac{\partial f}{\partial y}\right| \leqslant r\left(\rho + \frac{1}{1+\alpha}\right),$$

for $x, y \in S(\rho)$.

If we choose $\ell = r(2\alpha\rho + 2\rho + 1)$ as a Lipschitz constant, one can see that the conditions (C1)–(C3) are fulfilled for sufficiently small r. In addition, we assume that ℓ is sufficiently small so that the conditions (C5) and (C6) are satisfied.

Suppose that $\alpha \geqslant 1$ and $\rho < 1/(1+\alpha)$. Then for $V(x) = x^2$, $x \in S(\rho)$ and $t \neq \theta_i$, we have

$$\begin{aligned} V'(x(t), x(\beta(t))) &= -2rx(t)\left(x(t) + \frac{1}{1+\alpha}\right)(\alpha x(t) + x(\beta(t))) \\ &\leqslant -2r\left(x(t) + \frac{1}{1+\alpha}\right)(\alpha x^2(t) - |x(t)||x(\beta(t))|) \\ &\leqslant -2r\left(x(t) + \frac{1}{1+\alpha}\right)(\alpha - 1)x^2(t) \leqslant 0 \end{aligned}$$

whenever $V(x(\beta(t))) \leqslant V(x(t))$. Theorem 5.3 implies that the zero solution of (5.17) is uniformly stable. This in turn leads to uniform stability of the positive equilibrium N^* of (5.16).

To prove uniform asymptotic stability, we need to satisfy the condition (iii) in Theorem 5.4. In view of uniform stability, given $\rho_1 \in (0,\rho)$ we know that there exists a $\delta > 0$ such that $x(t) \in S(\rho_1)$ for all $t \geqslant t_0$ whenever $|x(t_0)| < \delta$. Let us take a constant q such that $1 < q < \alpha$, then for $\psi(s) = q^2 s$, $w(s) = 2r(\alpha - q)\eta s^2$, $\eta = 1/(1+\alpha) - \rho_1$ and $V(x) = x^2$ as before, we have

$$V'(x(t),x(\beta(t))) \leqslant -2r\left(x(t) + \frac{1}{1+\alpha}\right)(\alpha - q)x^2(t) \leqslant -w(|x(t)|),\ t \neq \theta_i,$$

whenever $V(x(\beta(t))) < \psi(V(x(t)))$. Hence the solution $x = 0$ (resp. $N = N^*$) of (5.17) (resp. (5.16)) is uniformly asymptotically stable by Theorem 5.4.

5.2 The method of Lyapunov functions

The results of this section have been developed through the concept of "total stability" [146, 275], which is stability under persistent perturbations of the right hand side of a differential equation, and they originate from a special theorem by Malkin [220]. Then, one can accept our approach as comparison of stability of equations with piecewise constant argument and ordinary differential equations. Finally, it deserves to emphasize that the direct method for differential equations with deviating argument necessarily utilizes functionals [150, 270], but we use only Lyapunov functions to determine criteria of the stability.

Let $\mathbb{N}$ and $\mathbb{R}^+$ be the set of natural numbers and nonnegative real numbers, respectively, i.e., $\mathbb{N} = \{0,1,2,\ldots\}$, $\mathbb{R}^+ = [0,\infty)$. Denote the n-dimensional real space by $\mathbb{R}^n$ for $n \in \mathbb{N}$, and the Euclidean norm in $\mathbb{R}^n$ by $\|\cdot\|$.

Let us introduce a special notation:

$$\mathscr{K} = \{\psi : \psi \in C(\mathbb{R}^+,\mathbb{R}^+) \text{ is a strictly increasing function and } \psi(0) = 0\}.$$

We fix a real-valued sequence θ_i, $i \in \mathbb{N}$, such that $0 = \theta_0 < \theta_1 < \cdots < \theta_i < \cdots$ with $\theta_i \to \infty$ as $i \to \infty$, and shall consider the following equation

$$x'(t) = f(t,x(t),x(\beta(t))), \tag{5.18}$$

where $x \in B(h) = \{x \in \mathbb{R}^n : \|x\| < h\}$, $t \in \mathbb{R}^+$, $\beta(t) = \theta_i$ if $t \in [\theta_i, \theta_{i+1})$, $i \in \mathbb{N}$.

We say that a continuous function $x(t)$ is a solution of equation (5.18) on $\mathbb{R}^+$ if it satisfies (5.18) on the intervals $[\theta_i, \theta_{i+1})$, $i \in \mathbb{N}$ and the derivative $x'(t)$ exists everywhere with the possible exception of the points θ_i, $i \in \mathbb{N}$, where one-sided derivatives exist.

In the rest of this section, we assume that the following conditions hold:

(C1) $f(t,u,v) \in C(\mathbb{R}^+ \times B(h) \times B(h))$ is an $n \times 1$ real valued function;

(C2) $f(t,0,0) = 0$ for all $t \geqslant 0$;

(C3) f satisfies a Lipschitz condition with constants ℓ_1, ℓ_2 :

$$\|f(t,u_1,v_1) - f(t,u_2,v_2)\| \leqslant \ell_1 \|u_1 - u_2\| + \ell_2 \|v_1 - v_2\| \tag{5.19}$$

for all $t \in \mathbb{R}^+$ and $u_1, u_2, v_1, v_2 \in B(h)$;

(C4) there exists a constant $\theta > 0$ such that $\theta_{i+1} - \theta_i \leqslant \theta, i \in \mathbb{N}$;

(C5) $\theta[\ell_2 + \ell_1(1 + \ell_2\theta)e^{\ell_1\theta}] < 1$;

(C6) $\theta(\ell_1 + 2\ell_2)e^{\ell_1\theta} < 1$.

The proof of the following assertion is similar to that of Lemma 5.1.

Lemma 5.2. *If the conditions* (C1)–(C5) *are fulfilled, then we have the estimation*

$$\|x(\beta(t))\| \leqslant m\|x(t)\| \tag{5.20}$$

for all $t \in \mathbb{R}^+$, *where* $m = \{1 - \theta[\ell_2 + \ell_1(1 + \ell_2\theta)e^{\ell_1\theta}]\}^{-1}$.

Next, we need the following theorem which provides conditions for the existence and uniqueness of solutions on $\mathbb{R}^+$. Since the proof of the assertion is almost identical to Theorem 2.3, we omit it here.

Theorem 5.5. *Suppose that conditions* (C1) *and* (C3)–(C6) *are fulfilled. Then for every* $(t_0, x_0) \in \mathbb{R}^+ \times B(h)$ *there exists a unique solution* $x(t) = x(t,t_0,x_0)$ *of* (5.18) *on* $\mathbb{R}^+$ *with* $x(t_0) = x_0$.

5.2.1 *Main results*

Let us consider the following system associated with (5.18):

$$y'(t) = g(t, y(t)), \tag{5.21}$$

where $g(t,y(t)) = f(t,y(t),y(t))$. This system is an ordinary differential equation. To understand the role of equation (5.21), particularly the effect of piecewise constant argument in the investigation of (5.18), the following example is important. Consider the following linear equation with piecewise constant argument:

$$x'(t) = ax(t) + bx([t]), \quad x(0) = x_0. \tag{5.22}$$

The solution of (5.22) on $\mathbb{R}^+$ is given by [317]

$$x(t) = \left(e^{a(t-[t])}\left(1 + \frac{b}{a}\right) - \frac{b}{a}\right)\left(e^a\left(1 + \frac{b}{a}\right) - \frac{b}{a}\right)^{[t]} x_0.$$

It is clear from the last expression that the zero solution of (5.22) is asymptotically stable if and only if

$$-\frac{a(e^a+1)}{e^a-1} < b < -a \tag{5.23}$$

holds. On the other side, when we consider the following ordinary differential equation

$$y'(t) = ay(t) + by(t) = (a+b)y(t), \tag{5.24}$$

which is associated with (5.22), we see that the trivial solution of (5.24) is asymptotically stable if and only if

$$b < -a. \tag{5.25}$$

When the insertion of the greatest integer function is regarded as a "perturbation" of the linear equation (5.24), it is seen for (5.22) that the stability condition (5.23) is necessarily stricter than the one given by (5.25) for the corresponding "non-perturbed" equation (5.24). If we discuss stability of equation (5.18) on the basis of (5.21), we expect that a comparison, similar to the relation of the conditions of (5.23) and (5.25), can be generalized. Furthermore, stability conditions for the ordinary differential equation (5.21) may not be enough for the issue system (5.18). By means of the following theorems, we demonstrate that stability of (5.18) depends on that of the corresponding ordinary differential equation (5.21).

Theorem 5.6. *Suppose that* (C1)–(C6) *are valid and there exist a continuously differentiable function* $V : \mathbb{R}^+ \times B(h) \to \mathbb{R}^+$, $V(t,0) = 0$, *and a positive constant* α *such that*

(i) $u(\|y\|) \leqslant V(t,y)$ *on* $\mathbb{R}^+ \times B(h)$, *where* $u \in \mathscr{K}$;
(ii) $V'_{(5.21)}(t,y) \leqslant -\alpha\ell_2(1+m)\|y\|^2$ *for all* $t \in \mathbb{R}^+$ *and* $y \in B(h)$;
(iii) $\left\|\dfrac{\partial V(t,y)}{\partial y}\right\| \leqslant \alpha\|y\|$.

Then the zero solution of (5.18) *is stable.*

Proof. Let $h_1 \in (0,h)$. Given $\varepsilon \in (0,h_1)$ and $t_0 \in \mathbb{R}^+$, choose $\delta > 0$ sufficiently small such that $V(t_0,x(t_0)) < u(\varepsilon)$ whenever $\|x(t_0)\| < \delta$. If we evaluate the time derivative of V with respect to (5.18), we get, for $t \neq \theta_i$,

$$\begin{aligned} V'_{(5.18)}(t,x(t),x(\beta(t))) &= V_t(t,x(t)) + \langle V_x(t,x(t)), f(t,x(t),x(\beta(t)))\rangle \\ &= V_t(t,x(t)) + \langle V_x(t,x(t)), g(t,x(t))\rangle + \langle V_x(t,x(t)), h(t,x(t),x(\beta(t)))\rangle \end{aligned}$$

where $h(t,x(t),x(\beta(t))) = f(t,x(t),x(\beta(t))) - f(t,x(t),x(t))$. Hence, we have

$$\begin{aligned} V'_{(5.18)}(t,x(t),x(\beta(t))) &\leqslant -\alpha\ell_2(1+m)\|x(t)\|^2 + \|V_x(t,x(t))\|\,\|h(t,x(t),x(\beta(t)))\| \\ &\leqslant -\alpha\ell_2(1+m)\|x(t)\|^2 + \alpha\ell_2(1+m)\|x(t)\|^2 = 0, \end{aligned}$$

which implies that $V(t,x(t)) \leqslant V(t_0,x(t_0)) < u(\varepsilon)$ for all $t \geqslant t_0$, proving that $\|x(t)\| < \varepsilon$. □

Theorem 5.7. *Suppose that* (C1)–(C6) *are valid and there exist a continuously differentiable function* $V : \mathbb{R}^+ \times B(h) \to \mathbb{R}^+$ *and a constant* $\alpha > 0$ *such that*

(i) $u(\|y\|) \leqslant V(t,y) \leqslant v(\|y\|)$ *on* $\mathbb{R}^+ \times B(h)$, *where* $u, v \in \mathscr{K}$;
(ii) $V'_{(5.21)}(t,y) \leqslant -\alpha\ell_2(1+m)\|y\|^2$ *for all* $t \in \mathbb{R}^+$ *and* $y \in B(h)$;
(iii) $\left\| \dfrac{\partial V(t,y)}{\partial y} \right\| \leqslant \alpha\|y\|$.

Then the zero solution of (5.18) *is uniformly stable.*

Proof. Let $h_1 \in (0,h)$. Fix $\varepsilon > 0$ in the range $0 < \varepsilon < h_1$ and choose $\delta > 0$ such that $v(\delta) < u(\varepsilon)$. If $t_0 \geqslant 0$ and $\|x(t_0)\| < \delta$, then as a consequence of the condition (i) we have $V(t_0,x(t_0)) < v(\delta) < u(\varepsilon)$. Using the same argument used in the proof of Theorem 5.6, one can obtain that $V(t,x(t)) \leqslant V(t,x(t)) < u(\varepsilon)$ for all $t \geqslant t_0$. Hence $\|x(t)\| < \varepsilon$ for all $t \geqslant t_0$. □

Theorem 5.8. *Suppose that* (C1)–(C6) *hold true and there exist a continuously differentiable function* $V : \mathbb{R}^+ \times B(h) \to \mathbb{R}^+$, *constants* $\alpha > 0$ *and* $\tau > 1$ *such that*

(i) $u(\|y\|) \leqslant V(t,y) \leqslant v(\|y\|)$ *on* $\mathbb{R}^+ \times B(h)$, *where* $u, v \in \mathscr{K}$;
(ii) $V'_{(5.21)}(t,y) \leqslant -\tau\alpha\ell_2(1+m)\|y\|^2$ *for all* $t \in \mathbb{R}^+$ *and* $y \in B(h)$;
(iii) $\left\| \dfrac{\partial V(t,y)}{\partial y} \right\| \leqslant \alpha\|y\|$.

Then the zero solution of (5.18) *is uniformly asymptotically stable.*

Proof. In view of the Theorem 5.6, the equilibrium $x = 0$ of (5.18) is uniformly stable. We need to show that it is asymptotically stable as well. For $t \neq \theta_i$,

$$\begin{aligned} V'_{(5.18)}(t,x(t),x(\beta(t))) &\leqslant -\tau\alpha\ell_2(1+m)\|x(t)\|^2 + \alpha\ell_2(1+m)\|x(t)\|^2 \\ &= -(\tau-1)\alpha\ell_2(1+m)\|x(t)\|^2. \end{aligned}$$

Denote $w(\|x\|) = (\tau-1)\alpha\ell_2(1+m)\|x\|^2$. Let $h_1 \in (0,h)$. Choose $\delta > 0$ such that $v(\delta) < u(h_1)$. We fix $\varepsilon \in (0,h_1)$ and pick $\eta \in (0,\delta)$ such that $v(\eta) < u(\varepsilon)$. Let $t_0 \in \mathbb{R}^+$ and $\|x(t_0)\| < \delta$. We define $T = u(h_1)/w(\eta)$. We shall show that $\|x(\bar{t})\| < \eta$ for some $\bar{t} \in [t_0, t_0+T]$. If this were not true, then we would have $\|x(t)\| \geqslant \eta$ for all $t \in [t_0, t_0+T]$. For $t \in [t_0,t_0+T]$, $t \neq \theta_i$, we have

$$V'_{(5.18)}(t,x(t),x(\beta(t))) \leqslant -w(\|x(t)\|) \leqslant -w(\eta).$$

Since the function $V(t,x(t))$ and the solution $x(t)$ are continuous, we obtain that

$$V(t_0+T,x(t_0+T)) \leqslant V(t_0,x(t_0)) - w(\eta)T < v(\delta) - w(\eta)\frac{u(h_1)}{w(\eta)} < 0,$$

which is a contradiction. Hence, $\bar{t}$ exists. Now for $t \geqslant \bar{t}$ we have

$$V(t,x(t)) \leqslant V(\bar{t},x(\bar{t})) < v(\eta) < u(\varepsilon).$$

Finally, it follows from the hypothesis (i) that $\|x(t)\| < \varepsilon$ for all $t \geqslant \bar{t}$ and in turn for all $t \geqslant t_0 + T$. □

Remark 5.1. Theorems 5.6-5.8 provide criteria for stability, which are entirely constructed on the basis of Lyapunov functions. As for the functionals, they appear only in the proofs of theorems. Although the equations include deviating arguments, and functionals are ordinarily used in the stability criteria [150], we see that the conditions of the section, which guarantee stability, are definitely formulated without functionals.

Next, we want to compare our present results, which are obtained by the method of Lyapunov functions with the ones proved in [25], see also Chapter 5 by employing the Lyapunov-Razumikhin technique. To this end, let us discuss the following linear equation with piecewise constant argument of generalized type taken from [25],

$$x'(t) = -a_0(t)x(t) - a_1(t)x(\beta(t)), \tag{5.26}$$

where a_0 and a_1 are bounded continuous functions on $\mathbb{R}^+$. We suppose that the sequence $\theta_i, i \in \mathbb{N}$, with $\ell_1 = \sup_{t\in\mathbb{R}^+} |a_0(t)|$, $\ell_2 = \sup_{t\in\mathbb{R}^+} |a_1(t)|$, satisfies the conditions (C4)–(C6). One can check easily that conditions (C1)–(C3) are also valid. Under the assumption

$$0 \leqslant a_0(t) + a_1(t) \leqslant 2a_0(t), \quad t \geqslant 0, \tag{5.27}$$

it was obtained via the Lyapunov-Razumikhin method in [25] that the trivial solution of (5.26) is uniformly stable. Let us consider this equation using the results obtained in the present section. We set

$$(1+m) \sup_{t\in\mathbb{R}^+} |a_1(t)| \leqslant a_0(t) + a_1(t), \ t \geqslant 0. \tag{5.28}$$

In order to apply our results, we need the following equation besides (5.26);

$$y'(t) = -(a_0(t) + a_1(t))y(t). \tag{5.29}$$

Let us define a Lyapunov function $V(y) = \frac{\alpha}{2}y^2$, $y \in B(h)$, $\alpha > 0$. It follows from (5.28) that the derivative of $V(y)$ with respect to equation (5.29) is given by

$$\begin{aligned} V'_{(5.29)}(y(t)) &= -\alpha(a_0(t) + a_1(t))y^2(t) \\ &\leqslant -\alpha\ell_2(1+m)y^2(t). \end{aligned}$$

Then, by Theorem 5.7, the zero solution of (5.26) is uniformly stable.
In addition, taking $(a_0(t)+a_1(t)) \geqslant \tau\ell_2(1+m)$, $\tau > 1$, one can show that the trivial solution of (5.26) is uniformly asymptotically stable by Theorem 5.8.
We can see that theorems obtained by Lyapunov-Razumikhin method provide larger class of equations with respect to (5.26). However, from the perspective of the constructive analysis, the present method may be more preferable, since, for example, from the proof of Theorem 5.7, we have $V'_{(5.26)}(t,x(t),x(\beta(t))) \leqslant 0$, which implies $|x(t)| \leqslant |x(t_0)|$, $t \geqslant t_0$, for our specific Lyapunov function. Thus, by using the present results, it is possible to evaluate the number δ needed for (uniform) stability in the Definition 5.4 as $\delta = \varepsilon$.
Besides Theorems 5.6, 5.7 and 5.8, the following assertions may be useful for analysis of the stability of differential equations with piecewise constant argument. These theorems are important and have their own distinctive values with the newly required properties of the Lyapunov function and can be proved similarly.

Theorem 5.9. *Suppose that* (C1)–(C6) *are valid and there exist a continuously differentiable function* $V : \mathbb{R}^+ \times B(h) \to \mathbb{R}^+$ *and a positive constant* M *such that*

(i) $u(\|y\|) \leqslant V(t,y)$ *on* $\mathbb{R}^+ \times B(h)$, *where* $u \in \mathscr{K}$;
(ii) $V'_{(5.21)}(t,y) \leqslant -M\ell_2(1+m)\|y\|$ *for all* $t \in \mathbb{R}^+$ *and* $y \in B(h)$;
(iii) $\left\| \dfrac{\partial V(t,y)}{\partial y} \right\| \leqslant M.$

Then the zero solution of (5.18) *is stable.*

Theorem 5.10. *Suppose that* (C1)–(C6) *are valid and there exist a continuously differentiable function* $V : \mathbb{R}^+ \times B(h) \to \mathbb{R}^+$ *and a positive constant* M *such that*

(i) $u(\|y\|) \leqslant V(t,y) \leqslant v(\|y\|)$ *on* $\mathbb{R}^+ \times B(h)$, *where* $u, v \in \mathscr{K}$;
(ii) $V'_{(5.21)}(t,y) \leqslant -M\ell_2(1+m)\|y\|$ *for all* $t \in \mathbb{R}^+$ *and* $y \in B(h)$;
(iii) $\left\| \dfrac{\partial V(t,y)}{\partial y} \right\| \leqslant M.$

Then the zero solution of (5.18) *is uniformly stable.*

Theorem 5.11. *Suppose that* (C1)–(C6) *hold true and there exist a continuously differentiable function* $V : \mathbb{R}^+ \times B(h) \to \mathbb{R}^+$, *constants* $M > 0$ *and* $\tau > 1$ *such that*

(i) $u(\|y\|) \leqslant V(t,y) \leqslant v(\|y\|)$ *on* $\mathbb{R}^+ \times B(h)$, *where* $u, v \in \mathscr{K}$;
(ii) $V'_{(5.21)}(t,y) \leqslant -\tau M\ell_2(1+m)\|y\|$ *for all* $t \in \mathbb{R}^+$ *and* $y \in B(h)$;
(iii) $\left\| \dfrac{\partial V(t,y)}{\partial y} \right\| \leqslant M.$

Then the zero solution of (5.18) *is uniformly asymptotically stable.*

5.2.2 *Applications to the logistic equation*

In this section, we are interested in the stability of the positive equilibrium $N^* = 1/(a+b)$ of the following logistically growing population subjected to a density-dependent harvesting;

$$N'(t) = rN(t)[1 - aN(t) - bN(\beta(t))], \quad t > 0, \tag{5.30}$$

where $N(t)$ denotes the biomass of a single species, and r, a, b are positive parameters. There exists an extensive literature dealing with sufficient conditions for global asymptotic stability of equilibria for the logistic equation with piecewise constant argument (see [131, 204, 205, 235] and the references therein). For example, Gopalsamy and Liu [131] showed that N^* is globally asymptotically stable if $a/b \geqslant 1$. In these papers, the initial moments are taken as integers owing to the method of investigation: reduction to difference equations. Since our approach makes it possible to take not only integers, but also all values from $\mathbb{R}^+$ as initial moments, we can consider the stability in uniform sense.

Let us also discuss the biological sense of the insertion of the piecewise constant delay [131,205,235]. The delay means that the rate of the population depends both on the present size and the memorized values of the population. To illustrate the dependence, one may think populations, which meet at the beginning of a season, e.g., in springtime, with their instinctive evaluations of the population state and environment and implicitly decide which living conditions to prefer and where to go in line with group hierarchy, communications and dynamics and then adapt to those conditions.

By means of the transformation $x = b(N - N^*)$, equation (5.30) can be simplified as

$$x'(t) = -r\left[x(t) + \frac{1}{1+\gamma}\right][\gamma x(t) + x(\beta(t))], \tag{5.31}$$

where $\gamma = a/b$. Let us specify for (5.31) general conditions of Theorems 5.6, 5.7 and 5.8. We observe that $f(x,y) := -r(x + \frac{1}{1+\gamma})(\gamma x + y)$ is a continuous function and has continuous partial derivatives for x, $y \in B(h)$. It can be found easily that

$$\ell_1 = r\left(2\gamma h + h + \frac{\gamma}{1+\gamma}\right), \quad \ell_2 = r\left(h + \frac{1}{1+\gamma}\right).$$

One can see that (C1), (C2) and (C3) hold if r is sufficiently small. Moreover, we assume that (C4), (C5) and (C6) are satisfied.

Consider the following equation associated with (5.31);

$$y'(t) = -r(1+\gamma)y(t)\left[y(t) + \frac{1}{1+\gamma}\right]. \tag{5.32}$$

Suppose h is smaller than $\frac{1}{1+\gamma}$ and consider a Lyapunov function defined by $V(y) = \frac{\alpha}{2}y^2$, $y \in B(h)$, $\alpha > 0$. Then,

$$\begin{aligned} V'_{(5.32)}(y(t)) &= -\alpha r(1+\gamma)y^2(t)\left[y(t)+\frac{1}{1+\gamma}\right] \\ &\leqslant -\alpha r[1-h(1+\gamma)]y^2(t). \end{aligned}$$

For sufficiently small h, we assume that

$$\varphi(h,m) \leqslant \gamma \tag{5.33}$$

where

$$\varphi(h,m) = \frac{1-h(3+m)-\sqrt{(h(1+m))^2-6h(1+m)+1}}{2h}.$$

It follows from (5.33) that

$$\left(h+\frac{1}{1+\gamma}\right)(1+m) \leqslant 1-h(1+\gamma),$$

which implies in turn

$$V'_{(5.32)}(y(t)) \leqslant -\alpha\ell_2(1+m)y^2(t).$$

By Theorem 5.7, the zero solution of (5.31) is uniformly stable.

Next, we consider uniform asymptotic stability. Assuming for $\tau > 1$;

$$\psi(h,m,\tau) \leqslant \gamma \tag{5.34}$$

where

$$\psi(h,m,\tau) = \frac{1-h\tau(3+m)-\sqrt{(h\tau(1+m))^2-6h\tau(1+m)+1}}{2h},$$

we obtain that

$$\tau\left(h+\frac{1}{1+\gamma}\right)(1+m) \leqslant 1-h(1+\gamma).$$

One can easily show that $\psi(h,m,\tau) \geqslant 1$ for small h. Then for $V(y) = \frac{\alpha}{2}y^2$, we have

$$V'_{(5.32)}(y(t)) \leqslant -\tau\alpha\ell_2(1+m)y^2(t).$$

That is, condition (iii) of Theorem 5.8 is satisfied. Thus, the trivial solution $x = 0$ of (5.31) is uniformly asymptotically stable.

In the light of the above reduction, we see that the obtained conditions are valid for the stability of the equilibrium $N = N^*$ of (5.30).

Finally, we see that the condition (5.34) is stronger than the one $\gamma \geqslant 1$ taken from [131]. However, our results allow us to take all values from $\mathbb{R}^+$ as initial moments, whereas [131] considers only integers. Moreover, the piecewise constant argument is of generalized type.

Notes

Stability as a standard object to analyze in any theory of differential equations. It has been investigated already in first papers about differential equations with piecewise constant argument [85, 286, 319] and developed further in [81, 131, 133, 211, 254, 256, 258, 291]. Nevertheless, it seems that the method of reduction to discrete systems narrowed the problem. For example, uniform stability could not be analyzed by the method and stability has been investigated with only the initial zero moment mainly. It is clear that there were not applications of continuous Lyapunov functions or functionals. Introducing the comparison of values of solutions at switching moments, as well as involving integral equations in the discussion have allowed us in [25] to consider the system in the proper way, as continuous dynamics. Consequently, various opportunities are visible, now. To the best of our knowledge, there have been no results on stability obtained by the method for differential equations with piecewise constant argument, despite they are delay differential equations. It is well known that the second Lyapunov method utilizes functionals or functions for differential equations with deviating argument [150, 270]. However, their derivatives along solutions are always functionals. From this point of view, one can search for more constructive criteria of the stability, which include the derivatives as functions. Our observations allow us to find the stability criteria applying Lyapunov functions with derivatives being functions, that is, to have equations close to ordinary differential equations in the stability analysis. Moreover, the effective Lyapunov-Razumikhin method has been used for the stability analysis. The main results of this chapter were published in [25, 29] for the first time.

Chapter 6

The state-dependent piecewise constant argument

6.1 Introduction

In previous chapters, the differential equations with piecewise constant argument of generalized type (differential equations with piecewise constant arguments) of the form

$$\frac{dx(t)}{dt} = f(t, x(t), x(\beta(t))), \tag{6.1}$$

are considered, where $\beta(t) = \theta_i$ if $\theta_i \leqslant t < \theta_{i+1}$, i are integers, is an identification function, θ_i is a strictly increasing sequence of real numbers.

In the current chapter, we generalize the equations to a new type of systems, differential equations with state-dependent piecewise constant argument, where intervals of constancy of the independent argument are not prescribed and they depend on the present state of a motion. The method of analysis for these equations was initiated in previous chapters.

The chapter consists of two main parts. In Section 6.2 we introduce the most general, for the present time, form of the equations. Basic properties of the ordinary differential equations, *constancy switching surfaces* are defined, which give a start of investigation. One of them is called *the extension property*. Moreover, the definition of solutions is given. In the remaining part of the chapter, we realize the general concepts for a particular type of equations, namely, the quasilinear systems. Existence and uniqueness theorem, periodicity, and stability of the zero solution are discussed.

6.2 Generalities

Let $\mathbb{N}$, $\mathbb{R}$, $\mathbb{Z}$ be the sets of all natural and real numbers, and integers, respectively. Denote by $\|\cdot\|$ the Euclidean norm in $\mathbb{R}^n$, $n \in \mathbb{N}$.

Let $\mathscr{I} = (a, b) \subseteq \mathbb{R}$ and $\mathscr{A} \subseteq \mathbb{Z}$ be nonempty intervals of real numbers, and integers, correspondingly. Let $\mathscr{G} \subseteq \mathbb{R}^n$ be an open connected region. Denote by $C(\mathscr{G}, \mathscr{I})$ and

$C^1(\mathscr{G},\mathscr{I})$ the set of all continuous and continuously differentiable functions from $\mathscr{G}$ to $\mathscr{I}$, respectively. Fix a sequence of real valued functions $\{\tau_i(x)\} \subset C(\mathscr{G},\mathscr{I})$, where $i \in \mathscr{A}$. The next assumption will be necessary in this section.

(A1) There exist two positive real numbers θ and $\overline{\theta}$ such that $\theta \leqslant \tau_{i+1}(x) - \tau_i(y) \leqslant \overline{\theta}$ for all $x, y \in \mathscr{G}$ and $i \in \mathscr{A}$.

Set the surfaces $S_i = \{(t,x) \in \mathscr{I} \times \mathscr{G} : t = \tau_i(x)\}, i \in \mathscr{A}$, in $\mathscr{I} \times \mathscr{G}$, and define the regions $D_i = \{(t,x) \in \mathscr{I} \times \mathscr{G} : \tau_i(x) \leqslant t < \tau_{i+1}(x)\}, i \in \mathscr{A}$, and $D_r = \{(t,x) \in \mathscr{I} \times \mathscr{G} : \tau_r(x) \leqslant t\}$ if $\max \mathscr{A} = r < \infty$. According to (A1), D_i's, $i \in \mathscr{A}$, are nonempty disjoint sets.
In this section, the equation

$$\frac{dx(t)}{dt} = f(t, x(t), x(\beta(t,x))), \tag{6.2}$$

is considered, where $t \in \mathscr{I}, x \in \mathscr{G}$, and $\beta(t,x)$ is a functional such that if $x(t) : \mathscr{I} \to \mathscr{G}$ is a continuous function, and $(t, x(t)) \in D_i$ for some $i \in \mathscr{A}$, then $\beta(t,x) = \eta_i$, where η_i satisfies the equation $\eta = \tau_i(x(\eta))$. From the description of the functions τ, it implies that one can call the surfaces $t = \tau_i(x)$ as *constancy switching surfaces,* since the solution's piecewise constant argument changes its value at the moment of meeting one of the surfaces.
The system (6.2) is named as *a system of differential equations with state-dependent piecewise constant argument.*
The following condition is required to define a solution of the equation (6.2) on $\mathscr{I}$.

(A2) For a given $(t_0, x_0) \in \mathscr{I} \times \mathscr{G}$, there is an integer $j \in \mathscr{A}$ such that $t_0 \geqslant \tau_j(x_0)$, and $j \geqslant k$ if $t_0 \geqslant \tau_k(x_0), k \in \mathscr{A}$.

It can be easily seen that the functional $\beta(t,x)$ satisfies the constraint $\beta(t,x) \leqslant t$ for all $t \in \mathscr{I}, x \in \mathscr{G}$. Indeed, to define system (6.2), the point (t,x) must belong to D_j, for some $j \in \mathscr{A}$.
Consider the ordinary differential equation

$$\frac{dy(t)}{dt} = f(t, y(t), z), \tag{6.3}$$

where z is a constant vector in $\mathscr{G}$, together with the assumption

(B0) For a given $(t_0, x_0) \in \mathscr{I} \times \mathscr{G}$, the solution $y(t) = y(t, t_0, x_0)$ of the equation (6.3) exists and is unique in any interval of existence, and it has an open maximal interval of existence such that any limit point of the set $(t, y(t))$, as t tends to the endpoints of the maximal interval of existence, is a boundary point of $\mathscr{I} \times \mathscr{G}$.

Let us remind that condition (B0) is valid, if, for example, the function f is continuous in t, and satisfies the local Lipschitz condition in y.

From now on, the following conditions are needed:

(A3) For a given $(t_0, x_0) \in \mathscr{I} \times \mathscr{G}$ satisfying (A2), there exists a solution $y(t) = y(t, t_0, x_0)$ of the equation (6.3) such that $\eta_j = \tau_j(y(\eta_j))$ for some $j \in \mathscr{A}$, and $\eta_j \leqslant t_0$.

(A4) For each $z \in \mathscr{G}$ and $j \in \mathscr{A}$ solution $y(t, \tau_j(z), z)$ of the equation (6.3) does not meet the surface S_j if $t > \tau_j(z)$.

(A5) For a given $(t_0, x_0) \in \mathscr{I} \times \mathscr{G}$ belonging to S_j, $j \in \mathscr{A}$, there exist a surface $S_{j-1} \subset \mathscr{I} \times \mathscr{G}$, and a solution $y(t) = y(t, t_0, x_0)$ of the equation (6.3) such that $\eta_{j-1} = \tau_{j-1}(y(\eta_{j-1}))$ for some $\eta_{j-1} < t_0$.

If a point $(t_0, x_0) \in \mathscr{I} \times \mathscr{G}$ satisfies (A2) and (A3), then this point is said to have *the extension property.*

Next, the definition of a solution of (6.2) is given.

Definition 6.1. A function $x(t)$ is said to be a solution of the equation (6.2) on an interval $\mathscr{J} \subseteq \mathscr{I}$ if:

(i) it is continuous on $\mathscr{J}$,

(ii) the derivative $x'(t)$ exists at each point $t \in \mathscr{J}$ with the possible exception of the points η_i, $i \in \mathscr{A}$, for which the equation $\eta = \tau_i(x(\eta))$ is satisfied, where the right derivative exists,

(iii) the function $x(t)$ satisfies the equation (6.2) on each interval (η_i, η_{i+1}), $i \in \mathscr{A}$, and it holds for the right derivative of $x(t)$ at the points η_i.

Now, the problem of global existence of the solution $x(t) = x(t, t_0, x_0)$ of (6.2), will be considered, where $(t_0, x_0) \in \mathscr{I} \times \mathscr{G}$ is a point with the extension property.

Let us investigate the problem for increasing t. The point (t_0, x_0) is either in S_j, or there is a ball $B((t_0, x_0); \varepsilon) \subset D_j$ for some real number $\varepsilon > 0$, and $j \in \mathscr{A}$. The solution $x(t)$ is defined on an interval $[\eta_j, t_0]$, $\eta_j \leqslant t_0$ by the extension property, and satisfies the initial value problem (IVP)

$$\begin{aligned} y'(t) &= f(t, y(t), y(\eta_i)), \\ y(\eta_i) &= x(\eta_i), \end{aligned} \tag{6.4}$$

such that $\eta_i = \tau_i(x(\eta_i))$ for $i = j$ (See Fig 6.1). In view of (A4) and (B0), there exists a solution $\psi(t) = \psi(t, \eta_j, x(\eta_j))$ of (6.4) defined on the right maximal interval of existence,

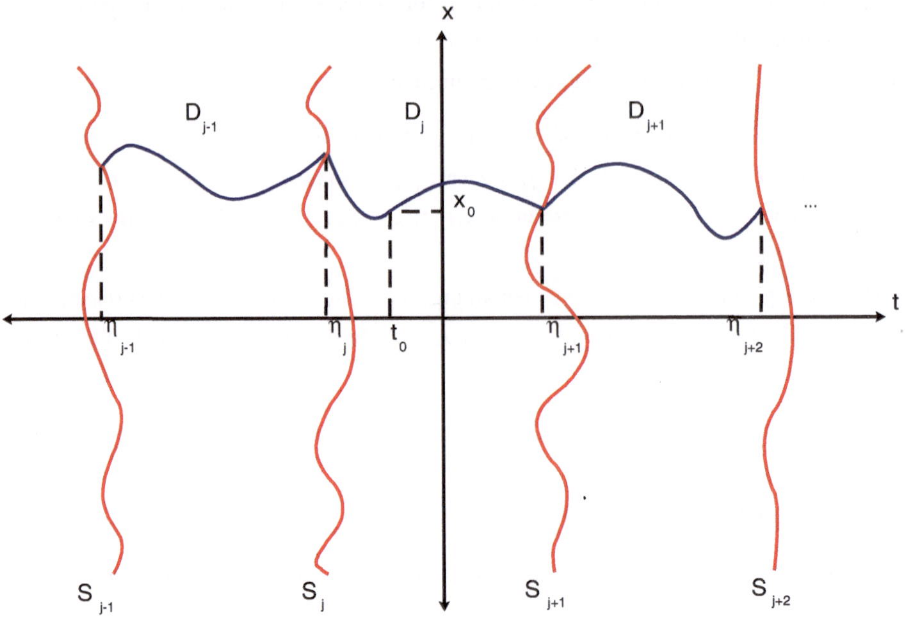

Fig. 6.1 A solution of a differential equation with state-dependent argument.

$[t_0, \beta)$. In the case of $\psi(t)$ does not intersect S_{j+1}, or the constancy switching surface S_{j+1} does not exist, the right maximal interval of $x(t)$ is $[t_0, \beta), \beta > t_0$. Otherwise, there is some $\xi \in \mathscr{I}$ such that $t_0 < \xi < \beta$, and $\xi = \tau_{j+1}(\psi(\xi))$. Then by denoting $\eta_{j+1} = \xi$, the solution $x(t)$ is defined as $\psi(t)$ on $[t_0, \eta_{j+1}]$. Now, the above discussion for (t_0, x_0) can be applied to the point $(\eta_{j+1}, x(\eta_{j+1}))$.

Proceeding in this way, we shall come either to the case when for some $k \in \mathscr{A}$, $k > j$, the solution $\psi(t) = \psi(t, \eta_k, x(\eta_k))$ has a right maximal interval $[\eta_k, \gamma)$ and this solution does not meet S_{k+1}, and then $[t_0, \gamma), \gamma > \eta_k$, is the right maximal interval of existence of $x(t)$. If there is no such k, then either $x(t)$ is continuable to $+\infty$ if the set $\mathscr{A}$ is unbounded from above, or the solution achieves the point $(\eta_r, x(\eta_r)), \eta_r = \tau_r(x(\eta_r))$ and then $x(t)$ has the right maximal interval $[t_0, \kappa), \kappa > \eta_r$ where $[\eta_r, \kappa)$ is the right maximal interval of the solution $\psi(t)$ of the equation (6.4) for $i = r$.

On the basis of the above discussion, if the extension property for (t_0, x_0) and the conditions (A4) and (B0) are valid, then the solution $x(t, t_0, x_0)$ of the equation (6.2) has a right maximal interval of existence, and it is open from the right.

Now, the case for decreasing t will be taken into account. Assume, again, that (t_0, x_0) satisfies the extension property. First, the situation when $(t_0, x_0) \in S_j$ should be considered. If the condition (A5) is not valid, then the solution $x(t, t_0, x_0)$ does not exist for $t \leqslant t_0$. Otherwise, it is continuable to η_{j-1} such that $\eta_{j-1} = \tau_{j-1}(x(\eta_{j-1}))$, and satisfies the equation (6.4) for $i = j-1$. Then, again, as for $(\eta_j, x(\eta_j))$, the same discussion should be made for the point $(\eta_{j-1}, x(\eta_{j-1}))$. Finally, we may conclude that either there exists η_k, $k \leqslant j$ such that the left maximal interval of $x(t)$ is $[\eta_k, t_0]$ (it is true also if there exists $k = \min \mathscr{A}$), or the solution is continuable up to $-\infty$. In the case when (t_0, x_0) is an interior point of D_j, and satisfies the extension property, $x(t, t_0, x_0)$ is continuable to the left till S_j, and the discussion made above can be repeated. Consequently, the left maximal interval of existence of $x(t)$ is either a closed interval $[\eta_k, t_0]$, $k \in \mathscr{A}$, or an infinite interval $(-\infty, t_0]$. By combining the left and right maximal intervals, the solution $x(t)$ is defined on the maximal interval of existence.

6.3 Quasilinear systems

In this section, the existence and uniqueness of solutions of quasilinear equations with state-dependent β function are investigated.

Let $\mathscr{I} = \mathbb{R}$, $\mathscr{G} = \mathbb{R}^n$ and $\mathscr{A} = \mathbb{Z}$. Fix a sequence of real numbers $\{\theta_i\} \subset \mathbb{R}$ such that $\theta_i < \theta_{i+1}$ for all $i \in \mathbb{Z}$. Take a sequence of functions $\xi_i(x) \in C(\mathbb{R}^n, \mathbb{R})$. Set $\tau_i(x) = \theta_i + \xi_i(x)$. Define the constancy switching surfaces $S_i = \{(t, x) \in \mathbb{R} \times \mathbb{R}^n : t = \theta_i + \xi_i(x)\}$, $i \in \mathbb{Z}$, and the regions $D_i = \{(t, x) \in \mathbb{R} \times \mathbb{R}^n : \theta_i + \xi_i(x) \leqslant t < \theta_{i+1} + \xi_{i+1}(x)\}$, $i \in \mathbb{Z}$.

The main subject of investigation is the quasilinear differential equation of the form

$$x'(t) = A(t)x(t) + F(t, x(t), x(\beta(t, x))), \tag{6.5}$$

where $t \in \mathbb{R}$, $x \in \mathbb{R}^n$, and $\beta(t, x)$ is a functional such that if $x(t) : \mathbb{R} \to \mathbb{R}^n$ is a continuous function, and $(t, x(t)) \in D_i$ for some $i \in \mathbb{Z}$, then $\beta(t, x) = \eta_i$, where η_i satisfies the equation $\eta = \theta_i + \xi_i(x(\eta))$.

Fix $H \in \mathbb{R}$, $H > 0$, and denote $K_H = \{x \in \mathbb{R}^n : \|x\| < H\}$. In the sequel, we make use of the following conditions:

(Q1) There exist positive real numbers c, d such that $c \leqslant \theta_{i+1} - \theta_i \leqslant d$, $i \in \mathbb{Z}$.

(Q2) There exists $l \in \mathbb{R}$, $0 \leqslant 2l < c$, such that $|\xi_i(x)| \leqslant l$, $i \in \mathbb{Z}$, for all $x \in K_H$.

(Q3) Functions $A : \mathbb{R} \to \mathbb{R}^{n \times n}$ and $F : \mathbb{R} \times \mathbb{R}^n \times \mathbb{R}^n \to \mathbb{R}^n$ are continuous.

(Q4) There exists a Lipschitz constant $L_1 > 0$ such that

$$\|F(t, x_1, y_1) - F(t, x_2, y_2)\| \leqslant L_1 \left[\|x_1 - x_2\| + \|y_1 - y_2\|\right]$$

for $t \in \mathbb{R}$ and $x_1, y_1, x_2, y_2 \in K_H$.

(Q5) $\sup_{t\in\mathbb{R}} \|A(t)\| = \kappa < \infty$.

(Q6) $\sup_{t\in\mathbb{R}} \|F(t,0,0)\| = N < \infty$.

(Q7) There exists a Lipschitz constant $L_2 > 0$ such that

$$|\xi_i(x) - \xi_i(y)| \leqslant L_2\|x-y\|$$

for all $x, y \in K_H$ and $i \in \mathbb{Z}$.

Explicitly, the conditions (Q1) and (Q2) imply (A1) with $\theta = c - 2l$ and $\overline{\theta} = d + 2l$. Also, the equation (6.3) for system (6.5) has the form

$$y'(t) = A(t)y(t) + F(t, y(t), z), \tag{6.6}$$

where $z \in \mathbb{R}^n$ is a constant vector. Hence, evidently, (A2) and (B0) are valid for the last equation, provided that the conditions (Q1)–(Q4) are satisfied.

Let $X(t)$ be a fundamental matrix solution of the homogeneous system, corresponding to the equation (6.6), that is,

$$x'(t) = A(t)x(t), \tag{6.7}$$

such that $X(0) = \mathscr{I}$, where $\mathscr{I}$ is an $n \times n$ identity matrix. Denote by $X(t,s) = X(t)X^{-1}(s)$, $t, s \in \mathbb{R}$, the transition matrix of (6.7). It can be verified that the inequalities

$$m \leqslant X(t,s) \leqslant M, \tag{6.8}$$

$$\|X(t,s) - X(\bar{t},s)\| \leqslant \kappa M|t-\bar{t}|, \tag{6.9}$$

are satisfied for the transition matrix $X(t,s)$, if $t, \bar{t}, s \in [\theta_j - l, \theta_{j+1} + l]$ for some $j \in \mathbb{Z}$, where $m = \exp(-\kappa\overline{\theta})$ and $M = \exp(\kappa\overline{\theta})$.

Let us fix $(t_0, x_0) \in \mathbb{R} \times \mathbb{R}^n$. The following lemma is the main auxiliary result of this chapter.

Lemma 6.1. *Suppose that* (Q1)–(Q3) *are fulfilled. Then, $x(t)$ is a solution of the equation* (6.5) *with $x(t_0) = x_0$ for $t \geqslant t_0$, if and only if it satisfies the equation*

$$x(t) = X(t,t_0)x_0 + \int_{t_0}^{t} X(t,s)F(s, x(s), x(\beta(s,x)))ds. \tag{6.10}$$

Proof. *Necessity.* Suppose that $x(t)$ is a solution of the equation (6.5) satisfying $x(t_0) = x_0$, where $(t_0, x_0) \in D_j$ for some $j \in \mathbb{Z}$. Define

$$\phi(t) = X(t,t_0)x_0 + \int_{t_0}^{t} X(t,s)F(s, x(s), x(\beta(s,x)))ds. \tag{6.11}$$

If $(t, x(t)) \in D_j \setminus S_j$, then there exists a moment $\eta_j \in \mathbb{R}$ such that $\beta(s,x) = \eta_j$ for all $(s, x(s)) \in D_j$. Also, the equations

$$\phi'(t) = A(t)\phi(t) + F(t, x(t), x(\eta_j)),$$

and

$$x'(t) = A(t)x(t) + F(t, x(t), x(\eta_j))$$

are satisfied, and this ends up with the equality

$$[\phi(t) - x(t)]' = A(t)[\phi(t) - x(t)].$$

Calculating the limit values at η_j, $j \in \mathbb{Z}$, one can find that

$$\phi'(\eta_j \pm 0) = A(\eta_j \pm 0)\phi(\eta_j \pm 0) + F(\eta_j \pm 0, x(\eta_j \pm 0), x(\beta(\eta_j \pm 0, x(\eta_j \pm 0)))),$$

$$x'(\eta_j \pm 0) = A(\eta_j \pm 0)x(\eta_j \pm 0) + F(\eta_j \pm 0, x(\eta_j \pm 0), x(\beta(\eta_j \pm 0, x(\eta_j \pm 0)))).$$

Consequently,

$$[\phi(t) - x(t)]'\big|_{t=\eta_j+0} = [\phi(t) - x(t)]'\big|_{t=\eta_j-0}.$$

Thus, $[\phi(t) - x(t)]$ is a continuously differentiable function defined for $t \geqslant t_0$ satisfying both the equation (6.7) and the initial condition $\phi(t_0) - x(t_0) = 0$. By means of uniqueness of solutions of the equation (6.7), it is clear that $\phi(t) - x(t) \equiv 0$ for $t \geqslant t_0$.

Sufficiency. Suppose that $x(t)$ is a solution of (6.10) for $t \geqslant t_0$. For a fixed integer j, if $(t, x(t)) \in D_j \setminus S_j$, then by differentiating (6.10) it is possible to see that $x(t)$ satisfies the equation (6.5). Moreover, it can be shown that $x(t)$ satisfies the equation (6.5) in D_j, by considering the limit $(t, x(t)) \to S_j$, and taking into account that $x(\beta(t, x))$ is a right-continuous function. The proof of the lemma is completed. □

The following example gives evidence about the difficulties for providing uniqueness of solutions of even simple differential equations with state-dependent piecewise constant argument.

Example 6.1. Consider the equation

$$x'(t) = -2x(\beta(t, x)), \tag{6.12}$$

where $\beta(t, x)$ is defined through the sequences $\theta_j = 2j$ and $\xi_j(x) = \cos x/4$, $j \in \mathbb{Z}$. Fix a point $(t_0, x_0) \in \mathbb{R} \times \mathbb{R}^n$, such that the equation $t_0 = (\cos x_0)/4$ is satisfied. The solution $x(t)$ of (6.12) with $x(t_0) = x_0$, is of the form $x(t) = (1 - 2(t - \cos x_0/4))x_0$ for $t \in [t_0, 5/4)$. Particularly, for $(t_0, x_0) = (1/4, 0)$ and $(1/4, 2\pi)$, the corresponding solutions are $x_1(t) = 0$ and $x_2(t) = \pi(3 - 4t)$, respectively, each of which passes through the point $(3/4, 0)$. Hence, the uniqueness is not attained.

Denote $\widetilde{M} = 2L_1H + N$. At this point we need the following assumption:

(Q8) $2M\overline{\theta}L_1 < \min\{1-2(\kappa H + M\widetilde{M})L_2,\ 1-N\overline{\theta}M/H\}$.

Let $h \in \mathbb{R}, 0 < h < \left(\frac{1-2ML_1\overline{\theta}}{M}H - N\overline{\theta}\right)$. The following lemma imposes sufficient conditions for the extension property to be satisfied by the equation (6.5).

Lemma 6.2. *Suppose that conditions* (Q1)–(Q8) *are fulfilled, and* $(t_0, x_0) \in D_j$ *for some integer* j *such that* $\|x_0\| < h$. *Then there exists a solution* $y(t) = y(t, t_0, x_0)$ *of the equation* (6.5) *such that*

$$\eta_j = \theta_j + \xi_j(y(\eta_j)) \text{ for some } \eta_j \leqslant t_0, \text{ and } y(t) \in K_H \text{ for all } t \in [\theta_j - l, \theta_{j+1} + l].$$

Proof. If $(t_0, x_0) \in S_j$, then by taking $\eta_j = t_0$ the result can be concluded directly. Suppose that $(t_0, x_0) \in D_j \setminus S_j$. Let us construct the following sequences. Take $\eta^0 = \theta_j$, $y_0(t) = X(t, t_0)x_0$, and define

$$\eta^{k+1} = \theta_j + \xi_j(y_k(\eta^k)), \tag{6.13}$$

$$y_{k+1}(t) = X(t, t_0)x_0 + \int_{t_0}^{t} X(t, s)F(s, y_k(s), y_k(\eta^k))ds \tag{6.14}$$

for all $k \in \mathbb{Z}, k \geqslant 0$.

Let $\|\cdot\|_0 = \max_{t\in[\theta_j - l, \theta_{j+1}+l]} \|\cdot\|$. It is straightforward to see that

$$\begin{aligned} \|y_{k+1}\|_0 &\leqslant M\|x_0\| + \left\| \int_{t_0}^{t} \|X(t, s)\|\|F(s, y_k(s), y_k(\eta^k))\|ds \right\|_0 \\ &\leqslant Mh + NM\overline{\theta} + 2ML_1\overline{\theta}\|y_k\|_0 \\ &\leqslant \frac{1-(2ML_1\overline{\theta})^{k+2}}{1-2ML_1\overline{\theta}}(Mh + NM\overline{\theta}). \end{aligned}$$

In view of (Q8), $y_k(t) \in K_H$ for all $t \in [\theta_j - l, \theta_{j+1} + l]$, $k \in \mathbb{Z}$, $k \geqslant 0$.

Now, we will deal with the uniform convergence of the sequence $\{y_k(t)\}$. The equations (6.13) and (6.14) imply that

$$\begin{aligned} |\eta^{k+1} - \eta^k| &= |\xi_j(y_k(\eta^k)) - \xi_j(y_{k-1}(\eta^{k-1}))| \\ &\leqslant L_2\|y_k(\eta^k) - y_{k-1}(\eta^{k-1})\|, \end{aligned}$$

$$\begin{aligned} \|y_{k+1} - y_k\|_0 &\leqslant \max_{t\in[\theta_j - l, \theta_{j+1}+l]} \left| \int_{t_0}^{t} M\|F(s, y_k(s), y_k(\eta^k)) - F(s, y_{k-1}(s), y_{k-1}(\eta^{k-1}))\|ds \right| \\ &\leqslant ML_1\overline{\theta}\left[\|y_k - y_{k-1}\|_0 + \|y_k(\eta^k) - y_{k-1}(\eta^{k-1})\|\right], \end{aligned}$$

$$
\begin{aligned}
\|y_{k+1}(\eta^{k+1})-y_k(\eta^k)\| \leqslant\ & \left\|X(\eta^{k+1},t_0)-X(\eta^k,t_0)\right\| \|x_0\| \\
& +\left|\int_{\eta_k}^{\eta_{k+1}} \|X(\eta^{k+1},s)F(s,y_k(s),y_k(\eta^k))\| ds\right| \\
& +\left|\int_{t_0}^{\eta_k} \|X(\eta^{k+1},s)F(s,y_k(s),y_k(\eta^k)) \right. \\
& \left. -X(\eta^k,s)F(s,y_{k-1}(s),y_{k-1}(\eta^{k-1}))\right\| ds\Big| \\
\leqslant\ & (\kappa h+\widetilde{M}(1+\kappa\overline{\theta}))M|\eta_{k+1}-\eta_k| \\
& +ML_1\overline{\theta}\left[\|y_k-y_{k-1}\|_0+\|y_k(\eta^k)-y_{k-1}(\eta^{k-1})\|\right] \\
\leqslant\ & M\big(L_2(\kappa h+\widetilde{M}(1+\kappa\overline{\theta})) \\
& +L_1\overline{\theta}\big)\left[\|y_k-y_{k-1}\|_0+\|y_k(\eta^k)-y_{k-1}(\eta^{k-1})\|\right] \\
\leqslant\ & \big(L_2(\kappa H+M\widetilde{M})+ML_1\overline{\theta}\big)\left[\|y_k-y_{k-1}\|_0\right. \\
& \left.+\|y_k(\eta^k)-y_{k-1}(\eta^{k-1})\|\right].
\end{aligned}
$$

Then,

$$|\eta^{k+1}-\eta^k| \leqslant \left[2\big(L_2(\kappa H+M\widetilde{M})+ML_1\overline{\theta}\big)\right]^{k-1}\overline{\theta}M\widetilde{M}, \tag{6.15}$$

$$\|y_{k+1}(\eta^{k+1})-y_k(\eta^k)\| \leqslant \left[2\big(L_2(\kappa H+M\widetilde{M})+ML_1\overline{\theta}\big)\right]^{k}\overline{\theta}M\widetilde{M}, \tag{6.16}$$

$$\|y_{k+1}-y_k\|_0 \leqslant \left[2\big(L_2(\kappa H+M\widetilde{M})+ML_1\overline{\theta}\big)\right]^{k}\overline{\theta}M\widetilde{M}. \tag{6.17}$$

Thus, there exist a unique moment η_j, and a solution $y(t)$ of the equation (6.5) with $y(t_0)=x_0$ such that $\eta_j=\theta_j+\xi_j(y(\eta_j))$, and η^k and y_k converge as $k\to\infty$, respectively. The lemma is proved. □

In what follows, we will consider the differential equations of type (6.5) such that the solutions intersect each constancy switching surface not more than once. In the previous section this assumption coincides with (A4). The following lemma defines the sufficient condition for this property.

From now on we shall need the following condition.

(Q9) $L_2\left[\kappa MH+M\widetilde{M}\right]<1.$

Lemma 6.3. *Suppose that* (Q1)–(Q7), (Q9) *hold. Then every solution $x(t)\in K_H$ of the equation* (6.5) *meets any constancy switching surface not more than once.*

Proof. Suppose the contrary. Then, there exist a solution $x(t)\in K_H$ of (6.5) and a surface S_j, $j\in\mathbb{Z}$ such that $x(t)$ meets this surface more than once. Let the first intersection be at $t = t_0$ and another intersection at $t = t^*$ so that we have $t_0 = \theta_j + \xi_j(x(t_0))$ and $t^* = \theta_j + \xi_j(x(t^*))$ for $t_0 < t^*$. Then, we have

$$\begin{aligned} |t^* - t_0| &\leqslant L_2 \|X(t^*, t_0)x(t_0) + \int_{t_0}^{t^*} X(t, s)F(s, x(s), x(\beta(s, x)))ds - x(t_0)\| \\ &\leqslant L_2 \left[\kappa MH + M\widetilde{M} \right] |t^* - t_0|, \end{aligned}$$

which contradicts (Q9). The lemma is proved. □

From the above lemmas we conclude the following theorem.

Theorem 6.1. *Assume that conditions* (Q1)–(Q9) *are fulfilled, and* $(t_0, x_0) \in D_j$ *for some* $j \in \mathbb{Z}$ *such that* $\|x_0\| < h$. *Then there exists a unique solution* $x(t) = x(t, t_0, x_0)$ *of the equation* (6.5) *on* $[\eta_j, \eta_{j+1}]$ *such that* $\eta_j = \theta_j + \xi_j(x(\eta_j)),\ \eta_{j+1} = \theta_{j+1} + \xi_{j+1}(x(\eta_{j+1}))$, *and* $x(t) \in K_H$.

6.4 Periodic solutions

In this section, we investigate periodic solutions of quasilinear differential equations with state-dependent piecewise constant argument of type (6.5).

Let ω and p be fixed positive real number and integer, respectively. We shall introduce the following assumptions:

(Q10) The functions $A(t)$ and $F(t, x, y)$ are ω-periodic in t.

(Q11) The sequence $\theta_i + \xi_i(x)$ satisfies (ω, p)-periodicity, that is, $\theta_{i+p} = \theta_i + \omega$ and $\xi_{i+p}(x) = \xi_i(x)$ for all $i \in \mathbb{Z}$ and $x \in \mathbb{R}^n$.

(Q12) $\det(I - X(\omega)) \neq 0$; that is, system (6.7) does not have any w-periodic solution.

We define, if (Q12) is fulfilled, the function

$$G(t, s) = \begin{cases} X(t)(I - X(\omega))^{-1}X^{-1}(s), & 0 \leqslant s \leqslant t \leqslant \omega, \\ X(t+\omega)(I - X(\omega))^{-1}X^{-1}(s), & 0 \leqslant t < s \leqslant \omega, \end{cases} \tag{6.18}$$

which is called *Green's function* [153]. Let $\max_{t,s \in [0,\omega]} \|G(t, s)\| = K$.

We need the following lemma to prove the main theorem. This lemma can be proved using Lemma 6.1.

Lemma 6.4. *Suppose that* (Q1)–(Q12) *are fulfilled. Then the solution* $x(t)$ *of the equation* (6.5) *is* w*-periodic if and only if it satisfies the integral equation*

$$x(t) = \int_0^{\omega} G(t, s)F(s, x(s), x(\beta(s, x)))ds. \tag{6.19}$$

Let $\|\cdot\|_\omega = \max_{t\in[0,\,\omega]} \|\cdot\|$. Denote by Φ the set of all continuous and piecewise continuously differentiable ω-periodic functions on $\mathbb{R}$ such that if $\phi \in \Phi$ then $\|\phi(t)\|_\omega < H$, $\|\frac{d\phi(t)}{dt}\|_\omega < N + (2L_1+\kappa)H$.

We introduce the following assumption to prove the next theorem.

(Q13) $(2KL_1\omega - 1)H + NK\omega < 0$;
$L_2(N+(2L_1+\kappa)H) < 1$;
$KL_1(2-L_2(N+(2L_1+\kappa)H))\omega + 2KHL_1L_2p + L_2(N+(2L_1+\kappa)H) < 1.$

Theorem 6.2. *Suppose that* (Q1)–(Q13) *hold. Then the equation* (6.5) *has a unique ω-periodic solution $\phi(t)$ such that $\phi(t) \in K_H$.*

Proof. Suppose that for all $x \in K_H$ and $k = j, \ldots, j+p-1$, for some $j \in \mathbb{Z}$ and $p > 1$, we have $0 \leqslant \theta_k + \xi_k(x) \leqslant \omega$. The other cases are similar. Define an operator T on Φ as

$$T[\phi] = \int_0^\omega G(t, s)F(s, \phi(s), \phi(\beta(s, \phi)))ds. \tag{6.20}$$

Using (Q13), it is easy to see that $\|T[\phi]\|_\omega < H$ and $\|\frac{dT[\phi]}{dt}\|_\omega < N + (2L_1+\kappa)H$. That is, $T[\phi] \in \Phi$.

Now, we will show that the operator T is contractive on Φ. Let $\phi_1, \phi_2 \in \Phi$. One can see that using (Q13), the function $\phi_i(t)$ intersects any constancy switching surface S_k exactly once at $t = \eta_k^i$ for all $i = 1, 2$ and $k = j, \ldots, j+p-1$. Without loss of generality suppose that $\eta_k^1 \leqslant \eta_k^2$.

Also, whereby the Mean Value Theorem and (Q13), it can be shown that the inequality

$$\|\phi_1(\eta_k^1) - \phi_2(\eta_k^2)\| \leqslant \frac{1}{1 - L_2(N + (2L_1+\kappa)H)}\|\phi_1 - \phi_2\|_\omega \tag{6.21}$$

holds.

By means of (6.20) and (Q11), the equation

$$\begin{aligned} T[\phi_i(t)] = &\int_0^{\eta_j^i} G(t, s)F(s, \phi_i(s), \phi_i(\eta_{j+p-1}^i))ds \\ &+ \sum_{k=j}^{j+p-2} \int_{\eta_k^i}^{\eta_{k+1}^i} G(t, s)F(s, \phi_i(s), \phi_i(\eta_k^i))ds \\ &+ \int_{\eta_{j+p-1}^i}^{\omega} G(t, s)F(s, \phi_i(s), \phi_i(\eta_{j+p-1}^i))ds \end{aligned}$$

is satisfied for $i = 1, 2$.

Then, by the help of (6.21), we have

$$\|T[\phi_1] - T[\phi_2]\|_\omega \leqslant K\Big[\int_0^{\eta_j^1} \|F(s, \phi_1(s), \phi_1(\eta_{j+p-1}^1)) - F(s, \phi_2(s), \phi_2(\eta_{j+p-1}^2))\|ds$$

$$+\sum_{k=j}^{j+p-2}\int_{\eta_k^2}^{\eta_{k+1}^1}\|F(s,\phi_1(s),\phi_1(\eta_k^1))-F(s,\phi_2(s),\phi_2(\eta_k^2))\|ds$$

$$+\int_{\eta_{j+p-1}^2}^{\omega}\|F(s,\phi_1(s),\phi_1(\eta_{j+p-1}^1))-F(s,\phi_2(s),\phi_2(\eta_{j+p-1}^2))\|ds$$

$$+\sum_{k=j}^{j+p-1}\int_{\eta_k^1}^{\eta_k^2}\|F(s,\phi_1(s),\phi_1(\beta(s,\phi_1)))-F(s,\phi_2(s),\phi_2(\beta(s,\phi_2)))\|ds\Big]$$

$$\leqslant\left[\frac{KL_1\omega(2-L_2(N+(2L_1+\kappa)H))+2KHL_1L_2p}{1-L_2(N+(2L_1+\kappa)H)}\right]\|\phi_1-\phi_2\|_\omega.$$

Hence, T is contractive and in view of the Lemma 6.4, we conclude that the fixed point is ω-periodic solution of the equation (6.5). This argument finishes the proof of the theorem. □

We carry out the last theorem in the following example.

Example 6.2. Consider the system of equations

$$x'(t)=-x(t)-a\sin(2\pi t+y(\beta(t,x,y))) \tag{6.22}$$

$$y'(t)=-2y(t)+a\sin(2\pi t+x(\beta(t,x,y))), \tag{6.23}$$

where $t,x,y\in\mathbb{R}$, and a is a positive real number. Here, $\beta(t,x,y)$ is defined by $\theta_j=j$, $\xi_j(x,y)=-a\cos(x+y)$. The numbers $L_1=a\sqrt{2}, L_2=a$, $\overline{\theta}=1+2a$, $\kappa=2$, $N=a\sqrt{2}$, $M=\mathrm{e}^{2+4a}$, $\widetilde{M}=(2H+1)a\sqrt{2}$, $\omega=1$, $p=1$, and $K=\mathrm{e}^2(1-\mathrm{e}^{-1})^{-1}$ are the corresponding parameters in the conditions of Theorem 6.2. One can show that the conditions (Q1)–(Q13) are fulfilled for $a=\mathrm{e}^{-4}$, $H=1$. Hence, by means of Theorem 6.2, we ensure that there is a 1-periodic solution of (6.23). Figure 6.2 shows the solution $(x(t),y(t))$ of (6.23) with an initial condition $(x(-\mathrm{e}^{-4}),y(-\mathrm{e}^{-4}))=(0.02,-0.02)$ that approaches to this periodic solution.

6.5 Stability of the zero solution

In this section, the sufficient conditions for stability of the zero solution are presented. Let us introduce the following conditions:

(Q14) $F(t,0,0)=0$ for all $t\in\mathbb{R}$.

(Q15) $M\left[(1+\overline{\theta}L_1)(\mathrm{e}^{ML_1\overline{\theta}}-1)+L_1\overline{\theta}\right]<1.$

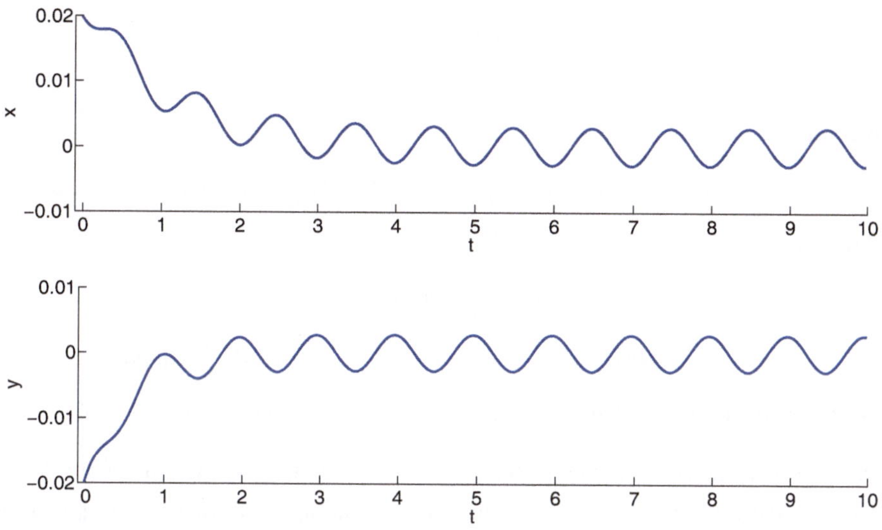

Fig. 6.2 A solution $(x(t), y(t))$ that approaches the 1-periodic solution as time increases.

Define

$$K(L_1, \overline{\theta}) = \frac{M}{1 - M\left[(1+\overline{\theta}L_1)(e^{ML_1\overline{\theta}} - 1) + L_1\overline{\theta}\right]}.$$

The following lemma plays a significant role in this section. The following lemma can be proved by using the technique in [9] and similar to [25, Lemma 1.2].

Lemma 6.5. *Suppose that* (Q1)–(Q9), (Q14), (Q15) *are fulfilled. Then, every solution* $x(t)$ *of the equation* (6.5) *satisfies the inequality*

$$\|x(\beta(t,x))\| \leqslant K(L_1, \overline{\theta})\|x(t)\| \tag{6.24}$$

for all $t \in \mathbb{R}$.

Proof. Fix $t \in \mathbb{R}$. Let $x(t)$ be a solution of (6.5). Then, there are $k \in \mathbb{Z}$, and $\eta_k \in \mathbb{R}$ such that $(t, x(t)) \in D_k$, and $\beta(t,x) = \eta_k$. Using Lemma 6.1, we have

$$x(t) = X(t, \eta_k)x(\eta_k) + \int_{\eta_k}^{t} X(t, s)F(s, x(s),\ x(\eta_k))ds.$$

Then,

$$\begin{aligned}\|x(t)\| &\leqslant M\|x(\eta_k)\| + ML_1\int_{\eta_k}^{t}(\|x(s)\| + \|x(\eta_k)\|)\,ds\\ &\leqslant M(1+\overline{\theta}L_1)\|x(\eta_k)\| + ML_1\int_{\eta_k}^{t}\|x(s)\|ds.\end{aligned}$$

Hence, by the help of the Gronwall-Bellman Lemma, the inequality

$$\|x(t)\| \leqslant M(1+\overline{\theta}L_1)e^{ML_1(t-\eta_k)}\|x(\eta_k)\|$$

is obtained.

Moreover,

$$x(\eta_k) = X(\eta_k,t)x(t) - \int_{\eta_k}^{t} X(\eta_k, s)F(s, x(s), x(\eta_k))ds,$$

and therefore,

$$\begin{aligned}\|x(\eta_k)\| &\leqslant M\|x(t)\| + ML_1\int_{\eta_k}^{t}(\|x(s)\| + \|x(\eta_k)\|)ds \\ &\leqslant M\|x(t)\| + M\left[(1+\overline{\theta}L_1)(e^{ML_1\overline{\theta}}-1) + L_1\overline{\theta}\right]\|x(\eta_k)\|.\end{aligned}$$

Thus, for $(t, x(t)) \in D_k$, we have $\|x(\eta_k)\| \leqslant K(L_1, \overline{\theta})\|x(t)\|$. The lemma is proved. □

Definition 6.2. The zero solution of (6.5) is said to be uniformly stable if for any $\varepsilon > 0$ and $t_0 \in \mathbb{R}$, there exists a $\delta = \delta(\varepsilon) > 0$ such that $\|x(t, t_0, x_0)\| < \varepsilon$ whenever $\|x_0\| < \delta$ for $t \geqslant t_0$.

Definition 6.3. The zero solution of (6.5) is said to be uniformly asymptotically stable if it is uniformly stable, and there is a number $b > 0$ such that for every $\zeta > 0$ there exists $T(\zeta) > 0$ such that $\|x_0\| < b$ implies that $\|x(t, t_0, x_0)\| < \zeta$ if $t > t_0 + T(\zeta)$.

Theorem 6.3. *Suppose that* (Q1)–(Q9), (Q14), (Q15) *hold. If the zero solution of the equation* (6.7) *is uniformly asymptotically stable, then for sufficiently small Lipschitz constant* L_1*, the zero solution of the equation* (6.5) *is uniformly asymptotically stable.*

Proof. Suppose that the zero solution of (6.7) is uniformly asymptotically stable. Then, there exist positive real numbers α and σ such that for $t > s$,

$$\|X(t, s)\| \leqslant \alpha e^{-\sigma(t-s)}. \tag{6.25}$$

Let $x(t)$ be a solution of (6.5) with the initial condition $x(t_0) = x_0$ such that $\|x_0\| \leqslant h$. For $t \geqslant t_0$, we have

$$\begin{aligned}\|x(t)\| &= \left\|X(t, t_0)x(t_0) + \int_{t_0}^{t} X(t, s)F(s, x(s), x(\beta(s, x)))ds\right\| \\ &\leqslant \alpha e^{-\sigma(t-t_0)}\|x_0\| + L_1\int_{t_0}^{t}\alpha e^{-\sigma(t-s)}(1+K(L_1, \overline{\theta}))\|x(s)\|ds,\end{aligned}$$

and

$$e^{\sigma t}\|x(t)\| \leqslant \alpha e^{\sigma t_0}\|x_0\| + \alpha L_1(1+K(L_1, \overline{\theta}))\int_{t_0}^{t} e^{\sigma s}\|x(s)\|ds.$$

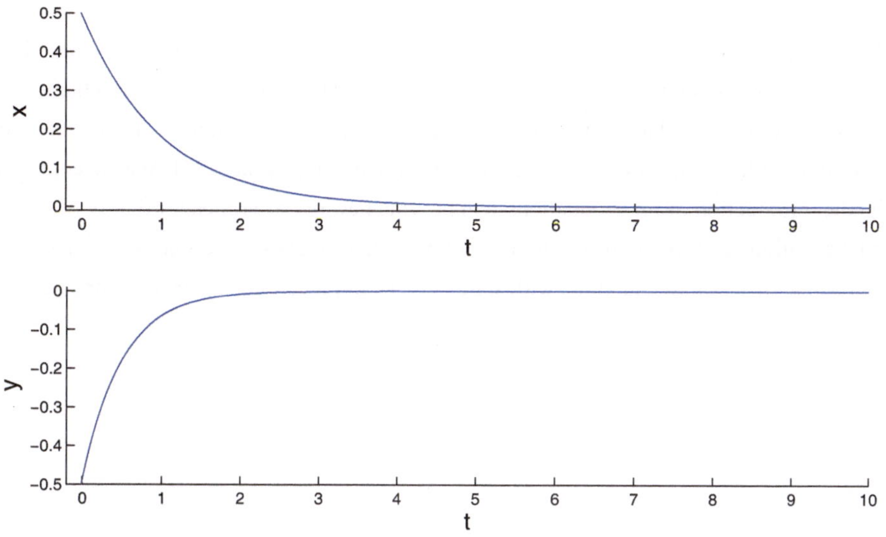

Fig. 6.3 A solution $(x(t), y(t))$ that approaches the zero solution as time increases.

Hence, the inequality

$$\|x(t)\| \leqslant \alpha e^{(\alpha L_1(1+K(L_1,\overline{\theta}))-\sigma)(t-t_0)} \|x_0\|$$

is obtained by the help of the Gronwall-Bellman Lemma. The proof of the theorem ends up by the fact that $\alpha L_1(1+K(L_1,\overline{\theta}))-\sigma<0$, for sufficiently small L_1. □

Next, we validate our last result with an example.

Example 6.3. In the present example, the equation

$$\begin{aligned} x'(t) &= -x(t) - a\sin^2(y(\beta(t,x,y))) \\ y'(t) &= -2y(t) + a\sin^2(x(\beta(t,x,y))) \end{aligned} \tag{6.26}$$

is taken into consideration, where $t, x, y \in \mathbb{R}$, and a is a positive real number. Here, $\beta(t,x,y)$ is defined through $\theta_j = j$ and $\xi_j(x,y) = -a\cos(x+y)$. The corresponding parameters in conditions of Theorem 6.3 are $L_1 = 2\sqrt{2}a$, $L_2 = a$, $\overline{\theta} = 1+2a$, $\kappa = 2$, $N = 0$, $M = e^{2+4a}$, $\widetilde{M} = 4\sqrt{2}aH$. It can be shown that the conditions (Q1)–(Q9), (Q14), (Q15) are satisfied for $a = e^{-4}$, $H = 1$. Hence, by Theorem 6.3, the zero solution is uniformly asymptotically stable. Figure 6.3 shows the solution $(x(t), y(t))$ of (6.26) with initial condition $(x(-e^{-4}), y(-e^{-4})) = (0.02, -0.02)$ that approaches to the zero solution.

Notes

The principal novelty of the chapter is the state-dependent constancy of arguments. This phenomenon has not been considered in literature, yet. The main results of this chapter are published in [33]. Similar results can be found for equations with impulses at variable moments of time [20] and differential equations on variable time scales [37]. We are confident that the introduction of these equations will provide new opportunities for theory of differential equations and for applications. Equations with γ functions as arguments can also be extended to differential equations with state-dependent piecewise constant argument.

Chapter 7

Almost periodic solutions

This chapter presents existence and stability of almost periodic solutions of the following system

$$\frac{dx(t)}{dt} = A(t)x(t) + f(t, x(\theta_{\upsilon(t)-p_1}), x(\theta_{\upsilon(t)-p_2}), \ldots, x(\theta_{\upsilon(t)-p_m})), \tag{7.1}$$

where $x \in \mathbb{R}^n, t \in \mathbb{R}$, $\upsilon(t) = i$ if $\theta_i \leqslant t < \theta_{i+1}, i = \ldots, -2, -1, 0, 1, 2, \ldots$, is an identification function, θ_i is a strictly ordered sequence of real numbers, unbounded on the left and on the right, p_j, $j = 1, 2, \ldots, m$, are fixed integers, and the linear homogeneous system associated with (7.1) satisfies exponential dichotomy. The problem of the existence is studied without any sign condition on deviations of the argument.

7.1 Introduction and Preliminaries

The existence of almost periodic solutions is one of the most interesting problems of the theory of differential equations (see, for example, [46, 87, 117] and the references cited there). This problem has been considered in the context of differential equations with piecewise constant argument in many papers, such as [44, 161, 170, 171, 283, 333, 336], [340–345].

In this chapter we will apply results of investigation of almost periodic discontinuous solutions [40–42], [148, 278]. From the literature one can see the connection between differential equations with piecewise constant arguments and impulsive differential equations. This was mentioned in [81] for differential equations with piecewise constant argument, and in [116] for differential equations with discontinuous right hand side.

Ideas on the spaces of discontinuous functions originated from [148,183,288,315]. Following these results, we introduce Bohner type discontinuous almost periodic functions [41]. We used a topology as well as a metric in the spaces of discontinuous functions and in the discrete spaces of sets of points.

Let $\mathbb{Z}$, $\mathbb{N}$, and $\mathbb{R}$ be the sets of all integers, natural and real numbers, respectively, and $\|\cdot\|$ be the Euclidean norm in $\mathbb{R}^n, n \in \mathbb{N}$. Let $s \in \mathbb{R}$ be a positive number. We denote $G_s = \{x \in \mathbb{R}^n \mid \|x\| \leqslant s\}$ and $G_s^m = G_s \times G_s \ldots \times G_s$ (that is, G_s^m is an $m-$ times Cartesian product of G_s). Let $C_0(\mathbb{R})$ (respectively $C_0(\mathbb{R} \times G_H^m)$ for a given $H \in \mathbb{R}$, $H > 0$) be the set of all bounded and uniformly continuous functions on $\mathbb{R}$ (respectively on $\mathbb{R} \times G_H^m$). For $f \in C_0(\mathbb{R})$ (respectively $C_0(\mathbb{R} \times G_H^m)$) and $\tau \in \mathbb{R}$, a translation of f by τ is a function $Q_\tau f = f(t+\tau)$, $t \in \mathbb{R}$ (respectively $Q_\tau f(t,z) = f(t+\tau, z)$, $(t,z) \in \mathbb{R} \times G_H^m$). A number $\tau \in \mathbb{R}$ is called an ε-translation number of a function $f \in C_0(\mathbb{R})$ $(C_0(\mathbb{R} \times G_H^m))$ if $\|Q_\tau f - f\| < \varepsilon$ for every $t \in \mathbb{R}$ $((t,z) \in \mathbb{R} \times G_H^m)$. A set $S \subset \mathbb{R}$ is said to be relatively dense if there exists a number $l > 0$ such that $[a, a+l] \cap S \neq \emptyset$ for all $a \in \mathbb{R}$.

Definition 7.1. A function $f \in C_0(\mathbb{R})(C_0(\mathbb{R} \times G_H^m))$ is said to be almost periodic (almost periodic in t uniformly with respect to $z \in G_H^m$) if for every $\varepsilon \in \mathbb{R}$, $\varepsilon > 0$, there exists a relatively dense set of ε-translation numbers of f.

Denote by $\mathscr{AP}(\mathbb{R})$ $(\mathscr{AP}(\mathbb{R} \times G_H^m))$ the set of all such functions.

The following assumptions will be needed throughout this chapter.

(A1) $A(t) \in \mathscr{AP}(\mathbb{R})$ is an $n \times n$ matrix;

(A2) $f \in \mathscr{AP}(\mathbb{R} \times G_s^m)$, for every $s \in \mathbb{R}$, $s \geqslant 0$;

(A3) $\exists l \in \mathbb{R}$, $l > 0$, such that

$$\|f(t,z_1) - f(t,z_2)\| \leqslant l \sum_{j=1}^{m} \|z_1^j - z_2^j\|,$$

where $z_i = (z_i^1, \ldots, z_i^m) \in \mathbb{R}^m$, $i = 1,2$.

Let

$$\frac{dx}{dt} = A(t)x \tag{7.2}$$

be the homogeneous linear system associated with (7.1), and $X(t)$ be a fundamental matrix of (7.2);

(A4) System (7.2) satisfies exponential dichotomy, that is, there exist a projection P and positive constants $\sigma_1, \sigma_2, K_1, K_2$, such that

$$\|X(t)PX^{-1}(s)\| \leqslant K_1 \exp(-\sigma_1(t-s)), \ t \geqslant s,$$
$$\|X(t)(I-P)X^{-1}(s)\| \leqslant K_2 \exp(\sigma_2(t-s)), \ t \leqslant s.$$

Denote by Θ a space of strictly ordered sequences $\{\theta_i\} \subset \mathbb{R}$, $i \in \mathbb{Z}$, such that $|\theta_i| \to \infty$, if $|i| \to \infty$. Let $\mathscr{PC}$ be the set of all piecewise continuous functions from $\mathbb{R}$ to $\mathbb{R}^n$ with discontinuities of the first type. Assume that the set of discontinuities of every function from $\mathscr{PC}$, numerated in a strict order, is an element of Θ. Moreover, these functions are uniformly continuous on the set $\bigcup_{i\in\mathbb{Z}}(\theta_i,\theta_{i+1})$, and they are left or right continuous at every point of discontinuity.

Denote by $\mathscr{PC}_r \subset \mathscr{PC}$ the set of all right continuous functions. If $\phi \in \mathscr{PC}$, then one can define a function $\phi_r \in \mathscr{PC}_r$, such that $\phi_r(t) = \phi(t)$ everywhere, except possibly at points $t = \theta_i$, that is,

$$\phi_r(t) = \begin{cases} \phi(t), & \text{if } t \neq \theta_i, \\ \phi(\theta_i+), \ i \in \mathbb{Z}, & \text{otherwise.} \end{cases}$$

We say that the function $\phi_r(t)$ is a right extension of the function $\phi(t) \in \mathscr{PC}$. The set of all left continuous functions, $\mathscr{PC}_l$, can be defined similarly. As the function $\upsilon(t)$ is right continuous, it is reasonable to consider only the space $\mathscr{PC}_r$, extending to the right a function from $\mathscr{PC}$, if necessary. Since functions from $\mathscr{PC}$ are assumed to be the derivatives or limits of the solutions of differential equations with piecewise constant arguments, no difficulty arises from this agreement. In what follows we assume that $\upsilon(t) \in \mathscr{PC}_r$.

It is obvious that the derivative of a solution $x(t)$ is a function from $\mathscr{PC}_r$, if we assume it to be right differentiable at $t = \theta_i$, $i \in \mathbb{Z}$.

Let

$$G(t,s) = \begin{cases} X(t)PX^{-1}(s), & \text{if } t \geqslant s, \\ X(t)(P-I)X^{-1}(s), & \text{if } t < s \end{cases}$$

be the Green's function of (7.2). Denote

$$F_\theta(\psi(t)) = f(t, \psi(\theta_{\upsilon(t)-p_1}), \psi(\theta_{\upsilon(t)-p_2}), \dots, \psi(\theta_{\upsilon(t)-p_m})),$$

where $\psi(t) \in C_0(\mathbb{R})$.

Lemma 7.1. *A function $x(t) \in C_0(\mathbb{R})$ is a solution of* (7.1) *if and only if*

$$x(t) = \int_{-\infty}^{\infty} G(t,s)F_\theta(x(s))ds. \tag{7.3}$$

Proof. *Necessity.* Let $x(t) \in C_0(\mathbb{R})$ be a solution of (7.1). Introduce a function

$$\phi(t) = \textstyle\int_{-\infty}^{\infty} G(t,s)F_\theta(x(s))ds. \tag{7.4}$$

One can easily see that the function $\phi(t)$ is bounded and continuous on $\mathbb{R}$.

If $t \neq \theta_i$, $i \in \mathbb{Z}$, then

$$\phi'(t) = A(t)\phi(t) + F_\theta(x(t))$$

and

$$x'(t) = A(t)x(t) + F_\theta(x(t)).$$

Consequently,

$$[\phi(t) - x(t)]' = A(t)[\phi(t) - x(t)].$$

Evaluating the limit values at $t = \theta_j$, $j \in \mathbb{Z}$, one can find that

$$\phi'(\theta_j \pm 0) = A(\theta_j \pm 0)\phi(\theta_j \pm 0) + F_\theta(x(\theta_j \pm 0)),$$

$$x'(\theta_j \pm 0) = A(\theta_j \pm 0)x(\theta_j \pm 0) + F_\theta(x(\theta_j \pm 0)).$$

Then,

$$[\phi(t) - x(t)]'|_{t=\theta_j+0} = [\phi(t) - x(t)]'|_{t=\theta_j-0}.$$

That is, $[\phi(t) - x(t)]$ is a continuously differentiable function on $\mathbb{R}$, satisfying (7.2), and $[\phi(t) - x(t)] = 0$ on $\mathbb{R}$.

Sufficiency. Assume that (7.3) is valid and $x(t) \in C_0(\mathbb{R})$. Fix $i \in \mathbb{Z}$ and consider the interval $[\theta_i, \theta_{i+1})$. If $t \in (\theta_i, \theta_{i+1})$, then by differentiating one can find that $x(t)$ satisfies (7.1). Moreover, considering $t \to \theta_i+$, and taking into account that $\upsilon(t)$ is a right-continuous function, we see that $x(t)$ satisfies (7.1) on $[\theta_i, \theta_{i+1})$. The lemma is proved. □

7.2 Wexler sequences

Fix a sequence $\theta \in \Theta$, and consider another sequence γ_i, $i \in \mathbb{Z}$, $\gamma_{i+1} \geqslant \gamma_i$, such that for every $\gamma_i \in \gamma$ there is an element $\theta_j \in \theta$ which satisfies $\gamma_i = \theta_j$. Let $m(i)$, $i \in \mathbb{Z}$, be the number of elements of γ which are equal to θ_i. We shall call this number the multiplicity of θ_i with respect to γ. Denote $m(\gamma) = \sup_i m(i)$. Denote by Γ the set of all sequences γ such that $m(\gamma) < \infty$. If $\gamma \in \Gamma$, then we say that $m(\gamma)$ is the maximal multiplicity of γ. It is obvious that $|\gamma_i| \to \infty$ if $|i| \to \infty$, for every $\gamma \in \Gamma$, and that $\Theta \subset \Gamma$. We say that θ is a support of γ, and γ a representative of θ. Introduce the following distance $\|\gamma^{(1)} - \gamma^{(2)}\| = \sup_i \|\gamma_i^{(1)} - \gamma_i^{(2)}\|$ if $\gamma^{(1)}, \gamma^{(2)} \in \Gamma$. We say that elements $\theta^{(1)}, \theta^{(2)} \in \Theta$ are $\varepsilon-$ equivalent and write $\theta^{(1)}\varepsilon\theta^{(2)}$, if there exist the representatives $\gamma^{(1)}$ and $\gamma^{(2)}$ of $\theta^{(1)}$ and $\theta^{(2)}$ in Γ, respectively, such that $\|\gamma^{(1)} - \gamma^{(2)}\| < \varepsilon$. Moreover, we shall say that these sequences are in the $\varepsilon-$ neighborhoods of each other.

The topology defined on the basis of all $\varepsilon-$ neighborhoods, $0 < \varepsilon < \infty$, of all elements of Θ is named as B^s- topology. Obviously, it is a Hausdorff topology.

Lemma 7.2. *If* $\theta^{(1)}\varepsilon_1\theta^{(2)}, \theta^{(2)}\varepsilon_2\theta^{(3)}$, *then* $\theta^{(1)}(\varepsilon_1 + \varepsilon_2)\theta^{(3)}$.

Proof. One can easily find that if $\theta^{(1)}\varepsilon_1\theta^{(2)}$, then for any element $\theta_i^{(1)}$ there exists at least one element $\theta_j^{(2)}$ such that $|\theta_j^{(2)}-\theta_i^{(1)}|<\varepsilon_1$, and vice versa. The same is true for elements of sequences $\theta^{(2)}$ and $\theta^{(3)}$, with ε_2. Consequently, it is easy to find that for any element $\theta_i^{(1)}$ there exists at least one element $\theta_j^{(3)}$ such that $|\theta_j^{(3)}-\theta_i^{(1)}|<\varepsilon_1+\varepsilon_2$, and vice versa. Using this fact, and that the sequences are strictly ordered, we shall construct, in what follows, two sequences $\overline{\gamma}^{(1)}$ and $\overline{\gamma}^{(3)}$ such that $\|\overline{\gamma}^{(1)}-\overline{\gamma}^{(3)}\|<\varepsilon_1+\varepsilon_2$. We start with $\theta_0^{(1)}$ and determine all those $\theta_i^{(3)}$, which are on the distance not less than ε from $\theta_0^{(1)}$. Let us say that they are $\theta_{j-k}^{(3)}<\theta_{j-k-1}^{(3)}<\cdots<\theta_j^{(3)}$. Define $\overline{\gamma}_i^{(1)}=\theta_0^{(1)}$ and $\overline{\gamma}_i^{(3)}=\theta_{j-k+i}^{(3)}, i=0,1,\ldots,k$. Next, consider elements $\theta_0^{(1)}<\theta_1^{(1)}<\cdots<\theta_l^{(1)}$, which are on the distance not less than ε from $\theta_j^{(3)}$. Introduce $\overline{\gamma}_i^{(3)}=\theta_j^{(3)},\overline{\gamma}_i^{(1)}=\theta_{-k+i}^{(1)}$, $i=k,\ldots,k+l$. Then, consider element $\theta_l^{(1)}$ with the similar procedure and so on. Let us built the sequences for decreasing indices. Consider elements $\theta_{-r}^{(1)}<\theta_{-r+1}^{(1)}<\cdots<\theta_0^{(1)}$, which are on the distance not less than ε from $\theta_{j-k}^{(3)}$. Set $\overline{\gamma}_i^{(3)}=\theta_{j-k}^{(3)}, \overline{\gamma}_i^{(1)}=\theta_i^{(1)}, i=-r,-r+1<\cdots,0$. Next, one can do the similar procedure for $\theta_{-r}^{(1)}$. Proceeding in this way we shall construct sequences $\overline{\gamma}^{(1)}$ and $\overline{\gamma}^{(3)}$ from Γ with the required property. The lemma is proved. □

Let a_i, $i\in\mathbb{Z}$, be a sequence in $\mathbb{R}^n$. An integer p is called an $\varepsilon-$almost period of the sequence, if $\|a_{i+p}-a_i\|<\varepsilon$ for any $i\in\mathbb{Z}$.

Definition 7.2. A sequence a_i, $i\in\mathbb{Z}$, is almost periodic, if for every $\varepsilon>0$ there exists a relatively dense set of its $\varepsilon-$almost periods.

Let $\gamma\in\Gamma$, i, $j\in\mathbb{Z}$. Denote $\gamma_i^j=\gamma_{i+j}-\gamma_i$ and define sequences $\gamma^j=\{\gamma_i^j\}\in\Gamma$, $j\in\mathbb{Z}$.

Definition 7.3 ([148, 278]). Sequences γ^j, $j\in\mathbb{Z}$, are equipotentially almost periodic if for any $\varepsilon>0$ there exists a relatively dense set of ε-almost periods that are common for all γ^j, $j\in\mathbb{Z}$.

Definition 7.4. We shall say that $\theta\in\Theta$ is a Wexler sequence if there exists a representative γ of θ with equipotentially almost periodic sequences γ^j, $j\in\mathbb{Z}$.

Let $\gamma\in\Gamma,\varepsilon>0$, be given. Denote by $T_\varepsilon\subset\mathbb{R}$ the set of numbers τ, for which there exists at least one integer q_τ such that

$$|\gamma_i^{q_\tau}-\tau|<\varepsilon,\;\; i\in\mathbb{Z}. \tag{7.5}$$

Denote by Q_τ the set of all numbers q_τ satisfying (7.5) for fixed ε and τ, and $Q_\varepsilon=\bigcup_{\tau\in T_\varepsilon}Q_\tau$. The following lemmas were proved in [148, 315] for $m(\gamma)=1$. But one can easily, repeating the proof in [148], verify that they are also valid if $1<m(\gamma)<\infty$.

Lemma 7.3. *The following statements are equivalent:*

(a) the sequences γ^j, $j \in \mathbb{Z}$, are equipotentially almost periodic;
(b) the set T_ε is relatively dense for each $\varepsilon > 0$;
(c) the set Q_ε is relatively dense for each $\varepsilon > 0$.

Lemma 7.4. *Assume that sequences γ^j, $j \in \mathbb{Z}$, are equipotentially almost periodic. Then for arbitrary $l > 0$ there exists $n_0 \in \mathbb{N}$ such that any interval of length l contains at most n_0 elements of γ.*

Fix $\theta \in \Theta$ and let $h = \{h_n\}$, $n \in \mathbb{N}$, be a sequence of real numbers. Assume that the sequence of shifts $\{\theta + h_n\}_n$ is convergent in B^s- topology. We shall denote the limit element as $Q_h\theta$.

Definition 7.5. An element $\theta \in \Theta$ has the Bohner property, if every sequence $\{h'_n\}$ contains a subsequence $h \subset h'$, such that $Q_h\theta$ exists.

Theorem 7.1. *If $\theta \in \Theta$ is a Wexler sequence, then it satisfies the Bohner property and $Q_h\theta$ is a Wexler sequence for arbitrary h.*

Proof. Let θ be a Wexler sequence and γ^j, $j \in \mathbb{Z}$, be equipotentially almost periodic, where γ is the representative of θ. The almost periodicity of γ^1 implies that there exists $\kappa > 0$ such that $0 \leqslant \theta_i^1 < \kappa$, $i \in \mathbb{Z}$. Hence, for arbitrary $n \in \mathbb{N}$ there exists i_n such that $\gamma_{i_n} + h_n \in [0, \kappa]$. Denote $\gamma^{(n)} = \{\gamma_{i+i_n}\}_i$. Clearly, $\gamma^{(n)} \in \Gamma$, $m(\gamma^{(n)}) = m(\gamma)$, and $\gamma^{(n)j}$, $j \in \mathbb{Z}$, are equipotentially almost periodic.
Using Theorem 1 in [148, p. 129] and the equipotentially almost periodicity of $\gamma^{(n)j}$, one can show that there exists a subsequence n_k, let us say it is n itself, such that for arbitrary $\varepsilon > 0$ there exists $n(\varepsilon) \in N$, such that

$$\|\gamma^{(m)j} - \gamma^{(p)j}\| < \frac{\varepsilon}{2}, \ \ j \in \mathbb{Z}, \ \text{if} \ m, p > n(\varepsilon). \tag{7.6}$$

Moreover, without loss of generality, we assume that $\gamma_0^{(n)} + h_n \to \gamma_0^{(0)} \in [0, \kappa]$. Consequently, for arbitrary $\varepsilon > 0$ there exists $n(\varepsilon)$ such that

$$\begin{aligned} &|\gamma^{(m)} + h_m - \gamma^{(p)} - h_p| < |\gamma_0^{(m)} + h_m - \gamma_0^{(p)} - h_p| + \\ &|\gamma^{(m)i} - \gamma^{(p)i}| < \frac{\varepsilon}{2} + \frac{\varepsilon}{2} = \varepsilon, \end{aligned} \tag{7.7}$$

if $m, p > n(\varepsilon)$. That is, if we fix $i \in \mathbb{Z}$, then $\{\gamma_i^{(n)} + h_n\}_n$ is a Cauchy sequence, and hence $\gamma_i^{(n)} + h_n \to \gamma_i^{(0)}, i \in \mathbb{Z}$. Furthermore, by (7.7) the convergence is uniform in i and $\gamma_{i+1}^{(0)} \geqslant \gamma_i^{(0)}, i \in \mathbb{Z}$. Finally, the condition $m(\gamma^{(n)}) = m(\gamma)$, and Lemma 7.4 imply that $m(\gamma^{(0)}) \leqslant$

$n_0 m(\gamma) < \infty$. It is obvious that $|\gamma_i^{(0)}| \to \infty$. Hence, $\gamma^0 \in \Gamma$. Assume that $\theta^{(0)}$ is the support of $\gamma^{(0)}$. Since $\gamma^{(n)j}$ are equipotentially almost periodic and $|\gamma_i^{(n)j} - \gamma_i^{(0)}| \to 0$ uniformly in i as $n \to \infty, j \in \mathbb{Z}$, one can show that $\gamma^{(0)j}, j \in \mathbb{Z}$, are equipotentially almost periodic in a similar manner as in the proof of the theorem on almost periodicity of a limit function [117]. Consequently, $\theta^{(0)}$ is a Wexler sequence. The theorem is proved. □

Theorem 7.2. *$\theta \in \Theta$ is a Wexler sequence if and only if it satisfies the Bohner property.*

Proof. *Necessity* is proved by Theorem 7.1.

Sufficiency. Assume that $\theta \in \Theta$ is not a Wexler sequence. Then θ^j, $j \in \mathbb{Z}$, are not equipotentially almost periodic, and by Lemma 7.3 there exist a number ε_0 and a sequence of sections $I_n = [h_n - l_n, h_n + l_n], n \in \mathbb{N}$, where l_1 is arbitrary, with $l_n > \max_{m<n} |h_m|$ such that the following inequality

$$\sup_{q,k \in \mathbb{Z}} |\theta_k^q - \xi| \geqslant \varepsilon_0, \xi \in \cup_n I_n, \tag{7.8}$$

is valid. Consider a sequence of shifts $\theta + h_n, n \in \mathbb{N}$, and set $h' = \{h_n\}$. For arbitrary $m, p \in \mathbb{N}, m > p$, we have that $h_m - h_p \in I_m$ and $\sup_{i,j \in \mathbb{Z}} |\theta_i^j - (h_m - h_p)| \geqslant \varepsilon_0$, or

$$\sup_{i,j \in \mathbb{Z}} |\theta_i + h_p - \theta_{i+j} - h_m| \geqslant \varepsilon_0. \tag{7.9}$$

The last inequality means that $\theta + h_m$ is not in the ε_0- neighborhood of $\theta + h_p$. Assume that there exists a subsequence $h \subset h'$ such that $\theta + h_{n_k}$ converges to $\theta^{(0)} \in \Theta$ uniformly in B^s- topology. Then there exist numbers n_m and $n_p, n_m > n_p$, such that $(\theta + h_{n_p}) \frac{\varepsilon_0}{2} \theta^{(0)}$ and $(\theta + h_{n_m}) \frac{\varepsilon_0}{2} \theta^{(0)}$. By Lemma 7.2, $(\theta + h_{n_p}) \, \varepsilon_0 \, (\theta + h_{n_m})$. The contradiction proves the theorem. □

Subsequences h and g are common subsequences of sequences h' and g' if $h_n = h'_{n(k)}$ and $g_n = g'_{n(k)}$ for some given function $n(k)$ [117]. The following theorem is an analogue of Theorem 1.17 from [117].

Theorem 7.3. *A sequence $\theta \in \Theta$ is a Wexler sequence if and only if for arbitrary h' and g' there are common subsequences $h \subset h'$ and $g \subset g'$ such that*

$$Q_{h+g}\theta = Q_h Q_g \theta. \tag{7.10}$$

Proof. *Necessity.* Let θ be a Wexler sequence. By the previous theorem there exists a subsequence $g'' \subset g'$ such that $Q_{g''}\theta$ exists and the limit is a Wexler sequence. Denote $\eta = Q_{g''}\theta$. Moreover, if $h'' \subset h'$ is common with g'', then one can find a sequence $h''' \subset h''$

such that $\mu = Q_{h'''}\eta$ is a Wexler sequence. If $g''' \subset g''$ is common with h''' then there exist common subsequences $h \subset h''', g \subset g'''$ such that $Q_{g+h}\theta = \zeta$. Since $h \subset h'', g \subset g'', Q_g\theta = \eta, Q_h\eta = \mu$. That is, if $\varepsilon > 0$ is fixed, then for sufficiently large n we have $\zeta\varepsilon(\theta + h_n + g_n)$, $\eta\varepsilon(\theta + g_n)$, and $\mu\varepsilon(\eta + h_n)$. Using Lemma 7.2, one can conclude that $\zeta\varepsilon\mu$. Hence, $\zeta = \mu$, as ε is arbitrarily small.

Sufficiency. Suppose that h' is the given sequence. Taking $g' = 0$ we see that the condition $Q_{g+h}\theta = Q_g Q_h\theta$ implies that $Q_h\theta$ exists, and hence by Theorem 7.2 the sequence θ is a Wexler sequence. The theorem is proved. □

7.3 Bohr-Wexler almost periodic functions

Definition 7.6. Let $u_1, u_2 \in \mathscr{PC}_r$, and $\theta^{(1)}, \theta^{(2)}$ be the sequences of the points of discontinuity of these functions, respectively. We shall say that u_1 is ε-equivalent to u_2, and denote $u_1\varepsilon u_2$, if $\theta^{(1)}\varepsilon\theta^{(2)}$ and $|u_1(t) - u_2(t)| < \varepsilon$ for all $t \in \mathbb{R} \setminus \bigcup_i[(\theta_i^{(1)} - \varepsilon, \theta_i^{(1)} + \varepsilon) \cup (\theta_i^{(2)} - \varepsilon, \theta_i^{(2)} + \varepsilon)]$. We also say that u_1 belongs to the ε-neighborhood of u_2, and vice versa, denoting $u_1 \in O(u_2, \varepsilon)$ and $u_2 \in O(u_1, \varepsilon)$, respectively.

Definition 7.7. The topology defined on the basis of all ε-neighborhoods of functions from $\mathscr{PC}_r$ is called $B-$ topology. It is clear that it is the Hausdorff topology.

Definition 7.8. A number τ is an $\varepsilon-$ translation number of $\phi \in \mathscr{PC}_r$ if $\phi(t+\tau) \in O(\phi(t), \varepsilon)$.

Definition 7.9. A function $\phi \in \mathscr{PC}_r$ is a Bohr-Wexler almost periodic function, if for any $\varepsilon > 0$ there exists a relatively dense set of ε-translation numbers of ϕ. If $\theta \in \Theta$ is a sequence of the moments of discontinuity of ϕ, then θ is a Wexler sequence.

We shall denote by $\mathscr{BWAP}$ the set of all Bohr-Wexler almost periodic functions. Let $h = \{h_n\}$ and $T_h\phi \in \mathscr{PC}_r$ be the limit of the sequence $\phi(t + h_n)$, $\phi \in \mathscr{PC}_r$, in B-topology, if it exists.

Definition 7.10. $\phi \in \mathscr{PC}_r$ satisfies the Bohner property if every sequence h' contains a subsequence $h \subseteq h'$ such that there exists $T_h\phi \in \mathscr{PC}_r$.

Lemma 7.5. *If $u_1\varepsilon_1 u_2$, and $u_2\varepsilon_2 u_3$, then $u_1(\varepsilon_1 + \varepsilon_2)u_3$.*

Proof. Lemma 7.1 and the relations $\theta^{(1)}\varepsilon_1\theta^{(2)}, \theta^{(2)}\varepsilon_2\theta^{(3)}$ imply that $\theta^{(1)}(\varepsilon_1+\varepsilon_2)\theta^{(3)}$. Moreover, one can easily obtain that if t belongs to the complement of the set

$$\bigcup_i[(\theta_i^{(1)}-(\varepsilon_1+\varepsilon_2), \theta_i^{(1)}+(\varepsilon_1+\varepsilon_2))\cup(\theta_i^{(3)}-(\varepsilon_1+\varepsilon_2), \theta_i^{(3)}+(\varepsilon_1+\varepsilon_2))] \quad (7.11)$$

then $t \in \mathbb{R}\setminus\bigcup_i[(\theta_i^{(j)}-\varepsilon_j, \theta_i^{(j)}+\varepsilon_j)\cup(\theta_i^{(2)}-\varepsilon_j, \theta_i^{(2)}+\varepsilon_j)]$, $j=1,3$.
That is why $|u_1(t)-u_3(t)| \leqslant |u_1(t)-u_2(t)|+|u_2(t)-u_3(t)| < \varepsilon_1+\varepsilon_2$ if (7.11) is valid. The lemma is proved. □

Theorem 7.4. *$\phi \in \mathscr{BWAP}$ if and only if ϕ satisfies the Bohner property. $T_h\phi \in \mathscr{BWAP}$ for $\phi \in \mathscr{BWAP}$, if the limit exists.*

Proof. *Necessity.* Consider a function $\phi \in \mathscr{BWAP}$, and $\theta \in \Theta$, a sequence of discontinuity points of ϕ, and $h' \subset \mathbb{R}$, a given sequence. By Theorem 7.1 there exists a subsequence of h' such that $T_{h'}\theta = \theta^0$ is a Wexler sequence, and without loss of generality we assume that it is h' itself. Consider a sequence ε_n such that $\varepsilon_n \to 0$, $n \to \infty$, and denote $A_n = \bigcup_i[\theta_i^0-\varepsilon_n, \theta_i^0+\varepsilon_n]$. Using the diagonal process [117], one can find a sequence $h^{(1)} \subseteq h'$, such that $\phi(t+h_n^{(1)})$ is uniformly convergent to $\phi^{(1)}$ on A_1. Then in the same way we can define a sequence $h^{(2)} \subseteq h^{(1)}$ such that $\phi(t+h_n^{(2)})$ is uniformly convergent to $\phi^{(2)}$ on A_2, and so on. Obviously, $\phi^{(i+1)} = \phi^{(i)}$ on A_i, $A_i \subset A_{i+1}$, $\bigcup_i A_i = R\setminus\theta^0$. Consequently, $\phi(t+h_n^{(n)})$ is convergent to $\phi^0 \in \mathscr{PC}_r$ in $B-$ topology.
Fix $\varepsilon > 0$. There exists $n(\varepsilon)$ such that for arbitrary $n > n(\varepsilon)$ the inequality $\phi(t+h_n^{(n)})\frac{\varepsilon}{3}\phi^0(t)$ is valid. If τ is an $\frac{\varepsilon}{3}-$ almost period of $\phi(t)$, then $\phi(t+h_n^{(n)}+\tau)\frac{\varepsilon}{3}\phi(t+h_n^{(n)})$ and $\phi(t+h_n^{(n)}+\tau)\frac{\varepsilon}{3}\phi^0(t+\tau)$. Now, using Lemma 7.5, one can obtain $\phi^0(t+\tau) \in O(\phi^0,\varepsilon)$.
Sufficiency. Let $\phi \notin \mathscr{BWAP}$. As similar to the continuous case, for some $\varepsilon_0 > 0$, we can find a sequence of sections $I_n = [h_n-l_n, h_n+l_n]$, $l_n \geqslant \max_{m<n}|h_m|$, l_1-arbitrary, such that if $\xi \in \bigcup_n I_n$, then $\phi(t+\xi) \notin O(\phi,\varepsilon_0)$. Denote $h' = \{h_n\}$, and assume that there exists $h \subset h'$ such that $T_{h'} = \phi^0 \in \mathscr{PC}_r$. Then there exists $n(\varepsilon)$ such that if $m > p > n(\varepsilon)$, then $\phi(t+h_m)\frac{\varepsilon_0}{2}\phi^0(t)$, and $\phi(t+h_p)\frac{\varepsilon_0}{2}\phi^0(t)$. By Lemma 7.5, $\phi(t+h_m)\varepsilon_0\phi(t+h_p)$. Hence, $\phi(t+(h_m-h_p))\varepsilon_0\phi(t)$, but $(h_m-h_p) \in I_m$. The theorem is proved. □
Let h', g' be given sequences. Subsequences $h \subset h'$, $g \subset g'$ are common subsequences of h', g', respectively, if $h_n = h'_{n(k)}$, $g_n = g'_{n(k)}$ for some function $n(k)$. There is analogue of Theorem 1.17 from [117], which can be proved similarly as Theorem 7.3.

Theorem 7.5. *$\phi \in \mathscr{BWAP}$ if and only if for arbitrary h', g' there exist common subsequences h, g such that $T_{h+g}\phi = T_hT_g\phi$.*

Lemma 7.6. *Assume that* $f \in \mathscr{PC}_r$, *and* (7.2) *satisfies exponential dichotomy. Then the system*

$$x' = A(t)x + f(t) \tag{7.12}$$

has a unique solution $x_0(t) \in C_0(t)$.

Proof. As similar to Lemma 7.1 one can find that

$$x_0(t) = \int_{-\infty}^{\infty} G(t,s)f(s)ds \tag{7.13}$$

is a solution of (7.12) and belongs to $C_0(\mathbb{R})$. Let $x_1(t)$ be another solution of (7.12), bounded on $\mathbb{R}$. One can see that the difference $x_1 - x_0$ is a continuously differentiable solution of system (7.2). Hence, it is a trivial solution of (7.2). The lemma is proved. □

Next, we assume that

(A5) θ is a Wexler sequence.

Using the Bohner property one can easily prove that the following assertion is valid.

Lemma 7.7. *Assume that* $\phi(t) \in \mathscr{AP}(\mathbb{R})$, *and condition* (A5) *is valid. Then* $\phi(\upsilon(t)) \in \mathscr{BWAP}$.

Consider functions $\phi(t) \in \mathscr{AP}$ and $\psi(t) \in \mathscr{BWAP}$.
For our convenience, following [46], we shall say that

(i) a sequence h is regular with respect to $\phi(t)$ if the sequence $\phi(t+h_n)$ is uniformly convergent on $\mathbb{R}$;
(ii) a sequence h is regular with respect to $\psi(t)$ if the sequence $\psi(t+h_n)$ is convergent in B-topology.

Let us denote by $\mathscr{L}(\phi)$ and $\mathscr{L}(\psi)$ the sets of all sequences regular with respect to ϕ and ψ, respectively.

7.4 Almost periodic solutions

Lemma 7.8. *Assume that* (7.2) *satisfies exponential dichotomy and* $f(t) \in \mathscr{BWAP}$. *Then there is a unique solution of* (7.12), $x_0(t) \in \mathscr{AP}(\mathbb{R})$, *such that* $\mathscr{L}(x_0) \subseteq \mathscr{L}(A, f)$, *and* $\|x_0\| \leqslant (\frac{K_1}{\sigma_1} + \frac{K_2}{\sigma_2})\|f\|$.

Proof. By Lemma 7.6, the function $x_0(t)$ defined by (7.13) is a solution of (7.12) and $x_0(t) \in C_0(\mathbb{R})$. It is easy to verify that $\|x_0\| \leqslant (\frac{K_1}{\sigma_1} + \frac{K_2}{\sigma_2})\|f\|$. Since every system in the hull of (7.2) satisfies exponential dichotomy [86], it has a unique bounded solution defined on $\mathbb{R}$. Let h' and g' be given sequences. Then there exist common subsequences $h \subset h'$ and $g \subset g'$ such that $T_{h+g}A = T_hT_gA$, $T_{h+g}f = T_hT_gf$, and there exist uniform limits on compact sets $y = T_{h+g}x_0$ and $z = T_hT_gx_0$. Since $y, z \in C_0(\mathbb{R})$ and they are solutions of the same equation, it follows that $y = z$. By Theorem 1.17 [117], $x_0(t)$ is an almost periodic function. Assume that for a given sequence h we have $T_hA = A^*$ and $T_hf = f^*$. We shall show that the limit T_hx_0 exists. Indeed, suppose to the contrary that the limit does not exist. Then there are two subsequences $h^{(1)} \subset h$ and $h^{(2)} \subset h$ such that

$$\|x_0(t+h_{1n}) - x_0(t+h_{2n})\| \geqslant \varepsilon_0 > 0,$$

or

$$\|x_0(t+h_{1n}-h_{2n}) - x_0(t)\| \geqslant \varepsilon_0 > 0$$

for all $n \in \mathbb{N}$. But $T_{h_1-h_2}A = A$, and $T_{h_1-h_2}f = f$. Hence, $T_{h_1-h_2}x_0 = x_0$. The theorem is proved. □

Using the Bohner property again and Lemma 7.7, one can prove that the following lemma is valid.

Lemma 7.9. *If $f \in \mathscr{AP}(\mathbb{R} \times G_H^m)$, and $\psi \in \mathscr{BWAP}, \psi : \mathbb{R} \to G_H$, then $F_\theta(\psi(t)) \in \mathscr{BWAP}$ and $\mathscr{L}(F_\theta(\psi(t))) \subseteq \mathscr{L}(f, \psi)$.*

Theorem 7.6. *Assume that conditions* (C1)–(C5) *are valid, and*

$$lm(\frac{K_1}{\sigma_1} + \frac{K_2}{\sigma_2}) < 1.$$

Then there exists a unique solution of (7.1), *$\phi(t) \in \mathscr{AP}(\mathbb{R})$, such that $\mathscr{L}(\phi) \subseteq \mathscr{L}(A, f, \upsilon)$.*

Proof. Let $\Psi = \{\psi \in \mathscr{AP}(\mathbb{R}) | \mathscr{L}(\psi) \subseteq \mathscr{L}(\mathscr{A}, \mathscr{F}, \upsilon)\}$ be a complete metric space with the sup-norm $\|\cdot\|_0$. Define an operator Π on Ψ such that

$$\Pi(\psi(t)) = \int_{-\infty}^{\infty} G(t,s)F_\theta(\psi(s))ds.$$

Lemma 7.9 implies that $F_\theta(\psi(s)) \in \Psi$ and $\Pi : \Psi \to \Psi$. If $\psi_1, \psi_2 \in \Psi$, then

$$\begin{aligned}
&\|\Pi(\psi_1(t)) - \Pi(\psi_2(t))\| \leqslant \|\int_{-\infty}^{t} X(t)PX^{-1}(s)(F_\theta(\psi_1(s)) - F_\theta(\psi_2(s)))ds\| + \\
&\|\int_{t}^{\infty} X(t)(I-P)X^{-1}(s)(F_\theta(\psi_1(s)) - F_\theta(\psi_2(s)))ds\| \leqslant \\
&lm(\frac{K_1}{\sigma_1} + \frac{K_2}{\sigma_2})\|\psi_1(t) - \psi_2(t)\|_0.
\end{aligned}$$

Thus, Π is a contractive operator and there exists a unique almost periodic solution of the equation

$$\psi(t) = \int_{-\infty}^{\infty} G(t,s)F_\theta(\psi(s))ds,$$

which is a solution of (7.1). The theorem is proved. □

Remark 7.1. Lemma 7.8 and Theorem 7.6 are analogous to the assertions which were obtained in [117] for ordinary differential equations.

7.5 Stability

This section is concerned with the problem of stability of the almost periodic solution of system (7.1). We consider a specific initial condition when values of solutions are evaluated only at points from sequence θ. This approach to the stability is natural for differential equations with piecewise constant argument [317]. Denote by $X(t,s) = X(t)X^{-1}(s)$ the transition matrix of (7.2). We will need the following assumptions:

(A6) $\exists\{\sigma,K\} \subset \mathbb{R}$, $K \geqslant 1$, $\sigma > 0$, such that $\|X(t,s)\| \leqslant K\exp(-\sigma(t-s))$, $t \geqslant s$;
(A7) $l < \dfrac{\sigma}{mK}$.

Assume that $p_j \geqslant 0$, $j = \overline{1,m}$ and denote $\tau = \max\{\sup_t(t-\theta_{\upsilon(t)-p_j}),\ j = \overline{1,m}\} > 0$, $\zeta(l) = 1-\exp(a\tau)Klm(\sigma-a)^{-1}$, where $a \in \mathbb{R}$, $0 < a < \sigma$, is fixed.

(A8) $\zeta(l) > 0$.

Conditions (A1)–(A7) and Theorem 7.6 imply that there exists a unique solution of (7.1), $\xi(t) \in \mathscr{A}\mathscr{P}(\mathbb{R})$.
Fix $\varepsilon > 0$ and denote $L(l,\delta) = \frac{K}{\zeta(l)}\delta$, where $\delta \in \mathbb{R}$, $\delta > 0$. Take δ so small that $L(l,\delta) < \varepsilon$. Assume that $t_0 \in \theta$. Moreover, without any loss of generality, assume that $t_0 = \theta_0 = 0$. Fix a sequence $\eta^j \in \mathbb{R}^n$, $j = \overline{1,m}$, $\max\|\eta^j\| < \delta$. Denote $p^0 = \max_{\overline{1,m}} p_j$, and let Ψ_η be the set of all continuous functions which are defined on $[\theta_{-p^0},\infty)$. And if $\psi \in \Psi_\eta$ then:

(1) $\psi(\theta_{-p_j}) = \eta^j$, $j = \overline{1,m}$;
(2) $\psi(t)$ is uniformly continuous on $[0,+\infty)$;
(3) $\|\psi(t)\| \leqslant L(l,\delta)\exp(-at)$ if $t \geqslant 0$.

Consider the following differential equations with piecewise constant arguments and the initial condition

$$\frac{dv}{dt} = A(t)v + w(t, v(\theta_{\upsilon(t)-p_1}), v(\theta_{\upsilon(t)-p_2}), \ldots, v(\theta_{\upsilon(t)-p_m})),$$
$$v(s) = \eta^j, \quad j = \overline{1,m}, \tag{7.14}$$

where

$$w(t, v(\theta_{\upsilon(t)-p_1}), v(\theta_{\upsilon(t)-p_2}), \ldots, v(\theta_{\upsilon(t)-p_m})) =$$
$$f(t, \xi(\theta_{\upsilon(t)-p_1}) + v(\theta_{\upsilon(t)-p_1}), \xi(\theta_{\upsilon(t)-p_2}) +$$
$$v(\theta_{\upsilon(t)-p_2}), \ldots, \xi(\theta_{\upsilon(t)-p_m}) + v(\theta_{\upsilon(t)-p_m})) -$$
$$f(t, \xi(\theta_{\upsilon(t)-p_1}), \xi(\theta_{\upsilon(t)-p_2}), \ldots, \xi(\theta_{\upsilon(t)-p_m})),$$

and w satisfies $w(t,0) = 0$,

$$\|w(t,v_1) - w(t,v_2)\| \leqslant l \sum_{j=1}^{m} \|v_1^j - v_2^j\|,$$

$v_i = (v_i^1, \ldots, v_i^m) \in R^m, i = 1,2$. The following definition is an adapted version of a definition from [81].

Definition 7.11. A function $v(t)$ is a solution of the initial value problem (7.14) on the interval $[\theta_{-p^0}, \infty)$ if the following conditions are fulfilled:

(i) $v(\theta_{-p_j}) = \eta^j$, $j = \overline{1,m}$;
(ii) $v(t)$ is continuous on $[\theta_{-p^0}, \infty)$;
(iii) the derivative $v'(t)$ exists at each point $t \in [0,\infty)$ with the possible exception of the points θ_j, $j \geqslant 0$, where one-sided derivatives exist;
(iv) equation (7.14) is satisfied by $v(t)$ on each interval $[\theta_j, \theta_{j+1})$, $j \geqslant 0$.

It is obvious that $v'(t)$ is the restriction on $[0,\infty)$ of a function from $\mathscr{PC}_r$ and the last definition can be used for differential equations with piecewise constant arguments (7.1), too.

Denote $\overline{\theta} = \sup_i(\theta_{i+1} - \theta_i)$. There exists a number $M > 0$ such that $\|X(t,s)\| \leqslant M$ if $|t-s| \leqslant \overline{\theta}$.

Assume additionally that

(A9) $M\overline{\theta}ml < 1$.

Theorem 7.7. *Assume that* (A1)–(A3), (A5)–(A9) *are valid. Then there exists a unique solution of the initial value problem* (7.14), *$v(t) \in \Psi_\eta$, defined on* $[0,\infty)$.

Proof. Similarly to the proof of Lemma 7.1 we can check that the initial value problem is equivalent to the following integral equation

$$v(t) = X(t,0)\eta^0 + \int_0^t X(t,s)F_w(v(s))ds,$$
$$v(\theta_{-p_j}) = \eta^j, \quad j = \overline{1,m}, \tag{7.15}$$

where $F_w(v(s)) = w(s, v(\theta_{\upsilon(s)-p_1}), v(\theta_{\upsilon(s)-p_2}), \ldots, v(\theta_{\upsilon(s)-p_m}))$. Define on Ψ_η an operator Π such that if $\psi \in \Psi_\eta$, then

$$\Pi\psi = \begin{cases} \psi, & t \in [\theta_{-p_j}, 0] \\ X(t,0)\eta^0 + \int_0^t X(t,s)F_\omega(\psi(s))ds, & t \geqslant 0. \end{cases}$$

We shall show that $\Pi : \Psi_\eta \to \Psi_\eta$. Indeed, for $t \geqslant 0$ it is true that

$$\|\Pi\psi\| \leqslant K\exp(-\sigma t)\delta + \int_0^t K\exp(-\sigma(t-s))lL(l,\delta)\sum_{j=0}^{m}\exp(-a\theta_{\upsilon(s)-p_j})ds \leqslant$$

$$\exp(-at)\left[K\delta + \frac{m\exp(a\tau)KlL(l,\delta)}{\sigma - a}\right] = L(l,\delta)\exp(-at).$$

Differentiating $\Pi\psi$ on $[0,\infty)$, it is easy to show that $[\Pi\psi]'$ exists on $[0,\infty)$ except possibly on a countable set of isolated points of discontinuity of the first kind, and that it is bounded on $[0,\infty)$. Hence, $\Pi\psi$ is a uniformly continuous function defined on $[0,\infty)$.
Let $\psi_1, \psi_2 \in \Psi_\eta$. Then

$$\|\Pi\psi_1 - \Pi\psi_2\| \leqslant \int_0^t b\exp(-a(t-s))lm\|\psi_1 - \psi_2\|_1 ds \leqslant \frac{Klm}{\sigma}\sup_{t\geqslant 0}\|\psi_1 - \psi_2\|.$$

Using a contraction mapping argument, one can conclude that there exists a unique fixed point $v(t,\eta)$ of the operator $\Pi:\ \Psi_\eta \to \Psi_\eta$ which is a solution of (7.14). To complete the proof we should show that there is not any other solution of the problem.
Consider first the interval $[\theta_0, \theta_1]$. Assume that on this interval, (7.14) has two different solutions v_1, v_2 of the problem. Obviously, their difference $w = v_1 - v_2$ is again a solution of the equation. Denote $\overline{m} = \max_{[\theta_0,\theta_1]}\|w(t)\|$, and assume, on contrary, that $m > 0$. We have that on the interval

$$\|w(t)\| = \|\int_0^t X(t,s)F_w(w(s))ds\| \leqslant Ml\overline{\theta}m\overline{m}.$$

The last inequality contradicts condition (A9). Now, using induction, one can easily prove the uniqueness for all $t \geqslant 0$. The theorem is proved. □
Denote $\phi = \{\phi^j\}$, $j = \overline{1,m}$, a sequence of vectors from $\mathbb{R}^n$. Let $x(t,\phi)$ be a solution of (7.1) such that $x(\theta_{-p_j}, \phi) = \phi^j$, $j = \overline{1,m}$.

Definition 7.12. The almost periodic solution $\xi(t)$ of (7.1) is said to be exponentially stable if there exists a number $a \in \mathbb{R}$, $a > 0$, such that for every $\varepsilon > 0$ there exists a number $\delta = \delta(\varepsilon)$, such that the inequality $\max_{j=\overline{1,m}} \|\phi^j - \xi(\theta_{-p_j})\| < \delta$ implies $\|x(t,\phi) - \xi(t)\| < \varepsilon \exp(-at)$ for all $t \geqslant 0$.

Consider now a solution $x(t,\phi)$ of (7.1) such that $\max_{j=\overline{1,m}} \|\phi^j - \xi(\theta_{-p_j})\| < \delta$. Since the solution $v(t)$ of the equation (7.14), satisfying $v(\theta_{-p_j}) = \phi^j - \xi(\theta_{-p_j})$, $j = \overline{1,m}$, exists, $x(t,\phi) = \xi(t) + v(t), t \in [\theta_{-p^0}, \infty)$, and $x(t,\phi)$ is uniquely continuable to ∞. Thus, the following theorem is proved.

Theorem 7.8. *Assume that* (A1)–(A3), (A5)–(A9) *are valid. Then the almost periodic solution* $\xi(t)$ *of* (7.1) *is exponentially stable.*

Notes

Almost periodic solutions of differential equations with piecewise constant arguments have been discussed in [44,161,170,171,283,333,336], [340–345] through solutions of discrete equations . That is why, only the reducible case were considered. In this chapter, basically results of the paper [12] are delivered. We investigate the problem in most general form, the case of exponentially dichotomous system, nonlinear with respect to solutions with non-deviated arguments. The theorem in [117], which is proved for ordinary differential equations, is generalized for differential equations with deviated argument of mixed type, delay and advanced. We also prove theorems for almost periodic sequences and discontinuous almost periodic functions such that they can be applied to impulsive differential equations [19,20], [40–42], [278].

Chapter 8

Stability of neural networks

In this chapter, the method of Lyapunov functions for differential equations with piecewise constant argument considered in Chapter 5 is applied to a model of recurrent neural networks (RNNs). The model includes both advanced and delayed arguments. We obtain sufficient conditions for global exponential stability of the equilibrium point. The feasibility of the results are illustrated by examples with numerical simulations.

8.1 Introduction

Recurrent neural networks have been deeply investigated in recent years for solving problems in engineering systems and computer sciences, such as associative memory, image processing, optimization, and pattern recognition problems [89–91, 162].
It is well known that the dynamical behavior of recurrent neural networks plays a crucial role in the applications of the networks. That is, stability and convergence of neural networks are prerequisites. However, the uniform asymptotic stability is not enough to design neural networks. In other words, the global exponential stability, which guarantees a neural network to converge fast enough in order to achieve fast response, is also needed. Moreover, in the analysis of dynamical neural networks for parallel computation and optimization, one requires to ensure a desired exponential convergence rate of the networks trajectories, starting from arbitrary initial states to the equilibrium point which corresponds to the optimal solution in order to increase the rate of convergence to the equilibrium point of the networks and reduce the neural computing time. Hence, from the mathematical and engineering points of view, the neural networks must contain a unique equilibrium point, which is globally exponentially stable. Therefore, many researchers paid great attention to the problem of stability analysis of recurrent neural networks and many results on this topic have been reported in the literature; see, e.g., [167, 257, 346, 347], and the references

therein.

The methods of Lyapunov functions and functionals are among the most popular tools in studying the problem of the stability for recurrent neural networks (see [69, 103, 167, 257, 346, 347]). However, there is not an easy way, yet, to construct Lyapunov functions or functionals for each problem that satisfies the required conditions in the classical stability theory. In this chapter, we obtain some stability conditions for recurrent neural networks model based on the Lyapunov's second method. Although the new model preserves both advanced and delayed arguments, it is worth to be mentioned that the stability conditions are given in terms of inequalities, and we know from the theory of differential equations with deviating arguments, this method necessarily utilizes functionals [150, 185, 317]. Moreover, one should emphasize that there is an opportunity to apply Lyapunov functions technique for estimating domains of attraction which has a particular interest to evaluate the efficiency of recurrent neural networks [337]. The crucial novelty of this chapter is that the argument is of mixed (advanced-delayed) type. In biology, the reasons for argument to be delayed are discussed well [237, 262]. However, the role of advanced argument is not analyzed properly. Nevertheless, the importance of anticipation for biology is mentioned by some authors. For example, in paper [63], it is supposed that synchronization of biological oscillators may request anticipation of counterparts behavior. Hence, one can assume that equations for neural networks may also need anticipation, which is usually reflected in models by advanced argument. Therefore, the systems posed in this chapter can be useful in the future analysis of recurrent neural networks.

Furthermore, the idea of involving both advanced and delayed arguments in the recurrent neural networks can be explained by the existence of advanced and retarded actions in a model of classical electrodynamics [104]. Moreover, mixed type of the argument may depend on traveling waves emergence in CNNs [314]. Understanding the structure of such traveling waves is significant due to their potential applications in engineering systems and computer sciences, such as image processing (see, for example, [89–92, 160]).

8.2 Preliminaries

Denote the n-dimensional real space by $\mathbb{R}^n$, $n \in \mathbb{N}$, and the norm of a vector $z \in \mathbb{R}^n$ by $\|z\| = \sum_{i=1}^{n} |z_i|$.

We fix two real valued sequences θ_i, ζ_i, $i \in \mathbb{N}$, such that $\theta_i < \theta_{i+1}$, $\theta_i \leqslant \zeta_i \leqslant \theta_{i+1}$ for all $i \in \mathbb{N}$, $\theta_i \to \infty$ as $i \to \infty$, and shall consider the following recurrent neural networks model described by differential equations with piecewise constant argument of generalized type:

$$z_i'(t) = -a_i z_i(t) + \sum_{j=1}^{n} b_{ij} f_j(z_j(t)) + \sum_{j=1}^{n} c_{ij} g_j(z_j(\gamma(t))) + I_i, \tag{8.1}$$
$$a_i > 0, \; i = 1, 2, \ldots, n,$$

where $\gamma(t) = \zeta_k$, if $t \in [\theta_k, \theta_{k+1})$, $k \in \mathbb{N}$, $t \in \mathbb{R}^+ = [0, \infty)$, n corresponds to the number of units in a neural network, $z_i(t)$ stands for the state vector of the i-th unit at time t, $f_j(z_j(t))$ and $g_j(z_j(\gamma(t)))$ denote, respectively, the measures of activation to its incoming potentials of the unit j at time t and $\gamma(t)$, b_{ij}, c_{ij}, I_i are real constants, b_{ij} means the strength of the j-th unit on the i-th unit at time t, c_{ij} infers the strength of the j-th unit on the i-th unit at time $\gamma(t)$, I_i signifies the external bias on the i-th unit and a_i represents the rate with which the i-th unit will reset its potential to the resting state in isolation when it is disconnected from the network and external inputs.

The following assumptions will be needed throughout this section:

(A1) the activation functions $f_j, g_j \in C(\mathbb{R}^n)$ satisfy $f_j(0) = 0$, $g_j(0) = 0$ for each $j = 1, 2, \ldots, n$;

(A2) there exist Lipschitz constants L_i^1, $L_i^2 > 0$ such that

$$|f_i(u) - f_i(v)| \leqslant L_i^1 |u - v|,$$

$$|g_i(u) - g_i(v)| \leqslant L_i^2 |u - v|$$

for all $u, v \in \mathbb{R}^n$, $i = 1, 2, \ldots, n$;

(A3) there exists a positive number θ such that $\theta_{i+1} - \theta_i \leqslant \theta$, $i \in \mathbb{N}$;

(A4) $\theta\,[m_1 + 2m_2]\,\mathrm{e}^{m_1\theta} < 1$;

(A5) $\theta\,\left[m_2 + m_1(1 + m_2\theta)\mathrm{e}^{m_1\theta}\right] < 1$,

where

$$m_1 = \max_{1 \leqslant i \leqslant n} \left(a_i + L_i^1 \sum_{j=1}^{n} |b_{ji}| \right), \quad m_2 = \max_{1 \leqslant i \leqslant n} \left(L_i^2 \sum_{j=1}^{n} |c_{ji}| \right).$$

In the following theorem, we obtain sufficient conditions for the existence of a unique equilibrium, $z^* = (z_1^*, \ldots, z_n^*)^T$, of (8.1).

Theorem 8.1. *Suppose* (A1) *and* (A2) *hold. If the neural parameters* a_i, b_{ij}, c_{ij} *satisfy*

$$a_i > L_i^1 \sum_{j=1}^{n} |b_{ji}| + L_i^2 \sum_{j=1}^{n} |c_{ji}|, \quad i = 1, \ldots, n,$$

then (8.1) *has a unique equilibrium* $z^* = (z_1^*, \ldots, z_n^*)^T$.

The proof of the theorem is almost identical to Theorem 2.1 in [230] and we omit it here. The next theorem provides conditions for the existence and uniqueness of solutions on $\mathbb{R}^+$. The proof of the assertion is similar to that of Theorem 2.3, but, for convenience of the reader we place the full proof of the assertion.

Theorem 8.2. *Assume that conditions* (A1)–(A4) *are fulfilled. Then, for every* $(t_0, z^0) \in \mathbb{R}^+ \times \mathbb{R}^n$, *there exists a unique solution* $z(t) = z(t,t_0,z^0) = (z_1(t),\ldots,z_n(t))^T$, $t \in \mathbb{R}^+$, *of* (8.1), *such that* $z(t_0) = z^0$.

Proof. *Existence.* Fix $k \in \mathbb{N}$. We assume without loss of generality that $\theta_k \leqslant \zeta_k < t_0 \leqslant \theta_{k+1}$. To begin with, we shall prove that for every $(t_0, z_0) \in [\theta_k, \theta_{k+1}] \times \mathbb{R}^n$, there exists a unique solution $z(t) = z(t,t_0,z^0) = (z_1(t),\ldots,z_n(t))^T$, of (8.1) such that $z(t_0) = z^0 = (z_1^0,\ldots,z_n^0)^T$. Let us denote for simplicity $\xi(t) = z(t,t_0,z^0)$, $\xi(t) = (\xi_1,\ldots,\xi_n)^T$, and consider the equivalent integral equation

$$\xi_i(t) = z_i^0 + \int_{t_0}^{T} \left[-a_i\xi_i(s) + \sum_{j=1}^{n} b_{ij} f_j(\xi_j(s)) + \sum_{j=1}^{n} c_{ij} g_j(\xi_j(\zeta_k)) + I_i \right] ds.$$

Define a norm $\|\xi(t)\|_0 = \max_{[\zeta_k,t_0]} \|\xi(t)\|$ and construct the following sequences $\xi_i^m(t)$, $\xi_i^0(t) \equiv z_i^0$, $i = 1,\ldots,n$, $m \geqslant 0$ such that

$$\xi_i^{m+1}(t) = z_i^0 + \int_{t_0}^{T} \left[-a_i\xi_i^m(s) + \sum_{j=1}^{n} b_{ij} f_j(\xi_j^m(s)) + \sum_{j=1}^{n} c_{ij} g_j(\xi_j^m(\zeta_k)) + I_i \right] ds.$$

One can find that

$$\|\xi^{m+1}(t) - \xi^m(t)\|_0 \leqslant [\theta(m_1+m_2)]^m \tau,$$

where

$$\tau = \theta \left[(m_1+m_2)\|z^0\| + \sum_{i=1}^{n} I_i \right].$$

Thus, there exists a unique solution $\xi(t) = z(t,t_0,z^0)$ of the integral equation on $[\zeta_k, t_0]$. Then, conditions (A1) and (A2) imply that $z(t)$ can be continued to θ_{k+1}, since it is a solution of ordinary differential equations

$$z_i'(t) = -a_i z_i(t) + \sum_{j=1}^{n} b_{ij} f_j(z_j(t)) + \sum_{j=1}^{n} c_{ij} g_j(z_j(\zeta_k)) + I_i, \;\; a_i > 0, \;\; i = 1,2,\ldots,n$$

on $[\theta_k, \theta_{k+1})$. Next, again, using the same argument we can continue $z(t)$ from $t = \theta_{k+1}$ to $t = \zeta_{k+1}$, and then to θ_{k+2}. Hence, the mathematical induction completes the proof.

Uniqueness. Denote by $z^1(t) = z(t,t_0,z^1)$, $z^2(t) = z(t,t_0,z^2)$, the solutions of (8.1), where $\theta_k \leqslant t_0 \leqslant \theta_{k+1}$. It is sufficient to check that for every $t \in [\theta_k, \theta_{k+1}]$, $z^2 = (z_1^2,\dots,z_n^2)^T$, $z^1 = (z_1^1,\dots,z_n^1)^T \in \mathbb{R}^m$, $z^2 \neq z^1$ implies $z^1(t) \neq z^2(t)$. Then, we have that

$$\|z^1(t) - z^2(t)\| \leqslant \|z^1 - z^2\| + \sum_{i=1}^{n} \left\{ \int_{t_0}^{T} \left[a_i|z_i^2(s) - z_i^1(s)| \right.\right.$$

$$\left.\left. + \sum_{j=1}^{n} L_i^1 |b_{ji}||z_i^2(s) - z_i^1(s)| + \sum_{j=1}^{n} L_i^2 |c_{ji}||z_i^2(\zeta_k) - z_i^1(\zeta_k)| \right] ds \right\} \leqslant$$

$$(\|z^1 - z^2\| + \theta m_2 \|z^1(\zeta_k) - z^2(\zeta_k)\|) + \int_{t_0}^{T} m_1 \|z^1(s) - z^2(s)\| ds.$$

The Gronwall-Bellman Lemma yields that

$$\|z^1(t) - z^2(t)\| \leqslant \left(\|z^1 - z^2\| + \theta m_2 \|z^1(\zeta_k) - z^2(\zeta_k)\|\right) \mathrm{e}^{m_1\theta}.$$

Particularly,

$$\|z^1(\zeta_k) - z^2(\zeta_k)\| \leqslant \left(\|z^1 - z^2\| + \theta m_2 \|z^1(\zeta_k) - z^2(\zeta_k)\|\right) \mathrm{e}^{m_1\theta}.$$

Thus,

$$\|z^1(t) - z^2(t)\| \leqslant \left(\frac{\mathrm{e}^{m_1\theta}}{1 - m_2\theta \mathrm{e}^{m_1\theta}}\right) \|z^1 - z^2\|. \tag{8.2}$$

On the other hand, assume on the contrary that there exists $t \in [\theta_k, \theta_{k+1}]$ such that $z^1(t) = z^2(t)$. Hence,

$$\begin{aligned}\|z^1 - z^2\| &= \sum_{i=1}^{n} \left| \int_{t_0}^{T} \left[-a_i\left(z_i^2(s) - z_i^1(s)\right) + \sum_{j=1}^{n} b_{ij}\left[f_j(z_j^2(s)) - f_j(z_j^1(s))\right] \right.\right. \\ &\quad \left.\left. + \sum_{j=1}^{n} c_{ij}\left[g_j(z_j^2(\zeta_k)) - g_j(z_j^1(\zeta_k))\right] \right] ds \right| \\ &\leqslant \sum_{i=1}^{n} \left\{ \int_{t_0}^{T} \left[a_i|z_i^2(s) - z_i^1(s)| + \sum_{j=1}^{n} L_i^1|b_{ji}||z_i^2(s) - z_i^1(s)| \right.\right. \\ &\quad \left.\left. + \sum_{j=1}^{n} L_i^2|c_{ji}||z_i^2(\zeta_k) - z_i^1(\zeta_k)| \right] ds \right\} \\ &\leqslant \theta m_2 \|z^1(\zeta_k) - z^2(\zeta_k)\| + \int_{t_0}^{T} m_1 \|z^1(s) - z^2(s)\| ds. \end{aligned} \tag{8.3}$$

Consequently, substituting (8.2) in (8.3), we obtain

$$\|z^1 - z^2\| \leqslant \theta(m_1 + 2m_2)\mathrm{e}^{m_1\theta} \|z^1 - z^2\|. \tag{8.4}$$

Thus, one can see that (A4) contradicts with (8.4). The uniqueness is proved for $t \in [\theta_k, \theta_{k+1}]$. The extension of the unique solution on $\mathbb{R}^+$ is obvious. Hence, the theorem is proved. □

System (8.1) can be simplified as follows. Substituting $y(t) = z(t) - z^*$ into (8.1) leads to

$$y_i'(t) = -a_i y_i(t) + \sum_{j=1}^{n} b_{ij}\varphi_j(y_j(t)) + \sum_{j=1}^{n} c_{ij}\psi_j(y_j(\gamma(t))), \tag{8.5}$$

where $\varphi_j(y_j(t)) = f_j(y_j(t) + z_j^*) - f_j(z_j^*)$ and $\psi_j(y_j(t)) = g_j(y_j(t) + z_j^*) - g_j(z_j^*)$ with $\varphi_j(0) = \psi_j(0) = 0$. From assumption (A2), $\varphi_j(\cdot)$ and $\psi_j(\cdot)$ are also Lipschitzian with L_j^1, L_j^2, respectively.

It is clear that the stability of the zero solution of (8.5) is equivalent to that of the equilibrium z^* of (8.1). Therefore, we restrict our discussion to the stability of the zero solution of (8.5). First of all, we give the following lemma which is one of the most important auxiliary results of the present section.

Lemma 8.1. *Let $y(t) = (y_1(t), \ldots, y_n(t))^T$ be a solution of* (8.5) *and* (A1)–(A5) *be satisfied. Then, the following inequality*

$$\|y(\gamma(t))\| \leqslant \lambda \|y(t)\| \tag{8.6}$$

holds for all $t \in \mathbb{R}^+$, where $\lambda = \left\{1 - \theta\left[m_2 + m_1(1 + m_2\theta)\mathrm{e}^{m_1\theta}\right]\right\}^{-1}$.

Proof. Fix $k \in \mathbb{N}$. Then for $t \in [\theta_k, \theta_{k+1})$,

$$y_i(t) = y_i(\zeta_k) + \int_{\zeta_k}^{T} \left[-a_i y_i(s) + \sum_{j=1}^{n} b_{ij}\varphi_j(y_j(s)) + \sum_{j=1}^{n} c_{ij}\psi_j(y_j(\zeta_k)) \right] ds,$$

where $\gamma(t) = \zeta_k$, if $t \in [\theta_k, \theta_{k+1})$, $t \in \mathbb{R}^+$. Taking norms of both sides for each $i = 1, 2, \ldots, n$ and adding all equalities, we obtain that

$$\begin{aligned}
\|y(t)\| &\leqslant \|y(\zeta_k)\| + \sum_{i=1}^{n} \left\{ \int_{\zeta_k}^{T} \left[a_i|y_i(s)| + \sum_{j=1}^{n} L_j^1|b_{ij}||y_j(s)| + \sum_{j=1}^{n} L_j^2|c_{ij}||y_j(\zeta_k)| \right] ds \right\} \\
&= \|y(\zeta_k)\| + \int_{\zeta_k}^{T} \left[\sum_{i=1}^{n} \left(a_i + L_i^1 \sum_{j=1}^{n} |b_{ji}| \right) |y_i(s)| + \sum_{i=1}^{n}\sum_{j=1}^{n} L_i^2|c_{ji}||y_i(\zeta_k)| \right] ds \\
&\leqslant (1 + m_2\theta)\|y(\zeta_k)\| + \int_{\zeta_k}^{T} m_1\|y(s)\| ds.
\end{aligned}$$

The Gronwall-Bellman Lemma yields

$$\|y(t)\| \leqslant (1 + m_2\theta)\mathrm{e}^{m_1\theta}\|y(\zeta_k)\|. \tag{8.7}$$

Furthermore, for $t \in [\theta_k, \theta_{k+1})$ we have

$$\begin{aligned}
\|y(\zeta_k)\| &\leqslant \|y(t)\| + \int_{\zeta_k}^{T} \left[\sum_{i=1}^{n} \left(a_i + L_i^1 \sum_{j=1}^{n} |b_{ji}| \right) |y_i(s)| + \sum_{i=1}^{n}\sum_{j=1}^{n} L_i^2|c_{ji}||y_i(\zeta_k)| \right] ds \\
&\leqslant \|y(t)\| + m_2\theta\|y(\zeta_k)\| + \int_{\zeta_k}^{T} m_1\|y(s)\| ds.
\end{aligned}$$

The last inequality together with (8.7) imply

$$\|y(\zeta_k)\| \leqslant \|y(t)\| + m_2\theta\|y(\zeta_k)\| + m_1\theta(1+m_2\theta)e^{m_1\theta}\|y(\zeta_k)\|.$$

Thus, it follows from condition (A4) that

$$\|y(\zeta_k)\| \leqslant \lambda\|y(t)\|, \quad t \in [\theta_k, \theta_{k+1}).$$

Hence, (8.6) holds for all $t \in \mathbb{R}^+$. This completes the proof. □

8.3 Main results

In this section we establish several criteria for global exponential stability of (8.5) based on the method of Lyapunov functions.

For convenience, we adopt the following notation in the sequel:

$$m_3 = \frac{1}{n}\min_{1\leqslant i\leqslant n}\left(a_i - \frac{1}{2}\sum_{j=1}^{n}\left(L_j^1|b_{ij}| + L_j^2|c_{ij}| + L_i^1|b_{ji}|\right)\right).$$

Theorem 8.3. *Suppose that* (A1)–(A5) *hold true. Assume, furthermore, that the following inequality is satisfied:*

$$m_3 > \frac{m_2\lambda^2}{2}. \tag{8.8}$$

Then the system (8.5) *is globally exponentially stable.*

Proof. We define a Lyapunov function by

$$V(y(t)) = \frac{1}{2}\sum_{i=1}^{n} y_i^2(t).$$

One can easily show that

$$\frac{1}{2n}\|y(t)\|^2 \leqslant V(y(t)) \leqslant \frac{1}{2}\|y(t)\|^2. \tag{8.9}$$

For $t \neq \theta_i$, $i \in \mathbb{N}$, the time derivative of V with respect to (8.5) is given by

$$V'_{(8.5)}(y(t)) = \sum_{i=1}^{n} y_i(t)y_i'(t) = \sum_{i=1}^{n} y_i(t)\left[-a_iy_i(t) + \sum_{j=1}^{n} b_{ij}\varphi_j(y_j(t)) + \sum_{j=1}^{n} c_{ij}\psi_j(y_j(\gamma(t)))\right] \leqslant$$

$$\sum_{i=1}^{n}\left[-a_iy_i^2(t) + \sum_{j=1}^{n} L_j^1|b_{ij}||y_i(t)||y_j(t)| + \sum_{j=1}^{n} L_j^2|c_{ij}||y_i(t)||y_j(\gamma(t))|\right] \leqslant$$

$$\sum_{i=1}^{n}\left[-a_iy_i^2(t) + \frac{1}{2}\sum_{j=1}^{n} L_j^1|b_{ij}|(y_i^2(t) + y_j^2(t)) + \frac{1}{2}\sum_{j=1}^{n} L_j^2|c_{ij}|(y_i^2(t) + y_j^2(\gamma(t)))\right] \leqslant$$

$$-\sum_{i=1}^{n}\left[\left(a_i-\frac{1}{2}\sum_{j=1}^{n}(L_j^1|b_{ij}|+L_j^2|c_{ij}|+L_i^1|b_{ji}|)\right)y_i^2(t)\right]+\frac{1}{2}\sum_{i=1}^{n}\sum_{j=1}^{n}L_i^2|c_{ji}|y_i^2(\gamma(t))\leqslant$$

$$-\min_{1\leqslant i\leqslant n}\left(a_i-\frac{1}{2}\sum_{j=1}^{n}(L_j^1|b_{ij}|+L_j^2|c_{ij}|+L_i^1|b_{ji}|)\right)\sum_{i=1}^{n}y_i^2(t)+$$

$$\frac{1}{2}\max_{1\leqslant i\leqslant n}\left(L_i^2\sum_{j=1}^{n}|c_{ji}|\right)\sum_{i=1}^{n}y_i^2(\gamma(t))\leqslant$$

$$-m_3\|y(t)\|^2+\frac{m_2}{2}\|y(\gamma(t))\|^2.$$

By using Lemma 8.1, we obtain

$$V'_{(8.5)}(y(t))\leqslant -m_3\|y(t)\|^2+\frac{m_2\lambda^2}{2}\|y(t)\|^2=-\left(m_3-\frac{m_2\lambda^2}{2}\right)\|y(t)\|^2.$$

Now, define β for convenience as follows:

$$\beta=m_3-\frac{m_2\lambda^2}{2}>0.$$

Then, we have for $t\neq\theta_i$

$$\frac{d}{dt}(e^{2\beta t}V(y(t)))=\mathrm{e}^{2\beta t}(2\beta)V(y(t))+e^{2\beta t}V'_{(8.5)}(y(t))\leqslant\beta\mathrm{e}^{2\beta t}\|y(t)\|^2-\beta\mathrm{e}^{2\beta t}\|y(t)\|^2=0.$$

From (8.9) and using the continuity of the function V and the solution $y(t)$, we obtain

$$\mathrm{e}^{2\beta t}(1/2n)\|y(t)\|^2\leqslant\mathrm{e}^{2\beta t}V(y(t))\leqslant\mathrm{e}^{2\beta t_0}V(y(t_0))\leqslant\mathrm{e}^{2\beta t_0}(1/2)\|y(t_0)\|^2,$$

which implies $\|y(t)\|\leqslant\sqrt{n}\|y(t_0)\|\mathrm{e}^{-\beta(t-t_0)}$. That is, the system (8.5) is globally exponentially stable. □

In the next theorem, we utilize the same technique, used in previous theorem, to find new stability conditions for recurrent neural networks by choosing a different Lyapunov function defined as

$$V(y(t))=\sum_{i=1}^{n}\alpha_i|y_i(t)|,\quad \alpha_i>0,\quad i=1,2,\ldots,n.$$

For simplicity of notation, let us denote

$$m_4=\min_{1\leqslant i\leqslant n}\left(a_i-L_i^1\sum_{j=1}^{n}|b_{ji}|\right).$$

Theorem 8.4. *Suppose that* (A1)–(A5) *hold true. Assume, furthermore, that the following inequality is satisfied:*

$$m_4>m_2\lambda. \tag{8.10}$$

Then the system (8.5) *is globally exponentially stable.*

The proof of the assertion is similar to that of Theorem 8.3, so we omit it here.

8.4 Illustrative examples

In this section, we give three examples with simulations to illustrate our results. In the sequel, we assume that the identification function $\gamma(t)$ is with the sequences $\theta_k = k/9$, $\zeta_k = (2k+1)/18$, $k \in \mathbb{N}$.

Example 8.1. Consider the following recurrent neural networks with the argument function $\gamma(t)$:

$$\frac{dz(t)}{dt} = -\begin{pmatrix} 2 & 0 \\ 0 & 1.5 \end{pmatrix}\begin{pmatrix} z_1(t) \\ z_2(t) \end{pmatrix} + \begin{pmatrix} 0.02 & 0.03 \\ 0.01 & 1 \end{pmatrix}\begin{pmatrix} \tanh(z_1(t)) \\ \tanh(z_2(t)) \end{pmatrix}$$
$$+\begin{pmatrix} 0.08 & 1 \\ 0.01 & 1 \end{pmatrix}\begin{pmatrix} \tanh(\frac{z_1(\gamma(t))}{7}) \\ \tanh(\frac{z_2(\gamma(t))}{6}) \end{pmatrix} + \begin{pmatrix} 1 \\ 1 \end{pmatrix}. \tag{8.11}$$

It is easy to verify that (8.11) satisfies the conditions of Theorem 8.3 with $L_1^1 = L_2^1 = 1$, $L_1^2 = 1/7$, $L_2^2 = 1/6$, $m_1 = 2.53$, $m_2 = 0.3333$, $m_3 = 0.6308$, $m_4 = 0.47$, $\lambda = 1.7337$. Thus, according to this theorem the unique equilibrium $z^* = (0.6011, 1.3654)^T$ of (8.11) is globally exponentially stable. However, the condition (8.10) of Theorem 8.4 is not satisfied. Let us simulate a solution of (8.11) with initial condition $z_1^1(0) = z_1^0$, $z_2^1(0) = z_2^0$. Since the equation (8.11) is of mixed type, the numerical analysis has a specific character and it should be described more carefully. One will see that this algorithm is in full accordance with the approximations made in the proof of Theorem 8.2.

We start with the interval $[\theta_0, \theta_1]$, that is; $[0, 1/9]$. On this interval the equation (8.11) has the form

$$\frac{dz(t)}{dt} = -\begin{pmatrix} 2 & 0 \\ 0 & 1.5 \end{pmatrix}\begin{pmatrix} z_1(0) \\ z_2(0) \end{pmatrix} + \begin{pmatrix} 0.02 & 0.03 \\ 0.01 & 1 \end{pmatrix}\begin{pmatrix} \tanh(z_1(0)) \\ \tanh(z_2(0)) \end{pmatrix} +$$
$$\begin{pmatrix} 0.08 & 1 \\ 0.01 & 1 \end{pmatrix}\begin{pmatrix} \tanh(\frac{z_1(1/18)}{7}) \\ \tanh(\frac{z_2(1/18)}{6}) \end{pmatrix} + \begin{pmatrix} 1 \\ 1 \end{pmatrix},$$

where $z_i(1/18)$, $i = 1, 2$, are still unknown. For this reason, we will arrange approximations in the following way. Consider the sequence of the equations

$$\frac{dz^{(m+1)}(t)}{dt} = -\begin{pmatrix} 2 & 0 \\ 0 & 1.5 \end{pmatrix}\begin{pmatrix} z_1^{(m)}(0) \\ z_2^{(m)}(0) \end{pmatrix} + \begin{pmatrix} 0.02 & 0.03 \\ 0.01 & 1 \end{pmatrix}\begin{pmatrix} \tanh(z_1^{(m)}(0)) \\ \tanh(z_2^{(m)}(0)) \end{pmatrix} +$$
$$\begin{pmatrix} 0.08 & 1 \\ 0.01 & 1 \end{pmatrix}\begin{pmatrix} \tanh(\frac{z_1^{(m)}(1/18)}{7}) \\ \tanh(\frac{z_2^{(m)}(1/18)}{6}) \end{pmatrix} + \begin{pmatrix} 1 \\ 1 \end{pmatrix},$$

where $m = 0, 1, 2 \ldots$, with $z_1^0(t) \equiv z_1^0, z_2^0(t) \equiv z_2^0$. We evaluate the solutions, $z^{(m)}(t)$, by using MATLAB 7.8 and stop the iterations at $(z_1^{(500)}(t), z_2^{500}(t))$. Then, we assign $z_1(t) = z_1^{(500)}(t)$, $z_2(t) = z_2^{(500)}(t)$ on the interval $[\theta_0, \theta_1]$. Next, similar operation is done on the interval $[\theta_1, \theta_2]$. That is, we construct the sequence $(z_1^{(m)}, z_2^{(m)})$ of solutions again for the system

$$\frac{dz^{(m+1)}(t)}{dt} = -\begin{pmatrix} 2 & 0 \\ 0 & 1.5 \end{pmatrix}\begin{pmatrix} z_1^{(m)}(0) \\ z_2^{(m)}(0) \end{pmatrix} + \begin{pmatrix} 0.02 & 0.03 \\ 0.01 & 1 \end{pmatrix}\begin{pmatrix} \tanh(z_1^{(m)}(0)) \\ \tanh(z_2^{(m)}(0)) \end{pmatrix} +$$

$$\begin{pmatrix} 0.08 & 1 \\ 0.01 & 1 \end{pmatrix}\begin{pmatrix} \tanh(\frac{z_1^{(m)}(3/18)}{7}) \\ \tanh(\frac{z_2^{(m)}(3/18)}{6}) \end{pmatrix} + \begin{pmatrix} 1 \\ 1 \end{pmatrix}$$

with $z_1^0(t) \equiv z_1^{(500)}(1/9)$, $z_2^0(t) \equiv z_2^{(500)}(1/9)$, and $z_1(t) = z_1^{(500)}(t)$, $z_2(t) = z_2^{(500)}(t)$ on $[\theta_1, \theta_2]$. Proceeding in this way, one can obtain a simulation which demonstrates the asymptotic property.

Specifically, simulation result with several random initial points is shown in Figure 8.1. We must explain that the non-smoothness at the switching points θ_k, $k \in \mathbb{N}$ is not seen by simulation. That is why we have to choose the Lipschitz constants and θ small enough to satisfy the conditions of the theorems. So, the smallness "hides" the non-smoothness.

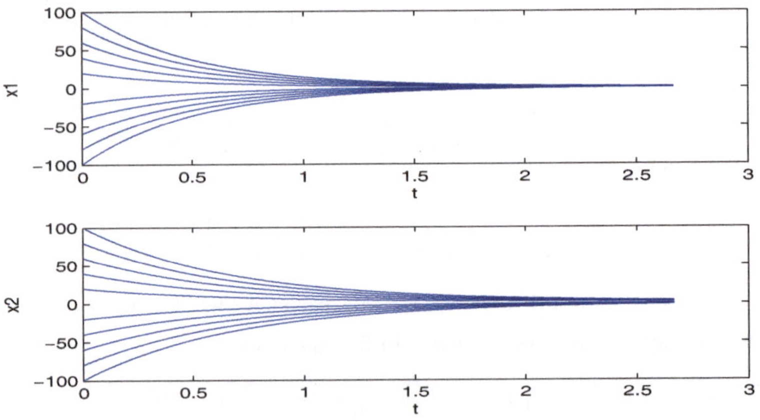

Fig. 8.1 Transient behavior of the recurrent neural networks in Example 8.1.

Let us now take the parameters such that the non-smoothness can be seen. Consider the

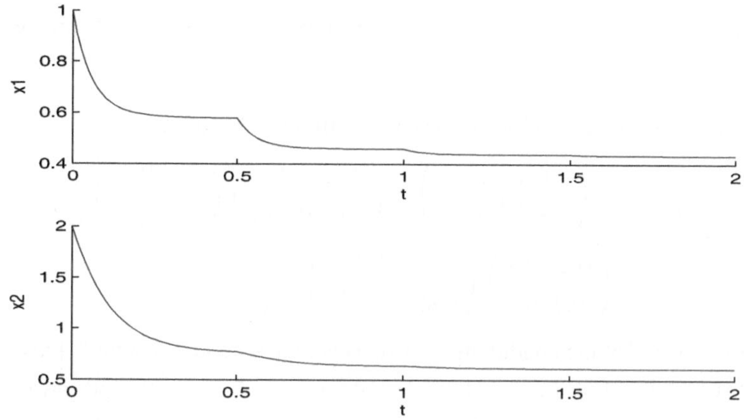

Fig. 8.2 The non-smoothness is seen at moments 0.5; 1; 1.5, which are switching points of the function $\gamma(t)$.

following recurrent neural networks:

$$\frac{dz(t)}{dt} = -\begin{pmatrix} 20 & 0 \\ 0 & 10 \end{pmatrix}\begin{pmatrix} z_1(t) \\ z_2(t) \end{pmatrix} + \begin{pmatrix} 2 & 1 \\ 8 & 0.2 \end{pmatrix}\begin{pmatrix} \tanh(z_1(t)) \\ \tanh(z_2(t)) \end{pmatrix} + \begin{pmatrix} 1 & 20 \\ 2 & 3 \end{pmatrix}\begin{pmatrix} \tanh(z_1(\gamma(t))) \\ \tanh(\frac{z_2(\gamma(t))}{2}) \end{pmatrix} + \begin{pmatrix} 1 \\ 1 \end{pmatrix}, \tag{8.12}$$

where $\theta_k = k/2$, $\zeta_k = (2k+1)/4$, $k \in \mathbb{N}$. One can see that θ and the Lipschitz coefficient are large this time. They do not satisfy the conditions of our theorems. It is illustrated in Figure 8.2 that the non-smoothness of the solution with the initial point $[1,2]^T$ can be seen at the switching points $\theta_k,\ k \in \mathbb{N}$.

This is important for us to see that the non-smoothness of solutions expected from the equations' nature is seen. Moreover, we can see that the solution converges to the unique equilibrium $z^* = (0.4325, 0.6065)^T$. It shows that the sufficient conditions which are found in our theorems can be elaborated further.

Example 8.2. Consider the following recurrent neural networks:

$$\frac{dz(t)}{dt} = -\begin{pmatrix} 2 & 0 \\ 0 & 2.5 \end{pmatrix}\begin{pmatrix} z_1(t) \\ z_2(t) \end{pmatrix} + \begin{pmatrix} 1 & 0.03 \\ 0.04 & 1 \end{pmatrix}\begin{pmatrix} \tanh(\frac{z_1(t)}{4}) \\ \tanh(z_2(t)) \end{pmatrix} + \begin{pmatrix} 1 & 0.04 \\ 0.02 & 0.07 \end{pmatrix}\begin{pmatrix} \tanh(\frac{z_1(\gamma(t))}{4}) \\ \tanh(\frac{z_2(\gamma(t))}{4}) \end{pmatrix} + \begin{pmatrix} 1 \\ 1 \end{pmatrix}. \tag{8.13}$$

It can be shown easily that (8.13) satisfies the conditions of the Theorem 8.4 if $L_1^1 = 1/4$, $L_2^1 = 1, L_1^2 = 1/4, L_2^2 = 1/4, m_1 = 3.53, m_2 = 0.2550, m_3 = 0.6181, m_4 = 1.47, \lambda = 2.6693,$

whereas the condition (8.8) of Theorem 8.3 does not hold. Hence, it follows from Theorem 8.4 that the unique equilibrium $z^* = (0.6737, 0.6265)^T$ of (8.13) is globally exponentially stable.

Example 8.3. Consider the following system of differential equations:

$$\frac{dz(t)}{dt} = -\begin{pmatrix} 3 & 0 \\ 0 & 3 \end{pmatrix}\begin{pmatrix} z_1(t) \\ z_2(t) \end{pmatrix} + \begin{pmatrix} 0.02 & 0.03 \\ 0.04 & 0.25 \end{pmatrix}\begin{pmatrix} \tanh(\frac{z_1(t)}{4}) \\ \tanh(\frac{z_2(t)}{4}) \end{pmatrix} +$$
$$\begin{pmatrix} 0.25 & 0.4 \\ 0.2 & 0.7 \end{pmatrix}\begin{pmatrix} \tanh(\frac{z_1(\gamma(t))}{4}) \\ \tanh(\frac{z_2(\gamma(t))}{4}) \end{pmatrix} + \begin{pmatrix} 1 \\ 1 \end{pmatrix}. \tag{8.14}$$

One can see easily that the conditions of both Theorem 8.3 and Theorem 8.4 are satisfied with $L_1^1 = 1/4$, $L_2^1 = 1/4$, $L_1^2 = 1/4$, $L_2^2 = 1/4$, $m_1 = 3.07$, $m_2 = 0.2750$, $m_3 = 1.4081$, $m_4 = 2.93$, $\lambda = 2.1052$, $\tau = 1.1$. Thus, according to Theorem 8.3 and Theorem 8.4 the unique equilibrium $z^* = (0.4172, 0.4686)^T$ of (8.14) is globally exponentially stable.

Notes

In this chapter the method of Lyapunov functions for differential equations with piecewise constant functions as argument is applied to the model of recurrent neural networks and new sufficient conditions which guarantee existence, uniqueness, and global exponential stability of the equilibrium point of the recurrent neural networks are obtained. The results can be extended to study more complex systems [16, 17]. On the basis of our results, one can estimate domains of attraction, which allows us to evaluate the performance of recurrent neural networks [337]. The results are published in [28]. More about stability of cellular neural networks with piecewise constant argument can be found in [27].

Chapter 9

The blood pressure distribution

9.1 The systemic arterial pressure

This section is organized in the following manner. In Subsection 9.1.1, we give a short background on systemic arterial pressure for those readers who come from a mathematical background and are not familiar with this subject. In Subsection 9.1.2, we build a model as a hybrid system with jumps equation and controlled switching moments and find conditions of oscillations in different types: periodic, almost periodic, ε-oscillations with their asymptotic properties and positiveness. In Subsection 9.1.3 we consider for the first time the regular behavior of systemic arterial pressure when the moments of jumps are not fixed, and a new jump occurs when the pressure value reaches a certain positive constant value. Finally, we present our conclusions.

9.1.1 *Background on systemic arterial pressure*

It is important to have a sense of the mechanisms describing systemic arterial pressure since they are the basis of the mathematical assumptions and derivations that we describe in this chapter. Therefore in this subsection, we introduce some of the control mechanisms behind systemic arterial pressure and its variation over time for those readers who are not familiar with human physiology.

Circulation of blood is very essential for functioning of the body since blood carries necessary substances such as oxygen, nutrients and hormones to the cells and collects waste products such as carbon dioxide and uric acid from the cells to be eliminated by the lungs and the kidneys. Circulating blood causes pressure against the walls of the arteries, the veins, and the heart chambers. Blood pressure in the systemic arteries as a result of pumping of oxygen rich blood by the left ventricle of the heart is called the *systemic arterial pressure* [166]. Any irregularities in the blood circulation, or deviations in the blood pressure may cause harm to the body, might even cause death. Therefore, there has been much research interest in understanding and modeling the mechanisms that control and regulate the blood pressure [53, 72, 78].

Systemic arterial pressure is regulated by the autonomic nervous system [297]. There are pressure sensors known as the baroreceptors located in great veins, the aortic arch and the carotid sinus that sense the changes in the blood pressure as the blood is pumped from the ventricles. The baroreceptors sense the rise in the arterial pressure and as a response, reflexes are initiated by the nervous system to slow down the heart rate to decrease the pressure. A decrease in the blood pressure is also sensed and heart rate is increased to increase the blood pressure. In addition to baroreceptors, there are chemical receptors known as chemoreceptors that are located in the carotid and aortic bodies. The ones in the carotid bodies sense decrease in the oxygen content, and the ones in the aortic bodies sense increase in the carbon dioxide content and hydrogen ion concentrations in the blood. As a result of the values sensed by these receptors, respiration rate is increased or decreased by the nervous system, heart rate is regulated with the changes in the respiration rate and blood pressure is thus controlled.

Circulation of blood in the body and regulation of blood pressure can be modeled by a sophisticated closed loop control system involving complex interactions between heart rate control and respiration rate, which is mediated by the autonomic nervous system. There has been much research interest in modeling this control system and estimating the corresponding model parameters [53, 72, 78]. These studies especially focused on the relation between blood pressure, heart rate and respiration.

Understanding the control mechanism that regulates blood pressure might find many applications in medicine that will help improving prosthetic device technologies. For example, in [276], systemic arterial model parameters are estimated using a modified recursive least squares algorithm for control of an electrically actuated total artificial heart. It is important to find the model parameters of the control system accurately because the artificial heart has to supply blood to the body in response to metabolic demands in a manner that is physiologically compatible with the body. In [72], the authors describe a study in which beat-to-beat blood pressure variability is recorded noninvasively to assess the hemodynamic effects of VVI and DDD pacing modes in pacemakers.

Heart rate variability refers to the beat-to-beat alterations in heart rate. Under resting conditions the electrocardiogram (ECG) of healthy individuals exhibits periodic variation in R-R intervals. The power spectrum of heart rate variability contains two major components:

the high frequency (0,18-0,4 Hz) component which is synchronous with respiration and is identical to respiratory sinus arrhythmia. The second is a low frequency (0,04 to 0,15 Hz) component that appears to be mediated by both the vagus and cardiac sympathetic nerves [154].

Blood pressure variability has a similar low frequency peak but much attenuated high frequency power. The high frequency peak is generally regarded as related to the respiratory sinus arrhythmia and therefore represents the modulation of vagal tone, while the low frequency component is thought to have its origins in variations in the blood pressure (so-called Mayer waves) related to sympathetic tone. Sympathetic blood pressure modulation is transduced into changes in heart rate through the actions of the vagal arm of the sinoaortic baroreflex. Thus the low frequency component relates both to parasympathetic and sympathetic factors. Blood pressure variability has been widely assessed by calculation of the standard deviation of 24 hour systolic, diastolic, and mean arterial pressure. Systolic blood pressure variability was obtained by calculating the standard deviation of the 24-hour mean, which was taken as the overall blood pressure variability. Several evidences collected during the past few years have shown that behavioral, neural, reflex and humoral factors participate in the phenomenon of blood pressure variability. Especially, blood pressure fluctuations are largely related to changes in respiration as well as to rhythmic alterations in central autonomic drive mediated by baroreflex mechanisms [221].

9.1.2 *The regular and irregular behavior of systemic arterial pressure when moments of the heart contraction are prescribed*

In the previous part, we had explained that the systemic arterial pressure has a periodical behavior. In the following, we study this periodic behavior (as well as non-periodic behavior) using the results of the theory of impulsive differential equations.

Let t_i, $i \in \mathbb{Z}$, be a sequence in $\mathbb{R}$, and $\mathscr{PC}$ be a space of piecewise continuous functions, defined on $\mathbb{R}$, with discontinuities of the first kind at the points t_i. Denote $P(t)-$ the value of systemic arterial pressure at time $t \geqslant 0$. It is shown in [166] that the regular change of systemic arterial pressure, after some simplifications, satisfies the following two equations

$$\frac{dP(t)}{dt} = -\alpha P(t) \text{ if } t \neq t_i, \tag{9.1}$$

where $t_{i+1} - t_i = T$, $i \in Z$, T-positive fixed number, and

$$P(t_i+) = P(t_i-) + I, \tag{9.2}$$

where $I > 0$ is fixed, $P(t_i+)$ and $P(t_i-)$ are values of systemic arterial pressure right after and before the heart contraction at time t_i.

It is not difficult to see that equations (9.1) and (9.2) can be written in the form

$$\frac{dP(t)}{dt} = -\alpha P(t), \quad t \neq t_i,$$
$$\Delta P|_{t=t_i} = I_i, \tag{9.3}$$

where $\Delta P|_{t=t_i} = P(t_i+) - P(t_i-)$, if we assume a particular case $I_{i+1} = I_i$, $i \in \mathbb{Z}$. But, in general, the last condition in this section is not fulfilled. Besides equation (9.3), we consider the following system:

$$\frac{dP(t)}{dt} = -\alpha P(t) + Q(t), \quad t \neq t_i,$$
$$\Delta P|_{t=t_i} = I_i, \tag{9.4}$$

where $Q(t) \in \mathscr{PC}$ and Q is not restricted by any sign assumption, and all I_i, $i \in \mathbb{Z}$, are positive. We shall call equation (9.4) as a *Q-perturbed system.*

Systems (9.3) and (9.4) are impulsive differential equations (IDE) [20, 192, 278]. Solutions of impulsive differential equations are piecewise continuous functions. We shall consider $P(t)$ as a solution of the impulsive differential equation for all $t \in \mathbb{R}$.

The concept of almost periodicity was introduced by H. Bohr [58] and plays a significant role in the theory of oscillations. D. Wexler started to investigate piecewise continuous almost periodic (PAP) functions [41, 42, 148, 278]. Below we shall give the description of the PAP functions.

Let a_i, $i \in \mathbb{Z}$, be a sequence in $\mathbb{R}$. An integer p is called an *ε-almost period* of the sequence a_i if $\|a_{i+p} - a_i\| < \varepsilon$, $i \in \mathbb{Z}$. If a sequence a_i is p-periodic, i.e. $a_{i+p} = a_i$ for all i, then for any positive ε, the number np, n is a natural number, is an *ε-almost period* of this sequence.

A set $S \subset \mathbb{R}$ is said to be relatively dense if there exists a number $l > 0$ such that $[a, a+l] \cap S \neq \emptyset$ for all $a \in \mathbb{R}$.

A sequence a_i is called almost periodic if for any positive ε there exists a relatively dense set of its ε-almost periods.

Let θ_i be a sequence of real numbers such that $\theta_i < \theta_{i+1}$ and $|\theta_i| \to \infty$ as $|i| \to \infty$. Denote $\theta_i^j = \theta_{i+j} - \theta_i$ for all integers i, j.

The family of sequences θ_i^j, i, $j \in \mathbb{Z}$, will be called equipotentially almost periodic if for an arbitrary $\varepsilon > 0$ there exists a relatively dense set of ε-almost periods, that are common to all the sequences $\{\theta_i^j\}$.

Assume that there exists a positive number θ such that $\sup_{i \in \mathbb{Z}}(\theta_{i+1} - \theta_i) > \theta$.

Definition 9.1 ([148]). We shall say that a function $\phi(t) \in \mathscr{PC}$ is a piecewise continuous almost periodic if:

a) *the sequence* t_i, $i \in \mathbb{Z}$, *is such that the derived sequences* t_i^j *are equipotentially almost periodic;*

b) for any positive ε there exists a positive number δ such that if the points t', t'' belong to the same interval of continuity and $|t' - t''| < \delta$, then $|\phi(t') - \phi(t'')| < \varepsilon$;

c) for any positive ε there exists a relatively dense set Γ of ε-almost periods such that if $\tau \in \Gamma$, then $|\phi(t+\tau) - \phi(t)| < \varepsilon$ for all $t \in \mathbb{R}$ which satisfy the condition $|t - \theta_i| > \varepsilon$, $i \in \mathbb{Z}$.

Definition 9.2. A function $\phi \in \mathscr{PC}$ is called ε-oscillatory for a given $\varepsilon > 0$, if there exists a piecewise continuous periodic or PAP function $\psi(t)$ such that $|\psi(t) - \phi(t)| < \varepsilon$ for all $t \geqslant 0$.

We shall call the periodic behavior of systemic arterial pressure as *regular* and call *almost periodic* or *ε-oscillatory* behavior of systemic arterial pressure as *irregular*, assuming for all cases that solutions are positive.

In what follows in this section we shall consider several impulsive differential equations which provide *regular* and *irregular* behavior of systemic arterial pressure.

a) *Existence of a positive and asymptotically stable T-periodic solution of the equation* (9.3). Let $t_i = iT$, $i \in \mathbb{Z}$, be the moments of jumps of solutions of the system (9.3), $I_i = I$, $i \in \mathbb{Z}$, and $P(t)$, $P(0) = P_0 > 0$, be a solution of (9.3). It is known [192,278] that the solution is equal to

$$P(t) = \exp(-\alpha t)\Big[P_0 + \sum_{0 \leqslant t_i < t} \exp(\alpha t_i) I\Big]. \tag{9.5}$$

Denote $P^0 = (\exp(\alpha T) - 1)^{-1}$, and let $P_0(t)$, $P_0(0) = P^0$, be a solution of (9.3). Then

$$\begin{aligned} P_0(t+T) &= \exp(-\alpha(t+T))\Big[P^0 + \sum_{0 \leqslant t_i < t+T} \exp(\alpha t_i) I\Big] \\ &= \exp(-\alpha t)\Big[P^0 \exp(-\alpha T) + \sum_{0 \leqslant t_i < t+T} \exp(\alpha(t_i - T)) I\Big] \\ &= \exp(-\alpha t)\Big[\sum_{0 \leqslant t_i < t} \exp(\alpha t_i) I\Big] \\ &= P_0(t). \end{aligned}$$

That is, $P_0(t)$ is a T-periodic solution, since the equality $P_0(t+T) = P_0(t)$ is valid for all t. Let $P(t)$ be an arbitrary solution of (9.3), $P(0) = P_0 > 0$. Then

$$P(t) - P_0(t) = \exp(-\alpha t)(P_0 - P^0).$$

Last equality implies that $P(t) - P_0(t) \to 0$ as $t \to \infty$. That is $P_0(t)$ is an asymptotically stable T-periodic solution of the system (9.3). The existence of an asymptotically stable T-periodic solution confirms the result of [166]. It is easy to see that $\min_{[0,T]} P_0(t) = P^0 \exp(-\alpha T) > 0$.

b) *Existence of a positive and asymptotically stable T-periodic solution of the periodic Q-perturbed system.*
In [166] some factors of systemic arterial pressure were neglected. Such as the systemic venous pressure, for example, but even a small perturbation may cause a significant change in a system's behavior. Hence, it is reasonable to consider a *Q-perturbed system.* Assume that $Q(t)$ is a T-periodic function. Let $P(t)$ be a solution of equation (9.4), $P(0) = P_0$. Then [192,278]

$$P(t) = \exp(-\alpha t)\left[P_0 + \int_0^t \exp(\alpha s)Q(s)ds + \sum_{0 \leqslant t_i < t} \exp(\alpha t_i)I\right]. \tag{9.6}$$

Denote $P^0 = (\exp(\alpha T) - 1)^{-1}\left(\int_0^T \exp(\alpha s)Q(s)ds + I\right)$. Let $P_0(t)$, $P_0(0) = P^0$, be a solution of (9.4). Similarly to the problem (a) one can show that $P_0(t)$ is T-periodic. Moreover, if

$$\int_0^T \exp(\alpha s)|Q(s)|ds < P^0 \exp(-\alpha T), \tag{9.7}$$

then solution $P_0(t)$ is positive-valued for all t.
Let $P(t)$ be a solution of (9.4) defined by (9.6). Then it is not difficult to obtain that

$$P(t) - P_0(t) = \exp(-\alpha t)(P_0 - P^0).$$

That is $P_0(t)$ is an asymptotically stable solution of the *Q-perturbed system.*

c) *Existence of an asymptotically stable ε-oscillatory solution of the Q-perturbed system, with arbitrary small $\varepsilon > 0$, when $Q(t)$ is not periodic, but is small in* sup*-norm, and the non-perturbed system is periodic.*
Assume that system (9.3) associated with (9.4) satisfies conditions of problem (a) and function $Q(t)$ is not T-periodic, but there exists a number $\delta > 0$ such that $|Q(t)| \leqslant \delta, t \geqslant 0$. We shall show that for a given $\varepsilon > 0$ there exists an ε-oscillation of (9.4), if δ is sufficiently small. Indeed, assume that $\delta < \alpha\varepsilon$. Let a function $P_0(t)$ ($P_0(t)$ is not solution!) be defined by the formula

$$P_0(t) = \exp(-\alpha t)\left[P^0 + \sum_{0 \leqslant t_i < t} \exp(\alpha t_i)I\right],$$

where $P^0 = (\exp(\alpha T) - 1)^{-1}$. We have proved already, that $P_0(t)$ is a periodic function. Let $\widetilde{P}(t)$, $\widetilde{P}(0) = P^0$, be a solution of (9.4). Then, using the formula (9.6) one can obtain that

$$\widetilde{P}(t) - P_0(t) = \int_0^t \exp(-\alpha(t-s))Q(s)ds,$$

and hence, $|\widetilde{P}(t) - P_0(t)| \leqslant \varepsilon$. Thus the solution $\widetilde{P}(t)$ is ε-oscillatory.
Denote $\beta = \min_{0 \leqslant t \leqslant T} P_0(t)$. Assume that ε is sufficiently small such that $\varepsilon < \beta$, then it is easily seen that $\widetilde{P}(t) > 0$ for all t.
Let $P(t)$ be a solution of (9.4) defined by (9.6), then $P(t) - \widetilde{P}(t) = \exp(-\alpha t)(P^0 - P_0)$. Hence, solution $\widetilde{P}(t)$ is asymptotically stable.

d) *Existence of an asymptotically stable almost periodic solution of the equation* (9.3) *such that all of its elements are almost periodic.*
We consider the problem of existence of PAP solutions. Let system (9.3) be given, where I_i is an almost periodic sequence, the sequences $t_i^j, j \in Z$, derivated from t_i, $i \in Z$, are equipotentially almost periodic. Moreover, we assume that there exists a positive number κ such that

$$\inf_{i \in \mathbb{Z}} I_i = \kappa. \tag{9.8}$$

Using Theorem 81 [278] one can see that the function

$$P^0(t) = \sum_{t_i < t} \exp(-\alpha(t - t_i))I_i \tag{9.9}$$

is a PAP function and asymptotically stable solution of the equation (9.3). It is obvious that the solution is positive. Moreover, one can easily see that $P^0(t) \geqslant \kappa \exp(-\alpha\theta) > 0$, for all $t \in \mathbb{R}$.

e) *Existence of an asymptotically stable almost periodic solution of the almost periodic Q-perturbed system.*
Assume that the behavior of systemic arterial pressure is described by the system (9.4), where all elements except $Q(t)$ are the same as in problem **d)**, and $Q(t)$ is a PAP function. Using Theorem 81 [278] again one can see that

$$P(t) = \int_{-\infty}^t \exp(-\alpha(t-s))Q(s)ds + \sum_{t_i < t} \exp(-\alpha(t - t_i))I_i \tag{9.10}$$

is PAP and asymptotically stable solution of equation (9.4).
It is easily seen that $P(t) > 0$, $t \in R$, if $\sup_R |Q(t)| < \alpha\kappa\exp(-\alpha\theta)$.
The idea of almost periodic nature of systemic arterial pressure also appears in [147].

f) *Existence of stable ε-oscillatory solution, with arbitrary small $\varepsilon > 0$, of the system* (9.4) *when $Q(t)$ is not periodic or almost periodic, but is small in* sup-*norm, and associated with* (9.4) *system* (9.3) *is almost periodic.*

Now, let $Q(t)$ satisfy the inequality $|Q(t)| < \delta, t \in R, \delta < \varepsilon\alpha$. Then the integral and the series in (9.10) are convergent and one can check by substitution that $P(t)$ is the solution of (9.4). Subtracting (9.9) from (9.10) we have that

$$|P(t) - P^0(t)| = \left| \int_{-\infty}^{t} \exp(-\alpha(t-s))Q(s)ds \right| < \frac{\delta}{\alpha} < \varepsilon.$$

Similarly to previous cases we can easily prove that $P(t)$ is asymptotically stable ε-oscillation of system (9.4), and it is positive if $\delta < \alpha\kappa\exp(-\alpha\theta)$.

9.1.3 *Regular behavior of systemic arterial pressure with non prescribed moments of the heart contraction*

In this part of our work we do not assume that the moments of jumps of systemic arterial pressure are fixed, but the discontinuities are committed when systemic arterial pressure achieves a special value $\overline{P} \in R, \overline{P} > 0$. The primary nervous mechanism for the control of cardiac output is the baroreceptor reflex. This reflex is initiated by stretch receptors, called baroreceptors, located in the walls of the carotid sinus and aortic arch, large arteries of the systemic circulation. The falling pressure at the baroreceptors elicits an immediate reflex, resulting in a strong sympathetic discharge. In this case one can suppose that an impulsive differential equation which describes the systemic arterial pressure has the form

$$\begin{aligned} &\frac{dP(t)}{dt} = -\alpha P(t), \quad P \neq \overline{P}, \\ &\Delta P|_{P=\overline{P}} = I, \end{aligned} \tag{9.11}$$

where $P \in \mathbb{R}, P \geqslant \overline{P}$, and $\alpha > 0, I > 0$, are fixed real numbers. We shall say that a solution $P(t)$ of (9.11) is ultimately periodic with a period $\omega > 0$, if there exists a number $T \in \mathbb{R}$ such that $P(t+\omega) = P(t), t \geqslant T$.

System (9.11) provides a discontinuous dynamical system [8], and to investigate these type of equations one can apply methods which are developed in [8–10].

Let us consider the following problem.

g) Prove that every solution of (9.11) $P(t) = P(t,0,P_0), P_0 \geqslant \overline{P}$, is ultimately periodic with one and the same period, and all of them are stable.

Fix $P_0 \in R, P_0 > \overline{P}$. Consider the solution $P_0(t) = P(t,0,P_0)$. The first moment of the jump $t = \theta_1^0$ can be found from the equation

$$P_0 e^{\alpha\theta_1^0} = \overline{P}. \tag{9.12}$$

That is $\theta_1^0 = \frac{1}{\alpha}\ln\frac{\overline{P}}{P_0}$. Then $P_0(\theta_1^0+) = \overline{P}+I$, and the second moment of the jump satisfies the equation

$$(\overline{P}+I)\mathrm{e}^{\alpha(\theta_2^0-\theta_1^0)} = \overline{P}, \tag{9.13}$$

then

$$\theta_2^0 - \theta_1^0 = \frac{1}{\alpha}\ln\frac{\overline{P}}{\overline{P}+I}.$$

We have $P_0(\theta_2^0+) = \overline{P}+I$ again. Proceeding the evaluation one can find that

$$\theta_{i+1}^0 - \theta_i^0 = \frac{1}{\alpha}\ln\frac{\overline{P}}{\overline{P}+I} \tag{9.14}$$

for all $i = 1,2,3,\ldots$. Denote $\omega = \frac{1}{\alpha}\ln\frac{\overline{P}}{\overline{P}+I}$, then

$$\theta_i^0 = \theta_1^0 + (i-1)\omega, \tag{9.15}$$

and for every $i = 1,2,3,\ldots$ we have that

$$P_0(t) = \mathrm{e}^{(t-\theta_i^0)} \text{ if } \theta_i^0 \leqslant t \leqslant \theta_{i+1}^0. \tag{9.16}$$

Now $\theta_i^0 \leqslant t \leqslant \theta_{i+1}^0$ implies that $\theta_{i+1}^0 \leqslant t+\omega \leqslant \theta_{i+2}^0$, and

$$P_0(t+\omega) = \mathrm{e}^{(t+\omega-(\theta_i^0-\omega))} = \mathrm{e}^{(t-\theta_i^0)},$$

if $t \geqslant \theta_1^0$. That is $P_0(t)$ is ultimately ω-periodic solution of (9.11).

9.1.4 *Investigation of stability*

In what follows we shall introduce the special topology in the space of piecewise continuous functions $\mathscr{PC}$ [8,9,11,20].

Let $u(t),\ v(t) : [0,\infty) \to R$, be piecewise functions, $\theta_i^u,\ \theta_i^v,\ i = 1,2,\ldots$, are strictly increasing sequences of discontinuity moments of u, v respectively, $\theta_i^u,\ \theta_i^v \to \infty$ as $i \to \infty$.

We shall say that v is in ε-neighborhood of v if:

1) $|\theta_i^u - \theta_i^v| < \varepsilon,\ \ \forall i \geqslant 1$;
2) $|u(t) - v(t)| < \varepsilon, t \in [0,\infty) \setminus \bigcup_{i=1}^{\infty}(\theta_i^u - \varepsilon, \theta_i^v - \varepsilon)$.

Definition 9.3. The solution $P_0(t)$ is stable if for arbitrary $\varepsilon > 0$ there exists $\delta = \delta(\varepsilon) > 0$ such that $|P_0 - P^0| < \delta$ implies that a solution $P(t) = P(t,0,P^0)$, $P^0 > \overline{P}$, of (9.11) is in ε-neighborhood of $P_0(t)$.

Fix $\varepsilon > 0$ such that $\varepsilon < \overline{P} + I$, and chose positive ε_1 satisfying

$$\varepsilon_1 = \min\left\{\frac{1}{\alpha}\ln\left(1 - \frac{\varepsilon}{\overline{P}+I}\right), \varepsilon\right\}. \tag{9.17}$$

Denote

$$\delta(\varepsilon_1) = \min\left(1 - \mathrm{e}^{-\varepsilon_1|\alpha|}, 1 - \mathrm{e}^{-\frac{\omega}{4}|\alpha|}, \varepsilon\right), \tag{9.18}$$

and let θ_i, $i \geqslant 1$, be a sequence of discontinuity points of $P(t)$. Then

$$|\theta_1 - \theta_1^0| = \left|\frac{1}{\alpha}\ln\frac{\overline{P}}{P_0} - \frac{1}{\alpha}\ln\frac{\overline{P}}{P^0}\right| = \frac{1}{|\alpha|}\left|\ln\frac{P^0}{P_0}\right| < \varepsilon_1. \tag{9.19}$$

Assume that $P^0 - P_0 > 0$. Last inequality implies that $-\varepsilon_1|\alpha| < \ln\frac{P^0}{P_0} < \varepsilon_1|\alpha|$, and, then $P_0(-1 + \mathrm{e}^{-\varepsilon_1|\alpha|}) < P^0 - P_0 < P_0(\mathrm{e}^{\varepsilon_1|\alpha|} - 1)$. One can see that $\mathrm{e}^{\varepsilon_1|\alpha|} - 1 \geqslant 1 - \mathrm{e}^{-\varepsilon_1|\alpha|}$. That is (9.18) implies (9.19).

Similarly one can check that the inequality is true if $P^0 - P_0 < 0$, and the inequality $|\theta_1 - \theta_1^0| < \frac{\omega}{4}$ follows (9.18). Assume, without loss of generality, that $\theta_1 < \theta_1^0$. Hence, $\theta_i < \theta_i^0$ for all i, and $\theta_i^0 - \theta_i = \theta_1^0 - \theta_1 < \min(\varepsilon_1, \frac{\omega}{4})$. If $t \in [\theta_i^0, \theta_{i+1}]$ then

$$\begin{aligned}|P_0(t) - P(t)| &= |(\overline{P}+I)\mathrm{e}^{\alpha(t-\theta_i^0)} - (\overline{P}+I)\mathrm{e}^{\alpha(t-\theta_i)}| \\ &= (\overline{P}+I)\mathrm{e}^{\alpha(t-\theta_i)}|\mathrm{e}^{-\alpha(\theta_i^0-\theta_i)} - 1| \\ &< (\overline{P}+I)[1 - \mathrm{e}^{\alpha\varepsilon_1}].\end{aligned} \tag{9.20}$$

Last inequality implies that

$$|P_0(t) - P(t)| < \varepsilon, \ t \in [\theta_i^0, \theta_{i+1}], \ i = 1, 2, 3, \ldots, \tag{9.21}$$

if $(\overline{P}+I)[1 - \mathrm{e}^{\alpha\varepsilon_1}] \leqslant \varepsilon$ or $\varepsilon_1 \leqslant \frac{1}{\alpha}\ln(1 - \frac{\varepsilon}{\overline{P}+I})$. Moreover, in $[0, \theta_1]$ we have that

$$|P_0(t) - P(t)| = |P^0\mathrm{e}^{\alpha t} - P_0\mathrm{e}^{\alpha t}| = \mathrm{e}^{\alpha t}|P^0 - P_0| \leqslant |P^0 - P_0| < \delta. \tag{9.22}$$

Comparing (9.18)–(9.22) we can find that $P(t)$ is lying in ε-neighborhood of $P_0(t)$ if $\delta > 0$ satisfies (9.18). That is $P_0(t)$ is stable.

9.1.5 *Conclusion*

In this section, we have modeled the behavior of systemic arterial pressure using the theory of impulsive differential equations and investigated these models. Problems (a)–(g) introduced in Subsections 9.1.2 and 9.1.3 have been considered for the first time in systemic arterial pressure modeling. Analysing the results of (a)–(g), one can conclude that the models permit regular (periodic) oscillations as well as nonregular oscillations. In the latter case they are almost periodic or ε-oscillations and might be useful to model irregular

arterial pulses. The approach we have proposed in this section might be used for investigation of arrhythmias and elaboration of the electrostimulation technique.
The models that we have described here do not reflect the true behavior of systemic arterial pressure but an approximate behavior in some sense. Further studies will include more realistic modeling of systemic arterial pressure . The results of the section have been announced in [32].

9.2 The global model

9.2.1 *Introduction and Preliminaries*

In this section we consider a system of differential equations the behavior of which solutions possesses several properties characteristic of the blood pressure distribution. The system can be used for a compartmental modeling [57, 175] of the cardiovascular system. It admits a unique bounded solution such that all coordinates of the solution are separated from zero by positive numbers, and which is periodic, eventually periodic or almost periodic depending on the moments of heart contraction. Appropriate numerical simulations are provided.
A living organism can be considered as a complex system of nonlinear oscillating structures of different origins. The system as an integrity is a "constellation" of all of its oscillations. The oscillators are connected with each other to build oscillating chains. The essential task of controlling processes in a living organism is to sustain the oscillating activity of certain significantly unstable elements in such a way, that the amplitudes of their oscillations do not become excessively large or excessively small. It is, therefore, natural to assume that the data of oscillations should be considered as one of the most important sources of information about the condition of the organism.
The movement of any fluid in a certain direction is the result of pressure differences in the starting and ending points of the system. The difference of pressure in heart chambers and in aorta, in aorta and outgoing blood vessels is the cause of blood flow in an organism. The last decades there has been a rise of intensive investigations of blood pressure in its connections with other parameters of the cardiovascular system [1,32], [50,57,71,120,124], [162,202,229], [243,247,261], [293]. At the same time we should recognize that due to the nonlinear properties of blood vessel walls and their variability along the length of a vessel the relation between actual oscillations of pressure and of blood volume [120], [229], [261] is very complex, and consequently, it is reasonable to consider the distribution of blood

pressure out of it's connection with other parameters, particularly blood volume. There are many different approaches to the problem of modeling cardiovascular dynamics including, for example, modeling of the coupling mechanisms as a system of differential equations with delay [71,129] or as a system of ordinary differential equations [166,202,293].

The idea of compartments makes analysis of a real world problem more abstract. Accordingly, we neither describe the physiological meaning of the compartments nor specify the number of these parts. The approach, which is based on introduction of compartments have been effectively applied to investigate different type processes in chemistry [244], medicine [47,139,175,191], epidemiology [237], ecology [223], pharmacokinetics [47,57,271,292].

On can easily see that all parts of the cardiovascular system should be involved if, let say, the blood flow problem is investigated, because the total volume of the blood is constant in the time, but the pressure could be discussed locally, as, for example, in [166, 202], where blood pressure for only compartment is studied. In our investigation we focus on the interactions of the blood pressures of several parts of the system. One can suppose that there exists, as a consequence of the blood float, a mutual interaction of blood pressures in different parts of the system. One of them, the systemic arterial pressure, should be singled out, as it is disturbed impulsively with large simultaneous inflow of blood in the time of heart contraction. Other compartments of the system do not directly undergo the impulsive action, so their change is continuous. We want to note that in this section we consider exclusively the aorta and the parts of the cardiovascular system interacting with the aorta through blood pressure, that is, only the compartments belonging to the systemic circulation are considered. The blood pressure in pulmonary artery also changes impulsively, and the differential equations system we propose can therefore be also used to model the pressure in this artery and its neighboring compartments. We, however, leave it out of our discussion.

The goal of our analysis is not to build a final model of cardiovascular system, but rather to introduce a system of differential equations, in which one coordinate is perturbed impulsively and other coordinates exhibit a continuous change, thus having certain properties characteristic of the actual blood pressure behavior. The obtained results now need qualitative experimental confirmation. We hope that our results will serve as a basis for further experimental and clinical studies, which can lead to better understanding of the cardiovascular system.

We denote C_0 the aorta and arteries, and P_0 blood pressure in C_0. Let C_i, $i = \overline{1,m}$, be other vessels and organs, which adjoin to C_0, and have an essential influence on fluctuations of

P_0, and P_i, $i=\overline{1,m}$, are the blood pressure values in compartments C_i, $\overline{1,m}$ (see Figure 1.3). We suppose that the pressure variables satisfy the following system of differential equations with impulses

$$\begin{aligned}
\frac{dP_0(t)}{dt} &= -k_0P_0 - g_0(P_0-P_1, P_0-P_2, \ldots, P_0-P_m),\\
\frac{dP_1(t)}{dt} &= -k_1P_1 + g_1(P_0-P_1),\\
\frac{dP_2(t)}{dt} &= -k_2P_2 + g_2(P_0-P_2),\\
&\ldots\ldots\\
\frac{dP_m(t)}{dt} &= -k_mP_m + g_m(P_0-P_m),\\
\Delta P_0|_{t=\theta_i} &= I_0 + J_0(P_0),
\end{aligned} \tag{9.23}$$

where $\Delta P_0|_{t=\theta_i} \equiv P_0(\theta_i+) - P_0(\theta_i)$, $P_0(\theta_i+) = \lim_{t\to\theta_i+} P_0(t)$.
The following assumptions for (9.23) will be needed throughout the section:

(C1) real constants $I_0, k_i, i=\overline{0,m}$, are positive;

(C2) J_0, g_i are real valued functions, $J_0(0)=0$, $g_i(0)=0$, $i=\overline{1,m}$, $J_0(z)>0$, $g_i(z)>0$, if $z>0$, $g_0(0,0,\ldots,0)=0$, $g_0(z_1,z_2,\ldots,z_m)>0$, if $z_j>0$, $j=\overline{1,m}$;

(C3) the functions J_0, g_i, $\overline{0,m}$, satisfy the Lipschitz condition

$$\begin{aligned}
&|g_i(z^1)-g_i(z^2)| \leqslant l_i|z^1-z^2|, \quad i=\overline{1,m},\\
&|g_0(z_1,\ldots,z_m)-g_0(w_1,\ldots,w_m)| \leqslant l_0\sqrt{\sum_{i=\overline{1,m}} |z_i-w_i|^2},\\
&|J_0(z^1)-J_0(z^1)| \leqslant l_J|z^1-z^2|;
\end{aligned} \tag{9.24}$$

(C4) there exists a number $\omega>0$, such that $i\omega \leqslant \theta_i < (i+1)\omega$, $i\in\mathbb{Z}$;

(C5) there exist positive real constants m_i, $i=\overline{0,m}, m_J$ such that

$$\sup_{z\geqslant 0} g_i(z) = m_i, \quad i=\overline{1,m}, \qquad \sup_{z_j\geqslant 0,\ j=\overline{1,m}} g_0(z_1,\ldots,z_m) = m_0, \quad \sup_{z\geqslant 0} J_0(z) = m_J. \tag{9.25}$$

We assume that atmospheric pressure has the zero value. Since pressures can not be negative, our main goal is to find the conditions which guarantee the existence of bounded positive solutions. Moreover, the coordinates of the solutions are separated from zero by some positive constants, and eventually the first coordinate must be larger than any other. The solutions should be exponentially stable under certain conditions, and their various oscillatory properties (periodicity, eventually periodicity, almost periodicity) correspondly to the particular properties of the sequence of moments θ_i.

Differential equations (9.23) as a model of the physiological process are developed using models, which can be found, for example, in [57, 71, 162, 202, 247] and many others. To clarify that, we shall describe the two following models.
In paper [202] the following differential equation is considered for the peripheric blood pressure P_p

$$\frac{dP_p}{dt} = -\frac{1}{RC}P_p + \frac{1}{C}Q(t), \tag{9.26}$$

where C is the total arterial compliance, $R-$ the arterial resistance, $Q(t)-$ the continuous blood flow. The formula for the continuous blood flow evaluation is given by $Q(t) = \frac{P_0(t)-P_p(t)}{r}$, where P_0- is the systemic arterial pressure, and $r-$ is the aortic impedance. As we suppose that the systemic arterial pressure is larger than the blood periferic pressure, $Q(t)$ is a positive function.
For the systemic arterial pressure the following differential equation with impulses has been discussed in book [166],

$$\begin{aligned} \frac{dP_0}{dt} &= -\frac{1}{RC}P_0, t \neq \theta_i, \\ P_0(\theta_i+) - P_0(\theta_i) &= \frac{V_j}{C}, \end{aligned} \tag{9.27}$$

where $\theta_i = iT$, $i \in \mathbb{Z}$, are prescribed moments, $P_0(\theta_i+)$ is the right limit's value, V_j are the stroke volumes, and C is the systemic arteria compliance.
It is natural to consider cardiovascular problems using discontinuous dynamics theory. One of the interesting approaches which should be mentioned is the method of circle mappings [50, 124] for cardiac arrhythmias. In [32] we consider the extended form of equation (9.27) for the systemic arterial pressure, and we investigate the problem of existence of positive periodic and almost periodic solutions, oscillations of the equation. The case when the moments of heart contraction are not prescribed, but are caused by a certain value of the systemic arterial pressure has also been investigated.
We have further developed the models of our predecessors in several ways:

1) The nonlinear perturbations caused by interactions are used in the model, both in differential equations and the impulsive part of the model. Function $g_0(P_0 - P_1, P_0 - P_2, \ldots, P_0 - P_m)$ is positive since the outflow from the aorta to neighboring compartments reduces the systemic arterial pressure, all differences in the function are assumed to be positive. Nonlinearities $g_i(P_0 - P_i)$, $i = \overline{1,m}$, are positive, when the differences again are positive, since the differences generate inflow into compartments, which naturally implies the rise of the pressure. Any function $g_i(P_0 - P_i)$ does not depend on P_j,

$j \neq i$, since we assume that there is no essential mutual interaction between compartments;

2) The equations for the compartments are derived from equation (9.26) for peripheric blood pressure. The nonnegative differences $\frac{P_0(t)-P_i(t)}{r_i}$ are replaced by the nonlinear functions $g_i(P_0 - P_i)$;
3) I_0 is a constant part of the instantaneous change of systemic arterial pressure;
4) Condition (C5) is chosen, assuming that the results obtained in our investigation can be interpreted for the blood pressure system activated by an artificial pacemaker, or for the case when fluctuations are regular. That is, we exclude the case when the set of points of discontinuity may have so called accumulation points.

We must also note that our results can be easily reconsidered when condition (C5) is replaced by condition $0 < \mu \leqslant \theta_{i+1} - \theta_i \leqslant \mu_1 < \infty$ or by the condition of existence of $\limsup_{T\to\infty} \frac{i(t_0,t_0+T)}{T} = p$, where $i(t_0,t_0+T)$ is the number of elements θ_i in the interval (t_0,t_0+T), and p is a nonnegative real number, uniform for all $t_0 \in \mathbb{R}$.
The section is organized in the following manner. In the next part we find the conditions for the equation such that there exists a unique solution bounded on the whole $\mathbb{R}$. The solution is periodic, almost periodic, or eventually periodic if the sequence of discontinuity moments has an appropriate property. Subsection contains the results on the stability and positiveness of the solutions. In the last part we specify the choice of the moments of discontinuity such that they may be determined artificially by a particular map, and by the initial moment. In this case the approach can be considered as a way to discuss the pacemaker's design. Appropriate numerical simulations are provided.

9.2.2 *Bounded, periodic, eventually periodic and almost periodic solutions*

In this section we shall prove that the system admits oscillating solutions. The existence of an eventually periodic solution is proved for the first time. The assertions about the discontinuous periodic and almost periodic solutions are specified versions of general theorems of the theory of impulsive differential equations [20,42,192,278].
Firstly, let us define the solutions of system (9.23).
A left continuous function $P(t): \mathbb{R} \to \mathbb{R}^{m+1}$ is in the set $\mathscr{PC}(\mathbb{R})$, if it is continuous on $\mathbb{R}$, except at the points θ_i, $i \in \mathbb{Z}$, where its first coordinate may have discontinuities of the first kind, and is left continuous.
A function $P(t) \in \mathscr{PC}(\mathbb{R})$ is a solution of (9.23) if:

(1) the differential equation is satisfied for $P(t), t \in \mathbb{R}$, except at the points θ_i, $i \in Z$, where it holds for the left derivative of $P(t)$;

(2) the jump equation is satisfied by $P_0(t)$ for every $i \in \mathbb{Z}$.

We shall also need the following set of functions.

A left continuous function $P(t)$ is in $\mathscr{PC}(\mathbb{R}_+), \mathbb{R}_+ = [t_0, \infty)$, $t_0 \in \mathbb{R}$, if it is continuous on $[t_0, \infty)$ except at the points θ_i, $t_0 \leqslant \theta_i < \infty$, where its first coordinate may have discontinuities of the first kind.

A solution $P(t) = (P_0(t), P_1(t), \ldots, P_m(t))$ of (9.23) on $[t_0, \infty)$ is a function in $\mathscr{PC}(\mathbb{R}_+)$ such that:

(1) the differential equation is satisfied for $P(t)$ on $\mathbb{R}_+$, except at the points θ_i, $\theta_i > t_0$, where it holds for the left derivative of $P(t)$;

(3) the jump equation is satisfied by $P_0(t)$ for every i, $\theta_i > t_0$.

One can easily show, using the theory of impulsive differential equations [20, 192, 278], that under the aforementioned conditions a solution $P(t, t_0, \pi_0)$, $(t_0, \pi_0) \in \mathbb{R} \times \mathbb{R}^{m+1}$, $\pi_0 = (\pi_0^0, \pi_0^1, \ldots, \pi_0^m)$, of (9.23) exists and is unique on $[t_0, \infty)$.

One can also find that the solution satisfies the following integral equation

$$
\begin{aligned}
P_0(t) &= \mathrm{e}^{-k_0(t-t_0)}\pi_0^0 - \int_{t_0}^{t} \mathrm{e}^{-k_0(t-s)} g_0(P_0(s)-P_1(s), P_0(s)-P_2(s), \ldots, P_0(s)-P_m(s))ds \\
&\quad + \sum_{t_0 \leqslant \theta_i < t} \mathrm{e}^{-k_0(t-\theta_i)}(I_0 + J_0(P_0(\theta_i))), \\
P_i(t) &= \mathrm{e}^{-k_i(t-t_0)}\pi_0^i + \int_{t_0}^{t} \mathrm{e}^{-k_i(t-s)} g_i(P_0(s)-P_i(s))ds, i = \overline{1,m}.
\end{aligned} \tag{9.28}
$$

We say that a sequence θ_i has the *p-property*, $p \in \mathbb{N}$, if $\theta_{i+p} = \theta_i + p\omega$ for all integers i. Moreover, we say that the sequence has the p-property eventually, if $\theta_{i+p} - \theta_i \to p\omega$ as $i \to \infty$.

Consider a strictly ordered sequence of real numbers $t_i, i \in \mathbb{Z}$. Denote $t_j^i = t_{i+j} - t_i$, $i,\ j \in \mathbb{Z}$, and define the sequences $\{t_j^i\}_i$, $i,\ j \in \mathbb{Z}$. Following [42, 148, 278] we call this family of sequences equipotentially almost periodic if for an arbitrary positive ε there exists a relatively dense set of $\varepsilon-$almost periods, common for all sequences $\{t_j^i\}$, $j \in \mathbb{Z}$.

It is proved in [42] (see also [278]) that the family θ_j^i, $i,\ j \in \mathbb{Z}$, is equipotentially almost periodic if the sequence $\theta_i - i\omega$, $i \in \mathbb{Z}$, is almost periodic. That is, if we chose an almost periodic sequence ξ_i, $i \in \mathbb{Z}$, such that $0 \leqslant \xi_i < \omega$, $i \in \mathbb{Z}$, then the sequence θ_i, where $\theta_i = i\omega + \xi_i$, $i \in \mathbb{Z}$, is equipotentially almost periodic.

Let $\widehat{[a,b]}$ be an oriented interval, that is $\widehat{[a,b]} = [a,b]$, if $a \leqslant b$, and $\widehat{[a,b]} = [b,a]$, if $a > b$. We say that a function $\phi(t)$ in $\mathscr{PC}(\mathbb{R})$ is eventually $p\omega$-periodic if the sequence of discontinuities θ_i has the p-property eventually, and

$$\lim_{t\to\infty}[\phi(t+p\omega)-\phi(t)] = 0, \tag{9.29}$$

for all t such that $t \notin \widehat{[\theta_i, \theta_{i+p} - p\omega]} \cap \widehat{[\theta_{i+1}, \theta_{i+p+1} - p\omega]}$.
Next, we shall give a definition from the last section (see, also [148,278]) which is slightly modified for our case.
A function $\phi(t)$ in $\mathscr{PC}(\mathbb{R})$ is a piecewise continuous almost periodic function if:

(a) the function is uniformly continuous on the union of all intervals of the continuity (θ_i, θ_{i+1}), $i \in \mathbb{Z}$;
(b) for every positive ε there exists a relatively dense set Γ of almost periods such that if $\gamma \in \Gamma$, then $\|\phi(t+\gamma) - \phi(t)\| < \varepsilon$ for all $t \in \mathbb{R}$ such that $|t - \theta_i| > \varepsilon$, $i \in \mathbb{Z}$.

We may assume that

(C6) $L(l_0, l_1, \ldots, l_m, l_J) = \sqrt{\left[\frac{l_0\sqrt{2m}}{k_0} + \frac{l_J e^{k_0\omega}}{1-e^{-k_0\omega}}\right]^2 + \sum_{i=1}^{m} \frac{l_i^2}{k_i^2}} < 1.$

Theorem 9.1. *If conditions* (C1)–(C6) *are fulfilled, then there exists a unique solution of* (9.23) *bounded on* $\mathbb{R}$ *and:*

(1) *the bounded solution has the period* $p\omega$ *if the sequence* θ_i *has the p-property for a fixed* $p \in \mathbb{N}$;
(2) *the bounded solution is an eventually* $p\omega$*-periodic function if the sequence* θ_i *has the p-property eventually for a fixed* $p \in \mathbb{N}$;
(3) *the bounded solution is a piecewise continuous almost periodic functionif the sequence* $\theta_i - i\omega$, $i \in \mathbb{Z}$, *is almost periodic.*

Proof. We shall use the norm $|\phi|_0 = \sup_{\mathbb{R}} |\phi(t)|$ for scalar valued functions defined on $\mathbb{R}$, and $\|\phi\|_0 = \sup_{\mathbb{R}} \|\phi(t)\|$ for vector-valued functions. Denote by Ω a subset of $\mathscr{PC}(\mathbb{R})$ such that if $\phi(t) \in \Omega$, $\phi = (\phi_0, \phi_1, \ldots, \phi_m)$, then $|\phi_0|_0 \leqslant (\frac{m_0}{k_0} + \frac{m_J e^{k_0\omega}}{1-e^{-k_0\omega}})$, $|\phi_i|_0 \leqslant \frac{m_i}{k_i}, i = \overline{1,m}$. One can easily verify that (9.23) has a bounded solution $P(t)$ if and only if $P(t)$ is a bounded on $\mathbb{R}$ solution of the following integral equation

$$P_0(t) = -\int_{-\infty}^{t} e^{-k_0(t-s)} g_0(P_0(s)-P_1(s),\dots,P_0(s)-P_m(s))ds$$
$$+\sum_{\theta_i<t} e^{-k_0(t-\theta_i)}(I_0+J_0(P_0(\theta_i))),$$
$$P_i(t) = \int_{-\infty}^{t} e^{-k_i(t-s)} g_i(P_0(s)-P_i(s))ds,\ i=\overline{1,m}. \tag{9.30}$$

Define on Ω an operator T such that

$$(T\phi)_0(t) = -\int_{-\infty}^{t} e^{-k_0(t-s)} g_0(\phi_0(s)-\phi_1(s),\dots,\phi_0(s)-\phi_m(s))ds$$
$$+\sum_{\theta_i<t} e^{-k_0(t-\theta_i)}(I_0+J_0(\phi_0(\theta_i))),$$
$$(T\phi)_i(t) = \int_{-\infty}^{t} e^{-k_i(t-s)} g_i(\phi_0(s)-\phi_i(s))ds,\ i=\overline{1,m}. \tag{9.31}$$

We can verify that $T:\Omega\to\Omega$. Take $\phi,\ \psi\in\Omega$. One can obtain that

$$\|T\phi-T\psi\|_0 \leqslant L(l_0,l_1,\dots,l_m,l_J)\|\phi-\psi\|_0,$$

and, consequently, condition (C6) implies that the operator is contractive. It is not difficult to check that the space Ω is complete. Thus, there exists a unique solution of (9.23) from Ω. Assume now that the sequence θ_i has the p-property. Then, using the standard method, starting with a $p\omega$-periodic function from Ω, we can, using the operator T, construct a sequence of $p\omega$-periodic approximations of the bounded solution, which is also $p\omega$-periodic. Consider the case where θ_i has the p-property eventually. In order to prove assertion (2) of the theorem, it is sufficient, according to the above discussion of boundedness of the solution, to prove that $T\phi$ is eventually $p\omega$-periodic, if ϕ is an eventually $p\omega$-periodic function.

Fix a positive number ε. Since θ_i has the p-property eventually, and ϕ is eventually $p\omega$-periodic function, one can find positive numbers $T_2>T_1$ such that:

(1) if $t\geqslant T_2$ and $\theta_i+\varepsilon<t<\theta_{i+1}-\varepsilon$ for some $i\in\mathbb{Z}$, then $\theta_{i+p}<t+p\omega<\theta_{i+1+p}$ and $\|\phi(t+p\omega)-\phi(t)\|<\varepsilon$;

(2) $2e^{-k_0(T_2-T_1)}[\frac{m_0}{k_0}+\frac{m_J e^{k_0\omega}}{1-e^{-k_0\omega}}]<\varepsilon,\ 2e^{-k_0(T_2-T_1)}\frac{2m_i}{k_i}<\varepsilon.$

We have that

$$|(T\phi)_0(t+p\omega)-(T\phi)_0(t)| = \Big|-\int_{-\infty}^{t+p\omega} e^{-k_0(t+p\omega-s)} g_0(\phi_0(s)-\phi_1(s),\dots,\phi_0(s)-\phi_m(s))ds$$

$$+\sum_{\theta_i<t+p\omega} e^{-k_0(t+p\omega-\theta_i)}(I_0+J_0(\phi_0(\theta_i)))+\int_{-\infty}^{t} e^{-k_0(t-s)}g_0(\phi_0(s)-\phi_1(s),\dots,\phi_0(s)-\phi_m(s))ds$$

$$-\sum_{\theta_i<t} e^{-k_0(t-\theta_i)}(I_0+J_0(\phi_0(\theta_i)))\Big| \leqslant \int_{-\infty}^{T_1} e^{-k_0(t-s)}2m_0 ds+$$

$$\sum_{\theta_i<T_1} e^{-k_0(t-\theta_i)}2m_J+\int_{T_1}^{t} e^{-k_0(t-s)}l_0\sqrt{2m}\|\phi(s+p\omega)-\phi(s)\|ds+$$

$$\sum_{T_1\leqslant\theta_i<t} e^{-k_0(t-\theta_i)}l_J\|\phi(\theta_{i+p})-\phi(\theta_i)\|+\sum_{\theta_i<t}\int_{\theta_i-\varepsilon}^{\theta_i+\varepsilon} e^{-k_0(t-s)}2m_0 ds \leqslant$$

$$\varepsilon\left[1+\frac{l_0\sqrt{2m}}{k_0}+\frac{e^{\kappa\omega}}{1-e^{-\kappa\omega}}(l_J+4m_0e^{k_0\varepsilon})\right].$$

Similarly, one can obtain that

$$|(T\phi)_i(t+p\omega)-(T\phi)_i(t)| \leqslant \varepsilon\left[1+\frac{l_i}{k_i}+4m_ie^{k_0\varepsilon}\frac{e^{\kappa\omega}}{1-e^{-\kappa\omega}}\right],\quad i=\overline{1,m}.$$

That is, there exists a unique eventually $p\omega$-periodic solution of (9.23).

Now assume that the sequence $\theta_i-i\omega$ is almost periodic. From the previous discussion we see that to prove the existence of a d.a.p. solution we need only to verify that the function $T\phi$ is piecewise continuous almost periodic if ϕ is a piecewise continuous almost periodic function. By Lemma 35 [278] (see also [42]), for a given positive ε there exist a real number ν, $0<\nu<\varepsilon$, and relatively dense sets of real numbers Γ and integers H, such that:

(1) $\|\phi(t+\gamma)-\phi(t)\|<\varepsilon$;
(2) $|\theta_k^h-\gamma|<\nu$, $k\in\mathbb{Z}$, $h\in H$, $\gamma\in\Gamma$.

Take the real number ν such that $\|\phi(t_1)-\phi(t_2)\|<\varepsilon$ if t_1, t_2 belong to the same interval of continuity of the function $\phi(t)$ and $|t_1-t_2|<\nu$. We have that if $|t-\theta_i|>\varepsilon$, $i\in\mathbb{Z}$, then

$$|(T\phi)_0(t+\gamma)-(T\phi)_0(t)| = |\int_{-\infty}^{t+\gamma} e^{-k_0(t+\gamma-s)}g_0(\phi_0(s)-\phi_1(s),\dots,\phi_0(s)-\phi_m(s))ds+$$

$$\sum_{\theta_i<t+\gamma} e^{-k_0(t+\gamma-\theta_i)}(I_0+J_0(\phi(\theta_i)))-\int_{-\infty}^{t} e^{-k_0(t-s)}g_0(\phi_0(s)-\phi_1(s),\dots,\phi_0(s)-\phi_m(s))ds-$$

$$\sum_{\theta_i<t} e^{-k_0(t-\theta_i)}(I_0+J_0(\phi(\theta_i)))| \leqslant \int_{-\infty}^{t} e^{-k_0(t-s)}l_0\sqrt{2m}\|\phi(s+\gamma)-\phi(s)\|ds+$$

$$\sum_{\theta_i<t} e^{-k_0(t-\theta_i)}l_J\|\phi(\theta_{i+h})-\phi(\theta_i)\|+\sum_{\theta_i<t}\int_{\theta_i-\varepsilon}^{\theta_i+\varepsilon} e^{-k_0(t-s)}2m_0 ds \leqslant$$

$$\varepsilon\left[\frac{l_0\sqrt{2m}}{k_0}+\frac{2e^{k_0\omega}}{1-e^{-k_0\omega}}(l_J+4m_0e^{k_0\varepsilon})\right]$$

and

$$|(T\phi)_i(t+\gamma)-(T\phi)_i(t)| \leqslant \varepsilon\left[\frac{l_i}{k_i}+4m_ie^{k_i\varepsilon}\frac{e^{k_i\omega}}{1-e^{-k_i\omega}}\right],\quad i=\overline{1,m}.$$

That is, $T\phi$ is a piecewise continuous almost periodic function. The theorem is proved. □

9.2.3 *Stability and positiveness*

In this section we show that the bounded solution is exponentially stable, and that every coordinate of the solution is separated from zero by some positive number. Moreover, the first coordinate of the solution, which is the value of the systemic arterial pressure, is higher than any other pressure value.

Let us denote the solution bounded on $\mathbb{R}$ as $\xi(t) = (\xi_0(t), \xi_1(t), \ldots, \xi_m(t))$. Next, we find the conditions for the positiveness of this solution. Assume additionally that

(C8) $\dfrac{m_0}{k_0} + \dfrac{m_i}{k_i} < I_0 \dfrac{\mathrm{e}^{-k_0\omega}}{1-\mathrm{e}^{-k_0\omega}}, \quad i = \overline{1,m}.$

Using (9.30) we obtain that

$$\xi_0(t) \geqslant I_0 \frac{\mathrm{e}^{-k_0\omega}}{1-\mathrm{e}^{-k_0\omega}} - \frac{m_0}{k_0} > 0, \quad t \in \mathbb{R},$$

and

$$|\xi_i(t)| \leqslant \frac{m_i}{k_i}, \quad t \in \mathbb{R}, \quad i = \overline{1,m}.$$

Hence,

$$\xi_0(t) - \xi_i(t) \geqslant I_0 \frac{\mathrm{e}^{-k_0\omega}}{1-\mathrm{e}^{-k_0\omega}} - \frac{m_0}{k_0} - \frac{m_i}{k_i} = \delta > 0, \quad t \in \mathbb{R}, \quad i = \overline{1,m}, \tag{9.32}$$

and using (9.30) again one can see that

$$\xi_i(t) \geqslant \frac{\overline{g}_i}{k_i} > 0, \quad t \in \mathbb{R}, \quad i = \overline{1,m},$$

where $\overline{g}_i$ is the minimal value of the function $g_i(z)$ for $z \in [\delta, \frac{m_0}{k_0} + \frac{m_J \mathrm{e}^{k_0\omega}}{1-\mathrm{e}^{-k_0\omega}} + \frac{m_i}{k_i}]$.

The following assertion is proved.

Theorem 9.2. *Assume that conditions* (C1)–(C8) *are valid. Then for the bounded solution* $\xi(t)$ *of* (9.23) *there exist positive constants* ν_j, $j = \overline{1,m}$, μ_i, $i = \overline{0,m+1}$ *such that the inequalities* $\xi_0(t) - \xi_i(t) > \nu_j$, $\xi_i(t) \geqslant \mu_i$ *are valid for all* i, j, *and* $t \in \mathbb{R}$.

Let us give the definition of uniform exponential stability of the solution. Denote $P(t) = P(t, t_0, P_0)$, a solution of (9.23).

Definition 9.4. The solution $\xi(t)$ is called uniformly exponentially stable if there exists a number $\alpha \in \mathbb{R}, \alpha > 0$, such that for every $\varepsilon > 0$ there exists a number $\delta = \delta(\varepsilon)$ such that the inequality $\|P(t) - \xi(t)\| < \varepsilon \exp(-\alpha(t-t_0))$, for all $t \geqslant t_0$, holds, if $\|P_0 - \xi(t_0)\| < \delta$.

Fix a positive number σ, $0 < \sigma < \min_{i=\overline{0,m}} k_i$, denote

$$m(l_0, l_1, \ldots, l_m, l_J) = 1 - \max\left\{ \frac{l_0\sqrt{2m}}{k_0 - \sigma} + \frac{l_J \mathrm{e}^{(k_0-\sigma)\omega}}{1-\mathrm{e}^{-(k_0-\sigma)\omega}}, \frac{2l_i}{k_i - \sigma}, \quad i = \overline{0,m} \right\},$$

and assume that the Lipschitz coefficients are small so that

(C9) $m(l_0,l_1,\dots,l_m,l_J)>0$.

Theorem 9.3. *Assume that conditions* (C1)–(C7), (C9) *are valid. Then the bounded solution* $\xi(t)$ *of* (9.23) *is uniformly exponentially stable.*

Proof. One can see that $v(t)=P(t)-\xi(t), v=(v_0,v_1,\dots,v_m)$, is a solution of the equation

$$\begin{aligned}&\frac{dv_0}{dt}=-k_0v_0+w_0(v),\\&\frac{dv_i}{dt}=-k_iv_i+w_i(v_0,v_i),\ \ i=\overline{1,m},\ \ t\neq\theta_j,\\&\Delta v_0|_{t=\theta_j}=u_0(v_0).\end{aligned}\tag{9.33}$$

where functions w_i, $i=\overline{0,m}$, and u_0 satisfy the following conditions:

$$|w_0(v)|\leqslant l_0\sqrt{2m}\|v\|;$$

$$|w_i(v_i,v_0)|\leqslant l_i(|v_i|+|v_0|),\ \ i\in\mathbb{Z};$$

$$|u_0(v_0)|\leqslant l_J|v_0|.$$

Thus, the problem of the stability of $\xi_0(t)$ is reduced to the stability of the zero solution of (9.33). Fix $\varepsilon>0$ and denote $K=K(l_0,l_1,\dots,l_m,l_J,\delta)=\frac{\delta}{m(l_0,l_1,\dots,l_m,l_J)}$, where $\delta\in\mathbb{R}$, $\delta>0$, and take δ so small that $K(l_0,l_1,\dots,l_m,l_J,\delta)<\varepsilon$. Assume, without loss of generality, that $t_0=0$. Let $v(t,v_0)$ be a solution of (9.33) such that $v(0,v_0)=v_0=(v^0,v^1,\dots,v^m)$.
Denote by Ψ a set of all functions $\psi=(\psi_0,\psi_1,\psi_2,\dots,\psi_m)$, defined on $\mathbb{R}_+=[0,\infty)$, such that:

1) $\psi(0)=v_0$;
2) $\psi(t)\in\mathscr{PC}(R_+)$;
3) $|\psi_i(t)|\leqslant K\exp(-\sigma t)$ if $t\geqslant 0$, $i=\overline{0,m}$.

Define an operator Π on Ψ such that if $\psi\in\Psi$, then

$$(\Pi\psi)_0=\mathrm{e}^{-k_0t}v^0+\int\limits_0^t\mathrm{e}^{-k_0(t-s)}w_0(\psi(s))ds+\sum_{0\leqslant\theta_i<t}\mathrm{e}^{-k_0(t-\theta_i)}u_0(v_0(\theta_i)),$$

$$(\Pi\psi)_i=\mathrm{e}^{-k_it}v^i+\int\limits_0^t\mathrm{e}^{-k_i(t-s)}w_i(v_0(s),v_i(s))ds,\ \ i=\overline{1,m}.\tag{9.34}$$

We shall show that $\Pi:\Psi\to\Psi$. Indeed, for $t\geqslant 0$ it is true that

$$|(\Pi\psi)_0|\leqslant\mathrm{e}^{-k_0t}\delta+\int_0^t\mathrm{e}^{-k_0(t-s)}l_0\sqrt{2m}K\mathrm{e}^{-\sigma s}ds$$

$$+\sum_{0\leqslant\theta_i<t}\mathrm{e}^{-k_0(t-\theta_i)}l_JK\mathrm{e}^{-\sigma\theta_i}\leqslant\mathrm{e}^{-\sigma t}\left[\delta+K\left(\frac{l_0\sqrt{2m}}{k_0-\sigma}+\frac{l_J\mathrm{e}^{(k_0-\sigma)\omega}}{1-\mathrm{e}^{-(k_0-\sigma)\omega}}\right)\right]\leqslant K\mathrm{e}^{-\sigma t}.$$

Similarly,

$$|(\Pi\psi)_i| \leqslant \exp(-\sigma t)\left[\delta + 2K\frac{l_i}{k_i-\sigma}\right] \leqslant K\exp(-\sigma t), \;\; i=\overline{1,m}.$$

Let $\psi_1, \psi_2 \in \Psi_\eta$. Then, we have that

$$\|\Pi\psi_1 - \Pi\psi_2\| \leqslant L(l_0, l_1, \ldots, l_m, l_J)\sup_{t\geqslant 0}\|\psi_1 - \psi_2\|,$$

where the coefficient is described in (C7). Using the contraction mapping argument one can conclude that there exists a unique fixed point $v(t, v_0)$ of the operator $\Pi : \Psi \to \Psi$ which is a solution of (9.33). Theorem is proved. □

From (9.32) it follows that the bounded solution has the first coordinate larger than any other coordinate, and the attractiveness of the solution implies that any other solution in its neighbourhood eventually has the first coordinate as its largest coordinate. Now, we can make some physiological conclusions: the normal state of the distribution of blood pressure is such that systemic arterial pressure is higher than any other pressure. The initial state may be odd, that is, the initial systemic arterial pressure may be lower than the pressure in a neighbouring compartment, but after a certain period of time the state becomes normal. We may call this period the time of stabilization of a solution.

In order to carry out numerical simulations of the obtained theoretical results we consider the following equation:

$$\begin{aligned} P_0' &= -0.5P_0 - 0.1(P_0 - P_1), \\ P_1' &= -0.7P_1 + 0.1(P_0 - P_1), \\ P_2' &= -1.2P_1 + 0.1(P_0 - P_2), \\ \Delta P_0|_{t=\theta_i} &= 0.07, \end{aligned} \tag{9.35}$$

where $\theta_i = i + \frac{1}{4}|\sin(i) - \sin(\sqrt{2}i)|$. It is proved in [42] that the sequence $\frac{1}{4}|\sin(i) - \sin(\sqrt{2}i)|$ is almost periodic. The theoretical part of the section implies that system (9.35) has a discontinuous almost periodic solution.

As can be seen from Figure 9.1, the solution $P(t)$, $P(0) = (0.08, 0.05, 0.03)$ is approaching the almost periodic solution as time increases.

Figure 9.2 shows that there exists a period of stabilization of a solution with initial value $P(0) = (0.03, 0.5, 0.2)$ so that the first coordinate is smaller than the second and third ones at the initial moment $t = 0$.

9.2.4 *Discussion*

As follows from the preceding discussion, if system (9.23) satisfies certain conditions, then systemic arterial pressure and the blood pressure in compartments oscillate, remaining

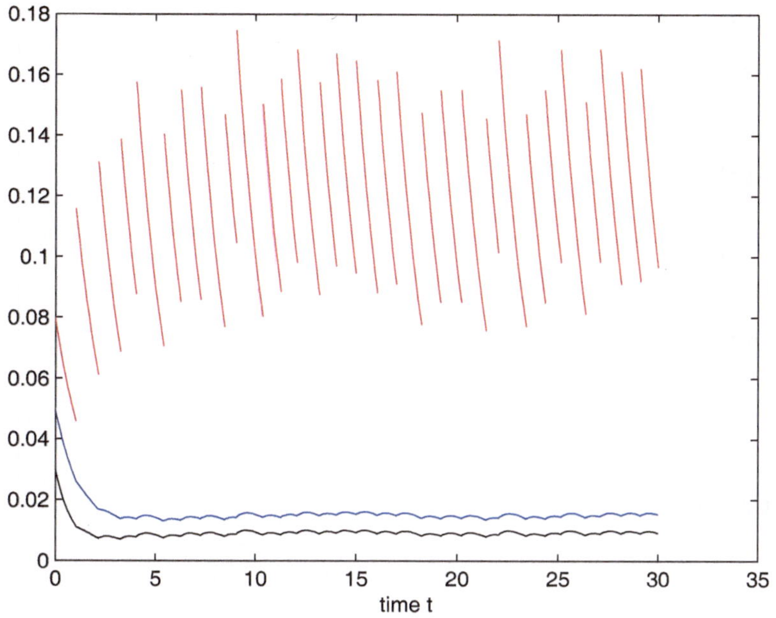

Fig. 9.1 The graph of the first coordinate $P_0(t)$ of the solution is shown in red, of the second coordinate $P_1(t)$ in blue, and the third coordinate $P_2(t)$ in black. The initial value of the first coordinate is larger than that of the second and third coordinates, and one can easily see from the graphs that our theoretical predictions are in full accordance with the properties represented by the figure: the solutions are separated from zero by some positive numbers, coordinate $P_0(t)$ is always larger than $P_1(t)$, $P_2(t)$, and we may suppose that the solution is approaching the discontinuous almost periodic solution of the system, as time increases.

positive. The shape of the oscillations depends on the behavior of the moments when contraction of the left ventricle takes place. The oscillations are asymptotically stable. That is, they do not react significantly to external perturbations.

From the results obtained above one can see that the strike type influence of the left ventricle contraction on the systemic arterial pressure is "softened" when it reaches the peripheral compartments through the connections of the system, so that the compartments' oscillations are continuous. On the other hand, this influence is sufficiently large to sustain the positiveness of all coordinates and keep the aortic pressure higher than in any other compartment.

In the next section we develop the proposed model further, to obtain additional features of the regular behavior of blood pressure, as well as to investigate irregular (chaotic) processes

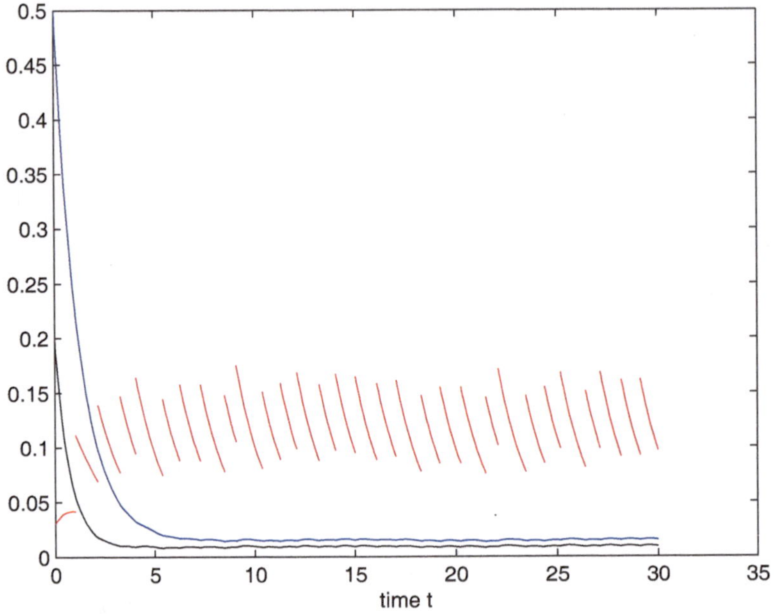

Fig. 9.2 We can see that despite the "odd" relation between the coordinates at the initial moment, the normal state is eventually achieved, when coordinate $P_0(t)$ is larger than $P_1(t)$ and $P_2(t)$. The stabilization time of the solution is no more than 5 units. Again, we have the consistency of the theoretical result with the simulation observations.

in the system. The problem of the period-doubling and intermittency routes to the chaos can serve as an important theoretical tool in the studies of the arythmia processes.

9.3 Chaotic dynamics

In this part of the chapter we consider the dynamics of blood pressure, initiated in the last section, concentrating on the interaction between systemic arterial pressure and peripheric blood pressure. A system of impulsive differential equations is applied as a model. The main result of the present section is the existence of Devaney's chaos ingredients: sensitivity of solutions, transitivity, and existence of infinitely many periodic solutions, in the case, when generation of moments of discontinuity is defined as a special initial value problem. The method of chaos creation proposed in [16–20] is applied. Appropriate examples are provided, including a simulation of the chaotic attractor.

9.3.1 *Introduction and Preliminaries*

The following system is a consequence of the modeling efforts of our predecessors [166, 202],

$$\begin{aligned}
\frac{dP_s(t)}{dt} &= -k_sP_s + g_s(P_s(t) - P_p(t)),\\
\frac{dP_p(t)}{dt} &= -k_pP_p + g_p(P_s(t) - P_p(t)), \ \ t \neq \theta_i,\\
\Delta P_s|_{t=\theta_i} &= I_s + J_s(P_s).
\end{aligned} \tag{9.36}$$

The general version of this system has been introduced in the last section.
We shall need the following assumptions throughout the current section:

(C1) real constants I_s, k_s, k_p are positive;

(C2) J_s, g_s, g_p are real valued continuous functions, $J_s(0) = g_s(0) = g_p(0) = 0$, $g_p(z) > 0$, $g_s(z) < 0$, if $z > 0$;

(C3) the function J_s is non-increasing;

(C4) the functions J_s, g_s, g_p satisfy the Lipschitz condition

$$\begin{aligned}
&|g_s(z^1) - g_s(z^2)| \leqslant l_s|z^1 - z^2|, |g_p(z^1) - g_p(z^2)| \leqslant l_p|z^1 - z^2|,\\
&|J_s(z^1) - J_s(z^1)| \leqslant l_J|z^1 - z^2|;
\end{aligned} \tag{9.37}$$

(C5) the sequence θ_i, $i \in \mathbb{Z}$, satisfies the following property: there exists a number $\omega > 0$, such that $i\omega \leqslant \theta_i < (i+1)\omega$, $i \in \mathbb{Z}$;

(C6) there exist positive real constants m_s, m_p such that

$$\sup_{z\geqslant 0}|g_s(z)| = m_s, \ \ \sup_{z\geqslant 0}|g_p(z)| = m_p. \tag{9.38}$$

Condition (C3) implies that $\sup_{z\geqslant 0}|J_0(z)| = m_J$, where m_J is a positive real number.
We consider the atmospheric pressure to be zero, so that, both of the two types pressures considered are negative. Consequently, our goal is to find conditions which guarantee the existence of bounded positive solutions. These solutions exponentially stable under certain conditions. They are periodic or eventually periodic depending on the particular properties of the sequence of moments θ_i when the jumps of the pressure in the aorta occur. Finally, we consider a special initial value problem, which can be thought of as a model of the cardiovascular system with an artificial pacemaker. Chaotic behavior is investigated for this initial value.
Below, we give a short motivation for introducing conditions (C1)–(C5).
The function $g_s(P_s - P_p)$ is non-positive since the outflow from the aorta to the neighboring compartment reduces systemic arterial pressure. Hence, we assume that P_s is always not

smaller than P_p. Nonlinearity $g_p(P_s - P_p)$ is non positive, since the difference generates the inflow to the compartment and, consequently, implies the rise in peripheric pressure.
The term I_0 in the impulsive part of the equation is caused by the constant stroke volume, which does not depend on systemic arterial pressure, but may depend on the distance ω, and the features of the left heart. The monotonicity of the jump function $J_s(P_s)$ has the following physiological interpretation. One can see that if the pressure P_s has a jump at the moment $t = \theta_p$ then the consequent volume of accumulated blood in the left heart is larger, the larger is the distance $\theta_{i+1} - \theta_p$, that is the time between two consecutive moments of heart contraction. And if the duration increases then systemic arterial pressure decreases because of the negative coefficient in the equation. Hence, to guarantee a larger inflow of blood we should assume that the function is decreasing.
Condition (C5) assumes that the results of our investigation can be interpreted as describing a blood pressure system activated by an artificial pacemaker, or the case when the change is close to regular.
The theoretical background of our investigation could be found in [20, 42, 278].
In the next section we give conditions for the equation such that there exists a unique bounded on the whole of $\mathbb{R}$ solution. The solution is periodic if the sequence of discontinuity moments has an appropriate property. The section contains, also, results on the stability and positiveness of the solutions. In the next section, a special initial value problem is considered, using the logistic map to create the sequence of the moments jumps. We introduce eventually periodic solutions and prove the existence of these solutions for certain values of the parameter of the logistic equation. All three ingredients of the Devaney's chaos for the problem are considered in this section. Appropriate examples are provided.

9.3.2 *Stability and positiveness of bounded and periodic solutions*

In this part we just repeat results of the previous section specified for equation (9.36).
We may assume that

$$\textbf{(C7)}\ L(l_s, l_p, l_J) = \sqrt{\left[\frac{l_s\sqrt{2}}{k_s} + \frac{l_J e^{k_s\omega}}{1 - e^{-k_s\omega}}\right]^2 + \frac{(l_p\sqrt{2})^2}{k_p^2}} < 1.$$

Theorem 9.4. *If conditions* (C1)–(C7) *are fulfilled, then there exists a unique bounded on* $\mathbb{R}$ *solution of* (9.36). *If the sequence* θ_p *has the p-property for a fixed* $p \in \mathbb{N}$, *then the bounded solution has period* $p\omega$.

The unique bounded on $\mathbb{R}$ solution satisfies the following integral equation

$$P_s(t) = \int_{-\infty}^{t} e^{-k_s(t-s)} g_0(P_s(s) - P_p(s))ds + \sum_{\theta_i<t} e^{-k_s(t-\theta_p)}(I_0 + J(P_s(\theta_i))),$$

$$P_p(t) = \int_{-\infty}^{t} e^{-k_p(t-s)} g_p(P_s(s) - P_p(s))ds. \tag{9.39}$$

Let us denote the bounded on $\mathbb{R}$ solution as $\xi(t) = (\xi_s(t), \xi_p(t))$. If we use the norm $|\phi|_0 = \sup_{\mathbb{R}} |\phi(t)|$ for scalar-valued functions defined on $\mathbb{R}$, and $\|\phi\|_0 = \sup_{\mathbb{R}} \|\phi(t)\|$ for $\phi \in \mathscr{PC}(\mathbb{R})$, then every bounded solution satisfies $|\xi_s|_0 \leqslant (\frac{m_s}{k_s} + \frac{m_J e^{k_s\omega}}{1-e^{-k_s\omega}}) = M_s$, $|\xi_p|_0 \leqslant \frac{m_p}{k_p} = M_p$.
Next we obtain conditions for the positiveness of this solution. Assume additionally that

(C8) $\dfrac{m_s}{k_s} + \dfrac{m_p}{k_p} < \|I_s\| \dfrac{e^{-k_s\omega}}{1-e^{-k_s\omega}}.$

Using (9.39) we find that

$$\xi_s(t) \geqslant \|I_s\| \frac{e^{-k_s\omega}}{1-e^{-k_s\omega}} - \frac{m_s}{k_s} > 0, \ t \in \mathbb{R},$$

and

$$|\xi_p(t)| \leqslant \frac{m_p}{k_p}, \ t \in \mathbb{R}, \ i = \overline{1,m}.$$

Hence,

$$\xi_s(t) - \xi_p(t) \geqslant \|I_s\| \frac{e^{-k_s\omega}}{1-e^{-k_s\omega}} - \frac{m_s}{k_s} - \frac{m_p}{k_p} = \delta > 0, \ t \in \mathbb{R}, \ i = \overline{1,m}.$$

Using condition (C3) and (9.39) again we obtain that

$$\xi_p(t) \geqslant \frac{\overline{g}_p}{k_p} > 0, \ t \in \mathbb{R},$$

where $\overline{g}_p$ is the minimal value of the function $g_p(z)$ for $z \in [\delta, \frac{m_s}{k_s} + \frac{m_J e^{k_s\omega}}{1-e^{-k_s\omega}} + \frac{m_p}{k_p}]$.
Thus, we have proved the existence of positive numbers μ_s, μ_p such that

$$\xi_s(t) \geqslant \mu_s, \ \xi_p(t) \geqslant \mu_p, \ t \in \mathbb{R}.$$

Fix a positive number σ, $0 < \sigma < \min\{k_s, k_p\}$, denote

$$m(l_s, l_p, l_J) = 1 - \max\left\{ \frac{2l_s}{k_s - \sigma} + \frac{l_J e^{(k_s-\sigma)\omega}}{1-e^{-(k_s-\sigma)\omega}}, \frac{2l_p}{k_p - \sigma} \right\},$$

and assume that the Lipshitz coefficients are sufficiently small so that

(C9) $m(l_s, l_p, l_J) > 0.$

Theorem 9.5. *Assume that* (C1)–(C9) *are valid. Then the bounded solution* $\xi(t)$ *of* (9.36) *is uniformly exponentially stable.*

9.3.3 *Blood pressure dynamics as a special initial value problem: chaotic behavior*

The phenomenal exploration of the irregular behavior of dynamical systems was done in [114, 156, 207, 214, 225, 277, 281, 289] Many papers considering the subject for specified models as well as for equations of general type have been published. One of the ways to discover chaos on the basis of qualitative theory is to look for topological components of chaos summarized in [98].

For the reminder in this section we assume that the sequence of discontinuity points θ_i is defined by a particular function and depends on the choice of the initial moment. More precisely, consider $h(t) = h(t,\mu) \equiv \mu t^2(1-t)$, the logistic map, where $\mu > 4$ is a parameter, and $\omega = 1, I = [0,1]$. It is known that there exists a positively invariant subset Λ of I.

For every $t_0 \in \Lambda$ one can construct a sequence $\kappa(t_0)$ of real numbers κ_i, $i \in \mathbb{Z}$, in the following way. If $i \geqslant 0$, then $\kappa_{i+1} = h(\kappa_i,\mu)$ and $\kappa_0 = t_0$.

Let us show, how the sequence is defined for negative i. Denote $s^0 = S(t_0)$, $s^0 = (s_0^0 s_1^0 \ldots)$. Consider elements $\underline{s} = (0 s_0^0 s_0^1 \ldots)$, $\overline{s} = (1 s_0^0 s_0^1 \ldots)$ of Σ_2, such that $\sigma(\underline{s}) = \sigma(\overline{s}) = s^0$ and $\underline{t} = S^{-1}(\underline{s}), \overline{t} = S^{-1}(\overline{s})$. The homeomorphism implies that $h(\overline{t},\mu) = h(\underline{t},\mu) = t_0$. Set $h^{-1}(t_0,\mu)$ may consist of not more than two elements $\overline{t}$, $\underline{t} \in \Lambda$. Each of these two values can be chosen as $\kappa_{-1}(t_0,\mu)$. Obviously, one can continue the process to $-\infty$, choosing always one element from the set h^{-1}. We have finalized the construction of the sequence, and, moreover, it is proved that $\kappa(t_0,\mu) \subset \Lambda$.

Thus, infinitely many sequences $\kappa(t_0,\mu)$ can be constructed for a given t_0. However, each of this type of sequence is unique for an increasing i. Fix one of the sequences and define a sequence $\theta(t_0) = \{\theta_i\}$, $\theta_i = i + \kappa_i$, $i \in \mathbb{Z}$. The sequence has a *periodicity property* if there exists $p \in \mathbb{N}$ such that $\theta_{i+p} = \theta_i + p$, for all $i \in \mathbb{Z}$. If we denote by Π the set of all such sequences $\{\theta_i\}$, $i \in \mathbb{Z}$, then a multivalued functional $w : I \to \Pi$ such that $\theta(t_0) = w(t_0)$ is defined.

Let us introduce the following special initial value problem for the impulsive differential equation (9.36)

$$\begin{aligned}
\frac{dP_s(t)}{dt} &= -k_s P_s + g_s(P_s - P_p), \\
\frac{dP_p(t)}{dt} &= -k_p P_p + g_1(P_s - P_p), t \neq \theta_i(t_0), \\
\Delta P_s|_{t=\theta_i(t_0)} &= I_s + J_s(P_s), \\
P(t_0) &= P_0, t_0 \in I,
\end{aligned} \tag{9.40}$$

where P_0 is a vector from $\mathbb{R}^2$.

Our main goal for this section is to investigate various regular and irregular dynamics of solutions of impulsive differential equations. The detailed investigation of the dynamics is introduced in [16–20], where all chaotic properties are discussed.

9.3.4 *Periodic solutions revisited. Eventually periodic solutions*

In what follows we assume that conditions (C1)–(C9) are valid with $\omega = 1$. Then by previous results for every $t_0 \in I$ such that $\kappa(t_0)$ is p-periodic, $p \in \mathbb{N}$, there exists a p-periodic uniformly exponentially stable positive solution of the impulsive differential equation. Let us denote the periodic solution as $\xi(t,t_0)$. Moreover, in what follows we assume a bounded solution as a bounded on $\mathbb{R}$ function, and we denote these solutions of (9.40) as $P(t,t_0)$.
It is known [225] that there exists an infinite sequence of the parameter μ values, $3 < \mu_1 < \mu_2 < \cdots < \mu_k \cdots < 3.8284\ldots$, such that $\kappa(t,\mu_i), i \geqslant 1$, has an asymptotically stable prime period-2^i point $t_i^* \in I$ with a region of attraction $(t_i^* - \delta_i, t_i^* + \delta_i)$. Beyond the value $3.8284\ldots$, there are cycles with every integer period [207].
Let $\widehat{[a,b]}$ be an oriented interval, that is $\widehat{[a,b]} = [a,b]$, if $a \leqslant b$, and $\widehat{[a,b]} = [b,a]$, otherwise. Fix a number j such that $h(t,\mu_j)$ has a period-2^j point t_j^* with a region of attraction $(t_j^* - \delta_j, t_j^* + \delta_j) \in I$. Denote $p = 2^j$, $t^* = t_j^*$, $\delta = \delta_j$. In what follows we shall investigate the dynamics of the IVP near the point t^*. It follows from Theorem 9.4 that there exists a p-periodic solution $\xi(t,t^*)$ of (9.40).
We shall need the following defintion.

Definition 9.5. We say that a function $\phi(t)$ from $\mathscr{PC}(\mathbb{R})$ with a sequence of discontinuities θ_i is eventually $p\omega$-periodic if for every positive ε there exists a moment $T > 0$, such that

(1) $|\theta_{i+p} - \theta_i - p\omega| < \varepsilon$, if $\theta_i > T$;
(2) $\|\phi(t+p\omega) - \phi(t)\| < \varepsilon$ for all $t > T$, such that $|t - \theta_i| > \varepsilon$, $i \in \mathbb{R}$.

Solution $\xi(t,t^*)$ attracts all other solutions of (9.40) that have the same initial moment t^*. Hence, it is obvious that all solutions of IVP (9.40) with the same initial moment t^* are eventually p-periodic. But our main interest lies in bounded on $\mathbb{R}$ solutions $P(t,t_0)$, $t_0 \neq t^*$, which have starting moments in $(t^* - \delta, t^* + \delta)$. Let us show that they are eventually p-periodic. We may assume that

(C10) $l_s + l_p + \ln(1 + l_J) < \min\{k_s, k_p\}$.

Theorem 9.6. *If conditions* (C1)–(C10) *are fulfilled, then solution* $P(t,t_0)$ *is eventually*

p-periodic.

Proof. Since t_0 belongs to the basin of attractiveness of t^*, $\theta(t_0)$ satisfies the first condition of Definition 9.5. Let us check if the other condition of the Definition holds for $P(t,t_0)$. One can easily see that it is fulfilled if

$$\|P(t,t_0) - \xi(t,t^*)\| \to 0, \text{ as } t \to \infty, \tag{9.41}$$

for all $t \notin [\widehat{\theta_i(t^*), \theta_i(t_0)}]$.

It is difficult to evaluate the difference between $P(t,t_0)$ and $\xi(t,t^*)$ since their moments of discontinuity do not coincide. For this reason let us apply the method of B-equivalence [20]. Let us consider the following system

$$\begin{aligned}
\frac{dQ_s(t)}{dt} &= -k_sQ_s + g_s(Q_s - Q_p),\\
\frac{dQ_p(t)}{dt} &= -k_pQ_p + g_p(Q_s - Q_p), t \neq \theta_i(t^*),\\
\Delta Q_s|_{t=\theta_i(t^*)} &= I_s + J_s(Q_s) + W_i(Q).\\
Q(t_0) &= P_0, t_0 \in I.
\end{aligned} \tag{9.42}$$

The two IVPs (9.40) and (9.42) are B-equivalent [20], if their solutions with the same initial data coincide in their common domain only if $t \notin [\widehat{\theta_i(t_0), \theta_i(t^*)}], \imath \in \mathbb{Z}$.

Next we shall define a function W for the last IVP such that it is $B-$equivalent to (9.40). Introduce the following system of ordinary differential equations

$$\begin{aligned}
\frac{dR_s(t)}{dt} &= -k_sR_s + g_s(R_s - R_p),\\
\frac{dR_p(t)}{dt} &= -k_pR_p + g_p(R_s - R_p),
\end{aligned} \tag{9.43}$$

and assume, without any loss of generality, that $\theta_i(t_0) \leqslant \theta_i(t^*)$, $i \in \mathbb{Z}$. Let $R(t,u,R_0)$ be a solution of (9.43) with the initial data u, R_0. Fix $Q = (Q_s, Q_p)$, and denote

$$\begin{aligned}
r &= (r_s, r_p),\\
r_s &= I_s + J_s(R_s(\theta_i(t_0), \theta_i(t^*), Q)) + R_s(\theta_i(t_0), \theta_i(t^*), Q),\\
r_p &= R_s(\theta_i(t_0), \theta_i(t^*), Q).
\end{aligned}$$

If $W_i(Q) = R(\theta_i(t^*), \theta_i(t_0), r) - I_s - J_s(Q)$, then one can verify that IVPs (9.40) and (9.42) are $B-$equivalent [20]. Moreover, every $W_i(Q)$ is a continuous function, and if $\theta_i(t_0) \to \theta_i(t^*)$ as $i \to \infty$, then $W_i(Q) \to 0$ uniformly on every bounded set from $\mathbb{R}^2$.

Introduce the norm $\|a\|_1 = |a_1| + |a_2|$, if $a = (a_1, a_2)$, and denote $M = M_s + M_p$, $\kappa_1 = \max\{k_s, k_p\}$, $\kappa_2 = \min\{k_s, k_p\}$. Fix a positive $\varepsilon < 2M$, and choose positive integers k, i_0 and a number $\varepsilon_1 > 0$, such that

(1) $\varepsilon_1 e^{\kappa_2} \sum_{i=0}^{k} e^{\kappa_2 i} < \frac{\varepsilon}{4}$;
(2) $2Me^{-(\kappa_2 - l_s - l_p - \ln(1+l_J))k} < \frac{\varepsilon}{4}$;
(3) $\|W_i(Q)\| < \varepsilon_1$ if $\|Q\| < M, i \geqslant i_0$.

Then for $t \geqslant i_0$

$$\xi_s(t) = e^{-k_s(t-i_0)}\xi_s(i_0) + \int_{i_0}^{t} e^{-k_s(t-u)} g_s(\xi_s(u) - \xi_p(u))du + \sum_{i_0 \leqslant \theta_i < t} e^{-k_s(t-\theta_i)}(I_s + J_s(\xi_s(\theta_i))),$$

$$\xi_p(t) = e^{-k_p(t-t_0)}\xi_p(i_0) + \int_{i_0}^{t} e^{-k_p(t-u)} g_p(\xi_s(u) - \xi_p(u))du, \tag{9.44}$$

and

$$Q_s(t) = e^{-k_s(t-i_0)}Q_s(i_0) + \int_{i_0}^{t} e^{-k_s(t-u)} g_s(Q_s(u) - Q_p(u))du + \sum_{i_0 \leqslant \theta_i < t} e^{-k_s(t-\theta_i)}(I_s + J_s(Q_s(\theta_i)) + W_i(Q(\theta_i)))$$

$$Q_p(t) = e^{-k_p(t-t_0)}Q_p(i_0) + \int_{i_0}^{t} e^{-k_p(t-u)} g_p(Q_s(u) - Q_p(u))du. \tag{9.45}$$

Then

$$\|\xi(t) - Q(t)\|_1 \leqslant e^{-\kappa_2(t-i_0)}\|\xi(i_0) - Q(i_0)\|_1 + \int_{i_0}^{t} e^{-\kappa_2(t-u)}(l_s + l_p)\|\xi(u) - Q(u)\|_1 du + \sum_{i_0 \leqslant \theta_i < t} e^{-\kappa_2(t-\theta_i)}(l_J\|\xi(\theta_i) - Q(\theta_i)\|_1 + \varepsilon_1),$$

and for $i_0 \leqslant t \leqslant i_0 + k$,

$$\|\xi(t) - Q(t)\|_1 \leqslant \|\xi(i_0) - Q(i_0)\|_1 e^{-\kappa_2(t-i_0)} + \int_{i_0}^{t} e^{-\kappa_2(t-u)}(l_s + l_p)\|\xi(u) - Q(u)\|_1 du + \sum_{i_0 \leqslant \theta_i < t} e^{-\kappa_2(t-\theta_i)}(l_J\|\xi(\theta_i) - Q(\theta_i)\|_1 + \varepsilon_1).$$

If we denote $v(t) = \|\xi(t) - Q(t)\|_1 e^{\kappa_2 t}$, then

$$v(t) \leqslant \left[\|\xi(i_0) - Q(i_0)\|_1 e^{\kappa_2 i_0} + \varepsilon_1 e^{\kappa_2}\sum_{i=1}^{k} e^{\kappa_2 i}\right] + \int_{i_0}^{t} (l_s + l_p)v(u)du + \sum_{i_0 \leqslant \theta_i < t} l_J v(\theta_i).$$

Now, applying the Gronwall-Bellman Lemma for piecewise continuous functions [278] one can find that

$$v(t) \leqslant \left[\|\xi(i_0) - Q(i_0)\|_1 \mathrm{e}^{\kappa_2 i_0} + \varepsilon_1 \mathrm{e}^{\kappa_2} \sum_{i=1}^{k} \mathrm{e}^{\kappa_2 i}\right] \mathrm{e}^{(l_s + l_p + \ln(1+l_J))(t - i_0)},$$

and

$$\|\xi(i_0+k) - Q(i_0+k)\|_1 \leqslant \|\xi(i_0) - Q(i_0)\|_1 \mathrm{e}^{-(\kappa_2 - l_s - l_p - \ln(1+l_J))k} + \mathrm{e}^{\kappa_2}\varepsilon_1 \sum_{i=1}^{k} \mathrm{e}^{\kappa_2 i} \mathrm{e}^{-(\kappa_2 - l_s - l_p - \ln(1+l_J))k} < \frac{\varepsilon}{4} + \frac{\varepsilon}{4} = \frac{\varepsilon}{2}. \tag{9.46}$$

Similarly to (9.46), we can obtain that for $t \in [i_0 + k, i_0 + 2k]$

$$\|\xi(t) - Q(t)\|_1 \leqslant \frac{\varepsilon}{2} \mathrm{e}^{-(\kappa_2 - l_s - l_p - \ln(1+l_J))(t - i_0)} + \mathrm{e}^{\kappa_2}\varepsilon_1 \sum_{i=1}^{k} \mathrm{e}^{\kappa_2 i} \mathrm{e}^{-(\kappa_2 - l_s - l_p - \ln(1+l_J))(t - i_0)} < \frac{\varepsilon}{2} + \frac{\varepsilon}{4} \leqslant \varepsilon, \tag{9.47}$$

and

$$\|\xi(i_0+2k) - Q(i_0+2k)\|_1 \leqslant \|\xi(i_0+k) - Q(i_0+k)\|_1 \mathrm{e}^{-(\kappa_2 - l_s - l_p - \ln(1+l_J))k} + \mathrm{e}^{\kappa_2}\varepsilon_1 \sum_{i=1}^{k} \mathrm{e}^{\kappa_2 i} \mathrm{e}^{-(\kappa_2 - l_s - l_p - \ln(1+l_J))k} < \frac{\varepsilon}{4} + \frac{\varepsilon}{4} = \frac{\varepsilon}{2}. \tag{9.48}$$

Hence, by mathematical induction, for arbitrary $t \geqslant i_0 + k$ we have that $\|\xi(t) - Q(t)\|_1 < \varepsilon$, and since ε can be arbitrarily small, $P(t, t_0)$, $t_0 \in (t^* - \delta, t^* - \delta)$, is eventually p-periodic. The theorem is proved. □

Since every bounded solution $P(t, t_0)$ is an attractor of all solutions of the system (9.43), it follows that $\xi(t, t^*)$ is an attractor of all solutions of (9.40) with $t_0 \in (t^* - \delta, t^* + \delta)$.

9.3.5 *Chaos*

It is known that a chaotic process takes place on a compact domain. Hence, it is natural to discuss the irregular behavior of the system only for the union of positive-valued and bounded solutions, that is positive solutions $P(t, t_0)$, which satisfy the inequalities $|P_s(t)|_0 < M_s$, $|P_p(t)|_0 < M_p$ for all $t \in \mathbb{R}$. Below, we say that a solution is bounded if it is bounded on $\mathbb{R}$. We should remark that, since any of these bounded solutions is an attractor of all solutions, that have the same initial moment, the chaotic properties are appropriate for all solutions, not only the bounded ones. So, to describe the chaos in the model we can consider only bounded solutions.

In this section we assume that $\mu > 4$. Then [272] there exists an invariant Cantor set $\Lambda \subset I$ such that $h(x, \mu)$ is chaotic on Λ. That is, h has sensitive dependence on initial conditions, periodic points are dense in Λ and h is topologically transitive. We may also point out that there are infinitely many orbits of h with different periods, that for each $p \in \mathbb{N}$ there exists a solution with period p, and that topological transitivity means existence of positive trajectory of h, dense in Λ.

Consider the sequence space [98]

$$\Sigma_2 = \{s = (s_0 s_1 s_2 \ldots) : s_j = 0 \text{ or } 1\}$$

with the metric

$$d[s,t] = \sum_{i=0}^{\infty} \frac{|s_i - t_i|}{2^i},$$

where $t = (t_0 t_1 \ldots) \in \Sigma_2$, and the shift map $\sigma : \Sigma_2 \to \Sigma_2$, such that $\sigma(s) = (s_1 s_2 \ldots)$. The map σ is continuous, $\text{card}(\text{Per}_n(\sigma)) = 2^n$, $\text{Per}(\sigma)$ is dense in Σ_2, and there exists a dense orbit in Σ_2.

If we denote

$$I_0 = \left[0, \frac{1}{2} - \sqrt{\frac{1}{4} - \frac{1}{\mu}}\right], \ A_0 = \left(\frac{1}{2} - \sqrt{\frac{1}{4} - \frac{1}{\mu}}, \frac{1}{2} + \sqrt{\frac{1}{4} - \frac{1}{\mu}}\right), \ I_1 = \left[\frac{1}{2} + \sqrt{\frac{1}{4} - \frac{1}{\mu}}, 1\right].$$

then $I = I_0 \cup A_0 \cup I_1$, $\Lambda \subset I_0 \cup I_1$, $h(I_0) = h(I_1) = I$, $h(A_0) \cap I = \emptyset$.

Consider the itinerary of x, $S(x) = (s_0 s_1 \ldots)$, where $s_j = 0$, if $h^j(x) \in I_0$, and $s_j = 1$, if $h^j(x) \in I_1$. The function $S(x)$ is a homeomorphism between Λ and Σ_2, and $S \circ h = \sigma \circ S$. That is, h and σ are topologically conjugate.

Next, let us consider some useful properties of the elements of Π. We shall formulate and prove three very important consequences of the topological conjugacy of the symbolical dynamics and of the dynamics generated by the logistic map [98] in the following assertion. Despite their simplicity we have not find them in a appropriate literature. So, we decided to give the full proof of these assertions.

Let $J \subseteq \mathbb{R}$ be an open interval. Introduce the following distance $\|\theta(t_0) - \theta(t_1)\|_J = \sup_{\theta_i(t_0), \theta_i(t_1) \in J} |\theta_i(t_0) - \theta_i(t_1)|$. One can read about the possible uses of this distance in [10].

Lemma 9.1. *If $\mu > 4$, then*

(a) *for each $\theta(t_0) \in \Pi$, arbitrarily small $\varepsilon > 0$, and arbitrarily large positive number E there exists a periodical sequence $\theta(t_1) \in \Pi$ such that $|\theta(t_0) - \theta(t_1)|_J < \varepsilon$, where $J = (0,E)$;*

(b) *there exists a sequence $\theta(t^*) \in \Pi$ such that for each $t_0 \in \Lambda$, and for arbitrarily small $\varepsilon > 0$, and arbitrarily large positive number E there exists an integer m such that $|\theta(t_0) - \theta(t^*, m)|_J < \varepsilon$, where $J = (0,E)$.*

Proof. (a) Fix numbers $t_0 \in \Lambda, \varepsilon > 0, E > 0$, and denote $S(t_0) = s^0 = (s_0^0 s_1^0 \ldots)$. Since S is a homeomorphism and Σ_2 is compact, there exists a number $\delta > 0$ such that $|S^{-1}(s^1) - S^{-1}(s^2)| < \varepsilon$ if $d[s^1, s^2] < \delta$, where $s^1, s^2 \in \Sigma_2$.

Next we define a periodic sequence $s = (s_0 s_1 s_3 \ldots)$ from Σ_2 to conform with ε and E. Take a number $l \in \mathbb{N}$ such that $l > E$ and $\frac{1}{2^{l-1}} < \delta$. Assume that $s_i = s_i^0, i = 0, 1, \ldots, 2l-1$, and $s_{i+2l} = s_i$, $i \in \mathbb{Z}$. Consider the sequence $\kappa_i(t_0) = h^i(t_0)$, $i \geqslant 0$. Since $h^i(t_0) = S^{-1} \circ \sigma^i \circ S$, and $d[\sigma^i s^0, \sigma^i s] < \delta$, $i = 0, 1, \ldots, l$, we have that $|h^i(t_0) - h^i(S^{-1}(s)| < \varepsilon$. So, if we denote $t = S^{-1}(s)$ then $|\kappa_i(t_0) - \kappa_i(t)| < \varepsilon$, $i = 0, 1, \ldots, l$. In other words, $\|\theta(t_0) - \theta(t)\|_J < \varepsilon$, where $J = (0, E)$. The assertion is proved.

(b) Consider the sequence $s^* \in \Sigma_2$ such that

$$s^* = \underbrace{01}_{\text{1 element blocks}} \quad | \underbrace{00011011}_{\text{2 element blocks}} |\ldots,$$

that is s^* is constructed by successively listing all blocks of $0's$ and $1's$ of length 1, then length 2, etc. The sequence s^* is dense in Σ_2 [98]. Denote $t^* = S^{-1}(s^*)$. Let us fix $t_0 \in \Lambda$, $\varepsilon > 0$ and $E > 0$. Similar to the previous proof, fix a number $\delta > 0$ such that $|S^{-1}(s^1) - S^{-1}(s^2)| < \varepsilon$ if $d[s^1, s^2] < \delta$, where $s^1, s^2 \in \Sigma_2$. Take $l \in \mathbb{N}$ such that $l > E + 1$ and $\frac{1}{2^{l-1}} < \delta$. We can find a block of length $2l$, $s_m^* s_{m+1}^* \cdots s_{m+2l}^*$ such that the i-th element of $S(t_0)$, $i = 0, 1, \ldots, 2l$, is equal to s_{i+m}^*. One can see that $\|\theta(t_0) - \theta(t^*, m)\|_J < \varepsilon$, where $J = (0, E)$. The lemma is proved. □

Since the sequence θ depends on $t_0 \in I$, in (9.40) we shall denote the space of piecewise continuous $\mathscr{PC}(t_0, \mathbb{R})$, instead of $\mathscr{PC}(\mathbb{R})$. Let us fix an interval $J \subset \mathbb{R}$, and t_0, $t_1 \in I$. We shall say that a function $\xi(t) \in \mathscr{PC}(t_0, \mathbb{R})$ is ε-equivalent to a function $\psi(t) \in \mathscr{PC}(t_1, \mathbb{R})$ on J and write $\xi(t)(\varepsilon, J)\psi(t)$ if $\|\theta(t_s) - \theta(t_p)\|_J < \varepsilon$ and $\|\xi(t) - \psi(t)\| < \varepsilon$ for all t from J such that $t \notin \bigcup_{\theta_p(t_s), \theta_p(t_p) \in J} [\theta_p(\widehat{t_s), \theta_p}(t_p)]$. More information about topologies in the set of discontinuous functions can be found in [8–10], [183, 278, 288].

Definition 9.6. We shall say that a bounded solution $P(t) = P(t, t_0, P_0)$, $t_0 \in \Lambda$, of (9.40) is sensitive with respect to the initial data if there exist positive real numbers ε_0, ε_1 such that for every $\delta > 0$ one could find a pair $(t_1, P_1) \in \Lambda \times \mathbb{R}^n$, $|t_0 - t_1| + \|P_0 - P_1\| < \delta$, and an interval J_1 in $[t_0, \infty)$ of length not less than ε_1 such that $\|P_1(t) - P(t)\| > \varepsilon_0$, $t \in J_1$, where $P_1(t)$, $P_1(t_1) = P_1$, is a bounded solution of (9.40), and there is no point of discontinuity of $P_1(t)$ and $P(t)$ in J_1.

Definition 9.7. A bounded solution $P(t) = P(t, t_0, P_0)$, $t_0 \in \Lambda$, $t \geqslant t_s$, of (9.40) is called dense in the set of all bounded solutions which start on Λ if for each bounded solution

$P_1(t) = P(t,t_1,P_1)$, $t_1 \in \Lambda$, of (9.40) and arbitrarily large positive number E there exist an interval J of length E and a real number s such that $P(t+s)(\varepsilon,J)P_1(t)$.

Definition 9.8. The set of all periodic solutions $\xi(t,t_0)$, $t_0 \in \Lambda$, of (9.40) is called dense in the set of all bounded solutions which start on Λ if for each bounded solution $P(t) = P(t,t_1)$, $t_1 \in \Lambda$, of (9.40) and each $\varepsilon > 0$, $E > 0$, there exists a periodic solution $\xi(t)$ and an interval $J \subset [t_1,\infty)$ with length E such that $\phi(t)(\varepsilon,J)P(t)$.

Theorem 9.7. *Assume that conditions* (C1)–(C8) *are fulfilled. Then there exists a dense in the sense of Definition 9.7 solution of* (9.40).

Proof. By Lemma 9.1 (b) there exists $t^* \in \Lambda$ such that $\theta(t^*)$ is dense in Π, such that for each $t_0 \in \Lambda$, for arbitrarily small $\varepsilon > 0$, and arbitrarily large positive number E there exists a positive integer m such that $|\theta(t_0) - \theta(t^*,m)|_J < \varepsilon$, where $J = (0,E)$. By Theorem 9.4 there exists a unique bounded solution $\xi^*(t) = \xi(t,t^*)$. Let us prove that $\xi^*(t)$ is the dense solution.

Consider an arbitrary solution $\xi(t) = \xi(t,t_0)$, $t_0 \in \Lambda$, of (9.40). Then for $t \geqslant 0$, we have that

$$\begin{aligned}
&\xi_s^*(t+m) = \mathrm{e}^{-k_s(t+m-m)}\xi_s^*(m) + \int_m^{t+m} \mathrm{e}^{-k_s(t+m-u)} g_s(\xi_s(u) - \xi_p(u))du + \\
&\sum_{m \leqslant \theta_i < t+m} \mathrm{e}^{-k_s(t+m-\theta_i)}(I_s + J_s(\xi_s(\theta_i))) = \\
&\mathrm{e}^{-k_s t}\xi_s^*(m) + \int_0^t \mathrm{e}^{-k_s(t-u)} g_s(\xi_s^*(u+m) - \xi_p^*(u+m))du + \\
&\sum_{m \leqslant \theta_i < t+m} \mathrm{e}^{-k_s(t+m-\theta_i)}(I_s + J_s(\xi_s^*(\theta_i))).
\end{aligned} \tag{9.49}$$

Similarly,

$$\xi_p^*(t+m) = \mathrm{e}^{-k_p t}\xi_p^*(m) + \int_0^t \mathrm{e}^{-k_p(t-u)} g_p(\xi_s^*(u) - \xi_p^*(u))du. \tag{9.50}$$

Moreover,

$$\begin{aligned}
&\xi_s(t) = \mathrm{e}^{-k_s t}\xi_s(0) + \int_0^t \mathrm{e}^{-k_s(t-u)} g_s(\xi_s(u) - \xi_p(u))du + \sum_{0 \leqslant \theta_i < t} \mathrm{e}^{-k_s(t-\theta_i)}(I_s + J_s(\xi_s(\theta_i))), \\
&\xi_p(t) = \mathrm{e}^{-k_p t}\xi_p(0) + \int_0^t \mathrm{e}^{-k_p(t-u)} g_p(\xi_s(u) - \xi_p(u))du.
\end{aligned} \tag{9.51}$$

Now, using the last three formulas and using the B-equivalence technique, similarly to the proof of Theorem 9.6, we can find a number E_1 sufficiently large so that there exists a subinterval J of J_1 of length E such that $\xi_*(t+m)$ and $\xi(t)$ are ε-equivalent on J. The theorem is proved. □

We shall need the following assumption.

(C8) $L < \dfrac{I_s e^{-k_s}}{16M}$.

Theorem 9.8. *Assume that conditions* (C1)–(C8) *are fulfilled. Then every bounded solution of* (9.40) *is sensitive with respect to the initial data.*

Proof. Let $S(t_0) = s^0 = (s_0^0, s_1^0, \ldots)$. Fix a number $t_1 \in \Lambda$ such that $S(t_1) = s^1 = (s_0^0, s_1^0, \ldots, s_{n-1}^0, s_n^1, s_{n+1}^0, s_{n+2}^0, \ldots)$, $s_n^1 \neq s_n^0$, for some $n > 0$. We have that

$$d[\sigma^i s^0, \sigma^i s^1] = \begin{cases} \frac{1}{2^{n-i}} & \text{if } 0 \leqslant i \leqslant n, \\ 0 & \text{if } i > n. \end{cases}$$

Take a positive number δ. Assume that $n \geqslant 8$, it is an even number, and that it is sufficiently large so that $|t_0 - t_1| = |S^{-1}(s^0) - S^{-1}(s^1)| < \frac{\delta}{2}$.

Since S is a homeomorphism and the set Σ_2 is compact, for a given i, $0 \leqslant i \leqslant n$, the set

$$d_i = \{(\overline{s}, \widetilde{s}) \in \Sigma_2 \times \Sigma_2 : d[\overline{s}, \widetilde{s}] \geqslant \frac{1}{2^{n-i}}\}$$

is compact, and

$$\min_{(\overline{s}, \widetilde{s}) \in P_i} |S^{-1}(\overline{s}) - S^{-1}(\widetilde{s})| = \mu_i > 0,$$

$d_{i+1} \subseteq d_i$, $\mu_{i+1} \geqslant \mu_i$, $0 \leqslant i < n$. Fix i_0, $\frac{n}{2} \leqslant i_0 < n-1$. Then $|\kappa_i(t_0) - \kappa_i(t_1)| > \mu_{i_0}$ if $i = i_0$, $i_0 + 1$.

We also have that $|\kappa_i(t_0) - \kappa_i(t_1)| \leqslant \sqrt{1 - \frac{4}{\mu}}$ if $0 \leqslant i < n$.

Without loss of generality, assume that $\kappa_i(t_0) < \kappa_i(t_1)$ for all i. Thus, there is a number k among i_0, i_0+1, such that $\kappa_k(t_1) - \kappa_k(t_0) > \mu_{i_0}$ and $\kappa_k(t_0) - \kappa_{k-1}(t_1) \geqslant \frac{1}{2}(1 - \sqrt{1 - \frac{4}{\mu}})$.

It is obvious that a solution is sensitive if its first coordinate is.

Fix a solution $P(t) = P(t, t_0)$. We shall show that the constants ε_0, ε_1 of Definition 9.6 can be taken equal to $\varepsilon_0 = \frac{I_s e^{-2k_s}}{8}$, $\varepsilon_1 = \min\{\mu_{i_0}, \frac{1}{2}(1 - \sqrt{1 - \frac{4}{\mu}})\}$.

Let us fix a solution $P^1(t) = (P_s^1(t), P_p^1(t))$, $\|P_1 - P_0\| < \frac{\delta}{2}$.

Below, we consider two alternative cases. Assume, first, that $|P_s((\theta_k(t_0)) - P_s^1((\theta_k(t_0))| < \frac{I_s e^{-2k_s}}{4}$.

Then on the interval $[\theta_k(t_0), \theta_k(t_1)]$

$$P_s(t) = e^{-k_s(t - \theta_k(t_0))} P_s(\theta_k(t_0)) + \int_{\theta_k(t_0)}^{t} e^{-k_s(t-u)} g_s(P_s(s) - P_p(s)) du +$$

$$e^{-k_s(t-\theta_k(t_0))}(I_s + J_s(P_s(\theta_k(t_0)))),$$

and

$$P_s^1(t) = e^{-k_s(t-\theta_k(t_0))}P_s^1(\theta_k(t_0)) + \int_{\theta_k(t_0)}^{t} e^{-k_s(t-u)}g_s(P_s^1(s) - P_p^1(s))du.$$

Hence,

$$|P_s(t) - P_s^1(t)| \geqslant |e^{-k_s(t-\theta_k(t_0))}(I_s + J_s(P_s(\theta_k(t_0))))| -$$

$$e^{-k_s(t-\theta_k(t_0))}|P_s(\theta_k(t_0)) - P_s^1(\theta_k(t_0))| - \int_{\theta_k(t_0)}^{t} e^{-k_s(t-u)}2LMdu \geqslant \varepsilon_0.$$

In the case when $|P_s(\theta_k(t_0)) - P_s^1(\theta_k(t_0))| \geqslant \frac{I_s e^{-k_s}}{4}$, we consider the interval $[\theta_{k-1}(t_1), \theta_k(t_0)]$, where

$$P_s(t) = e^{-k_s(t-\theta_k(t_0))}P_s(\theta_k(t_0)) + \int_{\theta_k(t_0)}^{t} e^{-k_s(t-u)}g_s(P_s(s) - P_p(s))du,$$

and

$$P_s^1(t) = e^{-k_s(t-\theta_k(t_0))}P_s^1(\theta_k(t_0)) + \int_{\theta_k(t_0)}^{t} e^{-k_s(t-u)}g_s(P_s^1(s) - P_p^1(s))du.$$

Consequently,

$$|P_s(t) - P_s^1(t)| \geqslant e^{-k_s(t-\theta_k(t_0))}|P_s(\theta_k(t_0)) - P_s^1(\theta_k(t_0))| - 2LM \geqslant \varepsilon_0.$$

The theorem is proved. □

One must say that under the additional condition $|g_p(z_1) - g_p(z_2)| \leqslant |z_1 - z_2|$, $z_1, z_2 \in \mathbb{R}$, one can easily prove that a bounded solution is sensitive in the second coordinate, too.

The next theorem can be proved similar to the verification of Theorem 9.7 using B-equivalence technique and Lemma 9.1 (a).

Theorem 9.9. *Assume that conditions* (C1)–(C8) *are fulfilled. Then the set of all periodic solutions* $\xi(t, t_0)$, $t_0 \in \Lambda$, *of* (9.40) *is dense in the set of all bounded solutions.*

9.3.6 *Example*

Hammel *et all.* [151] have given a computer-assisted proof that an approximate trajectory of the logistic map can be shadowed by a true trajectory for a long time. We can apply this result to make reliable the following simulations.

It is not easy task to find initial moments which can illustrate the chaos of a system, if $\mu > 4$. For this reason we propose to consider values of parameter μ, which are not discussed, but they can help us demonstrate the chaotic nature of the considered initial value problem. We construct a simulation to show that the dynamics of blood pressure can exhibit sensitivity if $\mu = 4$, and intermittency for $\mu = 3.8282$. Moreover, if $\mu = 4$ we observe the chaos attractor by a stroboscopic sequence on $\mathbb{R}^2$.

Let the following system be given

$$\begin{aligned} P_0' &= -2P_0 - g_1(P_0 - P_1), \\ P_1' &= -3P_1 + g_2(P_0 - P_1), \\ \Delta P_0|_{t=\theta_i} &= 0.5, \end{aligned} \tag{9.52}$$

where $\theta_i = i + \xi_i$, the sequence ξ is defined recursively, $\xi_i = 4\xi_{i-1}(1 - \xi_{i-1})$, $\xi_0 = t_0$, $t_0 \in [0,1]$, $i \geqslant 0$, $g_1(u) = g_2(u) = l \sin^2 u$, $W(s) = 1 + s^2$, if $|s| \leqslant l$. One can easily see that all the functions are Lipschitzian with a constant proportional to l. In what follows we assume that $l = 10^{-4}$.

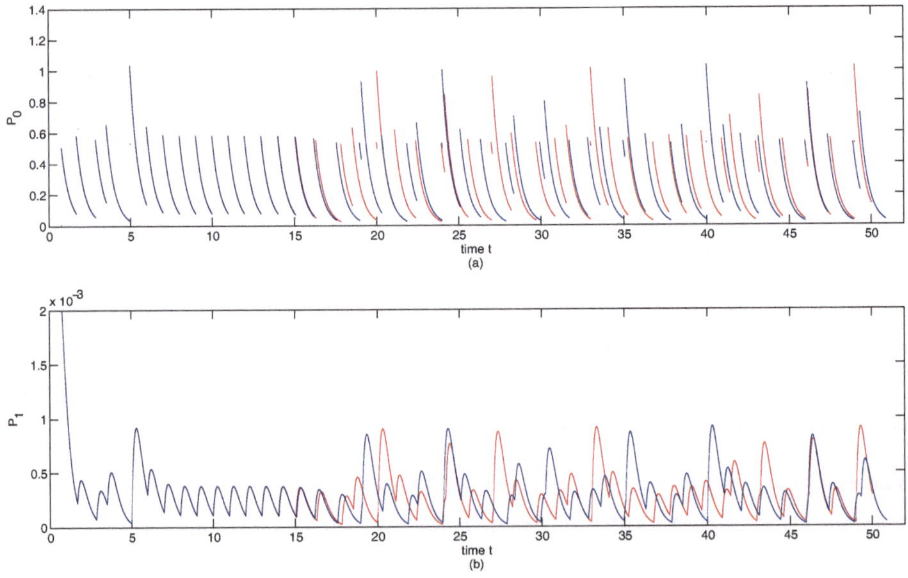

Fig. 9.3 Graphs of coordinates $P_0(t)$, $P_1(t)$ are in the blue, and of coordinates $\overline{P}_0(t)$, $\overline{P}_1(t)$ are red. The coordinates abruptly become significantly different when t is near 15, while coinciding for all t in the interval $(t_1, 15)$.

Consider two solutions $P(t) = (P_0, P_1)$, $\overline{P}(t) = (\overline{P}_0, \overline{P}_1)$, with initial moments $t_0 = 7/9$ and

$\bar{t}_0 = 7/9 + 3^{-12}$, respectively. That is, we take the initial values close to each other, and, moreover, the solutions with identical initial values, $P(t_0) = \overline{P}(\bar{t}_0) = (0.005, 0.002)$. The graphs of the coordinates of these solutions (Figure 9.3) show that the solutions abruptly become different when t is between 15 and 20, despite being very close to each other for all t in the interval $(t_1, 15)$. One can conclude that the phenomenon of sensitivity is numerically observable. Next, in Figure 9.4 the chaotic attractor is shown by using points $P(n)$, $n = 1, 2, 3, \ldots, 75000$, in P_1, P_2-plane.
A more detailed, magnified, view of the attractor in Figure 9.5 is given.
If one consider (9.52) with the sequence $\theta_i = i + \xi_i$, $\xi_i = 3.8282\xi_{i-1}(1 - \xi_{i-1})$, $\xi_0 = t_0$, $t_0 \in [0, 1]$, $i \geqslant 0$, then the phenomenon of intermittency, i.e. irregular switching between periodic and chaotic behavior, for a solution $P(t) = (P_0, P_1)$ can be observed in Figure 9.6. The coefficient's value of 3.8282 is such that the logistic map admits intermittency [98].

Notes

In this chapter we consider the dynamics of blood pressure, concentrating on the interaction between systemic arterial pressure and peripheric blood pressure. A system of impulsive differential equations is applied as a model.
In Section 9.1 a simple two-dimensional model is considered. We find conditions of oscillations in different types: periodic, almost periodic, ε-oscillations with their asymptotic properties and positiveness. Apparently, ε-oscillations are introduced at first time in literature. In Subsection 9.1.3 we consider for the first time the regular behavior of systemic arterial pressure when the moments of jumps are not fixed, and a new jump occurs when the pressure value reaches a certain positive constant value. The bouncing ball [280] has dynamics similar to the model discussed in this section.
In Section 9.2 we introduce a multidimensional system of differential equations the behavior of which solutions possesses several properties characteristic of the blood pressure distribution. The system is a compartmental model of the cardiovascular system. It admits a unique bounded solution such that all coordinates of the solution are separated from zero by positive numbers, and which is periodic, eventually periodic or almost periodic depending on the moments of heart contraction.
The main result of Section 9.3 is the existence of Devaney's chaos ingredients: sensitivity of solutions, transitivity, and existence of infinitely many periodic solutions, in the case, when generation of moments of discontinuity is defined as a special initial value problem. The

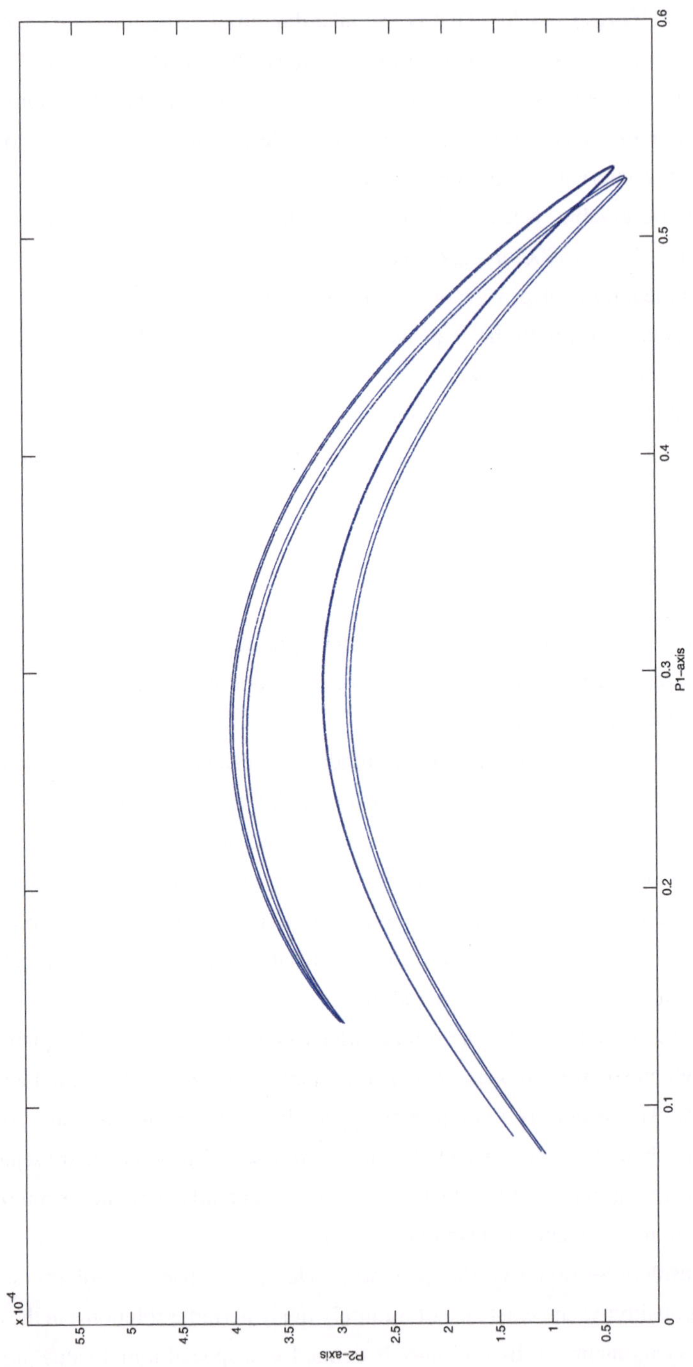

Fig. 9.4 The chaotic attractor by a stroboscopic sequence $P(n)$, $1 \leqslant n \leqslant 75000$, is observable.

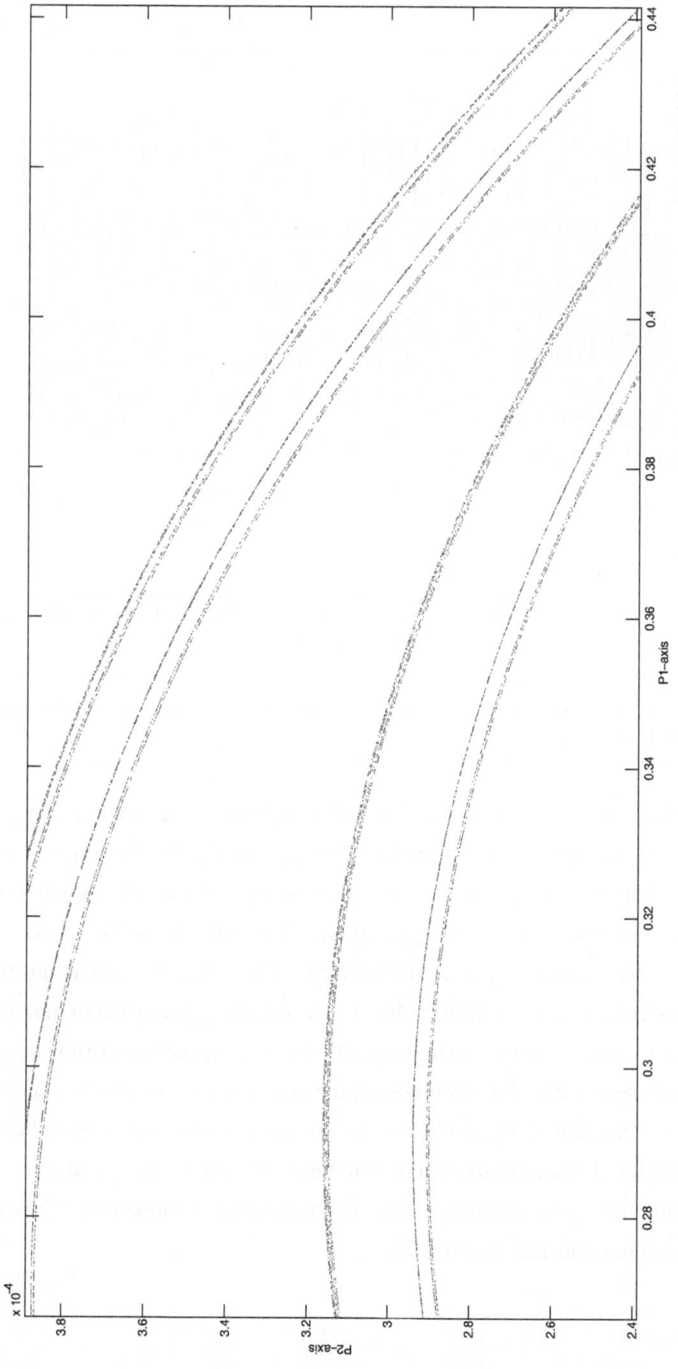

Fig. 9.5 An enlargement of the small rectangular of Figure 9.4. The characteristic fine structure of the attractor begins to emerge.

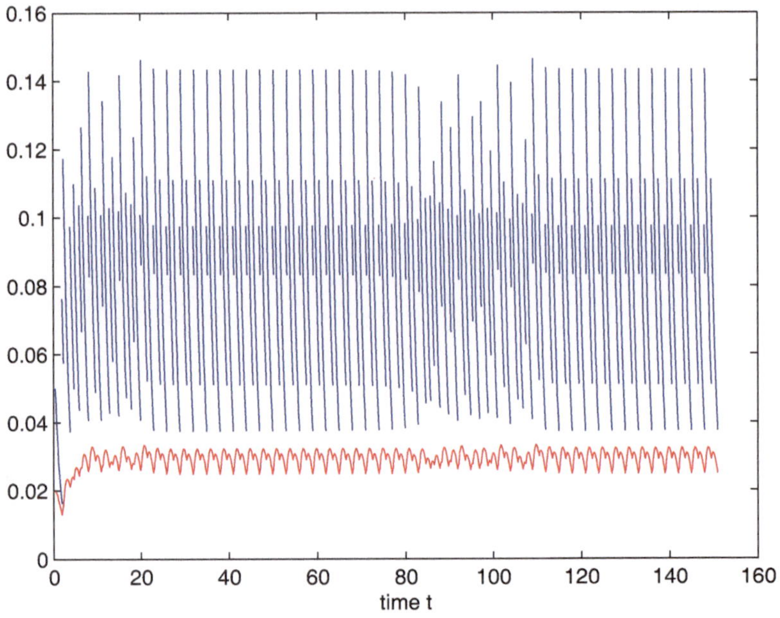

Fig. 9.6 Both coordinates $P_0(t), P_1(t)$, which are in red and blue colors respectively, have intermittency bouts in the intervals $(0,20)$ and $(90,110)$.

method of chaos creation proposed in [16–20] is applied. In this section we have continued investigation of qualitative characteristics of the system (9.36), Section 9.2. We introduce a new initial boundary value problem for the system, and consider equation (9.40). The main modeling novelty, useful for applications, is that the moments of discontinuity are prescribed by the choice of the initial moment. Thus, the moment is an initial for the impulsive system, as well is an initial value for the discrete component of the hybrid system. Apparently, this assumption could be accepted as an appropriate deterministic condition for many real life processes. We have defined the complex discontinuous dynamics proving that sensitivity and transitivity, as well as existence of countable set of periodic solutions are proper for the set of solutions of the problem bounded on the whole real axes. Appropriate numerical simulations are provided. The chaotic attractor is observable. The main results of the chapter are published in [30–32].

Chapter 10

Integrate-and-fire biological oscillators

10.1 Introduction

In paper [262] C. Peskin develops the integrate-and-fire model of the cardiac pacemaker [181] to a population of identical pulse-coupled oscillators. Thus, it was proposed to consider a model of cardiac pacemaker, where signal of fire arises not from an outside stimuli, but in the population of cells itself. Then, the well known conjectures of self-synchronization were formulated. Solution of these conjectures for identical oscillators [216,262], and a wide discussion of the subject stimulated mathematicians as well as biologists for the intensive investigations in the field [60,109,122,188,222,285,296,298,299]. It is natural that the problem has been considered in the more general form. In [216] the method of phase diagrams effectively is used to discuss the models. In the paper [24] we suggest a special map, which helped us to solve the synchronization problem for non-identical oscillators. A version of the model is considered such that perturbations can be evaluated still to save the synchronization.

One has to say that the collective behavior of biological and chemical oscillators is a fascinating topic that has attracted a lot of attention in the last 50 years [61,62,95], [106–108], [109,122–124,127,129,130,152,163–165,184,187,203,216,259,296,300,301,331,332]. An exceptional place in the analysis belongs to synchronization, which in its general sense is understood as phase locking, frequency locking, and synchrony itself, that is motion in unison [62,106], [63,108,109,122–125,127,129,130,152,163–165,184,187,203,216,259, 260,266,296,300,301,331,332].

It was conjectured in [262] that the cardiac pacemaker model self-synchronizes such that:

(C1) For arbitrary initial conditions, the system approaches a state in which all the oscillators are firing synchronously.

(C2) This remains true even when the oscillators are not quite identical.

The conjecture (C1) is solved in [262] for a system with two oscillators, and in [216] for the generalized model of two and more oscillators. The last paper gave start to an intensive and productive investigation of the problem and its applications [60,109,122,188,222,285, 296,298,299]. As far as we know the conjecture (C2) remains unsolved. Even a developed non-identity concept has not been found in the literature. Nowadays, the conjectures are developed in relation to real world questions, biological ones in particular. Let us mention some of such problems important for our paper.

(i) What differences may exist between non-identical oscillators to still allow for synchronization?
(ii) What will be changed in the dynamics if one replaces all-to-all coupling with more local interactions?
(iii) Is the dynamics is affected by the choice of pulse strengths?
(iv) Do continuous or piece-wise continuous couplings synchronize the model?
(v) Will the synchrony of firing in unison be achieved if the coupling effect delays?

In the present analysis we address questions (i), (iii).
Results, that concern questions (iv), (v) are considered in our papers [22,23]. In the present chapter we generalize the model, and propose a version of non-identity. The model is considered such that perturbations save the synchronization. These oscillators are not only pulse-coupled, but connected during the time between moments of firing. That is, the modeling differential equations are not separated. One can see that this approach may provide more biological sense to this theory. The chapter consists of main results, simulations and discussion of possible extensions. The main role in our analysis is played by a specially defined map. It is not a Poincaré map, since it transforms the coordinate of one oscillator to that of another, and the two interchange roles in the course of the mapping. If there are more than two oscillators, they are used in pairs to shape the map with the interference of other oscillators acting as perturbation. The difficulties arising due to the perturbations of the identity are solved with a technique that was developed for differential equations with discontinuities at non-fixed moments [12,39,99,113,192].
The main object of the present investigation is an integrate-and-fire model, which consists of n non-identical pulse-coupled oscillators, x_i, $i = 1,2,\ldots,n$. If the system does not fire the oscillators satisfy the following equations

$$x_i' = f(x_i) + \phi_i(x). \tag{10.1}$$

The domain consists of all points $x = (x_1, x_2, \ldots, x_n)$ such that $0 \leqslant x_i \leqslant 1 + \zeta_i(x)$ for all $i = 1,2,\ldots,n$. When the oscillator x_j increases from zero, and meets the surface such that

$x_j(t) = 1 + \zeta_j(x(t))$, then it fires, $x_j(t+) = 0$. This firing changes the values of all oscillators with $i \neq j$,

$$x_i(t+) = \begin{cases} 0, \text{ if } x_i(t) + \varepsilon + \varepsilon_i \geqslant 1 + \zeta_i(x), \\ x_i(t) + \varepsilon + \varepsilon_i, \text{ otherwise.} \end{cases} \tag{10.2}$$

Thus, it is assumed that if $x_i(t) \geqslant 1 + \zeta_i(x) - \varepsilon - \varepsilon_i$, then the oscillator fires, too. It is assumed also that there exist positive constants μ_i and ξ_i such that $|\phi_i(x)| < \mu_i$ and $|\zeta_i(x)| < \xi_i$, for all x and i. In what follows, we call the real numbers ε, μ_i, ξ_i, ε_i, *parameters*, assuming the first one is positive. Moreover, constants ξ_i, ε_i, μ_i, will be called *parameters of perturbation*. If all of them are zeros, then the model of identical oscillators is obtained. We assume that $\varepsilon + \varepsilon_i > 0$. That is, an exhibitory model is under discussion. The function f is positive valued and lipschitzian. Moreover, assume that ζ_i are continuous and ϕ_i are locally Lipschitzian for all i.

The coupling in the model is all-to-all such that each firing elicits jumps in all non-firing oscillators. If several oscillators fire simultaneously, then other oscillators react as if just one oscillator fires. In other words, any firing acts only as a signal which abruptly provokes a change of state. The intensity of the signal is not important, and pulse strengths are not additive. A system of oscillators is synchronized if all of them fire in unison.

In the present analysis we address synchronization as well as the existence of periodic solutions. Results that concern continuous and delayed couplings are considered in [22,23]. We believe that the approach proposed in this chapter will be useful for the investigation of a wide range of problems, focusing not only on synchrony and pulse-couplings, but also phase locking, frequency locking of systems, families of oscillators with continuous couplings. The method can be used to analyze inhibitory models as well as to evaluate the effects of coupling time deviations. Moreover, the model is suitable for the investigation of the existence of quasi-periodic and almost periodic motions. In the last section of the chapter delayed pulse-couplings are discussed.

10.2 The prototype map

In this section we shall define the map, which is the basic instrument of our investigation. It is constructed for a model more general, than is needed for this chapter, to be the basis for future investigations.

Let us consider two identical oscillators, $x_1(t)$, $x_2(t)$, $t \geqslant 0$, which satisfy the following differential equations

$$x_i' = f(x_i), \tag{10.3}$$

where $0 \leqslant x_i \leqslant 1$, $i = 1,2$. When the oscillator x_j fires at the moment t such that $x_j(t) = 1, x_j(t+) = 0$, then the value of another oscillator with $i \neq j$, changes so that

$$x_i(t+) = \begin{cases} 0, \text{if } \ x_i(t)+\varepsilon \geqslant 1, \\ x_i(t)+\varepsilon, \text{otherwise.} \end{cases} \tag{10.4}$$

Denote by $u(t,0,u_0)$, the solution of the equation

$$u' = f(u), \tag{10.5}$$

such that $u(0,0,u_0) = u_0$. Assume that the solution exists, is unique and continuable to the threshold for all u_0. Consider the solution $u(t) = u(t,0,v+\varepsilon)$ of (10.5). Denote by $s(v)$ the moment when $u(s) = 1$, and define the function $\overline{L}(v) = u(s,0,0)$ on $(0, 1-\varepsilon)$.
The following conditions will be needed throughout the chapter:

(A1) $\overline{L}(v)$ is a strictly decreasing continuous function;
(A2) $\eta = \lim_{v\to 0+} \overline{L}(v) > 1-\varepsilon$;
(A3) $\lim_{v\to 1-\varepsilon} \overline{L}(v) = 0$.

Conditions (A1), (A3) are valid, if, for example, f is a positive and Lipschitzian function. Another case will be considered in Example 10.3. It is obvious that there exists a unique fixed point, $v^*, \overline{L}(v^*) = v^*$.
Now, define a map $L : [0,1] \to [0,1]$, such that

$$L(v) = \begin{cases} \overline{L}(v), & \text{if } \ v \in (0, 1-\varepsilon), \\ \eta, & \text{if } \ v = 0, \\ 0, & \text{if } \ v \in [1-\varepsilon, 1]. \end{cases} \tag{10.6}$$

This newly defined function is continuous on $[0,1]$. The sketch of its graph is shown in Figure 10.1. The graph of the map is in red.
To make the following discussion constructive consider the sequence of maps $L^k(v)$, $k = 1,2,\dots,$ where $L^k(v) = L(L^{k-1}(v))$ if $k \geqslant 2$. Their graphs with $k = 1,2,3$ are shown in Figure 10.2. The graphs of L, L^2 and L^3 in red, blue and green respectively.
Denote $a_0 = 0$, $a_1 = 1-\varepsilon$, $a_2 = L^{-1}(1-\varepsilon)$, $a_3 = (L^2)^{-1}(1-\varepsilon), \dots$. The sequence can be obtained also through iterations $a_0 = 0$, $a_1 = 1-\varepsilon$, $a_{k+1} = L^{-1}(a_k)$, $k = 1,2,\dots,$ which are seen in Figure 10.1.It is clear that the sequences a_{2i} and a_{2i+1} are monotonic, increasing and decreasing respectively. One can verify existence of a fixed point $v^{**} \leqslant v^*$ of the map $L^2(v)$ such that $\widehat{v} = L(v^{**}) \geqslant v^*$, and there are no fixed points of L^2 in $(0, v^{**})$. Moreover,

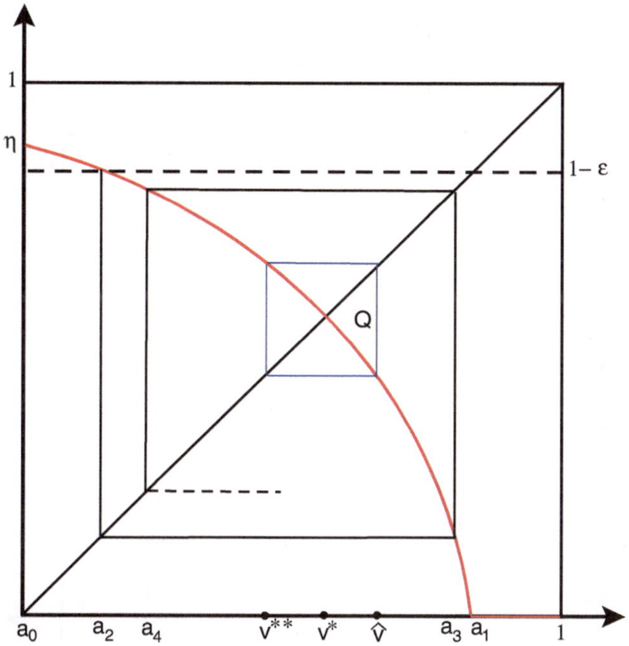

Fig. 10.1 The graph of function $w = L(v)$, in red, and the period−2 orbit in blue. The points $a_0 = 0$, $a_1 = 1-\varepsilon$, $a_{k+1} = L^{-1}(a_k)$, $k = 1,2,3$. (Color online)

$a_{2i} \to v^{**}$ and $a_{2i+1} \to \widehat{v}$ as $i \to \infty$. In the case $v^{**} = v^*$, there is no non-trivial period−2 points of L. In Figure 10.1 the period−2 orbit in blue. Let us show how iterations of L can be useful for the investigation of synchronization. Consider a motion $(x_1(t), x_2(t))$ and a firing moment $t_0 \geqslant 0$ such that $x_1(t_0) = 1$, $x_1(t_0+) = 0$, $x_2(t_0+) = v$, $v \in [0,1]$.

Lemma 10.1. *Motion $(x_1(t), x_2(t))$ synchronizes if and only if there exists a number k such that* $1-\varepsilon \leqslant L^k(v) \leqslant 1$.

Proof. Let us consider only necessity, since sufficiency is obvious. We shall consider the following two cases: (α) $0 \leqslant v < 1-\varepsilon$; (β) $1-\varepsilon \leqslant v \leqslant 1$.
(α) It is clear that the couple does not synchronize at the moment $t = t_0$. While it is not in synchrony, there exists a sequence $t_0 < t_1 < \ldots$, such that x_1 fires at moments t_i with even i and x_2 fires at t_i with odd indexes. Set $v_i = x_1(t_i)$ if i is odd, and $v_i = x_2(t_i)$ if it is even. Use the definition of L and identity of oscillators to obtain that $v_{i+1} = L(v_i), i \geqslant 1$. This demonstrates that map L evaluates alternatively the sequence of values x_1 and x_2 at firing moments.

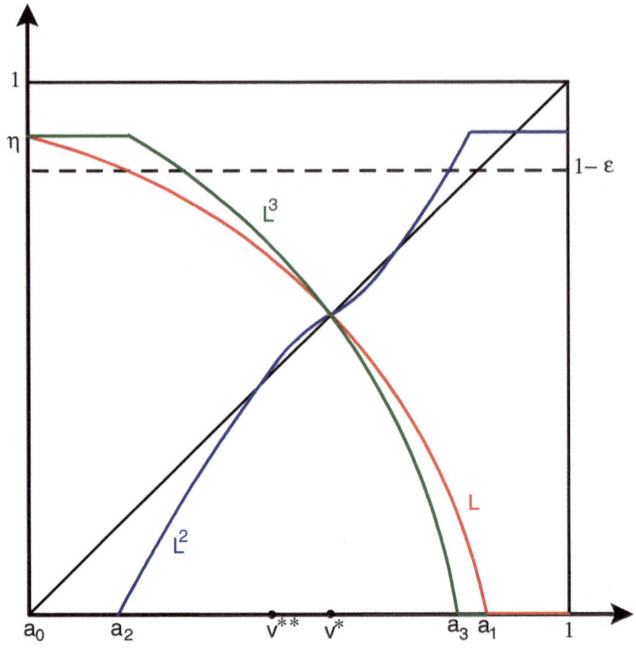

Fig. 10.2 The graphs of L, L^2 and L^3 in red, blue and green respectively. (Color online)

The pair synchronizes eventually if and only if there exists $k \geqslant 1$ such that $x_1(t) \neq x_2(t)$, if $t \leqslant t_k$, and $x_1(t) = x_2(t)$, for $t > t_k$. Both oscillators have to fire at t_k. That is, $1-\varepsilon \leqslant v_k < 1$.
(β) Consider $1-\varepsilon \leqslant v < 1$, as the case $v = 1$ is primitive. We have that t_0 is a common firing moment of both x_1 and x_2, and it is the synchronization moment. Moreover, $1-\varepsilon < L^2(x_2(t_1)) = \eta < 1$. The lemma is proved. □

Thus, it is confirmed that the analysis of synchronization is fully consistent with the dynamics of the introduced map $L(v)$ on $[0,1]$, and, therefore, L can be used as a valuable tool in further investigation of the topic.

Let us consider the rate of synchronization. We solve the problem by indicating initial points which synchronize after precisely k, $k \geqslant 0$, iterations of the map. Denote by S_k the region in $[0,1]$, where points v are synchronized after k iterations of map L. One can see that $S_0 = [1-\varepsilon, 1]$, $S_1 = [a_0, a_2]$, and $S_k = (a_{k-1}, a_{k+1}]$, if $k \geqslant 3$, is an odd positive integer, and $S_k = [a_{k+1}, a_{k-1})$, if $k \geqslant 2$, is an even positive integer.

From the discussion made above it follows that the closer v is to v^{**} from the left or to $\widehat{v}$ from the right, the later is the moment of synchronization.

Denote by T the natural period of oscillators, that is, the period, when there are no couplings, and by $\widetilde{T}$ the time needed for solution $u(t,0,v^*)$ of (10.5) to achieve threshold. Since each oscillator necessarily fires within any interval of length T and the distance between two firing moments of an oscillator are not less than $\widetilde{T}$, on the basis of the above discussion, the following assertion is valid.

Theorem 10.1. *Assume that conditions* (A1)–(A3) *are valid, and* $t_0 \geqslant 0$ *is a firing moment such that* $x_1(t_0) = 1,\ x_1(t_0+) = 0$. *If* $x_2(t_0+) \in S_m$, *m is a natural number, then the couple* $x_1,\ x_2$ *synchronizes within the time interval* $[t_0 + \frac{m}{2}\widetilde{T}, t_0 + mT]$.

One can easily see that whenever condition (A2) is not valid, the system does not synchronize.

Example 10.1. Consider the integrate-and-fire model of two identical oscillators $x_1,\ x_2$, with the differential equations

$$x_i' = x_i^2 + c, \tag{10.7}$$

where $i = 1,2$, and c is a positive constant. It is known [123] that the canonical type I phase model [107] can be reduced by a transformation to the form

$$u' = u^2 + c. \tag{10.8}$$

This time we investigate the model with the pulse-coupling.

Since the two equations are identical, we shall consider a solution $u(t)$ of equation (10.8) to construct map L. We have that $u(t,0,v+\varepsilon) = \sqrt{c}\tan\left(ct + \arctan\left(\frac{v+\varepsilon}{\sqrt{c}}\right)\right)$ and

$$\sqrt{c}\tan\left(cs + \arctan\left(\frac{v+\varepsilon}{\sqrt{c}}\right)\right) = 1. \tag{10.9}$$

Next, $u(s,0,0) = \sqrt{c}\tan(cs)$, and by applying (10.9) we find that

$$L(v) = c\frac{1-v-\varepsilon}{c+v+\varepsilon},$$

if $v \in (0, 1-\varepsilon)$, and the fixed point is equal to $v^* = \sqrt{(c+\varepsilon/2)^2 + c(1-\varepsilon)} - (c+\varepsilon/2)$. Evaluate

$$L(0) = c\frac{1-\varepsilon}{c+\varepsilon}$$

to see that $L(0) < 1-\varepsilon$, and condition (A2) is not valid. Moreover, one can verify that $L'(v) < 0$.

Thus, we obtain that the couple does not synchronize, and our simulations confirm this.

Example 10.2. Consider the following integrate-and-fire model of two identical oscillators, x_1, x_2, with the differential equations

$$x_i' = S - \gamma x_i, \tag{10.10}$$

where $i = 1,2$, positive constants S, γ satisfy $\kappa = \frac{S}{\gamma} > 1$. One can find that $u(t,0,v+\varepsilon) = (v+\varepsilon)\mathrm{e}^{-\gamma t} + \kappa(1-\mathrm{e}^{-\gamma t})$ and $u(s,0,0) = \kappa(1-\mathrm{e}^{-\gamma s})$. The last two expressions imply that

$$L(v) = \kappa \frac{1-(v+\varepsilon)}{\kappa-(v+\varepsilon)}, \tag{10.11}$$

if $0 < v < 1-\varepsilon$.

There is a unique fixed point of L and L^2, and it is equal to

$$v^* = \left(\kappa - \frac{\varepsilon}{2}\right) - \sqrt{\kappa^2 - \kappa + \frac{\varepsilon^2}{4}}. \tag{10.12}$$

Finally, $L(0) = \kappa\frac{1-\varepsilon}{\kappa-\varepsilon} > 1-\varepsilon$. That is, all conditions of the last theorem are valid, and the assertion in [262] is proved.

Remark 10.1. Map L is similar to that in [262], but the argument here is a coordinate before a jump, while in the paper the argument is a coordinate after a jump. This difference is not a critical one. The most important point is that C. Peskin uses it only as an auxiliary device to build the Poincaré map. We use L itself as the main map, with a newly defined continuous extension, which simplifies the discussion in this section and throughout our investigation. We revisit the problem of two identical oscillators, since L is the prototype map in our analysis. In addition to the main synchronization result, regions with equal time of synchronization are indicated, and the value of the fixed point, v^*, is evaluated.

Example 10.3. Consider the following system of integrate-and-fire and pulse-coupled oscillators, x_1, x_2, such that

$$\begin{aligned} x_1' &= f(x_1), \\ x_2' &= f(x_2), \end{aligned} \tag{10.13}$$

where

$$f(s) = \begin{cases} 4-3s & \text{if } 0 < s \leqslant 1/3; \\ 3 & \text{if } 1/3 < s \leqslant 2/3; \\ 4-3(s-2/3) & \text{if } 2/3 < s \leqslant 1. \end{cases}$$

We have found that map L for this system exists and is equal to

$$L(v) = \begin{cases} 2\dfrac{2-3v-3\varepsilon}{4-3v-3\varepsilon} & \text{if } 0 < v \leqslant \dfrac{1}{3} - \varepsilon; \\ 1-v-\varepsilon & \text{if } \dfrac{1}{3} - \varepsilon < v \leqslant \dfrac{2}{3} - \varepsilon; \\ \dfrac{4}{3}\dfrac{1-v-\varepsilon}{2-v-\varepsilon} & \text{if } \dfrac{2}{3} - \varepsilon < v \leqslant 1. \end{cases}$$

One can check that conditions (A1)–(A3) are fulfilled for this map. Moreover, the fixed points are equal to $v^* = (1-\varepsilon)/2$, $v^{**} = 1/3$ and $\widehat{v} = 2/3 - \varepsilon$. Finally, all the motions, which start outside of the periodic trajectory synchronize eventually, and all of them are periodic inside the trajectory.

The last example shows that the assumptions for the map L, including the existence of the non-trivial period-2 motion, can be realized for even the differential equations with discontinuous right-hand side. Moreover, in future investigations one can consider isolated periodic solutions, stable or unstable. The theoretical consequences of this research are clear, if one uses the mappings theory, but construction of examples requires additional time.

Example 10.4. Consider the model of two integrate-and-fire identical oscillators, x_1, x_2, which are pulse-coupled and

$$\begin{aligned} x_1' &= S - \gamma x_1 + \beta x_2, \\ x_2' &= S - \gamma x_2 + \beta x_1, \end{aligned} \tag{10.14}$$

where constants S, γ and β are positive numbers. One can easily see that the system is the extended Peskin's model in Example 10.2. The terms with coefficient β are newly introduced in the system. They reflect the permanent influence of the partners during the process. Eigenvalues associated to (10.14) are $\lambda_1 = -\gamma + \beta$ and $\lambda_2 = -\gamma - \beta$. We suppose that β is small so that both eigenvalues are negative. Moreover, it is assumed $\kappa = S/\gamma > 1$. Then, $\kappa_1 = -S/\lambda_1 > 1$ if β is sufficiently small. The solution of system (10.14) with value $(0, v+\varepsilon)$ at $t = 0$, is equal to

$$\begin{aligned} u_1(t) &= \frac{1}{2}\left[e^{\lambda_1 t} - e^{\lambda_2 t}\right](v+\varepsilon) - \kappa_1\left(e^{\lambda_1 t} - 1\right), \\ u_2(t) &= \frac{1}{2}\left[e^{\lambda_1 t} + e^{\lambda_2 t}\right](v+\varepsilon) - \kappa_1\left(e^{\lambda_1 t} - 1\right). \end{aligned}$$

By using these expressions one can obtain the equation

$$\frac{1}{2}\left[e^{\lambda_1 s} + e^{\lambda_2 s}\right](v+\varepsilon) + \kappa_1\left(1 - e^{\lambda_1 s}\right) = 1,$$

and construct $L(v) = 1 - (v+\varepsilon)e^{\lambda_2 s}$. Map L is too complex to analyze for properties (A1)–(A3). That is why we will compare this model with the couple in Example 10.2. The last two equations imply $L(0) = \kappa\frac{1-\varepsilon}{\kappa-\varepsilon} > 1-\varepsilon$, if $\beta = 0$ and $v = 0$. That is, if β is sufficiently small, then condition (A2) is valid. We have found, also, by direct evaluations that the derivative $L'(v)$ is negative if S and β are sufficiently large and small respectively. That is, condition (A1) is fulfilled. It is easy to verify that condition (A3) is also correct.

Now, using the continuity theorem in parameters [88, 153], one can find that map L may admit a period-2 point only if the orbit is as close to the fixed point v^* of Example 10.2 as β is small. Consequently, the measure of the set of points, which can not be synchronized diminishes as $\beta \to 0$. This result is a new one. In previous papers the differential equations were separated.

10.3 Non-identical oscillators

Consider the model of n non-identical oscillators given by relations (3.1) and (10.2). The domain of this model consists of points $x = (x_1, x_2, \ldots, x_n)$ such that $0 \leqslant x_i \leqslant 1 + \zeta_i(x)$ for all $i = 1, 2, \ldots, n$.

Fix two of the considered oscillators, let us say, x_ℓ and x_r.

Lemma 10.2. *Assume that condition* (A2) *is valid, and* $t_0 \geqslant 0$ *is a firing moment such that* $x_\ell(t_0) = 1 + \zeta_\ell(x(t_0)), x_\ell(t_0+) = 0$. *If parameters are sufficiently close to zero, and absolute values of parameters of perturbation sufficiently small with respect to* ε, *then the couple* x_ℓ, x_r *synchronizes within the time interval* $[t_0, t_0 + T]$ *if* $x_r(t_0+) \notin [a_0, a_1)$ *and within the time interval* $[t_0 + \frac{m-1}{2}\widetilde{T}, t_0 + (m+1)T]$, *if* $x_r(t_0+) \in S_m$, $m \geqslant 1$.

Proof. Denote by $x(t) = (x_1(t), x_2(t), \ldots, x_n(t))$, the motion of the oscillator. If $1 + \zeta_r(x(t_0)) - \varepsilon - \varepsilon_r \leqslant x_r(t_0) \leqslant 1 + \zeta_r(x(t_0))$, then these two oscillators fire simultaneously, and we only need to prove the persistence of synchrony which will be done later. So, fix another oscillator $x_r(t)$ such that $0 \leqslant x_r(t_0) < 1 + \zeta_r(x(t_0)) - \varepsilon - \varepsilon_r$.

While the pair does not synchronize, there exists a sequence of moments $0 < t_0 < t_1 < \cdots$, such that oscillator x_ℓ fires at t_i with even i, and x_r fires at t_i with odd i. For the sake of brevity let $u_i = x_\ell(t_i)$, $i = 2j+1$, $u_i = x_r(t_i)$, $i = 2j$, $j \geqslant 0$. In what follows we shall evaluate the difference $u_{i+1} - L(u_i)$.

Let us fix an even i and $u_i = x_r(t_i)$. If the parameters are sufficiently small, then there are $k \leqslant n-2$ distinct firing moments of the motion $x(t)$ on the interval (t_i, t_{i+1}). Denote by $t_i < \theta_1 < \theta_2 < \cdots < \theta_k < t_{i+1}$, the moments of firing, when at least one of the coordinates of $x(t)$ fires, and $v(t, t_0, v_0)$ the solution of the equation (3.1) with $v(t_0, t_0, v_0) = v_0$. We have that

$$x_r(\theta_1) = x_r(t_i) + \varepsilon + \int_{t_i}^{\theta_1} f(x_r(s))ds + \int_{t_i}^{\theta_1} \phi_r(x(s))ds, \tag{10.15}$$

where $x(t) = v(t, t_i, x(t_i+))$,

$$x_r(\theta_2) = x_r(\theta_1) + \varepsilon + \int_{\theta_1}^{\theta_2} f(x_r(s))ds + \int_{\theta_1}^{\theta_2} \phi_r(x(s))ds, \tag{10.16}$$

where $x(t) = v(t, \theta_1, x(\theta_1+))$,

$$\cdots\cdots\cdots\cdots\cdots\cdots\cdots\cdots\cdots$$

$$x_r(t_{i+1}) = x_r(\theta_k) + \varepsilon + \int_{\theta_k}^{t_{i+1}} f(x_r(s))ds + \int_{\theta_k}^{t_{i+1}} \phi_r(x(s))ds, \tag{10.17}$$

where $x(t) = v(t, \theta_1, x(\theta_k+))$.

The moment t_{i+1} satisfies

$$1 + \zeta_r(x(t_{i+1})) - \varepsilon - \varepsilon_r \leqslant x_r(t_{i+1}) \leqslant 1 + \zeta_r(x(t_{i+1})). \tag{10.18}$$

Similarly to the expressions for x_r one can obtain

$$x_\ell(\theta_1) = \int_{t_i}^{\theta_1} f(x_\ell(s))ds + \int_{t_i}^{\theta_1} \phi_\ell(x(s))ds,$$

$$x_\ell(\theta_2) = x_\ell(\theta_1) + \varepsilon + \int_{\theta_1}^{\theta_2} f(x_\ell(s))ds + \int_{\theta_1}^{\theta_2} \phi_\ell(x(s))ds,$$

$$\cdots\cdots\cdots\cdots\cdots\cdots\cdots\cdots\cdots$$

$$x_\ell(t_{i+1}) = x_\ell(\theta_k) + \varepsilon + \int_{\theta_k}^{t_{i+1}} f(x_\ell(s))ds + \int_{\theta_k}^{t_{i+1}} \phi_\ell(x(s))ds. \tag{10.19}$$

Formulas (10.15) to (10.19) define $u_{i+1} = x_\ell(t_{i+1})$. Similarly one can evaluate the number for odd i.

Let us now find the value of $L(u_i)$. With this aim, evaluate

$$\phi(\bar{t}_{i+1}) = x_r(t_i) + \varepsilon + \int_{t_i}^{\bar{t}_{i+1}} f(\phi(s))ds, \tag{10.20}$$

where $\bar{t}_{i+1}$ satisfies $\phi(\bar{t}_{i+1}) = 1$, and

$$\psi(\bar{t}_{i+1}) = \int_{t_i}^{\bar{t}_{i+1}} f(\psi(s))ds, \tag{10.21}$$

to find that $L(u_i) = \psi(\bar{t}_{i+1})$. Next, we will show that the difference $u_{i+1} - L(u_i)$ is small if the parameters are small.

First, one can find

$$\phi(t) - x_r(t) = \int_{t_i}^{t} [f(\phi(s)) - f(x_r(s))]ds - \int_{t_i}^{t} \phi_r(x(s))ds, \tag{10.22}$$

for $t \in [t_i, \theta_1]$.

Then, by applying the Gronwall-Bellman Lemma one can easily see

$$|\phi(\theta_1) - x_r(\theta_1)| \leqslant \mu_r(\theta_1 - t_i)e^{\ell(\theta_1 - t_i)}, \tag{10.23}$$

where ℓ is the Lipschitz constant of f. Next, we have

$$|\phi(\theta_2) - x_r(\theta_2)| \leqslant [\mu_r(\theta_1 - t_i)e^{\ell(\theta_1 - t_i)} + \mu_r(\theta_2 - \theta_1) + \varepsilon]e^{\ell(\theta_2 - \theta_1)}, \qquad (10.24)$$

if $t \in [\theta_1, \theta_2]$.

Without loss of generality, assume that $t_{i+1} > \bar{t}_{i+1}$. Proceeding the evaluations made above, we can obtain $|1 - x_r(\bar{t}_{i+1})| = |\phi(\bar{t}_{i+1}) - x_r(\bar{t}_{i+1})| = \phi_1(\varepsilon, \mu_r)$, where

$$\Phi_1(\varepsilon, \mu_r) \equiv \mu_r[(\theta_1 - t_i)e^{\ell(\bar{t}_{i+1} - t_i)} + \sum_{j=1}^{k-1}(\theta_{j+1} - \theta_j)e^{\ell(\bar{t}_{i+1} - \theta_j)} +$$

$$(\bar{t}_{i+1} - \theta_k)e^{\ell(\bar{t}_{i+1} - \theta_k)}] + \varepsilon \sum_{j=1}^{k} e^{\ell(\bar{t}_{i+1} - \theta_j)}.$$

There are positive numbers μ and M, which satisfy $\mu \leqslant f(s) \leqslant M$, if $0 \leqslant s \leqslant 1 + \max_i \xi_i$. One can request the following inequality: $\max_{i=1,\ldots,n} \mu_i < \mu$. We have that

$$|x_r(t_{i+1}) - x_r(\bar{t}_{i+1})| \leqslant |1 - x_r(t_{i+1})| + |1 - x_r(\bar{t}_{i+1})| \leqslant \phi_1(\varepsilon, \mu_r) + \xi_r.$$

Consequently,

$$|t_{i+1} - \bar{t}_{i+1}| < \frac{\Phi_1(\varepsilon, \mu_r) + \xi_r}{\mu - \mu_r} \equiv \phi_2(\varepsilon, \mu_r, \xi_r).$$

By applying (10.19) and (10.21), making similar evaluations for (10.23) and (10.24) one can find $|\psi(\bar{t}_{i+1}) - x_\ell(\bar{t}_{i+1})| \leqslant \phi_1(\varepsilon, \mu_\ell)$.

Then, we have that

$$|u_{i+1} - L(u_i)| = |\psi(\bar{t}_{i+1}) - x_\ell(t_{i+1})| \leqslant |\psi(\bar{t}_{i+1}) - x_\ell(\bar{t}_{i+1})| + |x_\ell(t_{i+1}) - x_\ell(\bar{t}_{i+1})|,$$

and, consequently,

$$|u_{i+1} - L(u_i)| \leqslant \phi(\varepsilon, \mu_r, \mu_\ell, \xi_r), \qquad (10.25)$$

where $\Phi \equiv \phi_1 + \phi_2(M + \mu_r)$. It is obvious that Φ tends to zero as the parameters do. This convergence is uniform with respect to u_0. We can also vary the number of points θ_i and their location in the intervals (t_j, t_{j+1}) between 0 and $n-1$. The convergence also is indifferent with respect to these variations.

Consider the sequence of inequalities

$$|u_i - L^i(u_0)| \leqslant |u_i - L(u_{i-1})| + |L(u_{i-1}) - L(L^{i-1}(u_0))|, \quad i = 1, 2, \ldots.$$

Then recurrently, by applying continuity of L, (10.25) and $L^m(u_0) \in [1 - \varepsilon, 1]$, conclude that either $1 + \xi_r - \varepsilon - \varepsilon_r \leqslant u_m < 1 + \xi_r$ or $1 + \xi_\ell - \varepsilon - \varepsilon_\ell \leqslant u_{m+1} < 1 + \xi_\ell$, if the parameters are sufficiently close to zero, and absolute values of the parameters of perturbation are sufficiently small with respect to ε. Both of these inequalities bring the pair to synchronization.

Since each of the iterations of the map L happens within an interval with length not more than T, we obtain that couple x_ℓ, x_r synchronizes no later than $t = t_0 + (m+1)T$. Similarly, the couple synchronizes not earlier than $t = t_0 + \frac{m-1}{2}\widetilde{T}$.

If two oscillators x_ℓ and x_r are non-identical and fire simultaneously at a moment $t = \theta$, how will they retain the state of firing in unison, despite being different? To find the required conditions, let us denote by $\tau, \tau > \theta$ a moment when one of them, let's say x_r, fires. We have that $x_\ell(\theta+) = x_r(\theta+) = 0$. Then $x_\ell(t) = x_r(t),\ \theta \leqslant t \leqslant \tau$. It is clear that to satisfy $x_\ell(\tau+) = x_r(\tau+) = 0$, we need $1 + \zeta_\ell(x(\tau)) - \varepsilon - \varepsilon_\ell \leqslant x_\ell(\tau)$. By applying formula (10.18) again, this time with $t_i = \theta, t_{i+1} = \tau$, one can easily obtain that the inequality is correct if parameters are close to zero, and absolute values of the parameters of perturbation are sufficiently small with respect to ε. Thus, one can conclude that if a couple of oscillators is synchronized at some moment of time then it persistently continues to fire in unison. The lemma is proved. □

Remark 10.2. The last lemma not only plays an auxiliary role for the next main theorem, but can also be considered a synchronization result for the model of two non-identical oscillators.

Let us extend the result of the lemma for the whole ensemble.

Theorem 10.2. *Assume that condition* (A2) *is valid, and* $t_0 \geqslant 0$ *is a firing moment such that* $x_j(t_0) = 1 + \zeta_j(x(t_0)), x_j(t_0+) = 0$. *If the parameters are sufficiently close to zero, and absolute values of parameters of perturbation are sufficiently small with respect to* ε, *then the motion* $x(t)$ *of the system synchronizes within the time interval* $[t_0, t_0 + T]$, *if* $x_i(t_0+) \notin [a_0, a_1), i \neq j$, *and within the time interval* $[t_0 + \frac{1}{2}(\max_{i\neq j} k_i - 1)\widetilde{T}, t_0 + (\max_{i\neq j} k_i + 1)T]$, *if there exist* $x_s(t_0+) \in [a_0, a_1)$ *for some* $s \neq j$ *and* $x_i(t_0+) \in S_{k_i}, i \neq j$.

Proof. Consider the non-trivial case. Applying the last lemma we can see that each pair $(x_j, x_i), i \neq j$, synchronizes within $[t_0 + \frac{1}{2}(\max_{i\neq j} k_i - 1)\widetilde{T}, t_0 + (\max_{i\neq j} k_i + 1)T]$. The theorem is proved. □

Now, replace coupling (10.2) by

$$x_i(t+) = \begin{cases} 0 & \text{if } x_i(t) + \varepsilon + \varepsilon_i \geqslant 1 + \zeta_i(x), \\ x_i(t) + \overline{\varepsilon} + \varepsilon_i & \text{otherwise}, \end{cases} \tag{10.26}$$

where $\overline{\varepsilon}$, $\overline{\varepsilon} + \varepsilon_i > 0$, is a new parameter, independent of ε. Consider system (3.1) with (10.26). One can find that the following assertion is valid.

Theorem 10.3. *Assume that condition* (A2) *is valid, and* $t_0 \geqslant 0$ *is a firing moment such that* $x_j(t_0) = 1 + \zeta_j(x(t_0))$, $x_j(t_0+) = 0$. *If parameters* $\overline{\varepsilon}$, μ_i, ξ_i, ε_i, *are sufficiently close to zero, then the motion* $x(t)$ *of the system synchronizes within the time interval* $[t_0, t_0 + T]$, *if* $x_i(t_0+) \notin [a_0, a_1)$, $i \neq j$, *and within the time interval* $[t_0 + \frac{1}{2}(\max_{i\neq j} k_i - 1)\widetilde{T},$ $t_0 + (\max_{i\neq j} k_i + 1)T]$, *if there exist* $x_s(t_0+) \in [a_0, a_1)$ *for some* $s \neq j$ *and* $x_i(t_0+) \in S_{k_i}$, $i \neq j$.

We can see that (10.26) changes the style of interaction in the model. It depends on distance of oscillators to thresholds. We use this to introduce delay and continuous couplings in papers [22] and [23], respectively.

To illustrate Theorem 10.2, consider a group of oscillators, x_i, $i = 1, 2, \ldots, 100$, with random uniform distributed start values in $[0, 1]$. It is supposed that they satisfy the equations $x_i' = (3 + 0.01\overline{\mu}_i) - (2 + 0.01\overline{\zeta}_i)x_i$. The constants $\overline{\mu}_i$, $\overline{\zeta}_i$, as well as $\overline{\xi}_i$ in the thresholds $1 + 0.005\overline{\xi}_i$, $i = 1, 2, \ldots, 100$, are uniform random distributed numbers from $[0, 1]$. In Figure 10.3 one can see the result of simulation with $\varepsilon = 0.08$, where the state of the system is shown before the first, twenty first, forty second and sixty third firing of the system. The flat sections of the graph are groups of synchronized oscillators. So, it is obvious that eventually the model shows synchrony.

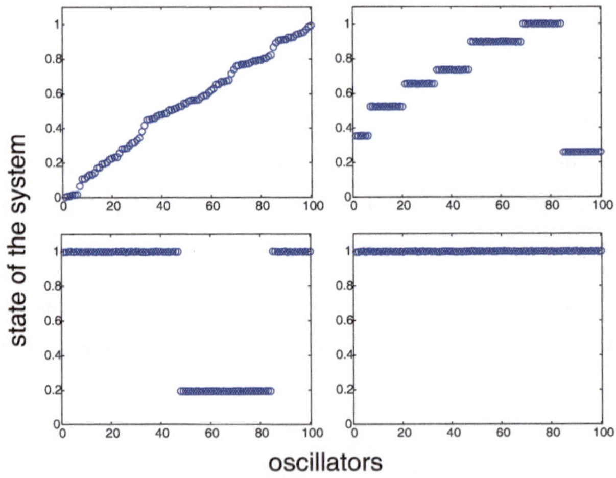

Fig. 10.3 The state of the model before the first, twenty first, forty second and sixty third firing of the system. The flat sections of the graph are groups of synchronized oscillators. (Color online)

Let us describe a more general system of oscillators such that Theorem 10.2 is still true. A system of n oscillators is given, such that if i-th oscillator does not fire or jump up, then it satisfies the i-th equation of system (3.1). If several oscillators x_{i_s}, $s = 1,2,\ldots,k$, fire so that $x_{i_s}(t) = 1+\zeta_{i_s}(x)$ and $x_{i_s}(t+) = 0$, then all other oscillators x_{i_p}, $p = k+1,k+1,\ldots,n$, change their coordinates by the law

$$x_{i_p}(t+) = \begin{cases} 0 & \text{if } x_{i_p}(t)+\varepsilon+\sum\limits_{s=1}^{k}\varepsilon_{i_p i_s} \geqslant 1+\zeta_{i_p}(x), \\ x_{i_p}(t)+\varepsilon+\sum\limits_{s=1}^{k}\varepsilon_{i_p i_s} & \text{otherwise.} \end{cases}$$

One can easily see that the last theorem is correct for the model just described, if $\varepsilon + \sum_{s=1}^{k}\varepsilon_{i_p i_s} > 0$, for all possible k, i_p and i_s.

Remark 10.3. The analysis of non-identical oscillators with non-small parameters may shed light on the investigation of arrhythmias, chaotic flashing of fireflies, etc. Namely, the dynamics in the neighborhood and inside the periodic trajectory, Q in Figure10.1, can be very complex. We do not exclude the possibility of chaos and fractals [298]. Bifurcation of periodic solutions can be discussed, if the parameters are small.

10.4 The Kamke condition

In this section we consider an integrate-and-fire model with a new type of continuous connection.

We believe that models of identical oscillators with more general differential equations,

$$\begin{aligned} x_1' &= g(x_1,x_2,\ldots,x_n), \\ x_2' &= g(x_2,x_3,\ldots,x_1), \\ &\cdots\cdots\cdots \\ x_n' &= g(x_n,x_1,\ldots,x_{n-1}), \end{aligned} \tag{10.27}$$

where $0 \leqslant x_i \leqslant 1$, $i = 1,2,\ldots,n$, are of both theoretical and applied interest. The positive valued function $g(y_1,y_2,\ldots,y_n)$ in (10.27) is continuously differentiable and indifferent with respect to permutations of coordinates y_2 to y_n.

When the oscillator x_j fires at the moment t such that $x_j(t) = 1$, $x_j(t+) = 0$, then the value of an oscillator with $i \neq j$, changes so that

$$x_i(t+) = \begin{cases} 0 & \text{if } x_i(t)+\varepsilon \geqslant 1, \\ x_i(t)+\varepsilon & \text{otherwise.} \end{cases} \tag{10.28}$$

Consider the cone $\mathbb{R}^n_+ \subset \mathbb{R}^n$ of all vectors with nonnegative coordinates. Introduce a partial order in the cone such that $a \leqslant b$ if $a_i \leqslant b_i, i = 1,2,\ldots,n$, [290].

We say that function g is of type $\mathscr{K}$ in $\mathbb{R}^n_+$ if $a \leqslant b$ implies that $g(a) \leqslant g(b)$. The sufficient condition for that is $\frac{\partial g(y)}{\partial y_i} \geqslant 0, i \neq 1$.

Let $u(t,t_0,u_0)$ and $u(t,t_0,u_1)$, u_0, $u_1 \in \mathbb{R}^n_+$, be solutions of (10.27). If g is of type $\mathscr{K}$, then [179, 290] the dynamics of (10.27) is monotone for $t \geqslant 0$. That is, $u(t,0,u_0) \leqslant u(t,0,u_1)$, if $u_0 \leqslant u_1$.

Consider first the model of two oscillators. Define the map L for this system in the following way. Take the solution $u(t) = u(t,0,(0,v+\varepsilon)) = (u_1,u_2)$. Denote by $s(v)$ the moment when $u_2(s) = 1$, and define the function $\overline{L}(v) = u_1(s)$ on $(0,1-\varepsilon)$. Then define map L through (10.6). Let us check if conditions (A1), (A3) are valid for this map. Indeed, the continuity of $\overline{L}$ is obvious. It is non-increasing since the monotonicity. Assume that there exist numbers $v_1, v_2 \in (0,1-\varepsilon)$ such that $v_1 < v_2$ and $s = s(v_1) = s(v_2)$. Then, we have a contradiction as the open interval (v_1,v_2) is mapped to the closed set $\{1\}$. That is, $\overline{L}$ satisfies (A1).

Condition (A3) is easily verifiable. Now, one can determine sets S_i, similarly to that in Section 10.1, and prove that the following theorem is valid.

Theorem 10.4. *Assume that g is of $\mathscr{K}$ type,* (A2) *is valid, and $t_0 \geqslant 0$ is a firing moment such that $x_1(t_0) = 1, x_1(t_0+) = 0$. If $x_2(t_0+) \in S_m$, m is a natural number, then the couple x_1, x_2 synchronizes within the time interval $[t_0 + \frac{m}{2}\widetilde{T}, t_0 + mT]$.*

We have, moreover, that, if g is of $\mathscr{K}$ type and (A2) is not true then the system does not synchronize.

Consider the multidimensional system of oscillators. Introduce the function $G(y,z) \equiv g(y,z,z,\ldots,z)$, and define the integrate-and-fire model of two identical oscillators y and z with the following system of differential equations

$$
\begin{aligned}
y' &= G(y,z),\\
z' &= G(y,z).
\end{aligned}
\tag{10.29}
$$

Denote by $u = (y,z), u(t) = u(t,0,(0,v+\varepsilon))$, the solution of (10.29), and by $s(v)$ the moment when $z(s) = 1$. Next, define the function $\overline{L}(v) = y(s)$ on $(0,1-\varepsilon)$. Then map L can be defined by (10.6) as well as correspond sets S_i. By applying the monotonicity of the dynamics, one can prove the following assertion, in a way very similar to that of Theorem 10.2.

Theorem 10.5. *Assume that* $0 \leqslant \frac{\partial g(y)}{\partial y_i} < \eta$, $i \neq 1$, *condition* (A2) *is valid, and* $t_0 \geqslant 0$ *is a firing moment such that* $x_j(t_0) = 1, x_j(t_0+) = 0$. *If parameter* η *is sufficiently small then the motion* $x(t)$ *of the system synchronizes within the time interval* $[t_0, t_0 + T]$, *if* $x_i(t_0+) \in S_0$, $i \neq j$, *and within the time interval* $[t_0 + \frac{1}{2}(\max_{i\neq j} k_i - 1)\widetilde{T}, t_0 + (\max_{i\neq j} k_i + 1)T]$, *if there exist* $x_s(t_0+) \in [a_0, a_1)$ *for some* $s \neq j$ *and* $x_i(t_0+) \in S_{k_i}$, $i \neq j$.

Example 10.4 shows that the analysis of the map L for Theorems 10.4 and 10.5 is not simple even with linear differential equations. The results of this section are therefore are provided for numerical application, as well as for future investigations.

10.5 The delayed pulse-coupling

In this section we consider the integrate-and-fire model of the cardiac pacemaker with delayed pulsatile coupling. Sufficient conditions of synchronization are obtained for identical and non-identical oscillators. We omit proof of all assertions in this section as they are similar to those in previous part of the chapter.
Delays arise naturally in many biological models [237]. In particular, they were considered in firefly models [62] as delay between stimulus and response, and in continuously coupled neuronal oscillators [182]. Authors of [109] considered the phenomenon for the Mirollo and Strogatz analysis, [216]. Identical oscillators were investigated. Two oscillators dynamics is discussed mathematically, and the multi-oscillatory system by computer simulations. It was found that the excitatory model of two units "can get only out-of-phase synchronization since in-phase synchronization proved to be not stable." In paper [122] a model without leakage was discussed, that is, oscillators increase at a constant rate between moments of firing. It was found that a periodic solution is reached after a finite time. Consequently, research of integrate-and-fire models, which admit delays and fire in unison is still on the agenda.

10.5.1 *The couple of identical oscillators*

Let us start to analyze two identical oscillators, which satisfy, if they do not fire, the following differential equations

$$x_i' = S - \gamma x_i, \tag{10.30}$$

where $0 \leqslant x_i \leqslant 1$, $i = 1, 2$. It is assumed that S, γ are positive numbers and $\kappa = \frac{S}{\gamma} > 1$.

When $x_j(t) = 1$, then the oscillator fires, $x_j(t+) = 0$. The firing changes value of the another oscillator, x_i, such that

$$x_i(t+) = 0 \ \text{ if } \ x_i(t) \geqslant 1 - \varepsilon, \tag{10.31}$$

and

$$x_i(t+\tau+) = x_i(t+\tau) + \varepsilon \ \text{ if } \ x_i(t) < 1 - \varepsilon. \tag{10.32}$$

We have that

$$x_i(s) = x_i(t)\mathrm{e}^{-\gamma(s-t)} + \int_t^s \mathrm{e}^{-\gamma(s-u)} S du$$

near t.

In what follows, assume that

$$\frac{\kappa - 1}{\kappa - 1 + \varepsilon} < \mathrm{e}^{-\gamma\tau}. \tag{10.33}$$

Then, from

$$\begin{aligned} \|x_i(s)\| &\leqslant \|x_i(t)\|\mathrm{e}^{-\gamma(s-t)} + \int_t^s \mathrm{e}^{-\gamma(s-u)} S du \\ &\leqslant (1-\varepsilon)\mathrm{e}^{-\gamma\tau} + \kappa(1 - \mathrm{e}^{-\gamma\tau}), \end{aligned}$$

and $x_i(t) < 1 - \varepsilon$, we obtain that $x_i(s) < 1$, for all $s \in [t, t+\tau]$. In other words oscillator x_i does not achieve the threshold within interval $[t, t+\tau]$, if the distance of $x_i(t)$ to threshold is more than ε. This is important for the construction of the prototype map, and makes a sense of condition (10.32).

One must emphasis that couplings of units are not only delayed in our model. By (10.31) oscillators interact instantaneously, if they are near threshold. This assumption is natural as firing provokes another oscillator, which being close to threshold "is ready" to react instantaneously. Otherwise, the interaction is retarded.

Next, we shall construct the prototype map. Fix a moment $t = \zeta$, when x_1 fires, and suppose that oscillators are not synchronized. In interval $[\zeta, \zeta+\tau]$ oscillator x_2 moves by law

$$x_2(t) = x_2(\zeta)\mathrm{e}^{-\gamma(t-\zeta)} + \int_\zeta^t \mathrm{e}^{-\gamma(t-u)} S du,$$

and

$$x_2(\zeta+\tau) = [x_2(\zeta) - \kappa]\mathrm{e}^{-\gamma\tau} + \kappa. \tag{10.34}$$

Denote $t = \eta$, the firing moment of x_2, then

$$x_2(\eta) = [x_2(\zeta+\tau) + \varepsilon]\mathrm{e}^{-\gamma(\eta-\zeta-\tau)} + \kappa[1 - \mathrm{e}^{-\gamma(\eta-\zeta-\tau)}].$$

The equation $x_2(\eta) = 1$ implies that

$$e^{-\gamma(\eta-\zeta)} = \frac{1-\kappa}{x_2(\zeta) - \kappa + \varepsilon_1}, \tag{10.35}$$

where $\varepsilon_1 = \varepsilon e^{\gamma\tau}$. Since $x_1(\eta) = \kappa[1 - e^{-\gamma(\eta-\zeta)}]$, we have that

$$x_1(\eta) = \kappa \frac{1-(x_2(\zeta)+\varepsilon_1)}{\kappa - (x_2(\zeta)+\varepsilon_1)}. \tag{10.36}$$

Introduce the following map

$$L_D(v,\varepsilon) = \kappa \frac{1-(v+\varepsilon_1)}{\kappa - (v+\varepsilon_1)}, \tag{10.37}$$

such that $x_1(\eta) = L_D(x_2(\zeta)), \varepsilon)$. If $t = \xi$ is the next to η firing moment of x_2, then one can similarly find that $x_2(\xi) = L_D(x_1(\eta), \varepsilon)$. One can see that the map L_D can be useful for our investigation, since it evaluates alternatively the sequence of values x_1 and x_2 at firing moments.

Take $\tau > 0$ so small that

$$e^{-\gamma\tau} > \varepsilon. \tag{10.38}$$

From (10.38) it implies that $\varepsilon_1 < 1$.

One can evaluate that

$$L_D(1-\varepsilon_1, \varepsilon) = 0.$$

and the derivatives of the map in $(0, 1-\varepsilon_1)$ satisfy

$$L'_D(v,\varepsilon) = \kappa \frac{1-\kappa}{(\kappa-(v+\varepsilon_1))^2} < 0, \tag{10.39}$$

and

$$L''_D(v,\varepsilon) = 2\kappa \frac{1-\kappa}{(\kappa-(v+\varepsilon))^3} < 0 \tag{10.40}$$

We can easily find that there is a fixed point of the map,

$$v^* = (\kappa - \frac{\varepsilon_1}{2}) - \sqrt{\kappa^2 - \kappa + \frac{\varepsilon_1^2}{4}}, \tag{10.41}$$

and

$$L'_D(v^*, \varepsilon) < -1. \tag{10.42}$$

That is, fixed point v^* is a repellor.

Now, we will define an extension of L_D on $[0,1]$ in the following way. Let

$$\omega = \kappa \frac{1-\varepsilon_1}{\kappa - \varepsilon_1}. \tag{10.43}$$

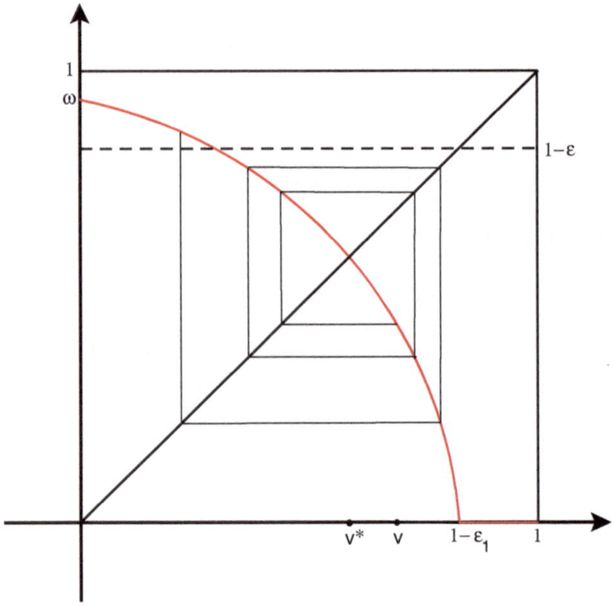

Fig. 10.4 The graph of map L_D in red, fixed point v^*, and stabilized trajectory are seen.

One can see that $1-\varepsilon<\omega<1$, if

$$e^{\gamma\tau}<\frac{\kappa}{\kappa-1+\varepsilon}. \tag{10.44}$$

In what follows, we assume that ε is sufficiently small such that (10.33) implies (10.44). We set $L_D(0,\varepsilon)=\omega$, and define $L_D(v,\varepsilon)=0$, if $1-\varepsilon_1\leqslant v\leqslant 1$. Since $L_D:[0,1]\to[0,1]$ is a monotonic continuous function and $[0,1]$ is an invariant set of this map, it is convenient for analysis by using iterations. The graph of this map is seen in Figure 10.4.

Now, by applying properties of L_D, and analyzing self-compositions of the map, one can easily obtain that for all $k\geqslant 0$ functions L_D^k have only one fixed point, v^*, and $|[L_D^k(v^*,\varepsilon)]'|>1$. We skip the discussion as it is respectively simple, and request a large place. Since all the maps L_D^k have one and the same fixed point, v^*, there is not a $k-$periodic motion, $k>1$, of the map. Consequently, for arbitrary point $v\neq v^*$ one has a stabilized trajectory present in Figure 10.4. The couple synchronizes when $L_D^k(v,\varepsilon)\geqslant 1-\varepsilon_1$.

Next, we investigate the rate of synchronization. Set $a_0=L_D^{-1}(\omega)=0$, $a_{k+1}=L_D^{-1}(a_k)$, $k=0,1,2,\ldots$.

Set $T=\frac{1}{\gamma}\ln\frac{\kappa}{\kappa-1}$ and denote by $\widetilde{T}$ the time needed for solution $u(t,0,v^*)$ of the equation $u'=S-\gamma u$, to achieve threshold. Since all oscillators fire within an interval of length T

and the distance between two firing moments of an oscillator are not less than $\widetilde{T}$, we can conclude that the following theorem is correct.

Theorem 10.6. *Assume that* (10.33) *and* (10.38) *are valid. If* $t_0 \geqslant 0$ *is a firing moment,* $x_1(t_0) = 1$, $x_1(t_0+) = 0$, *and* $x_2(t_0+) \in S_m$ *for some natural number* m, *then the couple* x_1, x_2 *of continuously coupled identical biological oscillators synchronizes within the time interval* $[t_0 + \frac{m}{2}\widetilde{T}, t_0 + Tm]$.

10.5.2 *Non-identical oscillators: the general case*

To make our investigation closer to the real world problems one has to consider an ensemble of non-identical oscillators. We will discuss the following system of equations

$$x_i' = (S + \mu_i) - (\gamma + \zeta_i)x_i, \tag{10.45}$$

where $0 \leqslant x_i \leqslant 1 + \xi_i$, $i = 1, 2, \ldots, n$. The constants S and γ are the same as in the last section such that $\kappa = \frac{S}{\gamma} > 1$. Moreover, constants μ_i and ζ_i are sufficiently small for $\kappa_i = \frac{S+\mu_i}{\gamma+\zeta_i} > 1$. When $x_j(t) = 1 + \xi_j$ then the oscillator fires, $x_j(t+) = 0$. The firing changes values of other oscillators x_i, $i \neq j$, such that

$$x_i(t+) = 0 \ \text{ if } \ x_i(t) \geqslant 1 - \varepsilon \tag{10.46}$$

and, if $x_i(t) < 1 - \varepsilon$, then

$$x_i(t + \tau +) = x_i(t + \tau) + \varepsilon + \varepsilon_i. \tag{10.47}$$

In what follows, we call real numbers ε, μ_i, ζ_i, ξ_i, ε_i, *parameters*, assuming the first one is positive. Moreover, constants μ_i, ζ_i, ξ_i, ε_i will be called *parameters of perturbation.* Assume that they are zeros to obtain the model of *identical oscillators.* We have that an exhibitory model is under discussion, that is $\varepsilon + \varepsilon_i > 0$ for all i. Coupling is all-to-all such that each firing elicits jumps in all non-firing oscillators. If several oscillators fire simultaneously, then other oscillators react as it just one oscillator fires. In other words, any firing acts only as a signal which abruptly provokes a state change, the intensity of the signal is not important, and pulse strengths are not additive. We have that

$$x_i(s) = x_i(t)\mathrm{e}^{-(\gamma+\zeta_i)(s-t)} + \int_t^s \mathrm{e}^{-(\gamma+\zeta_i)(s-u)}(S + \mu_i)du, \tag{10.48}$$

near t.

If one assume that condition (10.33) is valid, and constants μ_i and ζ_i are sufficiently small such that

$$\frac{\kappa_i - 1}{\kappa_i - 1 + \varepsilon} < \mathrm{e}^{-(\gamma+\zeta_i)\tau}, \tag{10.49}$$

then $x_i(s) < 1$ for all $s \in [t, t+\tau]$, if $x(t) < 1-\varepsilon$.

In this section we begin with analysis of a couple of oscillators of the ensemble of n oscillators, and find that the couple synchronizes if parameters close to zero. Then synchronization of the ensemble will be proved.

Consider the model of n non-identical oscillators given by relations (10.30) and (10.32). Fix two of them, let say, x_ℓ, x_r.

Lemma 10.3. *Assume that conditions* (10.33) *and* (10.38) *are valid,* $t_0 \geqslant 0$ *is a firing moment such that* $x_\ell(t_0) = 1+\xi_i$, $x_\ell(t_0+) = 0$. *If parameters are sufficiently close to zero, and absolute values of parameters of perturbation are sufficiently small with respect to* ε, *then the couple* x_ℓ, x_r *synchronizes within the time interval* $[t_0, t_0+T]$ *if* $x_r(t_0+) \notin [a_0, a_1)$ *and within the time interval* $[t_0 + \frac{m-1}{2}\widetilde{T}, t_0 + (m+1)T]$, *if* $x_r(t_0+) \in S_m$, $m \geqslant 1$.

Let us extend the result of the last Lemma for the whole ensemble.

Theorem 10.7. *Assume that* (10.33) *and* (10.38) *are valid,* $t_0 \geqslant 0$ *is a firing moment such that* $x_j(t_0) = 1+\xi_j$, $x_j(t_0+) = 0$. *If the parameters are sufficiently close to zero, and absolute values of parameters of perturbation are sufficiently small with respect to* ε, *then the motion* $x(t)$ *of the system synchronizes within the time interval* $[t_0, t_0+T]$, *if* $x_i(t_0+) \notin [a_0, a_1)$, $i \neq j$, *and within the time interval* $[t_0 + \frac{1}{2}(\max_{i\neq j} k_i - 1)\widetilde{T}, t_0 + (\max_{i\neq j} k_i + 1)T]$, *if there exist* $x_s(t_0+) \in [a_0, a_1)$ *for some* $s \neq j$ *and* $x_i(t_0+) \in S_{k_i}$, $i \neq j$.

Let us introduce a more general system of oscillators such that Theorem 10.7 is still true. Consider a system of n oscillators given such that if i-th oscillator does not fire or jump up, it satisfies i-th equation of system (10.30). If several oscillators x_{i_s}, $s = 1, 2, \ldots, k$, fire such that $x_{i_s}(t) = 1 + \phi(t, x(t), x(t-\tau_{i_s}))$, where $|\phi(t, x(t), x(t-\tau_i))| < \xi_i$, $i = 1, 2, \ldots, n$, and $x_{i_s}(t+) = 0$, then all other oscillators x_{i_p}, $p = k+1, k+1, \ldots, n$, change their coordinates by law

$$x_i(t+) = 0, \text{ if } x_i(t) \geqslant 1-\varepsilon \tag{10.50}$$

and, if $x_i(t) < 1-\varepsilon$, then

$$x_i(t+\tau+) = x_i(t+\tau) + \varepsilon + \sum_{s=1}^{k} \varepsilon_{i_p i_s}. \tag{10.51}$$

One can easily see that the last theorem is correct for the model just have been described, if $\varepsilon + \sum_{s=1}^{k} \varepsilon_{i_p i_s} > 0$, for all possible k, i_p and i_s, and we assume that ε_{ij} are also parameters of perturbation. Moreover, one can easily see that initial functions for thresholds conditions can be chosen arbitrarily with values in the domain of the system.

Remark 10.4. Our preliminary analysis shows that the dynamics in a neighborhood of v^* can be very complex. We do not exclude that a chaos appearance can be observed, and trajectories may belong to a fractal, if parameters are not small. It does not contradict the zero Lebesgue measure of non-synchronized points. Possibly, analysis of non-identical oscillators with not small parameters is of significant interests to explore arrhythmias, earthquakes, chaotic flashing of fireflies, etc.

Remark 10.5. The time of synchronization for a given initial point does not increase if number of oscillators increases (but the parameters needed to be closer to zero). This property, possibly, can be accepted as a small-world phenomenon.

10.5.3 *The simulation result*

To demonstrate our main result numerically, let us consider a model of 100 oscillators, which initial values are randomly uniform distributed in $[0,1]$. Their differential equations are of form

$$x_i' = (4.1 + 0.01 * \text{sort}(\text{rand}(1,n)) - (3.2 + 0.01 * \text{sort}(\text{rand}(1,n))x_i,$$

and thresholds

$$1 + 0.005 * \text{sort}(\text{rand}(1,n)), \quad i = 1,2,\ldots,100,$$

where deviations of coefficients the threshold are also uniformly random in $[0,1]$. We place the result of simulation with $\varepsilon = 0.06$ and $\tau = 0.002$ in Figure 10.5, where the state of the system is shown at the initial moment, before the 183-th jump, before the 366-th jump and the last is before the 549-th jump. That is, it is obvious that eventually all oscillators fire in unison.

Notes

A version of the integrate-and-fire model of pulse-coupled and non-identical oscillators is investigated in this chapter. We have solved the second Peskin's conjecture, though we have not showed that the measure of non-synchronized initial values is zero as it was shown in [216] for identical oscillators. However, we have located non-synchronized points, and showed that the time of synchronization infinitely increases as the region of points to be synchronized enlarges. Moreover, it is not necessary, in applications, for all points of the domain to fire in unison, and it is sufficient for the neighborhood of a motion to be synchronized.

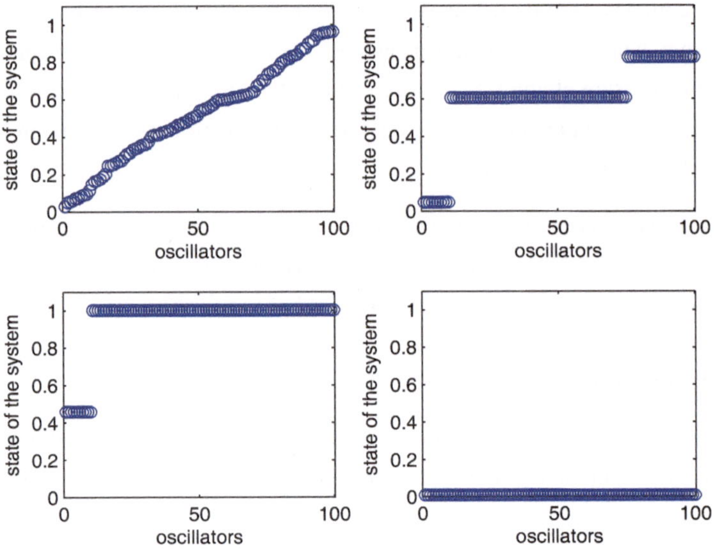

Fig. 10.5 The state of the model before the first, the 183-th, 366-th, and 549-th jump is seen. The flat fragments of the graph are groups of oscillators firing in unison.

A prototype map is introduced that helps to precise the results for a system of two identical oscillators and to solve the problem for the multi-non-identical-oscillators model. The map is, in fact a Poincaré map if one considers the identity of oscillators. The approach in this chapter is universal. For example, it can be applied if the thresholds are of the form $1+\phi(t,x)$ or $1+\phi(t)$, which represents an oscillating signal in physiology [127] and a variation of the threshold of the electronic relaxation oscillator [266]. Moreover, the jump's value ε as well as its perturbations may depend on x. The method can be easily extended for models, where differential equations have discontinuous right-hand sides, and only the existence of the map L with request properties are important to get appropriate results. Not only exhibitory, but inhibitory models of oscillators can be investigated, as well as delay of couplings [22, 62, 109, 122], and continuous couplings generated by firing [23, 301]. One can consider the models in this chapter as discontinuous *cooperative* systems [290]. Consequently, we expect that by applying methods of dynamical systems with variable moments of discontinuity, the results on monotone systems [158,159,290] can be extended for these models.

There is a rich collection of results on oscillators, obtained through experiments and simulations. The approach of the present chapter can give theoretical background for them and also form a basis for new ones. It can be applied not only to the problems of synchronization, but also to periodic, almost periodic motions, and the complex behavior of biological models. New small-world phenomena can be discovered. One can now request a certain property for the map L and then look for a system which meets the property. Thus, many new theoretical challenges can be brought under discussion. Conversely, if a system is given, then one can construct the corresponding map L, and by analyzing find new features which have not been mentioned in the present investigation.

The cardiac pacemaker model of identical and non-identical oscillators with delayed pulse-couplings is investigated in Section 10.5. We apply the method of investigation proposed in previous sections, which is based on a specially defined map. Sufficient conditions are found such that delay involvement in the Peskin's model does not change the synchronization result for identical and non-identical oscillators [22,216,262]. The result has a biological sense, since retardation is often presents in biological processes and if one proves that a phenomenon preserves even with delays, that makes us more confident that the model is adequate to the reality. Moreover, the method of treatment of models with delay can be useful for neural networks and earthquake faults [109,122,157,163,246] analysis. All proved assertions can be easily specified with $\tau = 0$, to obtain synchronization of the Peskin's model for identical [216,262] and nonidentical [22] oscillators. In particular, from the theorem of Subsection 10.5.1 one can obtain the result of Example 10.2. Main results of this chapter are published in [22,24].

Bibliography

[1] Abbiw-Jackson, R.M. and Langford, W.F., *Gain-induced oscillations in blood pressure*, Journal of Mathematical Biology, **37**, 203–234, (1998).

[2] Aftabizadeh, A. R. and Wiener, J., *Oscillatory and periodic solutions for systems of two first order linear differential equations with piecewise constant arguments*, Applcable Anal., **26**, 327–333, (1988).

[3] Aftabizadeh, A. R., Wiener, J. and Xu, J.-M., *Oscillatory and periodic solutions of delay differential equations with piecewise constant argument*, Proc. Amer. Math. Soc., **99**, 673–679, (1987).

[4] Aftabizadeh, A. R. and Wiener, J., *Differential inequalities for delay differential equations with piecewise constant argument*, Appl. Math. Comput., **24**, 183–194, (1987).

[5] Aftabizadeh, A. R. and Wiener, J., *Oscillatory and periodic solutions of advanced differential equations with piecewise constant argument*, Nonlinear Analysis and Applications, Dekker, New York, 31–38, (1987).

[6] Aftabizadeh, A. R. and Wiener, J., *Oscillatory and periodic solutions of an equation alternately of retarded and advanced type*, Appl. Anal., **23**, 219–231,(1986).

[7] Aftabizadeh, A. R. and Wiener, J., *Oscillatory properties of first order linear functional-differential equations*, Applicable Anal., **20**, 165–187, (1985).

[8] Akhmet M. U., *On the integral manifolds of the differential equations with piecewise constant argument of generalized type*, Proceedings of the Conference on Differential and Difference Equations at the Florida Institute of Technology, August 1-5, 2005,, Melbourne, Florida, Editors: R.P. Agarval and K. Perera, Hindawi Publishing Corporation, 2006, 11–20.

[9] Akhmet, M. U., *Integral manifolds of differential equations with piecewise constant argument of generalized type*, Nonlinear Anal.: TMA, **66**, 367–383, (2007).

[10] Akhmet, M. U., *On the reduction principle for differential equations with piecewise constant argument of generalized type*, J. Math. Anal. Appl., **336**, 646–663, (2007).

[11] Akhmet M. U., *Stability of differential equations with piecewise constant argument of generalized type*, Nonlinear Anal.: TMA, **68**, 794–803, (2008).

[12] Akhmet, M. U., *Almost periodic solutions of differential equations with piecewise constant argument of generalized type*, Nonlinear Anal.: HS, **2**, 456–467, (2008).

[13] Akhmet, M. U., *Asymptotic behavior of solutions of differential equations with piecewise constant arguments*, Appl. Math. Lett., **21**, 951–956, (2008).

[14] Akhmet M. U., *Almost periodic solutions of the linear differential equation with piecewise constant argument*, Discrete and Impulsive Systems, Series A, Mathematical Analysis, **16**, (2009) 743–753.

[15] Akhmet, M. U., *Existence and stability of almost-periodic solutions of quasi-linear differential*

equations with deviating argument, Appl. Math. Lett., **17**, 1177–1181, (2004).

[16] Akhmet, M. U., *Devaney's chaos of a relay system*, Communications in Nonlinear Science and Numerical Simulation, **14**, 1486–1493, (2009).

[17] Akhmet M. U., *Shadowing and dynamical synthesis*, International Journal of Bifurcation and Chaos, **19**, 1–8, (2009).

[18] Akhmet M. U., *Dynamical synthesis of quasi-minimal sets*, International Journal of Bifurcation and Chaos. 19, no. 7, 1–5, (2009).

[19] Akhmet M. U., *Li-Yorke chaos in the impact system*, J. Math. Anal. Appl., **351**, 804–810, (2009).

[20] Akhmet M. U., *Principles of discontinuous dynamical systems*, Springer, New York, (2010).

[21] Akhmet, M. U., *Perturbations and Hopf bifurcation of the planar discontinuous dynamical system*, Nonlinear Analysis, **60**, 163–178, (2005).

[22] Akhmet, M. U., *Synchronization of the cardiac pacemaker model with delayed pulse-coupling*, (submitted), arXiv:1102.4462v1.

[23] Akhmet, M. U., *Self-synchronization of the integrate-and-fire pacemaker model with continuous couplings*, (submitted).

[24] Akhmet, M. U., *Analysis of biological integrate-and-fire oscillators*, (submitted), arXiv:1011.5632v2.

[25] Akhmet, M. U., Aruğaslan, D. *Lyapunov-Razumikhin method for differential equations with piecewise constant argument* Discrete and Continuous Dynamical Systems, Series A, **25**, 457–466, (2009).

[26] Akhmet M. U., Arugaslan D. and Liu X., *Permanence of non-autonomous ratio-dependent predator-prey systems with piecewise constant argument of generalized type*, Dynamics of Continuous, Discrete and Impulsive Systems, Series A., **15**, 37–52, (2008).

[27] Akhmet, M. U., Aruğaslan, D. and Yılmaz, E., *Stability in cellular neural networks with piecewise constant argument*, Journal of computational and applied mathematics, **233**, 2365–2373, (2010).

[28] Akhmet, M. U., Aruğaslan, D. and Yılmaz, E., *Stability analysis of recurrent neural networks with piecewise constant argument of generalized type*, Neural Networks **23**, 805–811, (2010).

[29] Akhmet, M. U., Aruğaslan, D. and Yılmaz, E., *Method of Lyapunov functions for differential equations with piecewise constant delay*, Journal of computational and applied mathematics. (in press)

[30] Akhmet M. U., Bekmukhambetova G.A., A prototype compartmental model of the blood pressure distribution. Nonlinear Analysis: RWA, **11**, 1249–1257, (2010).

[31] Akhmet M. U., Bekmukhambetova G. A., On modeling of blood pressure distribution, Proceedings of the Fifth International Conference on Dynamic Systems and Applications held at Morehouse College, Atlanta, USA, May 30-June 2, 2007. Proceedings of Dynamic Systems and Applications, **5**, 17–20, (2008).

[32] Akhmet M. U., Bekmukhambetova G.A. and Y. Serinagaoglu, *Regular and Irregular Discontinuous Dynamics of the Systemic Arterial Pressure*, Extended abstracts of the International Conference on Numerical Analysis and Applied Mathematics 2005, 16-20 August, 2005, Rodos, pp. 44–46.

[33] Akhmet, M. U., Buyukadali, C., *Differential equations with a state-dependent piecewise constant argument*, Nonlinear Anal.: TMA, **72**, 4200–4210, (2010).

[34] Akhmet M. U. and Buyukadali C., *Periodic solutions of the system with piecewise constant argument in the critical case*, Comput. Math. Appl., 56, 2034–2042, (2008).

[35] Akhmet M. U., Buyukadali C. and Ergenc T., *Periodic solutions of the hybrid system with small parameter*, Nonlinear Analysis: HS, **2**, 532–543, (2008).

[36] Akhmet, M. U., Turan, M. *The differential equations on time scales through impulsive differen-*

tial equations, Nonlinear Anal.: TMA, **65**, 2043–2060, (2006).

[37] Akhmet,M. U., Turan, M., *Differential equations on variable time scales*, Nonlinear Anal.: TMA, **70**, 1175–1192, (2009).

[38] Akhmet M. U., Oktem H., Pickl S.W. and Weber G.-W., *An anticipatory extension of Malthusian model*, in: *Computing Anticipatory Systems*, CASYS'05, Seventh International Conference on Computing Anticipatory Systems (Liege, Belgium, August, 2005), D. Dubois, ed. AIP Conference Proceedings (Melville, New York), **839**, 260–264, (2006).

[39] Akhmetov, M. U., Perestyuk, N. A., *Differential properties of solutions and integral surfaces of nonlinear impulse systems*, Differential Equations, **28**, 445–453 (1992).

[40] Akhmetov, M. U., Perestyuk, N. A., *Periodic and almost-periodic solutions of strongly nonlinear impulse systems*, J. Appl. Math. Mech., **56**, 829–837, (1992).

[41] Akhmetov, M. U., Perestyuk, N. A., *Almost-periodic solutions of nonlinear impulse systems*, Ukrainian Math. J., **41**, 259–263 (1989).

[42] Akhmetov, M. U., Perestyuk, N. A., Samoilenko, A.M., *Almost-periodic solutions of differential equations with impulse action*, (Russian) Akad. Nauk Ukrain. SSR Inst., Mat. Preprint, no. 26, 49 pp, (1983).

[43] Alonso, A. and Hong, J., *Ergodic type solutions of differential equations with piecewise constant arguments*, Int. J. Math. Math. Sci., **28**, 609–619, (2001).

[44] Alonso, A., Hong, J. and Obaya, R., *Almost-periodic type solutions of differential equations with piecewise constant argument via almost periodic type sequences*, Appl. Math. Lett., **13**, 131–137, (2000).

[45] Altıntan, D., *Extension of the logistic equation with piecewise constant arguments and population dynamics*, MSc. Thesis, Middle East Technical University, (2006).

[46] Amerio, L. and Prouse, G., *Almost-periodic functions and functional equations*, Van Nostrand Reinhold Company, New York, (1971).

[47] Anderson, D.H., *Compartmental modeling and tracer kinetics*, Springer-Verlag, New-York, (1983).

[48] Andronov, A. A., Vitt, A. A. and Khaikin, C. E., *Theory of Oscillations*, Pergamon Press, Oxford, (1966).

[49] Apostol, T. *Mathematical Analysis*, Addison-Wesley Publishing Company, Massachusetts, London, (1971).

[50] Arnold V.I., *Cardiac arrhythmias and circle mappings*, Chaos, **1**, 20–24.

[51] Aulbach, B., Wanner, T., *Integral manifolds for Caratheodory type differential equations in Banach spaces*, in Six lectures on dynamical systems, (eds. B. Aulbach and F. Colonius), World Scientific, Singapore, 45–119, (1966).

[52] Balanter, B.J., Hanin, M.A. and Chernavsky, D.S., *Introduction to mathematical modelling of pathological processes*, (russian), Medicine, Moskow, (1980).

[53] Baselli, G., Cerutti, S., Civardi, S., Malliani, A., and Pagani, M., *Cardiovascular variability signals: Towards the identification of a closed-loop model of the neural control mechanisms*, IEEE Trans Biomed Eng., **35**, 1033–1046, (1988).

[54] Barbashin, E.A., *Introduction to the theory of stability*, Wolters-Noordhoff Publishing, Groningen, (1970).

[55] Bect, J., *A unifying formulation of the Fokker-Planck-Kolmogorov equation for general stochastic hybrid systems*, Nonlinear Anal. Hybrid Syst, **4**, 357–370, (2010).

[56] Blekhman, I. I., *Synchronization of dynamical systems* (Russian), Nauka, Moscow, (1971).

[57] Bellman, R., *Mathematical methods in medicine*, World Scientific, Singapore, (1983).

[58] Bohr, H., *Fastperiodische functionen*, Springer-Verlag, Berlin, (1932).

[59] Bogolyubov, N.N. *On some statistical methods in mathematical physics*, (Russian), Acad. Nauk U.R.S.R., (1945).

[60] Bottani, S., *Pulse-coupled relaxation oscillators: From biological synchronization to self-*

organized criticality, Phys. Rev. Lett., **74**, 4189–4182, (1995).

[61] P.C. Bressloff, S. Coombes, *Spike train dynamics underlying pattern formation in integrate-and-fire oscillator networks*, Phys. Rev. Lett.**81**, 2384–2387, (1998).

[62] J. Buck, *Synchronous Rhythmic Flashing of Fireflies. II.*, The Quarterly Review of Biology, **63**, no. 3, 265–290, (1988).

[63] Buck, J. and Buck, E., *Mechanism of rhythmic synchronous flashing of fireflies: Fireflies of Southeast Asia may use anticipatory time-measuring in synchronizing their flashing*, Science, **159**, 1319–1327, (1968).

[64] Burton, T. A. and Furumochi, T., *Fixed points and problems in stability theory for ordinary and functional differential equations*, Dynam. Systems Appl., **10**, 89–116, (2001).

[65] Busenberg, S. and Cooke, K. L., *Models of vertically transmitted diseases with sequential-continuous dynamics*, Nonlinear Phenomena in Mathematical Sciences, Academic Press, New York, 179–187, (1982).

[66] Byrne, H. M. and Gourley, S. A., *Global convergence in a reaction-diffusion equation with piecewise constant argument*, Math. Comput. Modelling, **34**, 403–409, (2001).

[67] Cabada, A., Ferreiro, J. B. and Nieto, J. J., *Green's function and comparison principles for first order periodic differential equations with piecewise constant arguments*, J. Math. Anal. Appl., **291**, 690–697, (2004).

[68] Carr, J., *Applications of center manifold theory*, Springer-Verlag, New York, (1981).

[69] Cao, J., *Global asymptotic stability of neural networks with transmission delays*, Int. J. Syst. Sci., **31**, 1313–1316, (2000).

[70] Carvalho, L. A. V. and Cooke, K. L., *A nonlinear equation with piecewise continuous argument*, Differential Integral Equations, **1** (3), 359–367, (1988).

[71] Cavalcanti, A. and Belardinelli, E. *Modelling cardiovascular variability using differential delay equation*, IEEE Trans. Biomed. Eng., **43**, 982, (1996).

[72] Channon, K. M., Hargreaves, M. R., Gardner, M. and Ormerod, O. J., *Noninvasive beat-to-beat arterial blood pressure measurement during VVI and DDD pacing: relationship to symptomatic benefit from DDD pacing*, Pacing Clin. Electrophysiol., **20**, 25–33, (1997).

[73] Chicone, C., Latushkin, Y., *Center manifolds for infinite-dimensional non -autonomous differential equations*, J. Differential Equations, **141**, 356–399, (1997).

[74] Chow, S.-N., Lu, K. C^k *centre unstable manifold*, Proc. Royal. Soc Edinburgh, **108A**, 285–317 (1988).

[75] Chueshov, I., *A reduction principle for coupled nonlinear parabolic-hyperbolic PDE*, J. Evol. Equ, **4**, 591–612, (2004).

[76] Ceragioli, F., *Finite valued feedback laws and piecewise classical solutions*, Nonlinear Anal., 65 (2006) 984–998.

[77] Coddington, E. A. and Levinson, N., *Theory of Ordinary Differential Equations*, New York, McGraw-Hill, (1955).

[78] Constant, I., Villain, E., Laude, D., Girard, A., Murat, I. and Elghozi, J.-L., *Heart rate control of blood pressure variability in children: A study in subjects with fixed ventricular pacemaker rhythm*, Clinical Science, **95**, 33–42, (1998).

[79] Cooke, K. L. and Györi, I., *Numerical approximation of the solutions of delay differential equations on an infinite interval using piecewise constant arguments*, Comput. Math. Appl., **28**, 81–92, (1994).

[80] K.L. Cooke, Turi, J., Turner, G. *Stabilization of hybrid systems in the presence of feedback delays*, Preprint Series 906, Inst. Math. and Its Appls., University of Minnesota, (1991).

[81] Cooke, K. L. and Wiener, J., *A survey of differential equation with piecewise continuous argument*, Lecture Notes in Math., **1475**, Springer, Berlin, 1–15, (1991).

[82] Cooke, K. L. and Wiener, J., *An equation alternately of retarded and advanced type*, Proc. Amer. Math. Soc., **99**, 726–732, (1987).

[83] Cooke, K. L. and Wiener, J., *Neutral differential equations with piecewise constant argument*, Boll. Un. Mat. Ital., **7**, 321–346, (1987).

[84] Cooke, K. L. and Wiener, J., *Stability for linear equations with piecewise continuous delay*, Comput. Math. Appl. Ser. A, **12**, 695–701, (1986).

[85] Cooke, K. L. and Wiener, J., *Retarded differential equations with piecewise constant delays*, J. Math. Anal. Appl., **99**, 265–297, (1984).

[86] Coppel, W.A. *Dichotomies in stability theory*, Lecture notes in Mathematics, Springer-Verlag, Berlin, Heidelberg, New York, (1978).

[87] Corduneanu, C., *Some remarks on functional equations with advanced-delayed operators*, Computing Anticipatory Systems, CASYS'03-Sixth International Conference, Liége, Belgium, 11-16 August 2003, American Institute of Physics, New York, (2004).

[88] Corduneanu, C., *Principles of Differential and Integral Equations*, Boston, Allyn and Bacon, (1971).

[89] Chua, L. O., *CNN: A Paradigm for Complexity*, World Scientific, Singapore, 1998.

[90] Chua, L. O., Yang, L, *Cellular neural networks: Theory*, IEEE Trans. Circuits Syst., **35**, 1257–1272, (1988).

[91] Chua, L. O., Yang, L, *Cellular neural networks: Applications*, IEEE Trans. Circuits Syst., **35**, 1273–1290, (1988).

[92] Chua, L. O., Roska, T., *Cellular neural networks with nonlinear and delay-type template elements*, in: Proc. 1990 IEEE Int. Workshop on Cellular Neural Networks and Their Applications, 12-25, (1990).

[93] Dai, L., *Nonlinear dynamics of piecewise constant systems and implementation of piecewise constant arguments*, World Scientific, Hackensack, NJ, (2008).

[94] Dai, L. and Singh, M. C., *On oscillatory motion of spring-mass systems subjected to piecewise constant forces*, J. Sound Vibration, **173**, 217–232, (1994).

[95] T. Danino, O. Mondragon-Palomino, L. Tsimring, J. Hasty, *A synchronized quorum of genetic clocks*, Nature **463**, 326–330, (2010).

[96] Dhage, B. and Lakshmikantham, V., *Basic results on hybrid differential equations*, Nonlinear Anal.: Hybrid Syst., **4**, 414–424, (2010).

[97] De la Sen, M. and Ibeas, A., *Stability results of a class of hybrid systems under switched continuous-time and discrete-time control*, Discrete Dyn. Nat. Soc. 2009, Art. ID 315713, 28 pp.

[98] Devaney, R., *An introduction to chaotic dynamical systems*, Addison-Wesley, Menlo Park California, (1990).

[99] Devi, Vasundara, J., Vatsala, A. S., *Generalized quasilinearization for an impulsive differential equation with variable moments of impulse*, Dynam. Systems Appl., **12**, 369–382, (2003).

[100] Diekmann, O., S.A. van Gils, Verduyn Lunel, S.M., Walther, H.-O., *Delay equations. Functional-, Complex,- and Nonlinear Analysis*, Springer Verlag, New York, (1995).

[101] Doi, S. and Kumagai, S., *From singularly perturbed system to continuous-discrete time hybrid dynamical system*, (Japanese) Systems Control Inform, **43**, 600–607, (1999).

[102] Doss, S. and Nasr, S. K., *On the functional equation* $y' = f(x, y(x), y(x+h))$, $h > 0$, Amer. J. Math., **75**, 713–716, (1953).

[103] Driessche, P. V. D. and Zou, X, *Global attractivity in delayed Hopfield neural network models*, SIAM J. Appl. Math., **58**, 1878–1890, (1998).

[104] Driver, R. D., *Can the future influence the present?*, Physical Review D, **19**, 1098–1107, (1979).

[105] Dubois, D. M., *Mathematical foundations of discrete and functional systems with strong and weak anticipations*, Anticipatory Behavior in Adaptive Learning Systems, Springer, LNAI **2684**, 110–132, (2003).

[106] Ermentrout, G.B. and Koppel, N., *Oscillator death in systems of coupled neural oscillators*, SIAM J. Appl. Math, **50**, no. 1, 125-146, (1990).

[107] Ermentrout, G.B. and Koppel, N., *Parabolic bursting in an excitable system coupled with a slow oscillators*, SIAM J. Math. Anal, **15**, 233–253, (1986).

[108] Ermentrout, G.B. and Terman, D.H., *Mathematical foundations of neuroscience*, Springer, New-York, (2010).

[109] Ernst, U., Pawelzik, K. and Geisel, T., *Delay-induced multistable synchronization of biological oscillators*, Phys. Rev. E, **57**, 2150–2162, (1998).

[110] Fan, M. and Wang, Q., *Periodic solutions of a class of nonautonomous discrete time semi-ratio-dependent predator-prey systems*, Discrete Contin. Dyn. Syst. Ser. B, **4**, 563–574, (2004).

[111] Fan, M. and Wang, K., *Periodic solutions of a discrete time nonautonomous ratio-dependent predator-prey system*, Math. Comput. Modelling, **35**, 951–961, (2002).

[112] Fan, Y.-H. and Li, W.-T., *Permanence for a delayed discrete ratio-dependent predator-prey system with Holling type functional response*, J. Math. Anal. Appl., **299**, 357–374, (2004).

[113] Feckan, M., *Bifurcation of periodic and chaotic solutions in discontinuous systems*, Arch. Math. (Brno), **34**, 73–82, (1998) 73–82.

[114] Feigenbaum, M., *Quantative universality for a class of nonlinear transformations*, J. Stat. Phys., **19**, 25–52, (1978).

[115] Feng, Y. and Dai, Y., *Oscillatory and asymptotic behavior of first order differential equation with piecewise constant deviating arguments*, Ann. Differential Equations, **15**, 345–351, (1999).

[116] Filippov, A. F., *Differential Equations with Discontinuous Right-Hand Sides*, Kluwer Academic Publishers, Dordrecht, (1988).

[117] Fink, A.M., *Almost-periodic differential quations*, Lecture notes in mathematics, Springer-Verlag, Berlin, Heidelberg, New York, (1974).

[118] Foias,C., Sell, G.R. and Temam, R., *Inertial manifolds for nonlinear evolutionary equations*, J. Differential Equations, **73**, 309–353, (1988).

[119] Frigon, M. and O'Regan, D., *Impulsive differential equations with variable times*, Nonlinear Anal.: TMA, **26**, 1913–1922, (1996).

[120] Fung, Y.C., *Biomechanics, Circulation*, Springer-Verlag, New-York, (1997).

[121] Furumochi, T., *Stability and oscillation for linear and nonlinear scalar equations with piecewise constant argument*, Appl. Anal., **44**, 113–125, (1992).

[122] Gerstner, W., *Rapid phase locking in systems of pulse-coupled oscillators with delays*, Phys. Rev. Lett., **76**, 1755–1758, (1996).

[123] Gerstner, W., Kistler, W.M., *Spiking neuron models: Single neurons, populations, plasticity*, Cambridge University Press, (2002).

[124] Glass, L., *Cardiac arrhythmias and circle maps - A classical problem*, Chaos, **1**, 13–19.

[125] Glass, L., Mackey, M.C., *From clocks to chaos: the rhythms of life*, Princeton University Press, (1988).

[126] Glass, L., Mackey, M.C., *A simple model for phase locking of biological oscillators*, J. Math. Biol, **7**, 339–367, (1979).

[127] Glass, L., *Synchronization and rhythmic processes in physiology*, Nature, **410**, 277–284, (2001).

[128] Glendinning, P. and Kowalczyk, P., *Micro-chaotic dynamics due to digital sampling in hybrid systems of Filippov type*, Phys. D, **239**, 58–71, (2010).

[129] Goel, P., Ermentrout, B., *Synchrony, stability, and firing patterns in pulse-coupled oscillators*, Physica D, **163**, 191–216, (2002).

[130] Golomb, D., Rinzel, J., *Dynamics of globally coupled inhibitory neurons with heterogeneity*,

Phys. Rev. E, **48**, 4810–4814, (1993).

[131] Gopalsamy, K. and Liu, P., *Persistence and global stability in a population model*, J. Math. Anal. Appl., **224**, 59–80, (1998).

[132] Gopalsamy, K., Liu, P., *Dynamics of social populations*, Nonlinear Anal.: TMA, **30**, 2595–2604, (1997).

[133] Gopalsamy, K., *Stability and Oscillation in Delay Differential Equations of Population Dynamics*, Kluwer Academic Publishers, Dordrecht, (1992).

[134] Gopalsamy, K., Kulenović, M. R. S. and Ladas, G., *On a logistic equation with piecewise constant arguments*, Differential Integral Equations, **4**, 215–223, (1991).

[135] Gopalsamy, K., Györi, I. and Ladas, G., *Oscillation of a class of delay equations with continuous and piecewise constant arguments*, Funkcial. Ekvac., **32**, 395–406, (1989).

[136] Grove, E. A., Ladas, G. and Zhang, S., *On a nonlinear equation with piecewise constant argument*, Comm. Appl. Nonlinear Anal., **4**, 67–79, (1997).

[137] Grove, E. A., Györi, I. and Ladas, G., *On the characteristic equation for equations with continuous and piecewise constant arguments*, Rad. Mat., **5**, 271–281, (1989).

[138] Guckenheimer, J. and Holmes, P., *Nonlinear oscillations, Dynamical systems, and Bifurcations of Vector Fields*, Springer-Verlag, New York, (1983).

[139] Györi, I. *On approximation of the solutions of delay differential equations by using piecewise constant argument*, Int. J. Math. Math. Sci., **14**, 111–126, (1991).

[140] Györi, I. and Hartung, F., *On numerical approximation using differential equations with piecewise-constant arguments*, Period. Math. Hungar., **56**, 55–69, (2008).

[141] Györi, I. and Hartung, F., *Numerical approximation of neutral differential equations on infinite interval*, J. Difference Equ. Appl., **8**, 983–999, (2002).

[142] Györi, I. and Eller, J., *Compartmental systems with pipes*, Mathematical Biosciences, **53**, 223–247, (1981).

[143] Györi, I. and Ladas, G., *Oscillation Theory of Delay Differential Equations with Applications*, Oxford University Press, New York, (1991).

[144] Györi, I. and Ladas, G., *Linearized oscillations for equations with piecewise constant arguments*, Differential Integral Equations, **2**, 123–131, (1989).

[145] Haddad, W.M., Chellaboina, V-S. and Nersesov, S.G., *Impulsive and Hybrid Dynamical Systems, Stability, Dissipativity, and Control*, Princeton University Press, Princeton, NJ, (2006).

[146] Hahn, W., *Stability of Motion*, Berlin, Springer-Verlag, (1967).

[147] Haidar, N.H.S., *An amplitude spectral tonometer*, Computers and Mathematics with applications, **46**, 959–969, (2003).

[148] A. Halanay, D. Wexler, *Qualitative theory of impulsive systems*, Edit. Acad. RPR, Bucuresti, (1968) (in Romanian).

[149] Hale, J. K., Lunel, S. M. V., *Introduction to Functional Differential Equations*, Springer-Verlag, New York, (1993).

[150] Hale, J. K., *Theory of Functional Differential Equations*, Springer-Verlag, New York, (1977).

[151] Hammel, S.M., Jorke, J.A. and Grebogi, C., *Do numerical orbits of chaotic dynamical processes represent true orbits?* Journal of Complexity, 136–145, **3**, (1987).

[152] Hanson, F.E., Case, J.F., Buck, E. and Buck, J., *Synchrony and flash entrainment in a New Guinea firefly*, Science, **174**, 161–164, (1971).

[153] Hartman, P., *Ordinary Differential Equations*, Philadelphia, Society for Industrial and Applied Mathematics, (2002).

[154] *Heart rate variability: Standards of measurement, physiological interpretation and clinical use. Task Force of the European Society of Cardiology and the North American Society of Pacing and Electrophysiology*, European Heart J., **17**, 1043-1065, (1996).

[155] Henry, D.B., *Geometric theory of semi-linear parabolic equations*, Springer-Verlag, New

York, (1981).

[156] Henon, M., *A two-dimensioanal mapping with a strange attractor*, Comm. Math. Phys., **50**, 69–77, (1976).

[157] Herz, A.V.M., Hopfield, J.J., *Earthquake cycles and neural perturbations: collective oscillations in systems with pulse-coupled thresholds elements*, Phys. Rev. Lett., **75**, 1222–1225, (1995).

[158] Hirsh, M., *Systems of differential equations which are competitive or cooperative 1: limit sets*, SIAM, J. Appl. Math., **13**, 167–179, (1982).

[159] Hirsh, M., *Systems of differential equations which are competitive or cooperative 1: convergence everywhere*, SIAM, J. Appl. Math., **16**, 423–439, (1985).

[160] Hsu, C. H., Lin, S. S., Shen, W.X., *Traveling waves in cellular neural networks*, International Journal of Bifurcation and Chaos, **9**, 1307–1319, (1999).

[161] Hong, J., Obaya, R. and Sanz, A., *Almost periodic type solutions of some differential equations with piecewise constant argument*, Nonlinear Anal.: TMA, **45**, 661–688, (2001).

[162] Hopfield, J. J., *Neurons with graded response have collective computational properties like those of two-stage neurons*, Proc. Nat. Acad. Sci. Biol., **81**, 3088–3092, (1984).

[163] Hopfield, J.J., *Neurons, dynamics and computation*, Physics Today, February, 40–46, (1994).

[164] Hopfield, J.J., Herz, A., *Rapid local synchronization of action potentials: Toward computation with coupled integrate-and-fire neurons*, PNAS., **92**, 6655–6662, (1995).

[165] Hoppensteadt F.C., Izhikevich, E.M., *Weakly connected neural networks*, Applied Mathematical Sciences, **126**, Springer, New York, Berlin, (1997).

[166] Hoppensteadt, F. C. and Peskin, C.S., *Mathematics in Medicine and in the Life Sciences*, Springer-Verlag, New York, Berlin, Heidelberg, (1992).

[167] Huang, H., Cao, J. and Wang, J, *Global exponential stability and periodic solutions of recurrent neural networks with delays*, Phys. Lett. A, **298**, 393–404, (2002).

[168] Huo, H.-F. and Li, W.-T., *Existence and global stability of periodic solutions of a discrete ratio-dependent food chain model with delay*, Appl. Math. Comput., **162**, 1333–1349, (2005).

[169] Huang, Z., Wang, X. and Xia, Y., *A topological approach to the existence of solutions for nonlinear differential equations with piecewise constant argument*, Chaos Solitons Fractals, **39**, 1121–1131, (2009).

[170] Huang, Z., Xia, Y. and Wang, X., *The existence and exponential attractivity of κ-almost periodic sequence solution of discrete time neural networks*, Nonlinear Dynam., **50**, 13–26, (2007).

[171] Huang, Z. Wang, X. and Gao, F., *The existence and global attractivity of almost periodic sequence solution of discrete-time neural networks*, Physics Letters A, **350**, 182–191, (2006).

[172] Huang, L. H. and Chen, X. X., *Oscillation of systems of neutral differential equations with piecewise constant argument*, (Chinese), J. Jiangxi Norm. Univ. Nat. Sci. Ed., **30**, 470–474, (2006).

[173] Huang, Y. K., *Oscillations and asymptotic stability of solutions of first order neutral differential equations with piecewise constant argument*, J. Math. Anal. Appl., **149**, 70–85, (1990).

[174] Huang, Y. K., *On a system of differential equations alternately of advanced and delay type*, Differential Equations and Applications, Ohio Univ. Press, Athens, 455–465, (1989).

[175] Jacquez, J.A., *Compartmental analysis in biology and medicine*, Elsevier, New York, (1972).

[176] Jayasree, K. N. and Deo, S. G., *On piecewise constant delay differential equations*, J. Math. Anal. Appl, **169**, 55–69, (1992).

[177] Elsamahy, E., Mahfouf, M. and Likens, D., *A hybrid intelligent closed-loop model for exploration of cardiovascular interactions*, Proc. of the 4th Ann. IEEE Conf. on Info. Tech.

Appl. In Biomedicine, 165–168. (2003).

[178] El'sgol'ts, L.E., *Introduction to the theory of differential equations with deviating arguments*, Holden-Day, Inc, San Francisco, London, Amsterdam, (1966).

[179] Kamke, E., *Zur Theorie der Systeme gewohnlicher Differentialgleichungen*. II, (German) Acta Math., **58**, 57–85, (1932).

[180] Kelley, A., *The stable, center-stable, center, center-unstable, unstable manifolds*, An appendix in Transversal mappings and flows, R. Abraham and J. Robbin, Benjamin, New York (1967).

[181] Knight, B.W., *Dynamics of encoding in a population of neurons*, J. Gen. Physiol., **59**, 734–766, (1972).

[182] Ko, T.-W., Ermentrout, G.B., *Effects of axonal time delay on synchronization and wave formation in sparsely coupled neuronal oscillators*, Phys. Rev. E, **76**, 1–8, (2007).

[183] Kolmogorov, A.N., *On the Skorokhod convergence*, (Russian. English summary), Teor. Veroyatnost. i Primenen., **1**, 239–247, (1956).

[184] Koppel, N., Ermentrout, G.B., Williams, T., *On chains of oscillators forced at one end*, SIAM J. Appl Math, **51**, 1397–1417, (1991).

[185] Krasovskii, N. N., *Stability of motion, applications of Lyapunov's second method to differential systems and equations with delay*, Stanford University Press, Stanford, California, (1963).

[186] Küpper, T. and Yuan, R., *On quasi-periodic solutions of differential equations with piecewise constant argument*, J. Math. Anal. Appl., **267**, 173–193, (2002).

[187] Kuramoto, Y. *Chemical oscillations*, Springer, Berlin, (1984).

[188] Kuramoto, Y. *Collective synchronization of pulse-coupled oscillators and excitable units*, Physica D, **50**, 15–30, (1991).

[189] Ladas, G., *Oscillations of equations with piecewise constant mixed arguments*, Differential Equations and Applications, Ohio Univ. Press, Athens, 64–69, (1989).

[190] Ladas, G., Partheniadis, E. C. and Schinas, J., *Oscillation and stability of second order differential equations with piecewise constant argument*, Rad. Mat., **5**, 171–178, (1989).

[191] Ladde, G.S. *Cellular systems, II, Stability of compartmental systems*, Math. Biosci., **30**, 1–21, (1976).

[192] Lakshmikantham, V., Bainov, D.D. and Simeonov, P. S., *Theory of Impulsive Differential Equations*, World Scientific, Singapore, (1989).

[193] Lakshmikantham, V., Leela, S. and Kaul, S., *Comparison principle for impulsive differential equations with variable times and stability theory*, Nonlin. Anal.: TMA, **22**, 499–503, (1994).

[194] Lakshmikantham, V. and Vasundhara D. J., *Hybrid systems with time scales and impulses*, Nonlin. Anal.: TMA, **65**, 2147–2152, (2006).

[195] Lakshmikantham, V. and Vatsala, A. S., *Hybrid systems on time scales. Dynamic equations on time scales*, J. Comput. Appl. Math., **141**, 227–235, (2002).

[196] Lakshmikantham, V. and Liu, X., *Impulsive hybrid systems and stability theory*, Dynam. Systems Appl., **7**, 1–9, (1998).

[197] Lakshmikantham, V. and Liu, X., *On quasistability for impulsive differential equations*, Nonlinear Anal.: TMA, **13**, 819–828, (1989).

[198] LaSalle, J. P., *The Stability and Control of Discrete Processes*, Springer-Verlag, New York, (1986).

[199] Lefschetz, S., *Differential equations, Geometric theory*, Wiley, New York, (1957).

[200] Lee, K. H., Ong, E. H., *A reduction principle for singular perturbation problems*, Appl. Math. Comput., **101**, 45–62, (1999).

[201] Leela, S. and Pandit, S. G., *On stability in terms of two measures of a hybrid system having partly invisible solutions*, Nonlinear Anal. Hybrid Syst., **3**, 713–718, (2009).

[202] Lerma, C., Minzoni, A., Infante O. and Jose, M.V., *A mathematical analysis for the cardiovascular control adaptations in chronic renal failure*, Artificial organs, **28**, 398–409, (2004).
[203] Lloyd, J.E., *Fireflies of Melanesia: biolumeniscence, mating behavior, and synchronous flashing (Coleoptera: Lampyridae)*, Eviron. Entomol., **2**, 991–1008, (1973).
[204] Li, H., Muroya, Y. and Yuan, R., *A sufficient condition for global asymptotic stability of a class of logistic equations with piecewise constant delay*, Nonlinear Anal. Real World Appl., **10**, 244–253, (2009).
[205] Li, H., Yuan, R., *An affirmative answer to Gopalsamy and Liu's conjecture in a population model*, J. Math. Anal. Appl., **338**, 1152–1168, (2008).
[206] Li, X. and Wang, Z., *Global attractivity for a logistic equation with piecewise constant arguments. Differences and differential equations*, Fields Inst. Commun., **42**, 215–222, (2004).
[207] Li, T. and Yorke, J., *Period three implies chaos*, Amer. Math. Monthly, **82**, 985–992, (1975).
[208] Lin, L. C. and Wang, G. Q., *Oscillatory and asymptotic behavior of first order nonlinear differential equations with retarded argument* $[t]$, Chinese Sci. Bull., **36**, 889–891, (1991).
[209] Liu, B. and Hill, D. J., *Stability of discrete impulsive hybrid systems via comparison principle*, IFAC 08, Seoul, Korea, July 6-11, 11520–11525, (2008).
[210] Liu, B. and Hill, D. J., *Comparison principle and stability of discrete impulsive hybrid systems*, IEEE Trans Circuits Syst. I Regul. Pap., **56**, 233–245, (2009).
[211] Liu, Y., *Global stability in a differential equation with piecewisely constant arguments*, J. Math. Res. Exposition, **21**, 31–36, (2001).
[212] Liu, P., Gopalsamy, K., *Global stability and chaos in a population model with piecewise constant arguments*, Appl. Math. Comp., **101**, 63–88, (1999).
[213] Liu, X., Shen, X. and Zhang, Y., *Stability analysis of a class of hybrid dynamic systems*, Dyn. Contin. Discrete Impuls. Syst. Ser. B Appl. Algorithms, **8**, 359–373, (2001).
[214] Lorenz, E.N., *Deterministic nonperiodic flow*, J. Atmos. Sci, **20**), 130–141, (1963).
[215] Lyapunov, A. M., *Problème général de la stabilité du mouvement*, Princeton Univ. Press, Princeton, N.J., (1949).
[216] Magnia, L., Scattolini R., *Stabilizing model predictive control of nonlinear continuous time systems*, Annual Reviews in Control, **28**, 1–11, (2004).
[217] Magnia, L., Scattolini R. *State-feedback MPS with piecewise continuous control for continuous-time nonlinear systems*, Proceedings of 41st IEEE conference on decision and control, Las Vegas, Nevada, USA, 4625–4630, (2002).
[218] Malkin, I.G. *Some problems in the theory of nonlinear oscillations*, GITTL, Moscow, (1956). (Russian) English Transl., U.S. Atomic Energy Commission Translation AEC-tr-3766, Books 1 and 2.
[219] Malkin, I.G., *Theory of stability of motion*, U.S. Atomic Energy Commission, Office of Technical Information, 1958.
[220] Malkin, I.G. *Stability in the case of constantly acting disturbances*, Appl. Math. Mech. [Akad. Nauk SSSR. Prikl. Mat. Mech.], **8**, 241–245, (1944).
[221] Mancia, G., Parati G., Di Rienzo, M. and Zanchetti, A., *Blood pressure variability*, In: A. Zanchetti and G. Mancia eds. Handbook of hypertension, vol. 17: Pathophysiology of hypertension, Elsevier Science, Amsterdam, 117–169, (1997).
[222] Mathar, R., Mattfeldt, J., *Pulse-coupled decentral synchronization*, SIAM J. Appl. Math., **56**, 1094–1106, (1996).
[223] Matis, J.M., Patten B.C. and White G.C. (eds), *Compartmental analysis of ecosystem models*, International Cooperative Pub. House, (1979).
[224] Matsunaga, H., Hara, T. and Sakata, S., *Global attractivity for a logistic equation with piecewise constant argument*, NoDEA, **8**, 45–52, (2001).
[225] May, R. *Simple mathematical models with very complicated dynamics*, Nature, **261**, 459–467,

(1976).

[226] May, R. M. and Oster, G. F., *Bifurcations and dynamic complexity in simple ecological models*, Amer. Natural, **110**, 573–599, (1976).

[227] Michel, A.N., *Recent trends in the stability analysis of hybrid dynamical systems*, IEEE Trans. Automat. Control, **45**, 120–134, (1999).

[228] Mirollo, R.E., Strogatz, S.H., *Synchronization of pulse-coupled biological oscillators*, SIAM Journal on Applied Mathematics, **50**, 1645–1662, (1990).

[229] McDonald, D. A., *The relation of pulsative pressure to flow in arteries*, J Physiol, **127**, 533–552, (1955).

[230] Mohamad, S. and Gopalsamy, K. *Exponential stability of continuous-time and discrete-time cellular neural networks with delays*, Applied Mathematics and Computation, **135**, 17–38, (2003).

[231] Mosterman, P.J. and Biswas, G., *A comprehensive methodology for building hybrid models of physical systems*, Artificial Intelligence, **121**, 171–209, (2000).

[232] Minh, N. V. and Dat, T. T., *On the almost automorphy of bounded solutions of differential equations with piecewise constant argument*, J. Math. Anal. Appl., **326**, 165–178, (2007).

[233] Minorsky, N., *Nonlinear oscillations*, Huntington, New York: Krieger Publishing, (1962).

[234] Murad, N. M. and Celeste, A., *Linear and nonlinear characterization of loading systems under piecewise discontinuous disturbances voltage: analytical and numerical approaches*, Power Electronics Systems and Applications, 291–297, (2004).

[235] Muroya, Y., *Persistence, contractivity and global stability in logistic equations with piecewise constant delays*, J. Math. Anal. Appl., **270**, 602–635, (2002).

[236] Muroya, Y. and Kato, Y., *On Gopalsamy and Liu's conjecture for global stability in a population model*, J. Comput. Appl. Math., **181**, 70–82, (2005).

[237] Murray, J. D., *Mathematical Biology: I. An Introduction*, Springer-Verlag, New York, Heidelberg, Berlin, (2002).

[238] Myshkis, A. D., *On certain problems in the theory of differential equations with deviating argument*, Russian Math. Surveys, **32**, 181–213, (1977).

[239] Myshkis, A. D., *Linear differential equations with retarded arguments*, Nauka, Moscow, in Russian (1972).

[240] Nagaev, R.F., *Dynamics of synchronising systems*, Springer-Verlag, Berlin, (2003).

[241] Nayfeh, A.H., Balachandran, B., *Applied Nonlinear Dynamics Analytical Computational and Experimental Methods*, Wiley, New York, (1995).

[242] Nemytskii, V. V., Stepanov, V. V., *Qualitative theory of Differential Equations*, Princeton University Press, Princeton, New Jersey, (1966).

[243] Nicols, W.W., O'Rourke, M.F., *McDonald's Blood flow in arteries: theoretical, experimental and clinical principles*, Third Edition, Lea & Febier, London, (1990).

[244] Nicolis G. and Prigogine, I., *Self-organization in non-equilibrium systems*, Wiley, New York, (1977).

[245] Nieto, J. J. and Rodriguez-Lopez, R., *Green's function for second order periodic boundary value problems with piecewise constant argument*, J. Math. Anal. Appl., **304**, 33–57, (2005).

[246] Olami, Z., Feder, H.J.S. and Christensen, K., *Self-organized criticality in a continuous, nonconservative cellular automaton modeling earthquakes*, Phys. Rev. Lett., **68**, 1244–1247, (1992).

[247] Ottesen, J.T. *Modelling of the baroreflex-feedback mechanism with time-delay*, J. Math. Biol., **36**, 41–63, (1997).

[248] Palmer, K.J., *Linearization near an integral manifold*, J. Math. Anal. Appl., **51**, 243–255, (1975).

[249] Papaschinopoulos, G., *Linearization near the integral manifold for a system of differential*

equations with piecewise constant argument, J. Math. Anal. Appl., **215**, 317–333, (1997).

[250] Papaschinopoulos, G., *On the integral manifold for a system of differential equations with piecewise constant argument*, J. Math. Anal. Appl., **201**, 75–90, (1996).

[251] Papaschinopoulos, G., *A linearization result for a differential equation with piecewise constant argument*, Analysis, **16**, 161–170, (1996).

[252] Papaschinopoulos, G., *Some results concerning second and third order neutral delay differential equations with piecewise constant argument*, Czechoslovak Math. J., **44** (119), 501–512, (1994).

[253] Papaschinopoulos, G., *Some results concerning a class of differential equations with piecewise constant argument*, Math. Nachr., **166**, 193–206, (1994).

[254] Papaschinopoulos, G., *On asymptotic behavior of the solutions of a class of perturbed differential equations with piecewise constant argument and variable coefficients*, J. Math. Anal. Appl., **185**, 490–500, (1994).

[255] Papaschinopoulos, G., *Exponential dichotomy, topological equivalence and structural stability for differential equations with piecewise constant argument*, Analysis, **14**, 239–247, (1994).

[256] Papaschinopoulos, G. and Schinas, J., *Existence stability and oscillation of the solutions of first order neutral delay differential equations with piecewise constant argument*, Appl. Anal., **44**, 99–111, (1992).

[257] Park, J. H., *Global exponential stability of cellular neural networks with variable delays*, Appl. Math. Comput., **183**, 1214–1219 (2006).

[258] Partheniadis, E. C., *Stability and oscillation of neutral delay differential equations with piecewise constant argument*, Differential Integral Equations, **1**, 459–472, (1988).

[259] Pavlidis, T., *A new model for simple neural nets and its application in the design of a neural oscillator*, Bull. Math. Biophys., **27**, 215–229, (1965).

[260] Pavlidis, T., *Biological oscillators: Their Mathematical Analysis*, Academic Press, (1973).

[261] Pedley, T., *The fluid mechanics of large blood vessels*, Cambridge University Press, London, New-York, (1980).

[262] Peskin, C.S., *Mathematical aspects of heart physiology*, Courant Institute of mathematical sciences, New York University, 268–278, (1975).

[263] Pinto, M., *Asymptotic equivalence of nonlinear and quasi linear differential equations with piecewise constant arguments*, Math. Comput. Modelling, **49**, 1750–1758, (2009).

[264] Pliss, V. A., *Integral Sets of Periodic Systems of Differential Equations*, Izdat. Nauka, Moscow, (1977).

[265] Pliss, V. A., *A reduction principle in the theory of stability of motion*, Izv. Akad. Nauk SSSR Ser. Mat., **28**, 1297–1324, (1964).

[266] Pikovsky, A., Rosenblum, M., Kurths, J., *Synchronization: A universal concept in nonlinear sciences*, Cambridge University Press, New York, (2001).

[267] Poincaré, H., *Les méthodes nouvelles de la mécanique céleste,* Gauthier-Villars, Paris, (1892).

[268] Pugh, C., Shub, M. *Linearization of normally hyperbolic diffeomorphisms and flows*, Invent. Math, **10**, 187–190, (1970).

[269] Qin, S. J., Badgwell, T. A., *An overview of industrial model predictive control technology*, in Kantor, J. C., Garcia, C. E., Carnahan, B. (eds.), Fifth International Conference on Chemical Process Control, 232–256, (1996).

[270] Razumikhin, B. S., *Stability of delay systems*, Prikl. Mat. Mekh., **20**, 500–512, (1956).

[271] Rescigno, A. and Segre, G., *Drug and tracer kinetics*, Waltham, Blaisdel, (1966).

[272] Robinson, C., *Dynamical Systems: stability, Symbolic dynamics, and Chaos*, CRC Press, Boca Raton, Ann Arbor, London, Tokyo, (1995).

[273] Roseau, M., *Vibrations non linéaires et théorie de la stabilité*, Springer-Verlag, Berlin-New York, (1966).

[274] Rosen, R., *Anticipatory Systems*, Pergamon Press, New York, (1985).

[275] Rouche, N.,Habets, P. and Laloy, M., *Stability Theory by Liapunov's Direct Method*, New York, Springer-Verlag, (1977).

[276] Ruchti, T.L., Brown, R. H., Feng, X. and Jeutter, D. C., *Estimation of systemic arterial parameters for control of an electrically actuated total artificial heart*, Proceedings, 32nd IEEE Midwest Symposium on Circuits and Systems, 640–643, (1989).

[277] Ruelle, D. *On the nature of turbulunce*, Commun. Math. Phys., **20**, 167–192, (1971).

[278] Samoilenko, A. M., Perestyuk, N. A., *Impulsive Differential Equations*, World Scientific, Singapore, (1995).

[279] Savkin, A.V. and Evans, R.J., *Hybrid dynamical systems: controller and sensor switching problems*, Birkhauser, Boston, Basel, Berlin, (2002).

[280] van der Schaft, A.J. and Schumacher, J.M., *An introduction to hybrid dynamical systems*, Springer, London, (2000).

[281] Sharkovskii, A.N., *Coexistence of cycles of a continuous map of a line into itself*, (Russian) Ukr. Mat. Zh., **16**, 61–71, (1964).

[282] Shcheglova, A. A., *Duality of the concepts of controllability and observability for degenerate linear hybrid systems*, Autom. Remote Control, **67**, 1445–1465, (2006).

[283] Seifert, G., *Almost periodic solutions of certain differential equations with piecewise constant delays and almost periodic time dependence*, J. Differential Equations, **164**, 451–458, (2000).

[284] Seifert, G., *On an interval map associated with a delay logistic equation with discontinuous delays*, Delay differential equations and dynamical systems (Claremont, CA, 1990), 243–249, Lecture Notes in Math., **1475**, Springer, Berlin, (1991).

[285] Senn, W., Urbanczik, R., *Similar non-leaky integrate-ans-fire neurons with instantaneous couplings always synchronize*, SIAM J. Appl. Math., **61**, 1143–1155, (2000).

[286] Shah, S. M. and Wiener, J., *Advanced differential equations with piecewise constant argument deviations*, Int. J. Math. Math. Sci., **6**, 671–703, (1983).

[287] Shen, J. H. and Stavroulakis, I. P., *Oscillatory and nonoscillatory delay equation with piecewise constant argument*, J. Math. Anal. Appl., **248**, 385–401, (2000).

[288] Skorokhod, A.V., *Limit theorems for random processes*, Theory Probab. Appl., **39**, 289-319, (1994).

[289] Smale, S., *Differentiable dynamical systems*, Bull. Amer. Math. Soc., **73**, 747–817, (1967).

[290] Smith, H.L., *Monotone dynamical systems: An introduction to the theory of competitive and cooperative systems*, AMS, (1995).

[291] So, J. W. H. and Yu, J. S., *Global stability in a logistic equation with piecewise constant arguments*, Hokkaido Math. J., **24**, 269–286, (1995).

[292] Solimano, F., Bischi, G.I., Bianchi, M., Rossi L. and Magnani, M., *A nonlinear three-compartment model for the administration of 2',3'- Dideoxycytidine by using red blood cells as bioreactors*, Bull. Math. Biol., **52**, 785–796, (1990).

[293] Stefanovska, A., Luchinsky, D.G. and McClintock, P., *Modeling couplings among the oscillators of the cardiovascular system*, Physiol. Meas., **22**, 551–564, (2001).

[294] Stokes, A. *Local coordinates around a limit cycle of a functional differential equation with applications*, J. Differential Equations, **24**, 153–172, (1977).

[295] Strogatz, S.H., *Nonlinear dynamics and chaos*, Perseus books, New York, (1994).

[296] Strogatz, S.H., *Sync: The Emerging Science of Spontaneous Order*, Hyperion, (2003).

[297] Talbot S. A., and Gessner, U., *Systems Physiology*, John Wiley and Sons, Inc., New York, (1973).

[298] Timme, M., Wolf, F. and Geisel, T., *Prevalence of Unstable Attractors in Networks of Pulse-Coupled Oscillators*, Phys. Rev. Lett., **89**, 154105, (2002).

[299] Timme, M., Wolf, F., *The simplest problem in the collective dynamics of neural networks: is*

synchrony stable? Nonlinearity, **21**, 1579–1599, (2008).

[300] Vasseur, D. A., Fox, J. W., *Phase-locking and environmental fluctuations generate synchrony in a predator-prey community*, Nature, **460**, 1007–1010, (2009).

[301] van Vreeswijk, C., *Partial synchronization in populations of pulse-coupled oscillators*, Phys. Rev. E, **54**, 5522–5537, (1996).

[302] Wang, G., *Periodic solutions of a neutral differential equation with piecewise constant arguments*, J. Math. Anal. Appl., **326**, 736–747, (2007).

[303] Wang, Q. and Liu, X., *Impulsive stabilization of delay differential systems via the Lyapunov-Razumikhin method*, Appl. Math. Lett., **20**, 839–845, (2007).

[304] Wang, Z. and Wu, J., *The stability in a logistic equation with piecewise constant arguments*, Differ. Eq. Dynam. Systems, **14**, 179–193, (2006).

[305] Wang, Y. and Yan, J., *Oscillation of a differential equation with fractional delay and piecewise constant argument*, Comput. Math. Appl., **52**, 1099–1106, (2006).

[306] Wang, G. Q. and Cheng, S. S., *Existence of periodic solutions for second order Rayleigh equations with piecewise constant argument*, Turkish J. Math., **30**, 57–74, (2006).

[307] Wang, G. Q. and Cheng, S. S., *Existence of periodic solutions for a neutral differential equation with piecewise constant argument*, Funkcial. Ekvac., **48**, 299–311, (2005).

[308] Wang, G. Q. and Cheng, S. S., *The set of periodic solutions of a neutral differential equation with constant delay and piecewise constant argument*, Port. Math., **62**, 295–302, (2005).

[309] Wang, G. Q. and Cheng, S. S., *Oscillation of second order differential equation with piecewise constant argument*, Cubo, **6**, 55–63, (2004).

[310] Wang, G. Q. and Cheng, S. S., *Note on the set of periodic solutions of a delay differential equation with piecewise constant argument*, Int. J. Pure Appl. Math., **9**, 139–143, (2003).

[311] Wang, Y. and Yan, J., *A necessary and sufficient condition for the oscillation of a delay equation with continuous and piecewise constant arguments*, Acta Math. Hungar., **79**, 229–235, (1998).

[312] Wang, Y. and Yan, J., *Necessary and sufficient condition for the global attractivity of the trivial solution of a delay equation with continuous and piecewise constant arguments*, Appl. Math. Lett., **10**, 91–96, (1997).

[313] Wang, L., Yuan, R. and Zhang, C., *Corrigendum to: "On the spectrum of almost periodic solution of second order scalar functional differential equations with piecewise constant argument"*, [J. Math. Anal. Appl. 303 (2005), 103–118, by Yuan, R.], J. Math. Anal. Appl., **349**, 299, (2009).

[314] Weng, P. and Wu, J., *Deformation of traveling waves in delayed cellular neural networks*, International Journal of Bifurcation and Chaos, **13**, 797–813, (2003).

[315] Wexler, D., *Solutions périodiques et presque-périodiques des systémes d'équations difféDretielles linéaires en distributions*, J. Differ. Eq., **2**, 12–32, (1966).

[316] Wiggins, S., *Introduction to Applied Nonlinear Dynamical Systems and Chaos*, New York, Springer, (2003).

[317] Wiener, J., *Generalized Solutions of Functional Differential Equations*, World Scientific, Singapore, (1993).

[318] Wiener, J., *A second-order delay differential equation with multiple periodic solutions*, J. Math. Anal. Appl, **229**, 659–676, (1999).

[319] Wiener, J., *Pointwise initial value problems for functional-differential equations*, Differential Equations, North-Holland, Amsterdam, 571–580, (1984).

[320] Wiener, J., *Differential equations with piecewise constant delays*, Trends in the Theory and Practice of Nonlinear Differential Equations, Marcel Dekker, New York, 547–552, (1983).

[321] Wiener, J. and Aftabizadeh, A. R., *Differential equations alternately of retarded and advanced type*, J. Math. Anal. Appl., **129**, 243–255, (1988).

[322] Wiener, J. and Cooke, K. L., *Oscillations in systems of differential equations with piecewise constant argument*, J. Math. Anal. Appl., **137**, 221–239, (1989).

[323] Wiener, J. and Debnath, L., *Boundary value problems for the diffusion equation with piecewise continuous time delay*, Int. J. Math. Math. Sci, **20**, 187–195, (1997).

[324] Wiener, J. and Debnath, L., *A survey of partial differential equations with piecewise continuous arguments*, Int. J. Math. Math. Sci., **18**, 209–228, (1995).

[325] Wiener, J. and Debnath, L., *A parabolic differential equation with unbounded piecewise constant delay*, Internat. J. Math. Math. Sci., **15**, 339–346, (1992).

[326] Wiener, J. and Heller, W., *Oscillatory and periodic solutions to a diffusion equation of neutral type*, Int. J. Math. Math. Sci, **22**, 313–348, (1999).

[327] Wiener, J. and Lakshmikantham, V., *A damped oscillator with piecewise constant time delay*, Nonlinear Stud., **7**, 78–84, (2000).

[328] Wiener, J. and Lakshmikantham, V., *Differential equations with piecewise constant argument and impulsive equations*, Nonlinear Stud., **7**, 60–69, (2000).

[329] Wiener, J. and Lakshmikantham, V., *Excitability of a second-order delay differential equation*, Nonlinear Anal. Real World Appl., **38**, 1–11, (1999).

[330] Wiener, J. and Shah, S. M., *Continued fractions arising in a class of functional-differential equations*, J. Math. Phys. Sci., **21**, 527–543, (1987).

[331] Winfree, A. T., *The Geometry of Biological Time*, Springer, New York, (1980).

[332] Winfree, A. T., *Biological rhythms and the behavior of populations of biological oscillators*, J. Theor. Biol., **16**, 15–42, (1967).

[333] Xia, Y., Huang, Z. and Han, M., *Existence of almost periodic solutions for forced perturbed systems with piecewise constant argument*, J. Math. Anal. Appl., **333**, 798–816, (2007).

[334] Yan, J., Zhao, A. and Nieto, J.J., *Existence and global activity of positive periodic solutions of periodic single-species impulsive Lotka-Volterra systems*, Math. Comput. Model., **40**, 509–518, (2004).

[335] Yang, P., Liu, Y. and Ge, W., *Green's function for second order differential equations with piecewise constant argument*, Nonlinear Anal., **64**, 1812–1830, (2006).

[336] Yang, X., *Existence and exponential stability of almost periodic solution for cellular neural networks with piecewise constant argument*, Acta Math. Appl. Sin., **29**, 789–800, (2006).

[337] Yang, X., Liao, X, Li, C. and Evans, D. J., *New estimate on the domains of attraction of equilibrium points in continuous Hopfield neural networks*, Phys. Lett. A, **351**, 161–166, (2006).

[338] Ye, H., Michel, A. N. and Hou, L., *Stability theory for hybrid dynamical systems. Hybrid control systems*, IEEE Trans. Automat. Control, **43**, 461–474, (1998).

[339] Yoshizawa, T., *Stability Theory by Lyapunov's Second Method*, The Mathematical Society of Japan, Tokyo, (1966).

[340] Yuan, R., *On the spectrum of almost periodic solution of second order scalar functional differential equations with piecewise constant argument*, J. Math. Anal. Appl., **303**, 103–118, (2005).

[341] Yuan, R., *The existence of almost periodic solutions of retarded differential equations with piecewise argument*, Nonlinear Anal., **48**, 1013–1032, (2002).

[342] Yuan, R., *Almost periodic solutions of a class of singularly perturbed differential equations with piecewise constant argument*, Nonlinear Anal., **37**, 841–859, (1999).

[343] Yuan, R. and Hong, J., *The existence of almost periodic solutions for a class of differential equations with piecewise constant argument*, Nonlinear Anal., **28** (8), 1439–1450, (1997).

[344] Yuan, R. and Hong, J., *Existence of almost periodic solutions of neutral differential equations with piecewise-constant argument*, Sci. China Ser. A, **39**, 1164–1177, (1996).

[345] Yuan, R. and Hong, J., *Almost periodic solutions of differential equations with piecewise con-*

stant argument, Analysis, **16**, 171–180, (1996).

[346] Zhang, Q., Wei, X. and Xu, J, *Stability of delayed cellular neural networks*, Chaos Solitons Fractals, **31**, 514–520, (2007).

[347] Zeng, Z. G., Wang, J., *Improved conditions for global exponential stability of recurrent neural networks with time-varying delays*, IEEE Trans. Neural Networks, **17**, 623–635, (2006).

Atlantis Studies in Probability and Statistics

Volume 4

Series editor

Chris P. Tsokos, Tampa, USA

For further volumes:
http://www.atlantis-press.com

Aims and scope of the series

The series 'Atlantis Studies in Probability and Statistics' publishes studies of high quality throughout the areas of probability and statistics that have the potential to make a significant impact on the advancement in these fields. Emphasis is given to broad interdisciplinary areas at the following three levels:

(I) Advanced undergraduate textbooks, i.e., aimed at the 3rd and 4th years of undergraduate study, in probability, statistics, biostatistics, business statistics, engineering statistics, operations research, etc.;

(II) Graduate-level books, and research monographs in the above areas, plus Bayesian, nonparametric, survival analysis, reliability analysis, etc.;

(III) Full Conference Proceedings, as well as selected topics from Conference Proceedings, covering frontier areas of the field, together with invited monographs in special areas.

All proposals submitted in this series will be reviewed by the Editor-in-Chief, in consultation with Editorial Board members and other expert reviewers.

For more information on this series and our other book series, please visit ourwebsite at: www.atlantis-press.com/Publications/books

AMSTERDAM—PARIS—BEIJING
ATLANTIS PRESS
Atlantis Press
29, avenue Laumière
75019 Paris, France

Mohammad Ahsanullah · B. M. Golam Kibria
Mohammad Shakil

Normal and Student's *t* Distributions and Their Applications

ATLANTIS
PRESS

Mohammad Ahsanullah
Department of Management Sciences
Rider University
Lawrenceville, NJ
USA

Mohammad Shakil
Department of Mathematics
Miami Dade College
Hialeah, FL
USA

B. M. Golam Kibria
Department of Mathematics and Statistics
Florida International University
Miami, FL
USA

ISSN 1879-6893
ISSN 1879-6907 (electronic)
ISBN 978-94-6239-060-7
ISBN 978-94-6239-061-4 (eBook)
DOI 10.2991/978-94-6239-061-4

Library of Congress Control Number: 2013957385

Printed on acid-free paper

To my wife, Masuda

M. Ahsanullah

To my late parents, Mr. Abdur Rahim Bhuiyan and Mrs. Ayesha Bhuiyan, and my wife, Orchi

B. M. Golam Kibria

To my late parents, Mr. M. A. Rab and Mrs. Humaira Khatoon, and my wife, Nausheen

M. Shakil

Preface

The normal and Student's t distributions are two of the most important continuous probability distributions, and are widely used in statistics and other fields of sciences. The distributions of the sum, product, and ratio of two independent random variables arise in many fields of research, for example, biology, computer science, control theory, economics, engineering, genetics, hydrology, medicine, number theory, statistics, physics, psychology, reliability, risk management, etc. This has increased the need to explore more statistical results on the sum, product, and ratio of independent random variables. The aim of this book is to study the *Normal and Student's t Distributions and Their Applications*. First, the distributions of the sum, product, and ratio of two independent normal random variables, which play an important role in many areas of research, are presented, and some of the available results are surveyed. The distributions of the sum, product, and ratio of independent Student's t random variables, which are of interest in many areas of statistics, are then discussed. The distributions of the sum, product, and ratio of independent random variables belonging to different families are also of considerable importance and one of the current areas of research interest. This book introduces and develops some new results on the distributions of the sum of the normal and Student's t random variables. Some properties of these distributions are also discussed. A new symmetric distribution has been derived by taking the product of the probability density functions of the normal and Student's t distributions. Some characteristics of the new distributions are presented. Before a particular probability distribution model is applied to fit the real-world data, it is necessary to confirm whether the given probability distribution satisfies the underlying requirements by its characterization. Thus, characterization of a probability distribution plays an important role in probability and statistics. We have also provided some characterizations of the family of normal and Student's t distributions.

We hope the findings of the book will be useful for the advanced undergraduate and graduate students, and practitioners in various fields of sciences.

As a preparation to study this book, the readers are assumed to have knowledge of calculus and linear algebra. In addition, they need to have taken first courses in probability and statistical theory.

We wish to express our gratitude to Dr. Chris Tsokos for his valuable suggestions and comments about the manuscript, which certainly improved the quality and presentation of the book. The first author thanks Z. Karssen and K. Jones of

Atlantis Press for the interesting discussions at a meeting in Athens, Greece, for the publication of this book. Summer research grant and sabbatical leave from Rider University enabled the first author to complete his part of the work. Part of the book is from the independent study of the third author with Dr. Kibria. The book was partially written while the second author was on sabbatical in 2010–2011, and he gratefully acknowledges the excellent research facilities of Florida International University. The third author is grateful to Miami Dade College for all the support, including STEM grants. Last but not least, the authors would like to express their deep regret for any error or omission or misprint or mistake, which is very likely to occur in any textbook of this type. We have endeavored our best that our book be typo free (which is impossible but our intention). All suggestions in this regard for improvement in the future are welcome, and will be highly appreciated and gratefully acknowledged.

Contents

Chapter 1
Introduction

The normal and Student's t distributions are two of the most important distributions in statistics. These distributions have been extensively studied and used by many researchers since their discoveries. This book reviews the normal and Student's t distributions, and their applications. These studies involve some preliminaries on random variables and distribution functions, which are defined below (in Sect. 1.1), (for details, see Lukacs 1972, Dudewicz and Mishra 1988, Rohatgi and Saleh 2001, Severini 2005, and Mukhopadhyay 2006, among others). Some special functions and mathematical results will also be needed, which, for the sake of completeness, are given below (in Sect. 1.2), (for details, see Abramowitz and Stegun 1970, Lebedev 1972, Prudnikov et al. 1986, and Gradshteyn and Ryzhik 2000, among others).

1.1 Some Preliminaries on Random Variables and Distributions

Definition 1.1.1 (Random Variable): Let $(\Omega,\ T,\ P)$ be a probability space, where $\Omega = \{w\}$ is a set of simple events, T is a σ-algebra of events, and P is a probability measure defined on $(\Omega,\ T)$. Let B be an element of the Borel σ-algebra of subsets of the real line R. A random variable $X = X(w)$ is defined as a finite single-valued function $X : \Omega \to R$ such that $X^{-1}(B) = \{w : X\ (w) \in B\} \in T,\ \forall$ Borel set $B \in R$. Thus, a random variable X is a real-valued function with domain Ω, that is, $X(w) \in R = \{y :\ -\infty < y < +\infty\},\ \forall\, w \in \Omega$.

Definition 1.1.2 (Cumulative Distribution Function): Let $B = (-\infty,\ x]$ in the above definition 1.1. Then the cumulative distribution function (cdf) or distribution function (df) of the random variable $X = X\ (w)$ is defined by $F_X(x) = P\ [X \leq x], \forall x \in (-\infty,\ +\infty)$, with the following properties:

(i) $F_X(x)$ is a non-decreasing function of x.
(ii) $F_X(-\infty) = 0, \quad F_X(+\infty) = 1$.
(iii) $F_X(x)$ is right continuous.

M. Ahsanullah et al., *Normal and Student's t Distributions and Their Applications*,
Atlantis Studies in Probability and Statistics 4, DOI: 10.2991/978-94-6239-061-4_1,

Definition 1.1.3 (Absolutely Continuous Distribution Function): The distribution function $F_X(x)$ of a random variable X is said to be absolutely continuous (with respect to Lebesgue measure) if $\exists$ a function $f_X(x) \geq 0$ such that $F_X(x) = \int_{-\infty}^{x} f_X(t)dt$.

Definition 1.1.4 (Probability Density Function): The function $f_X(x)$ in the above definition 1.1.3 is called the probability density function (pdf) or density function of the random variable X if it satisfies the following condition:

$$\int_{-\infty}^{\infty} f_X(x)dx = 1.$$

Definition 1.1.5 (Moments): If a random variable X has an absolutely continuous (with respect to Lebesgue measure) distribution with a pdf $f_X(x)$, then the nth moment about zero and the nth central moment of X are respectively defined by the following expressions:

$$\alpha_n = E\left(X^n\right) = \int_{-\infty}^{\infty} x^n f_X(x)dx,$$

when

$$E|X|^n = \int_{-\infty}^{\infty} |x|^n f_X(x)dx < \infty.$$

and

$$\beta_n = E[X - E(X)]^n = \int_{-\infty}^{\infty} (x - E(X))^n f_X(x)dx,$$

when

$$E|X - E(X)|^n = \int_{-\infty}^{\infty} |x - E(X)|^n f_X(x)dx < \infty.$$

Note that, in the above definitions, $\alpha_1 = E(X)$, and $\beta_2 = E[X - E(X)]^2$ are respectively called the expected value (or mean or mathematical expectation) and variance of the random variable X.

Definition 1.1.6 (Entropy): An entropy provides an excellent tool to quantify the amount of information (or uncertainty) contained in a random observation regarding its parent distribution (population). A large value of entropy implies the greater uncertainty in the data. As proposed by Shannon (1948), if a random variable X has an absolutely continuous distribution with a pdf $f_X(x)$, then the entropy of X is defined as

$$H_X[f_X(X)] = E[-\ln(f_X(X)] = -\int_{-\infty}^{\infty} f_X(x)\ \ln[f_X(x)]\ dx.$$

1.2 Some Useful Mathematical Results

The following special functions and mathematical results will be useful in our analyses.

(a) **Special Functions**

(i) The series

$${}_pF_q(\alpha_1, \alpha_2, \ldots, \alpha_p; \beta_1, \beta_2, \ldots, \beta_q; z) = \sum_{k=0}^{\infty}\left\{\frac{(\alpha_1)_k\ (\alpha_2)_k \cdots (\alpha_p)_k}{(\beta_1)_k\ (\beta_2)_k \cdots (\beta_q)_k}\frac{z^k}{k!}\right\},$$

is called a generalized hypergeometric series of order (p, q), where $(\alpha)_k$ and $(\beta)_k$ represent Pochhammer symbols and

$$(x)_k = x(x-1)......(x-k+1), \text{ and } (x)_0 = 1.$$

(ii) For $p = 1$ and $q = 2$, we have generalized hypergeometric function ${}_1F_2$ of order $(1, 2)$, given by ${}_1F_2(\alpha_1; \beta_1, \beta_2; z) = \sum_{k=0}^{\infty}\left\{\frac{(\alpha_1)_k}{(\beta_1)_k(\beta_2)_k}\frac{z^k}{k!}\right\}$.

(iii) For $p = 2$ and $q = 1$, the series given by ${}_2F_1(\alpha, \beta; \gamma; z) \equiv F(\alpha, \beta; \gamma; z) \equiv F(\beta, \alpha; \gamma; z) = \sum_{k=0}^{\infty}\left\{\frac{(\alpha)_k(\beta)_k z^k}{(\gamma)_k k!}\right\}$ is called generalized hypergeometric function ${}_2F_1$ of order $(2, 1)$. Also, we have $F(\alpha, \beta; \gamma; z) = (1-z)^{-\beta}F\left(\beta, \gamma-\alpha; \gamma; \frac{z}{z-1}\right)$.

(iv) For $p = 2$ and $q = 0$, the function defined by

$$\begin{aligned}\psi(\alpha, \gamma; z) &\equiv z^{-\alpha}{}_2F_0\left(\alpha, 1+\alpha-\gamma; \frac{-1}{z}\right) \equiv U(\alpha, \gamma, z)\\ &= \frac{1}{\Gamma(\alpha)}\int_0^{\infty} e^{-zt}t^{\alpha-1}(1+t)^{\gamma-\alpha-1}dt, \quad Re(\alpha) > 0\end{aligned}, \tag{1.1}$$

is called Kummer's function. Here $\mathrm{Re}(\alpha) > 0$, $\mathrm{Re}(z) > 0$, (see Prudnikov et al., 1986, volume 3, equation 7.2.2.7, page 435, or Abramowitz and Stegun, 1970, equation 13.2.5, page 505).

(v) The function defined by

$$\Phi_1(a, b; c; w, z) = \sum_{k,\, l=0}^{\infty} \frac{(a)_{k+l}\,(b)_k}{(c)_{k+l}} \frac{w^k z^l}{(k\,!)\,(l\,!)}$$
$$= \Gamma\begin{bmatrix} c \\ a,\ c-a \end{bmatrix} \int_0^1 u^{a-1}(1-u)^{c-a-1}(1-u\,w)^{-b} e^{u\,z}\,du, \tag{1.2}$$

where Re (a), Re $(c-a) > 0, |w| < 1$, and $(a)_k$ denote the Pochhammer symbol, is called a generalized (or confluent) hypergeometric function of two variables.

(vi) The integrals given by

$$\Gamma(\alpha) = \int_0^{\infty} t^{\alpha-1}e^{-t}dt,\ \alpha > 0,\ \gamma(\alpha, x) = \int_0^{x} t^{\alpha-1}e^{-t}dt\ ,\text{ and } \Gamma(\alpha,\ x) = \int_x^{\infty} t^{\alpha-1}e^{-t}dt,\ \alpha > 0,$$

are called (complete) gamma, incomplete gamma and complementary incomplete gamma functions, respectively. Note that $\Gamma(\alpha,\ x) + \gamma(\alpha, x) = \Gamma(\alpha)$.

(vii) The functions defined by $erf(x) = \frac{2}{\sqrt{\pi}} \int_0^x e^{-u^2} du$, and $erfc(x) = \frac{2}{\sqrt{\pi}} \int_x^{\infty} e^{-u^2} du = 1 - erf(x)$, are called error and complementary error functions respectively.

(viii) The function defined by $B(p,\ q) = \int_0^1 t^p(1-t)^{q-1}dt = \frac{\Gamma(p)\Gamma(q)}{\Gamma(p+q)}$, $p > 0,\ q > 0$, is known as beta function (or Euler's function of the first kind).

(ix) **Struve Function,** $H_\nu(x)$**:** It is defined as

$$H_\nu(x) = \frac{2x^{\nu+1}}{\sqrt{\pi}2^{\nu+1}\Gamma(\nu+3/2)} \sum_{k=0}^{\infty} \frac{1}{(3/2)_k(\nu+3/2)_k} \left(-\frac{x^2}{4}\right)^k.$$

(x) **Bessel Function of the First Kind,** $J_\nu(x)$**:** It is defined as

$$J_\nu(x) = \frac{x^\nu}{2^\nu\Gamma(\nu+1)} \sum_{k=0}^{\infty} \frac{1}{(\nu+1)_k\,(k!)} \left(-\frac{x^2}{4}\right)^k.$$

(xi) **Bessel Function of the Second Kind,** $Y_\nu(x)$**:** It is defined as

$$Y_\nu(x) = \frac{\cos(\nu\pi)\ J_\nu(x) - J_{-\nu}(x)}{\sin(\nu\pi)},$$

with $Y_0(.)$ interpreted as the limit $Y_0(x) = \lim_{\nu\to 0} Y_\nu(x)$.

(xii) **Modified Bessel Function of the First Kind,** $I_\nu(x)$**:** It is defined as

$$I_\nu(x) = \frac{x^\nu}{2^\nu \Gamma(\nu+1)} \sum_{k=0}^{\infty} \frac{1}{(\nu+1)_k\,(k!)} \left(\frac{x^2}{4}\right)^k.$$

(xiii) **Modified Bessel Function of the second kind** (note we need later to find the pdf of the product of two independent normal random variables). Modified Bessel function of the Second Kind, $K_\alpha(x)$ is defined as

$$K_\alpha(x) = \frac{J_\alpha(x)\cos(\alpha x) - J_{-\alpha}(x)}{\mathrm{Sin}(\alpha x)},$$

where

$$J_\alpha(x) = \frac{(x/2)^\alpha}{\Gamma(\alpha+1)} \alpha F_1(\alpha+1, -\frac{x^2}{4}).$$

In the case of integer order n, the function is defined by the limit as non-integer α tends to n. For n = 0

$$K_0(x) = \frac{1}{2}\int_{-\infty}^{\infty} \frac{\cos tx}{\sqrt{1+t^2}} dt$$

(xiv) **Modified Bessel Function of the Third Kind,** $K_\nu(x)$**:** It is defined as

$$K_\nu(x) = \frac{\pi\,\{I_{-\nu}(x) - I_\nu(x)\}}{2\sin(\nu\pi)},$$

with $K_0(.)$ interpreted as the limit $K_0(x) = \lim_{\nu\to 0} K_\nu(x)$.

(xv) **Meijer** ***G*****-Function:** It is defined as

$$G^{m,n}_{p,q}\left(x\Big|^{a_1,\dots,a_p}_{b_1,\dots,b_q}\right) = \frac{1}{2\pi i}\int_L \frac{x^{-t}\Gamma(b_1+t)\cdots\Gamma(b_m+t)\Gamma(1-a_1-t)\cdots\Gamma(1-a_n-t)}{\Gamma(a_{n+1}+t)\cdots\Gamma(a_p+t)\Gamma(1-b_{m+1}-t)\cdots\Gamma(1-b_q-t)} dt,$$

where $(e)_k = e(e+1)\cdots(e+k-1)$ denotes the ascending factorial and L denotes an integration path (for details on Meijer G-Function, see, Gradshteyn and Ryzhik (2000), Sect. 9.3, Page 1068).

(xvi) $\Gamma\left(n+\frac{1}{2}\right) = \frac{\sqrt{\pi}(2n)!}{2^{2n}(n!)}$, where $n > 0$ is an integer.

(xvii) For negative values, gamma function can be defined as

$$\Gamma\left(-n+\frac{1}{2}\right) = \frac{(-1)^n 2^n \sqrt{\pi}}{1.3.5.\dots(2n-1)}, \text{ where } n > 0 \text{ is an integer.}$$

The organization of this book is as follows. In Chap. 2 , some basic ideas, definitions and some detailed properties along with applications of the normal distributions have been presented. Some basic ideas, definitions and some detailed properties along with applications of the Student's t distributions have been presented in Chap. 3. Chapter 4 presents the distributions of the sum, product and ratio of normal random variables. In Chap. 5, sum, product and ratio for Student's t random variables have been given. Chapter 6 discusses the sum, product and ratio for random variables X and Y having the normal and Student's t distributions respectively and distributed independently of each other. In Chap. 7, a new symmetric distribution and its properties have been presented by taking the product of the probability density functions of the normal and Student's t distributions for some continuous random variable X. The characterizations of normal distributions are presented in Chap. 8. In Chap. 9, we presented the characterizations of Student's t distribution. Some concluding remarks and some future research on the sum, product and ration of two random variables are provided in Chap. 10.

Chapter 2
Normal Distribution

The normal distribution is one of the most important continuous probability distributions, and is widely used in statistics and other fields of sciences. In this chapter, we present some basic ideas, definitions, and properties of normal distribution, (for details, see, for example, Whittaker and Robinson (1967), Feller (1968, 1971), Patel et al. (1976), Patel and Read (1982), Johnson et al. (1994), Evans et al. (2000), Balakrishnan and Nevzorov (2003), and Kapadia et al. (2005), among others).

2.1 Normal Distribution

The normal distribution describes a family of continuous probability distributions, having the same general shape, and differing in their location (that is, the mean or average) and scale parameters (that is, the standard deviation). The graph of its probability density function is a symmetric and bell-shaped curve. The development of the general theories of the normal distributions began with the work of de Moivre (1733, 1738) in his studies of approximations to certain binomial distributions for large positive integer $n > 0$. Further developments continued with the contributions of Legendre (1805), Gauss (1809), Laplace (1812), Bessel (1818, 1838), Bravais (1846), Airy (1861), Galton (1875, 1889), Helmert (1876), Tchebyshev (1890), Edgeworth (1883, 1892, 1905), Pearson (1896), Markov (1899, 1900), Lyapunov (1901), Charlier (1905), and Fisher (1930, 1931), among others. For further discussions on the history of the normal distribution and its development, readers are referred to Pearson (1967), Patel and Read (1982), Johnson et al. (1994), and Stigler (1999), and references therein. Also, see Wiper et al. (2005), for recent developments. The normal distribution plays a vital role in many applied problems of biology, economics, engineering, financial risk management, genetics, hydrology, mechanics, medicine, number theory, statistics, physics, psychology, reliability, etc., and has been has been extensively studied, both from theoretical and applications point of view, by many researchers, since its inception.

M. Ahsanullah et al., *Normal and Student's t Distributions and Their Applications*,
Atlantis Studies in Probability and Statistics 4, DOI: 10.2991/978-94-6239-061-4_2,

2.1.1 Definition (Normal Distribution)

A continuous random variable X is said to have a normal distribution, with mean μ and variance σ^2, that is, $X \sim N(\mu, \sigma^2)$, if its pdf $f_X(x)$ and cdf $F_X(x) = P(X \leq x)$ are, respectively, given by

$$f_X(x) = \frac{1}{\sigma\sqrt{2\pi}} e^{-(x-\mu)^2/2\sigma^2}, \quad -\infty < x < \infty, \tag{2.1}$$

and

$$F_X(x) = \frac{1}{\sigma\sqrt{2\pi}} \int_{-\infty}^{x} e^{-(y-\mu)^2/2\sigma^2} dy$$

$$= \frac{1}{2}\left[1 + erf\left(\frac{x-\mu}{\sigma\sqrt{2}}\right)\right], \quad -\infty < x < \infty, \ -\infty < \mu < \infty, \ \sigma > 0, \tag{2.2}$$

where $erf(.)$ denotes error function, and μ and σ are location and scale parameters, respectively.

2.1.2 Definition (Standard Normal Distribution)

A normal distribution with $\mu = 0$ and $\sigma = 1$, that is, $X \sim N(0, 1)$, is called the standard normal distribution. The pdf $f_X(x)$ and cdf $F_X(x)$ of $X \sim N(0, 1)$ are, respectively, given by

$$f_X(x) = \frac{1}{\sqrt{2\pi}} e^{-x^2/2}, \ -\infty < x < \infty, \tag{2.3}$$

and

$$F_X(x) = \frac{1}{\sqrt{2\pi}} \int_{-\infty}^{x} e^{-t^2/2} dt, \quad -\infty < x < \infty,$$

$$= \frac{1}{2}\left[1 + erf\left(\frac{x}{\sqrt{2}}\right)\right], \quad -\infty < x < \infty. \tag{2.4}$$

Note that if $Z \sim N(0, 1)$ and $X = \mu + \sigma Z$, then $X \sim N(\mu, \sigma^2)$, and conversely if $X \sim N(\mu, \sigma^2)$ and $Z = (X - \mu)/\sigma$, then $Z \sim N(0, 1)$. Thus, the pdf of any general $X \sim N(\mu, \sigma^2)$ can easily be obtained from the pdf of $Z \sim N(0, 1)$, by using the simple location and scale transformation, that is, $X = \mu + \sigma Z$. To describe the shapes of the normal distribution, the plots of the pdf (2.1) and cdf (2.2),

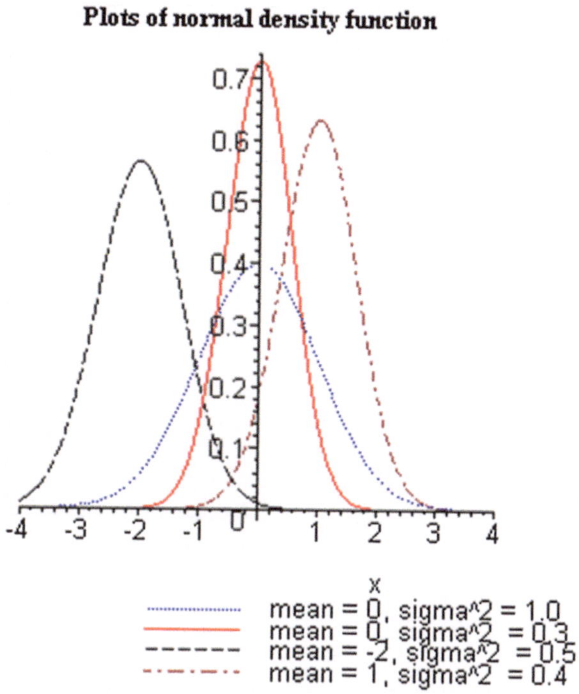

Fig. 2.1 Plots of the normal pdf, for different values of μ and σ^2

for different values of μ and σ^2, are provided in Figs. 2.1 and 2.2, respectively, by using Maple 10. The effects of the parameters, μ and σ^2, can easily be seen from these graphs. Similar plots can be drawn for other values of the parameters. It is clear from Fig. 2.1 that the graph of the pdf $f_X(x)$ of a normal random variable, $X \sim N(\mu,\ \sigma^2)$, is symmetric about mean, μ, that is $f_X(\mu + x) = f_X(\mu - x)$, $-\infty < x < \infty$.

2.1.3 Some Properties of the Normal Distribution

This section discusses the mode, moment generating function, cumulants, moments, mean, variance, coefficients of skewness and kurtosis, and entropy of the normal distribution, $N(\mu,\ \sigma^2)$. For detailed derivations of these, see, for example, Kendall and Stuart (1958), Lukacs (1972), Dudewicz and Mishra (1988), Johnson et al. (1994), Rohatgi and Saleh (2001), Balakrishnan and Nevzorov (2003), Kapadia et al. (2005), and Mukhopadhyay (2006), among others.

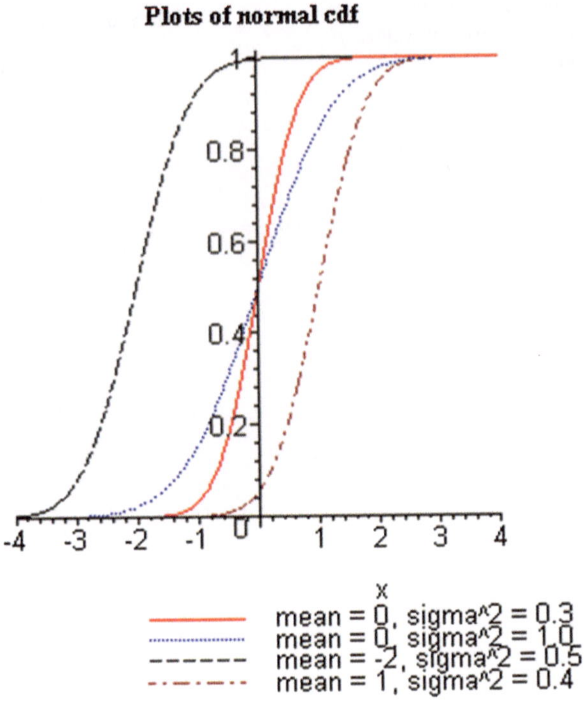

Fig. 2.2 Plots of the normal cdf for different values of μ and σ^2

2.1.3.1 Mode

The mode or modal value is that value of X for which the normal probability density function $f_X(x)$ defined by (2.1) is maximum. Now, differentiating with respect to x Eq. (2.1), we have

$$f'_X(x) = -\sqrt{\frac{2}{\pi}}\left[\frac{(x-\mu)\,e^{-(x-\mu)^2/2\sigma^2}}{\sigma^3}\right],$$

which, when equated to 0, easily gives the mode to be $x = \mu$, which is the mean, that is, the location parameter of the normal distribution. It can be easily seen that $f''_X(x) < 0$. Consequently, the maximum value of the normal probability density function $f_X(x)$ from (2.1) is easily obtained as $f_X(\mu) = \frac{1}{\sigma\sqrt{2\pi}}$. Since $f'(x) = 0$ has one root, the normal probability density function (2.1) is unimodal.

2.1.3.2 Cumulants

The cumulants k_r of a random variable X are defined via the cumulant generating function

$$g(t) = \sum_{r=1}^{\infty} k_r \frac{t^r}{r!}, \text{ where g(t)} = \ln\left(\mathrm{E}(\mathrm{e}^{\mathrm{t}X})\right).$$

For some integer $r > 0$, the rth cumulant of a normal random variable X having the pdf (2.1) is given by

$$\kappa_r = \begin{cases} \mu, & when\ r = 1; \\ \sigma^2, & when\ r = 2; \\ 0, & when\ r > 2 \end{cases}$$

2.1.3.3 Moment Generating Function

The moment generating function of a normal random variable X having the pdf (2.1) is given by (see, for example, Kendall and Stuart (1958), among others)

$$M_X(t) = E\left(e^{tX}\right) = e^{t\mu + \frac{1}{2}t^2\sigma^2}.$$

2.1.3.4 Moments

For some integer $r > 0$, the rth moment about the mean of a normal random variable X having the pdf (2.1) is given by

$$E\left(X^r\right) = \mu_r = \begin{cases} \frac{\sigma^r\,(r!)}{2^{\frac{r}{2}}\,[(r/2)!]}, & \text{for r even;} \\ 0, & \text{for r odd} \end{cases} \tag{2.5}$$

We can write $\mu_r = \sigma^r(r!!)$, where $m!! = 1.3.5.\ldots(m-1)$ for m even.

2.1.3.5 Mean, Variance, and Coefficients of Skewness and Kurtosis

From (2.5), the mean, variance, and coefficients of skewness and kurtosis of a normal random variable $X \sim N(\mu, \sigma^2)$ having the pdf (2.1) are easily obtained as follows:

(i) **Mean:** $\alpha_1 = E(X) = \mu$;
(ii) **Variance:** $Var(X) = \sigma^2, \quad \sigma > 0$;
(iii) **Coefficient of Skewness:** $\gamma_1(X) = \frac{\mu_3}{\mu_2^{3/2}} = 0$;
(iv) **Coefficient of Kurtosis:** $\gamma_2(X) = \frac{\mu_4}{\mu_2^2} = 3$.

where μ_r has been defined in Eq. (2.5).

Since the coefficient of kurtosis, that is, $\gamma_2(X) = 3$, it follows that the normal distributions are mesokurtic distributions.

2.1.3.6 Median, Mean Deviation, and Coefficient of Variation of $X \sim N(\mu, \sigma^2)$

These are given by

(i) **Median:** μ

(ii) **Mean Deviation:** $\left(\frac{2\sigma^2}{\pi}\right)^{\frac{1}{2}}$

(iii) **Coefficient of Variation:** $\frac{\sigma}{\mu}$

2.1.3.7 Characteristic Function

The characteristic function of a normal random variable $X \sim N(\mu, \sigma^2)$ having the pdf (2.1) is given by (see, for example, Patel et al. (1976), among others)

$$\phi_X(t) = E\left(e^{itX}\right) = e^{it\mu - \frac{1}{2}t^2\sigma^2}, \ \mathrm{i} = \sqrt{-1}.$$

2.1.3.8 Entropy

For some $\sigma > 0$, entropy of a random variable X having the pdf (2.1) is easily given by

$$\begin{aligned} H_X[f_X(x)] &= E[-\ln(f_X(X)] \\ &= -\int_{-\infty}^{\infty} f_X(x) \ln [f_X(x)]\, dx, \\ &= \ln\left(\sqrt{2\pi e}\sigma\right) \end{aligned}$$

(see, for example, Lazo and Rathie (1978), Jones (1979), Kapur (1993), and Suhir (1997), among others). It can be easily seen that $\frac{d(H_X[f_X(x)])}{d\sigma} > 0$, and $\frac{d^2(H_X[f_X(x)])}{d\sigma^2} < 0, \forall\sigma > 0, \forall\mu$. It follows that that the entropy of a random variable X having the normal pdf (2.1) is a monotonic increasing concave function of $\sigma > 0, \ \forall\mu$. The possible shape of the entropy for different values of the parameter σ is provided below in Fig. 2.3, by using Maple 10. The effects of the parameter σ on entropy can easily be seen from the graph. Similar plots can be drawn for others values of the parameter σ.

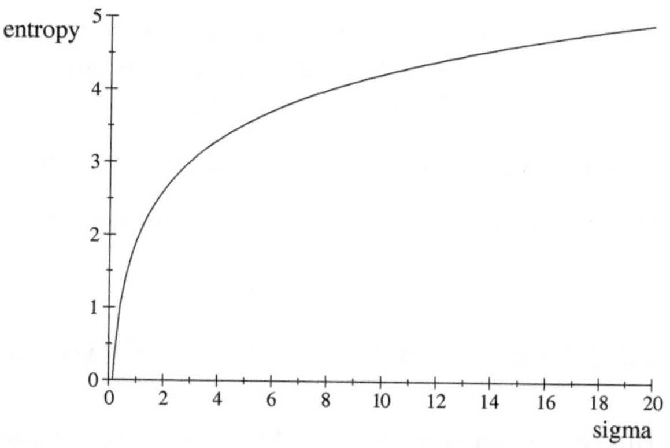

Fig. 2.3 Plot of entropy

2.1.4 Percentiles

This section computes the percentiles of the normal distribution, by using Maple 10. For any $p(0 < p < 1)$, the $(100p)th$ percentile (also called the quantile of order p) of $N(\mu,\ \sigma^2)$ with the pdf $f_X(x)$ is a number z_p such that the area under $f_X(x)$ to the left of z_p is p. That is, z_p is any root of the equation

$$\Phi(z_p) = \int_{-\infty}^{z_p} f_X(u)du = p.$$

Using the Maple program, the percentiles z_p of $N(\mu,\ \sigma^2)$ are computed for some selected values of p for the given values of μ and , which are provided in Table 2.1, when $\mu = 0$ and $\sigma = 1$. Table 2.1 gives the percentile values of z_p for $p \geq 0.5$. For $p < 0.5$, use $1 - Z_{1-p}$.

Table 2.1 Percentiles of $N(0,\ 1)$

p	z_p
0.5	0.0000000000
0.6	0.2533471031
0.7	0.5244005127
0.75	0.6744897502
0.8	0.8416212336
0.9	1.281551566
0.95	1.644853627
0.975	1.959963985
0.99	2.326347874
0.995	2.575829304
0.9975	2.807033768
0.999	3.090232306

Suppose $X_1, X_2, \ldots X_n$ are n independent N (0, 1) random variables and $M(n) = \max(X_1, X_2, \ldots X_n)$. It is known (see Ahsanullah and Kirmani (2008) p.15 and Ahsanullah and Nevzorov (2001) p.92) that

$P(M(n) \leq a_n + b_n x) \rightarrow e^{-e^{-x}}$, for all x as $n \rightarrow \infty$.

where $a_n = \beta_n - \frac{D_n}{2\beta_n}$, $D_n = \ln\ln n + \ln 4\pi$, $\beta_n = (2\ln n)^{1/2}$, $b_n - (2\ln n)^{-1/2}$.

2.2 Different Forms of Normal Distribution

This section presents different forms of normal distribution and some of their important properties, (for details, see, for example, Whittaker and Robinson (1967), Feller (1968, 1971), Patel et al. (1976), Patel and Read (1982), Johnson et al. (1994), Evans et al. (2000), Balakrishnan and Nevzorov (2003), and Kapadia et al. (2005), among others).

2.2.1 *Generalized Normal Distribution*

Following Nadarajah (2005a), a continuous random variable X is said to have a generalized normal distribution, with mean μ and variance $\frac{\sigma^2\Gamma\left(\frac{3}{s}\right)}{\Gamma\left(\frac{1}{s}\right)}$, where $s > 0$, that is, $X \sim N\left(\mu, \frac{\sigma^2\Gamma\left(\frac{3}{s}\right)}{\Gamma\left(\frac{1}{s}\right)}\right)$, if its pdf $f_X(x)$ and cdf $F_X(x) = P(X \leq x)$ are, respectively, given by

$$f_X(x) = \frac{s}{2\sigma\Gamma\left(\frac{1}{s}\right)} e^{-\left|\frac{x-\mu}{\sigma}\right|^s}, \tag{2.6}$$

and

$$F_X(x) = \begin{cases} \frac{\Gamma\left(\frac{1}{s}, \left(\frac{\mu - x}{\sigma}\right)^s\right)}{2\Gamma\left(\frac{1}{s}\right)}, & if\ x \leq \mu \\ 1 - \frac{\Gamma\left(\frac{1}{s}, \left(\frac{x - \mu}{\sigma}\right)^s\right)}{2\Gamma\left(\frac{1}{s}\right)}, & if\ x > \mu \end{cases} \tag{2.7}$$

where $-\infty < x < \infty$, $-\infty < \mu < \infty$, $\sigma > 0$, $s > 0$, and $\Gamma(a, x)$ denotes complementary incomplete gamma function defined by $\Gamma(a, x) = \int_x^\infty t^{a-1} e^{-t} dt$. It is easy to see that the Eq. (2.6) reduces to the normal distribution for $s = 2$, and Laplace distribution for $s = 1$. Further, note that if has the pdf given by (2.6), then the pdf of the standardized random variable $Z = (X - \mu)/\sigma$ is given by

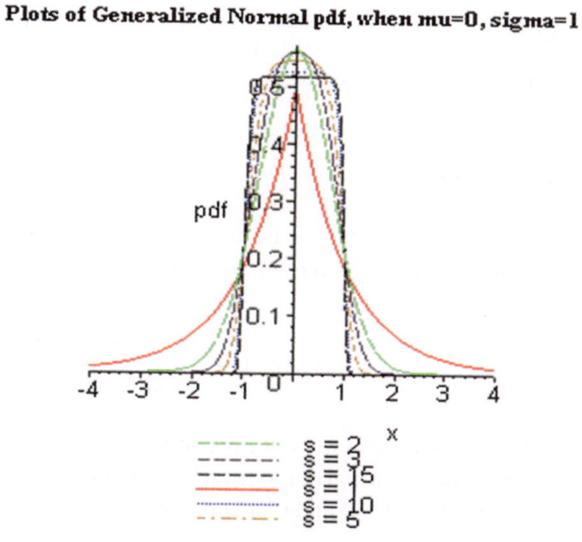

Fig. 2.4 Plots of the generalized normal pdf for different values of s

$$fz(z) = \frac{s}{2\Gamma\left(\frac{1}{s}\right)} e^{-|z|^s} \tag{2.8}$$

To describe the shapes of the generalized normal distribution, the plots of the pdf (2.6), for $\mu = 0, \sigma = 1$, and different values of s, are provided in Fig. 2.4 by using Maple 10. The effects of the parameters can easily be seen from these graphs. Similar plots can be drawn for others values of the parameters. It is clear from Fig. 2.4 that the graph of the pdf $f_X(x)$ of the generalized normal random variable is symmetric about mean, μ, that is

$$f_X(\mu + x) = f_X(\mu - x), \quad -\infty < x < \infty.$$

2.2.1.1 Some Properties of the Generalized Normal Distribution

This section discusses the mode, moments, mean, median, mean deviation, variance, and entropy of the generalized normal distribution. For detailed derivations of these, see Nadarajah (2005).

2.2.1.2 Mode

It is easy to see that the mode or modal value of x for which the generalized normal probability density function $f_X(x)$ defined by (2.6) is maximum, is given by $x = \mu$, and the maximum value of the generalized normal probability density function (2.6) is given by $f_X(\mu) = \frac{s}{2\sigma\Gamma\left(\frac{1}{s}\right)}$. Clearly, the generalized normal probability density function (2.6) is unimodal.

2.2.1.3 Moments

(i) For some integer $r > 0$, the rth moment of the generalized standard normal random variable Z having the pdf (2.8) is given by

$$E\left(Z^r\right) = \frac{1+(-1)^r}{2\Gamma\left(\frac{1}{s}\right)}\Gamma\left(\frac{r+1}{s}\right) \tag{2.9}$$

(i) For some integer $n > 0$, the nth moment and the nth central moment of the generalized normal random variable X having the pdf (2.6) are respectively given by the Eqs. (2.10) and (2.11) below:

$$E\left(X^n\right) = \frac{(\mu^n)\sum\limits_{k=0}^{n}\binom{n}{k}\left(\frac{\sigma}{\mu}\right)^k\left[1+(-1)^k\right]\Gamma\left(\frac{k+1}{s}\right)}{2\Gamma\left(\frac{1}{s}\right)} \tag{2.10}$$

and

$$E\left[(X-\mu)^n\right] = \frac{(\sigma^n)\left[1+(-1)^n\right]\Gamma\left(\frac{n+1}{s}\right)}{2\Gamma\left(\frac{1}{s}\right)} \tag{2.11}$$

2.2.1.4 Mean, Variance, Coefficients of Skewness and Kurtosis, Median and Mean Deviation

From the expressions (2.10) and (2.11), the mean, variance, coefficients of skewness and kurtosis, median and mean deviation of the generalized normal random variable X having the pdf (2.6) are easily obtained as follows:

(i) Mean: $\alpha_1 = E(X) = \mu$;

(ii) Variance: $Var(X) = \beta_2 = \frac{\sigma^2\Gamma\left(\frac{3}{s}\right)}{\Gamma\left(\frac{1}{s}\right)}, \quad \sigma > 0,\ s > 0;$

(iii) Coefficient of Skewness: $\gamma_1(X) = \frac{\beta_3}{\beta_2^{3/2}} = 0;$

(iv) Coefficient of Kurtosis: $\gamma_2(X) = \frac{\beta_4}{\beta_2^2} = \frac{\Gamma\left(\frac{1}{s}\right)\Gamma\left(\frac{5}{s}\right)}{\left[\Gamma\left(\frac{3}{s}\right)\right]^2}, \quad s > 0;$

(v) Median (X): μ;

(vi) Mean Deviation: $E\,|X - \mu| = \dfrac{\sigma\,\Gamma\left(\frac{2}{s}\right)}{\Gamma\left(\frac{1}{s}\right)}, \quad s > 0.$

2.2.1.5 Renyi and Shannon Entropies, and Song's Measure of the Shape of the Generalized Normal Distribution

These are easily obtained as follows, (for details, see, for example, Nadarajah (2005), among others).

(i) **Renyi Entropy:** Following Renyi (1961), for some reals $\gamma > 0,\ \gamma \neq 1$, the entropy of the generalized normal random variable X having the pdf (2.6) is given by

$$\Im_R(\gamma) = \frac{1}{1-\gamma}\ln\int_{-\infty}^{+\infty}[f_X(X)]^{\gamma}\,dx$$

$$= \frac{\ln(\gamma)}{s(\gamma-1)} - \ln\left[\frac{s}{2\sigma\Gamma\left(\frac{1}{s}\right)}\right], \quad \sigma > 0,\ s > 0,\ \gamma > 0,\ \gamma \neq 1.$$

(ii) **Shannon Entropy:** Following Shannon (1948), the entropy of the generalized normal random variable X having the pdf (2.6) is given by

$$H_X[f_X(X)] = E[-\ln(f_X(X)] = -\int_{-\infty}^{\infty} f_X(x)\,\ln\,[f_X(x)]\,dx,$$

which is the particular case of Renyi entropy as obtained in (i) above for $\gamma \to 1$. Thus, in the limit when $\gamma \to 1$ and using L'Hospital's rule, Shannon entropy is easily obtained from the expression for Renyi entropy in (i) above as follows:

$$H_X[f_X(X)] = \frac{1}{s} - \ln\left[\frac{s}{2\sigma\Gamma\left(\frac{1}{s}\right)}\right], \sigma > 0,\ s > 0.$$

(iii) **Song's Measure of the Shape of a Distribution:** Following Song (2001), the gradient of the Renyi entropy is given by

$$\Im'_R(\gamma) = \frac{d}{d\gamma}[\Im_R(\gamma)] = \frac{1}{s}\left\{\frac{1}{\gamma(\gamma-1)} - \frac{\ln(\gamma)}{(\gamma-1)^2}\right\} \tag{2.12}$$

which is related to the log likelihood by

$$\Im'_R(1) = -\frac{1}{2} Var\,[\ln f(X)]\,.$$

Thus, in the limit when $\gamma \to 1$ and using L'Hospital's rule, Song's measure of the shape of the distribution of the generalized normal random variable X having the pdf (2.6) is readily obtained from the Eq. (2.12) as follows:

$$-\,2\,\Im'_R(1) = \frac{1}{s},$$

which can be used in comparing the shapes of various densities and measuring heaviness of tails, similar to the measure of kurtosis.

2.2.2 Half Normal Distribution

Statistical methods dealing with the properties and applications of the half-normal distribution have been extensively used by many researchers in diverse areas of applications, particularly when the data are truncated from below (that is, left truncated,) or truncated from above (that is, right truncated), among them Dobzhansky and Wright (1947), Meeusen and van den Broeck (1977), Haberle (1991), Altman (1993), Buckland et al. (1993) , Chou and Liu (1998), Klugman et al. (1998), Bland and Altman (1999), Bland (2005), Goldar and Misra (2001), Lawless (2003), Pewsey (2002, 2004), Chen and Wang (2004) and Wiper et al. (2005), Babbit et al. (2006), Coffey et al. (2007), Barranco-Chamorro et al. (2007), and Cooray and Ananda (2008), are notable. A continuous random variable X is said to have a (general) half-normal distribution, with parameters μ (location) and σ (scale), that is, $X|\mu,\ \sigma \sim HN(\mu,\ \sigma)$, if its pdf $f_X(x)$ and cdf $F_X(x) = P(X \le x)$ are, respectively, given by

$$f_X(x|\mu,\ \sigma) = \sqrt{\frac{2}{\pi}}\frac{1}{\sigma}e^{-\frac{1}{2}\left(\frac{x-\mu}{\sigma}\right)^2}, \tag{2.13}$$

and

$$F_X(x) = erf\left(\frac{x-\mu}{\sqrt{2}\sigma}\right) \tag{2.14}$$

where $x \ge \mu,\ -\infty < \mu < \infty,\ \sigma > 0$, and $erf\,(.)$ denotes error function, (for details on half-normal distribution and its applications, see, for example, Altman (1993), Chou and Liu (1998), Bland and Altman (1999), McLaughlin (1999), Wiper et al. (2005), and references therein). Clearly, $X = \mu + \sigma\,|Z|$, where $Z \sim N(0,\ 1)$ has a standard normal distribution. On the other hand, the random variable $X = \mu - \sigma\,|Z|$ follows a negative (general) half- normal distribution. In particular, if $X \sim N\left(0,\ \sigma^2\right)$, then it is easy to see that the absolute value $|X|$ follows a half-normal distribution, with its pdf $f_{|X|}(x)$ given by

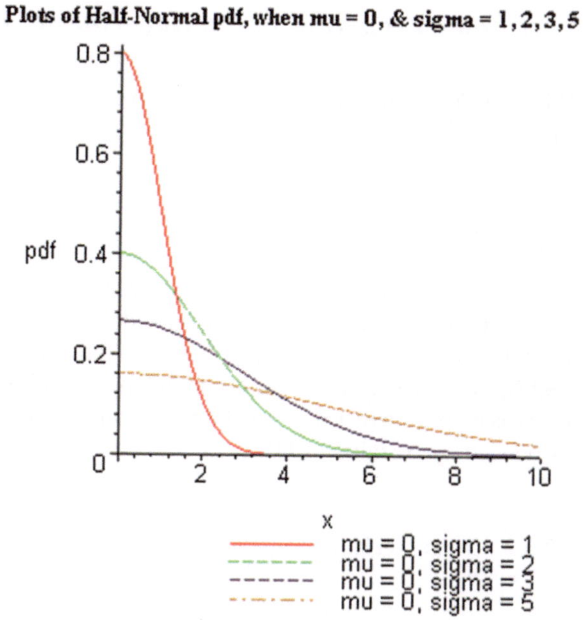

Fig. 2.5 Plots of the half-normal pdf

$$f_{|X|}(x) = \begin{cases} \frac{2}{\sqrt{2\pi}\,\sigma} e^{-\frac{1}{2}\left(\frac{x}{\sigma}\right)^2} & if\ x \geq 0 \\ 0 & if\ x < 0 \end{cases} \tag{2.15}$$

By taking $\sigma^2 = \frac{\pi}{2\theta^2}$ in the Eq. (2.15), more convenient expressions for the pdf and cdf of the half-normal distribution are obtained as follows

$$f_{|X|}(x) = \begin{cases} \frac{2\theta}{\pi} e^{-\left(\frac{x\theta}{\sqrt{\pi}}\right)^2} & if\ x \geq 0 \\ 0 & if\ x < 0 \end{cases} \tag{2.16}$$

and

$$F_{|X|}(x) = erf\left(\frac{\theta x}{\sqrt{\pi}}\right) \tag{2.17}$$

which are implemented in *Mathematica* software as HalfNormalDistribution[theta], see Weisstein (2007). To describe the shapes of the half-normal distribution, the plots of the pdf (2.13) for different values of the parameters μ and σ are provided in Fig. 2.5 by using Maple 10. The effects of the parameters can easily be seen from these graphs. Similar plots can be drawn for others values of the parameters.

2.2.3 Some Properties of the Half-Normal Distribution

This section discusses some important properties of the half-normal distribution, $X|\mu, \sigma \sim HN(\mu, \sigma)$.

2.2.3.1 Special Cases

The half-normal distribution, $X|\mu, \sigma \sim HN(\mu, \sigma)$ is a special case of the Amoroso, central chi, two parameter chi, generalized gamma, generalized Rayleigh, truncated normal, and folded normal distributions (for details, see, for example, Amoroso (1925), Patel and Read (1982), and Johnson et al. (1994), among others). It also arises as a limiting distribution of three parameter skew-normal class of distributions introduced by Azzalini (1985).

2.2.3.2 Characteristic Property

If $X \sim N(\mu, \sigma)$ is folded (to the right) about its mean, μ, then the resulting distribution is half-normal, $X|\mu, \sigma \sim HN(\mu, \sigma)$.

2.2.3.3 Mode

It is easy to see that the mode or modal value of x for which the half-normal probability density function $f_X(x)$ defined by (2.13) is maximum, is given at $x = \mu$, and the maximum value of the half-normal probability density function (2.13) is given by $f_X(\mu) = \frac{1}{\sigma}\sqrt{\frac{2}{\pi}}$. Clearly, the half-normal probability density function (2.13) is unimodal.

2.2.3.4 Moments

(i) kth **Moment of the Standardized Half-Normal Random Variable:** If the half-normal random variable X has the pdf given by the Eq. (2.13), then the standardized half-normal random variable $|Z| = \frac{X-\mu}{\sigma} \sim HN(0, 1)$ will have the pdf given by

$$f_{|Z|}(z) = \begin{cases} \frac{2}{\sqrt{2\pi}} e^{-\frac{1}{2}z^2} & if\ z \geq 0 \\ 0 & if\ z < 0 \end{cases} \tag{2.18}$$

For some integer $k > 0$, and using the following integral formula (see Prudnikov et al. Vol. 1, 1986, Eq. 2.3.18.2, p. 346, or Gradshteyn and Ryzhik

1980, Eq. 3.381.4, p. 317)

$$\int_0^{\infty} t^{\alpha - 1} e^{-\rho t^{\mu}} dt = \frac{1}{\mu} \rho^{-\frac{\alpha}{\mu}} \Gamma\left(\frac{\alpha}{\mu}\right), \quad \text{where } \mu, \ \text{Re}\,\alpha, \ \text{Re}\,\rho > 0,$$

the kth moment of the standardized half-normal random variable Z having the pdf (2.18) is easily given by

$$E\left(Z^k\right) = \frac{1}{\sqrt{\pi}} 2^{\frac{k}{2}} \Gamma\left(\frac{k+1}{2}\right), \tag{2.19}$$

where Γ (.) denotes gamma function.

(ii) **Moment of the Half-Normal Random Variable:** For some integer $n > 0$, the nth moment (about the origin) of the half-normal random variable X having the pdf (2.13) is easily obtained as

$$\begin{aligned} \mu_n' &= E\left(X^n\right) = E\left[(\mu + z\sigma)^n\right] = \sum_{k=0}^{n} \binom{n}{k} \mu^{n-k} \sigma^k E\left(Z^k\right) \\ &= \frac{1}{\sqrt{\pi}} \sum_{k=0}^{n} \binom{n}{k} 2^{\frac{k}{2}} \mu^{n-k} \sigma^k \Gamma\left(\frac{k+1}{2}\right) \end{aligned} \tag{2.20}$$

From the above Eq. (2.20), the first four moments of the half-normal random variable X are easily given by

$$\mu_1' = E[X] = \mu + \sigma\sqrt{\frac{2}{\pi}}, \tag{2.21}$$

$$\mu_2' = E\left[X^2\right] = \mu^2 + 2\sqrt{\frac{2}{\pi}}\mu\sigma + \sigma^2, \tag{2.22}$$

$$\mu_3' = E\left[X^3\right] = \mu^3 + 3\sqrt{\frac{2}{\pi}}\mu^2\sigma + 3\mu\sigma^2 + 2\sqrt{\frac{2}{\pi}}\sigma^3, \tag{2.23}$$

and

$$\mu_4' = E\left[X^4\right] = \mu^4 + 4\sqrt{\frac{2}{\pi}}\mu^3\sigma + 6\mu^2\sigma^2 + 8\sqrt{\frac{2}{\pi}}\mu\sigma^3 + 3\sigma^4. \tag{2.24}$$

(iii) **Central Moment of the Half-Normal Random Variable:** For some integer $n > 0$, the nth central moment (about the mean $\mu_1' = E(X)$) of the half-normal random variable X having the pdf (2.13) can be easily obtained using the formula

$$\mu_n = E\left[(X - \mu_1')^n\right] = \sum_{k=0}^{n} \binom{n}{k} (-\mu_1')^{n-k}\, E\left(X^k\right), \tag{2.25}$$

where $E\left(X^k\right) = \mu_k'$ denotes the kth moment, given by the Eq. (2.20), of the half-normal random variable X having the pdf (2.13).

Thus, from the above Eq. (2.25), the first three central moments of the half-normal random variable X are easily obtained as

$$\mu_2 = E\left[(X - \mu_1')^2\right] = \mu_2' - (\mu_1')^2 = \frac{\sigma^2(\pi - 2)}{\pi},$$

$$\mu_3 = \beta_3 = E\left[(X - \mu_1')^3\right] \tag{2.26}$$

$$= \mu_3' - 3\mu_1'\mu_2' + 2(\mu_1')^3 = \sqrt{\frac{2}{\pi}}\frac{\sigma^3(4 - \pi)}{\pi}, \tag{2.27}$$

and

$$\mu_4 = \beta_4 = E\left[(X - \mu_1')^4\right] = \mu_4' - 4\mu_1'\mu_3' + 6(\mu_1')^2\mu_2' - 3(\mu_1')^4$$
$$= \frac{\sigma^4(3\pi^2 - 4\pi - 12)}{\pi^2}. \tag{2.28}$$

2.2.3.5 Mean, Variance, and Coefficients of Skewness and Kurtosis

These are easily obtained as follows:

(i) **Mean** : $\alpha_1 = E(X) = \mu + \sigma\sqrt{\frac{2}{\pi}}$;

(ii) **Variance** : $Var(X) = \mu_2 = \sigma^2\left(1 - \frac{2}{\pi}\right), \quad \sigma > 0$;

(iii) **Coefficient of Skewness** : $\gamma_1(X) = \frac{\mu_3}{\mu_2^{3/2}} = \frac{\sqrt{2}(4 - \pi)}{\sqrt{(\pi - 2)^3}} \approx 0.9953$;

(iv) **Coefficient of Kurtosis** : $\gamma_2(X) = \frac{\mu_4}{\mu_2^2} = \frac{8(\pi - 3)}{(\pi - 2)^2} \approx 0.7614$;

2.2.3.6 Median (i.e., 50th Percentile or Second Quartile), and First and Third Quartiles

These are derived as follows. For any $p(0 < p < 1)$, the $(100p)th$ percentile (also called the quantile of order p) of the half-normal distribution, $X|\mu, \sigma \sim HN(\mu, \sigma)$, with the pdf $f_X(x)$ given by (2.13), is a number z_p such that the area under $f_X(x)$ to the left of z_p is p. That is, z_p is any root of the equation

$$F(z_p) = \int_{-\infty}^{z_p} f_X(t)dt = p. \tag{2.29}$$

For $p = 0.50$, we have the 50th percentile, that is, $z_{0.50}$, which is called the median (or the second quartile) of the half-normal distribution. For $p = 0.25$ and $p = 0.75$, we have the 25th and 75th percentiles respectively.

2.2.3.7 Derivation of Median(X)

Let m denote the median of the half-normal distribution, $X|\mu, \sigma \sim HN(\mu, \sigma)$, that is, let $m = z_{0.50}$. Then, from the Eq. (2.29), it follows that

$$0.50 = F(z_{0.50}) = \int_{-\infty}^{z_{0.50}} f_X(t)dt = \sqrt{\frac{2}{\pi}}\frac{1}{\sigma}\int_{-\infty}^{z_{0.50}} e^{-\frac{1}{2}\left(\frac{t-\mu}{\sigma}\right)^2} dt. \tag{2.30}$$

Substituting $\frac{t-\mu}{\sqrt{2}\,\sigma} = u$ in the Eq. (2.30), using the definition of error function, and solving for $z_{0.50}$, it is easy to see that

$$\begin{aligned} m = \text{Median}(X) = z_{0.50} &= \mu + \left(\sqrt{2}\right) erf^{-1}(0.50)\,\sigma \\ &= \mu + (\sqrt{2})(0.476936)\sigma \\ &\approx \mu + 0.6745\sigma, \quad \sigma > 0, \end{aligned}$$

where $erf^{-1}[0.50] = 0.476936$ has been obtained by using *Mathematica*. Note that the inverse error function is implemented in *Mathematica* as a *Built-in Symbol*, Inverse Erf[s], which gives the inverse error function obtained as the solution for z in $s = erf(z)$. Further, for details on Error and Inverse Error Functions, see, for example, Abramowitz and Stegun (1972, pp. 297–309), Gradshteyn and Ryzhik (1980), Prudnikov et al., Vol. 2 (1986), and Weisstein (2007), among others.

2.2.3.8 First and Third Quartiles

Let Q_1 and Q_3 denote the first and third quartiles of $X \sim HN(\mu, \sigma)$, that is, let $Q_1 = z_{0.25}$ and $Q_3 = z_{0.75}$. Then following the technique of the derivation of the Median(X) as in 2.2.3.7, one easily gets the Q_1 and Q_3 as follows.

(i) **First Quartile:** $Q_1 = \mu - 0.3186\sigma, \quad \sigma > 0;$
(ii) **Third Quartile:** $Q_3 = \mu + 1.150\sigma, \quad \sigma > 0.$

2.2.3.9 Mean Deviations

Following Stuart and Ord, Vol. 1, p. 52, (1994), the amount of scatter in a population is evidently measured to some extent by the totality of deviations from the mean and median. These are known as the mean deviation about the mean and the mean deviation about the median, denoted as δ_1 and δ_2, respectively, and are defined as follows:

$$\textbf{(i)}\ \delta_1 = \int_{-\infty}^{+\infty} |x - E(X)| f(x)dx,$$

$$\textbf{(ii)}\ \delta_2 = \int_{-\infty}^{+\infty} |x - M(X)| f(x)dx.$$

Derivations of δ_1 and δ_2 for the Half-Normal distribution, $X|\mu, \sigma \sim HN(\mu, \sigma)$: To derive these, we first prove the following Lemma.

Lemma 2.2.1: Let $\delta = \frac{\omega-\mu}{\sigma}$. Then

$$\int_{\mu}^{\infty} \frac{1}{\sigma}|x-\omega|\sqrt{\frac{2}{\pi}}e^{-(1/2)(\frac{x-\mu}{\sigma})^2}dx$$
$$= \sigma\sqrt{\frac{2}{\pi}}\left(-1 - \delta\sqrt{\frac{\pi}{2}} + e^{-\frac{\delta^2}{2}} + \delta\sqrt{\frac{\pi}{2}}erf(\frac{\delta}{\sqrt{2}})\right),$$

where $erf(z) = \int_0^z \frac{2}{\sqrt{\pi}}e^{-t^2}dt$ denotes the error function.

Proof: We have

$$\int_{\mu}^{\infty} \frac{1}{\sigma}|x-\omega|\sqrt{\frac{2}{\pi}}e^{-(1/2)(\frac{x-\mu}{\sigma})^2}dx$$

$$= \int_{\mu}^{\infty} \frac{|x-\mu-(\omega-\mu|}{\sigma}\sqrt{\frac{2}{\pi}}e^{-(1/2)(\frac{x-\mu}{\sigma})^2}dx$$

$$= \sigma\int_{0}^{\infty} |u-\delta|\sqrt{\frac{2}{\pi}}\, e^{-(1/2)u^2}du,$$

Substituting $\frac{x\ -\ \mu}{\sigma} = u,\ and\ \delta = \frac{\omega - \mu}{\sigma}$

$$= \sigma\int_{0}^{\delta} (\delta-u)\sqrt{\frac{2}{\pi}}e^{-(1/2)u^2}du + \sigma\int_{\delta}^{\infty} (u-\delta)\sqrt{\frac{2}{\pi}}e^{-(1/2)u^2}du$$

$$= \frac{\sigma}{\sqrt{\pi}}\left(\delta\sqrt{\pi}erf(\frac{\delta}{\sqrt{2}}) + \sqrt{2}e^{-\frac{\delta^2}{2}} - \sqrt{2}\right)$$

$$+ \frac{\sigma}{\sqrt{\pi}}\left(\delta\sqrt{\pi}erf(\frac{\delta}{\sqrt{2}}) + \sqrt{2}e^{-\frac{\delta^2}{2}} - \delta\sqrt{\pi}\right)$$

$$= \sigma\sqrt{\frac{2}{\pi}}\left(-1 - \delta\sqrt{\frac{\pi}{2}} + e^{-\frac{\delta^2}{2}} + \delta\sqrt{\frac{\pi}{2}}erf(\frac{\delta}{\sqrt{2}})\right).$$

This completes the proof of Lemma. □

Theorem 2.1: For $X|\mu,\ \sigma\ \sim\ HN(\mu,\ \sigma)$, the mean deviation, δ_1, about the mean, μ_1, is given by

$$\delta_1 = E\,|X - \mu_1| = \int_{0}^{\infty} |x - \mu_1|\, f(x)dx$$

$$= 2\sigma\sqrt{\frac{2}{\pi}}\left(-1 + e^{-\pi^{-1}} + erf(\pi^{-1/2})\right) \tag{2.31}$$

Proof: We have

$$\delta_1 = \int_{0}^{\infty} |x - \mu_1|\, f(x)dx$$

From Eq. (2.21), the mean of $X|\mu,\ \sigma\ \sim\ HN(\mu,\ \sigma)$ is given by

$$\mu_1 = E\,[X] = \mu + \sigma\sqrt{\frac{2}{\pi}}.$$

Taking $\omega = \mu_1$, we have

$$\delta = \frac{\omega - \mu}{\sigma} = \sqrt{\frac{2}{\pi}}.$$

Thus, taking $\omega = \mu_1$ and $\delta = \sqrt{\frac{2}{\pi}}$ in the above Lemma, and simplifying, we have

$$\delta_1 = 2\sigma\sqrt{\frac{2}{\pi}}\left(-1 + e^{-\pi^{-1}} + erf(\pi^{-1/2})\right),$$

which completes the proof of Theorem 2.1. □

Theorem 2.2: For $X|\mu, \sigma \sim HN(\mu, \sigma)$, the mean deviation, δ_2, about the median, m, is given by

$$\delta_2 = E|X - m| = \int_0^\infty |x - m| f(x)dx$$
$$= \sigma\sqrt{\frac{2}{\pi}}\left(k\sqrt{\pi} - 1 + 2e^{-k^2} + 2k\sqrt{\pi}erf(k)\right), \quad (2.32)$$

where $k = erf^{-1}(0.50)$.

Proof: We have

$$\delta_2 = \int_0^\infty |x - m| f(x)dx$$

As derived in Sect. 2.2.3.7 above, the median of $X|\mu, \sigma \sim HN(\mu, \sigma)$ is given by

$$m = \text{Median}(X) = \mu + \sqrt{2}erf^{-1}(0.50)\sigma = \mu + \sigma\sqrt{2}k,$$

where $k = erf^{-1}(0.50)$.
Taking $\omega = m$, we have

$$\delta = \frac{\omega - \mu}{\sigma} = \frac{m - \mu}{\sigma} = \frac{\mu + \sigma\sqrt{2}k - \mu}{\sigma} = \sqrt{2}k$$

Thus, taking $\omega = m$ and $\delta = \sqrt{2}k$ in the above Lemma, and simplifying, we have

$$\delta_2 = \sigma\sqrt{\frac{2}{\pi}}\left(k\sqrt{\pi} - 1 + 2e^{-k^2} + 2k\sqrt{\pi}erf(k)\right),$$

where $k = erf^{-1}(0.50)$. This completes the proof of Theorem 2.2. □

2.2.3.10 Renyi and Shannon Entropies, and Song's Measure of the Shape of the Half-Normal Distribution

These are derived as given below.

(i) **Renyi Entropy:** Following Renyi (1961), the entropy of the half- normal random variable X having the pdf (2.13) is given by

$$\Im_R(\gamma) = \frac{1}{1-\gamma} \ln \int_0^\infty [f_X(X)]^\gamma \, dx,$$
$$= \frac{\ln(\gamma)}{2(\gamma-1)} - \ln\left[\sqrt{\frac{2}{\pi}}\frac{1}{\sigma}\right], \quad \sigma > 0, \ \gamma > 0, \ \gamma \neq 1. \tag{2.33}$$

(ii) **Shannon Entropy:** Following Shannon (1948), the entropy of the half-normal random variable X having the pdf (2.13) is given by

$$H_X[f_X(X)] = E[-\ln(f_X(X)] = -\int_0^\infty f_X(x) \ln [f_X(x)] \, dx,$$

which is the particular case of Renyi entropy (2.31) for $\gamma \to 1$. Thus, in the limit when $\gamma \to 1$ and using L'Hospital's rule, Shannon entropy is easily obtained from the Eq. (2.33) as follows:

$$H_X[f_X(X)] = E[-\ln(f_X(X)] == \frac{1}{2} - \ln\left[\sqrt{\frac{2}{\pi}}\frac{1}{\sigma}\right], \quad \sigma > 0.$$

(iii) **Song's Measure of the Shape of a Distribution:** Following Song (2001), the gradient of the Renyi entropy is given by

$$\Im'_R(\gamma) = \frac{d}{d\gamma}[\Im_R(\gamma)] = \frac{1}{2}\left\{\frac{1}{\gamma(\gamma-1)} - \frac{\ln(\gamma)}{(\gamma-1)^2}\right\} \tag{2.34}$$

which is related to the log likelihood by

$$\Im'_R(1) = -\frac{1}{2} Var[\ln f(X)].$$

Thus, in the limit when $\gamma \to 1$ and using L'Hospital's rule, Song's measure of the shape of the distribution of the half-normal random variable X having the pdf (2.13) is readily obtained from the Eq. (2.33) as follows:

$$\Im'_R(1) = -\frac{1}{8} \; (< 0),$$

the negative value of Song's measure indicating herein a "flat" or "platykurtic" distribution, which can be used in comparing the shapes of various densities and measuring heaviness of tails, similar to the measure of kurtosis.

2.2.3.11 Percentiles of the Half-Normal Distribution

This section computes the percentiles of the half-normal distribution, by using Maple 10. For any $p(0 < p < 1)$, the $(100p)th$ percentile (also called the quantile of order p) of the half-normal distribution, $X|\mu, \sigma \sim HN(\mu, \sigma)$, with the pdf $f_X(x)$ given by (2.13), is a number z_p such that the area under $f_X(x)$ to the left of z_p is p. That is, z_p is any root of the equation

$$F(z_p) = \int_{-\infty}^{z_p} f_X(t)dt = p. \qquad (2.35)$$

Thus, from the Eq. (2.35), using the Maple program, the percentiles z_p of the half-normal distribution, $X|\mu, \sigma \sim HN(\mu, \sigma)$ can easily been obtained.

2.2.4 *Folded Normal Distribution*

An important class of probability distributions, known as the folded distributions, arises in many practical problems when only the magnitudes of deviations are recorded, and the signs of the deviations are ignored. The folded normal distribution is one such probability distribution which belongs to this class. It is related to the normal distribution in the sense that if Y is a normally distributed random variable with mean μ (location) and variance σ^2 (scale), that is, if $Y \sim N(\mu, \sigma^2)$, then the random variable $X = |Y|$ is said to have a folded normal distribution. The distribution is called folded because the probability mass (that is, area) to the left of the point $x = 0$ is folded over by taking the absolute value. As pointed out above, such a case may be encountered if only the magnitude of some random variable is recorded, without taking into consideration its sign (that is, its direction). Further, this distribution is used when the measurement system produces only positive measurements, from a normally distributed process. To fit a folded normal distribution, only the average and specified sigma (process, sample, or population) are needed. Many researchers have studied the statistical methods dealing with the properties and applications of the folded normal distribution, among them Daniel (1959), Leon et al. (1961), Elandt (1961), Nelson (1980), Patel and Read (1982),

Sinha (1983), Johnson et al. (1994), Laughlin (http://www.causascientia.org/math_stat/Dists/Compendium.pdf,2001), and Kim (2006) are notable.

Definition: Let $Y \sim N\left(\mu,\ \sigma^2\right)$ be a normally distributed random variable with the mean μ and the variance σ^2. Let $X = |Y|$. Then X has a folded normal distribution with the pdf $f_X(x)$ and cdf $F_X(x) = P(X \le x)$, respectively, given as follows.

$$f_X(x) = \begin{cases} \frac{1}{\sqrt{2\pi}\sigma}\left[e^{-\frac{(x-\mu)^2}{2\sigma^2}} + e^{-\frac{(-x-\mu)^2}{2\sigma^2}}\right], & x \ge 0 \\ 0, \quad x < 0 \end{cases} \tag{2.36}$$

Note that the μ and σ^2 are location and scale parameters for the parent normal distribution. However, they are the shape parameters for the folded normal distribution. Further, equivalently, if $x \ge 0$, using a hyperbolic cosine function, the pdf $f_X(x)$ of a folded normal distribution can be expressed as

$$f_X(x) = \frac{1}{\sigma}\sqrt{\frac{2}{\pi}}\cosh\left(\frac{\mu x}{\sigma^2}\right)e^{-\frac{(x^2+\mu^2)}{2\sigma^2}}, \quad x \ge 0.$$

and the cdf $F_X(x)$ as

$$F_X(x) = \frac{1}{\sqrt{2\pi}\sigma}\int_0^x \left(e^{-\frac{(y-\mu)^2}{2\sigma^2}} + e^{-\frac{(-y-\mu)^2}{2\sigma^2}}\right)dy,$$
$$x \ge 0,\ |\mu| < \infty,\ \sigma > 0. \tag{2.37}$$

Taking $z = \frac{y-\mu}{\sigma}$ in (2.37), the cdf $F_X(x)$ of a folded normal distribution can also be expressed as

$$F_X(x) = \frac{1}{\sqrt{2\pi}}\int_{-\mu/\sigma}^{(x-\mu)/\sigma}\left(e^{-\frac{1}{2}z^2} + e^{-\frac{1}{2}\left(z+\frac{2\mu}{\sigma}\right)^2}\right)dz,$$
$$z \ge 0,\ |\mu| < \infty,\ \sigma > 0, \tag{2.38}$$

where μ and σ^2 are the mean and the variance of the parent normal distribution. To describe the shapes of the folded normal distribution, the plots of the pdf (2.36) for different values of the parameters μ and σ are provided in Fig. 2.6 by using Maple 10. The effects of the parameters can easily be seen from these graphs. Similar plots can be drawn for others values of the parameters.

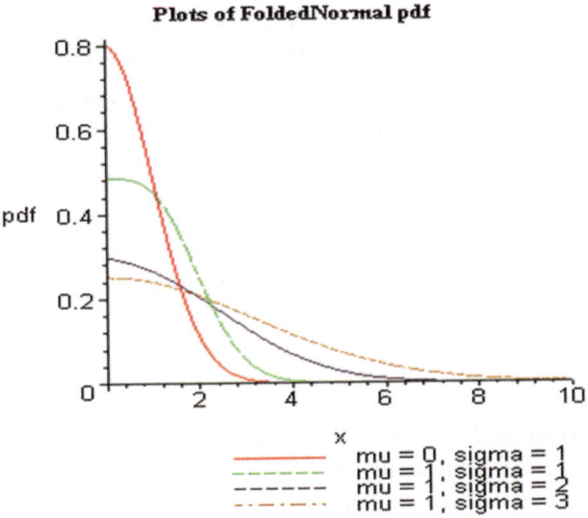

Fig. 2.6 Plots of the folded normal pdf

2.2.4.1 Some Properties of the Folded Normal Distribution

This section discusses some important properties of the folded normal distribution, $X \sim FN(\mu, \sigma^2)$.

2.2.4.2 Special Cases

The folded normal distribution is related to the following distributions (see, for example, Patel and Read 1982, and Johnson et al. 1994, among others).

(i) If $X \sim FN(\mu, \sigma^2)$, then (X/σ) has a non-central chi distribution with one degree of freedom and non-centrality parameter $\frac{\mu^2}{\sigma^2}$.

(ii) On the other hand, if a random variable U has a non-central chi distribution with one degree of freedom and non-centrality parameter $\frac{\mu^2}{\sigma^2}$, then the distribution of the random variable $\sigma\sqrt{U}$ is given by the pdf $f_X(x)$ in (2.36).

(iii) If $\mu = 0$, the folded normal distribution becomes a half-normal distribution with the pdf $f_X(x)$ as given in (2.15).

2.2.4.3 Characteristic Property

If $Z \sim N(\mu, \sigma)$, then $|Z| \sim FN(\mu, \sigma)$.

2.2.4.4 Mode

It is easy to see that the mode or modal value of x for which the folded normal probability density function $f_X(x)$ defined by (2.36) is maximum, is given by $x = \mu$, and the maximum value of the folded normal probability density function (2.35) is given by

$$f_X(\mu) = \frac{1}{\left(\sqrt{2\pi}\right)\sigma}\left[1 + e^{-\frac{2\mu^2}{\sigma^2}}\right]. \tag{2.39}$$

Clearly, the folded normal probability density function (2.36) is unimodal.

2.2.4.5 Moments

(i) rth **Moment of the Folded Normal Random Variable:** For some integer $r > 0$, a general formula for the rth moment, $\mu'_{f(r)}$, of the folded normal random variable $X \sim FN\left(\mu, \sigma^2\right)$ having the pdf (2.36) has been derived by Elandt (1961), which is presented here. Let $\theta = \frac{\mu}{\sigma}$. Define $\mathrm{I}_r(a) = \frac{1}{\sqrt{2\pi}}\int_a^\infty y^r e^{-\frac{1}{2}y^2}dy$, $r = 1, 2, \ldots$, which is known as the "incomplete normal moment." In particular,

$$\mathrm{I}_0(a) = \frac{1}{\sqrt{2\pi}}\int_a^\infty e^{-\frac{1}{2}y^2}dy = 1 - \Phi(a), \tag{2.40}$$

where $\Phi(a) = \frac{1}{\sqrt{2\pi}}\int_{-\infty}^a e^{-\frac{1}{2}y^2}dy$ is the CDF of the unit normal $N(0, 1)$. Clearly, for $r > 0$, $\mathrm{I}_r(a) = \left(\frac{1}{\sqrt{2\pi}}\right)a^{r-1}e^{-\frac{1}{2}a^2} + (r-1)\,\mathrm{I}_{r-2}(a)$. Thus, in view of these results, the rth moment, $\mu'_{f(r)}$, of the folded normal random variable X is easily expressed in terms of the I_r function as follows.

$$\begin{aligned}\mu'_{f(r)} &= E\left(X^r\right) = \int_0^\infty x f_X(x)\,dx \\ &= \left(\sigma^r\right)\sum_{j=0}^{r}\binom{r}{j}\theta^{r-j}\left[\mathrm{I}_j(-\theta) + (-1)^{r-j}\,\mathrm{I}_j(\theta)\right].\end{aligned} \tag{2.41}$$

From the above Eq. (2.41) and noting, from the definition of the I_r function, that $\mathrm{I}_2(-\theta) - \mathrm{I}_2(\theta) = -\left[\left(\frac{2}{\sqrt{2\pi}}\right)\theta e^{-\frac{1}{2}\theta^2} + \{1 - 2\mathrm{I}_0(-\theta)\}\right]$, the first four moments of the folded normal random distribution are easily obtained as follows.

$$\mu'_{f(1)} = E[X] = \mu_f = \left(\frac{2}{\sqrt{2\pi}}\right)\sigma e^{-\frac{1}{2}\theta^2} - \mu[1 - 2\mathrm{I}_0(-\theta)]$$
$$= \left(\frac{2}{\sqrt{2\pi}}\right)\sigma e^{-\frac{1}{2}\theta^2} - \mu[1 - 2\Phi(\theta)],$$
$$\mu'_{f(2)} = E\left[X^2\right] = \sigma_f^2 = \mu^2 + \sigma^2,$$
$$\mu'_{f(3)} = E\left[X^3\right] = \left(\mu^2 + 2\sigma^2\right)\mu_f - \mu\sigma^2[1 - 2\Phi(\theta)],$$

and

$$\mu'_{f(4)} = E\left[X^4\right] = \mu^4 + 6\mu^2\sigma^2 + 3\sigma^4. \tag{2.42}$$

(ii) **Central Moments of the Folded Normal Random Variable:** For some integer $n > 0$, the nth central moment (about the mean $\mu'_{f(1)} = E(X)$) of the folded normal random variable X having the pdf (2.36) can be easily obtained using the formula

$$\mu_{f(n)} = E\left[\left(X - \mu'_{f(1)}\right)^n\right] = \sum_{r=0}^{n}\binom{n}{r}\left(-\mu'_{f(1)}\right)^{n-r}E\left(X^r\right), \tag{2.43}$$

where $E\left(X^r\right) = \mu'_{f(r)}$ denotes the rth moment, given by the Eq. (2.41), of the folded normal random variable X. Thus, from the above Eq. (2.43), the first four central moments of the folded normal random variable X are easily obtained as follows.

$$\mu_{f(1)} = 0,$$
$$\mu_{f(2)} = \mu^2 + \sigma^2 - \mu_f^2,$$
$$\mu_{f(3)} = \beta_3 = 2\left[\mu_f^3 - \mu^2\mu_f - \left(\frac{\sigma^3}{\sqrt{2\pi}}\right)e^{-\frac{1}{2}\theta^2}\right],$$

and

$$\mu_{f(4)} = \beta_4 = \left(\mu^4 + 6\mu^2\sigma^2 + 3\sigma^4\right) + \left(\frac{8\sigma^3}{\sqrt{2\pi}}\right)e^{-\frac{1}{2}\theta^2}\mu_f + 2\left(\mu^2 - 3\sigma^2\right)\mu_f^2 - 3\mu_f^4. \tag{2.44}$$

2.2.4.6 Mean, Variance, and Coefficients of Skewness and Kurtosis of the Folded Normal Random Variable

These are easily obtained as follows:

(i) **Mean:** $E\ (X) = \alpha_1 = \mu_f = \left(\frac{2}{\sqrt{2\pi}}\right)\sigma e^{-\frac{1}{2}\theta^2} - \mu[1 - 2\Phi(\theta)]$,

(ii) **Variance:** $Var\ (X) = \beta_2 = \mu_{f(2)} = \mu^2 + \sigma^2 - \mu_f^2, \quad \sigma > 0$,

(iii) **Coefficient of Skewness:** $\gamma_1\ (X) = \frac{\mu_3}{[\mu_2]^{\frac{3}{2}}}$,

(iv) **Coefficient of Kurtosis:** $\gamma_2\ (X) = \frac{\mu_{f(4)}}{[\mu_{f(2)}]^2}$,

where the symbols have their usual meanings as described above.

2.2.4.7 Percentiles of the Folded Normal Distribution

This section computes the percentiles of the folded normal distribution, by using Maple 10. For any $p(0 < p < 1)$, the $(100p)th$ percentile (also called the quantile of order p) of the folded normal distribution, $X \sim FN\left(\mu, \sigma^2\right)$, with the pdf $f_X(x)$ given by (2.36), is a number z_p such that the area under $f_X(x)$ to the left of z_p is p. That is, z_p is any root of the equation

$$F(z_p) = \int_{-\infty}^{z_p} f_X(t)dt = p. \tag{2.45}$$

Thus, from the Eq. (2.45), using the Maple program, the percentiles z_p of the folded normal distribution can be computed for some selected values of the parameters.

Note: For the tables of the folded normal cdf $F_X(x) = P(X \le x)$ for different values of the parameters, for example, $\frac{\mu_f}{\sigma_f} = 1.3236, 1.4(0.1)3$, and $x = 0.1(0.1)7$, the interested readers are referred to Leon et al. (1961).

Note: As noted by Elandt (1961), the family of the folded normal distributions, $N_f\left(\mu_f, \sigma_f\right)$, is included between the half-normal, for which $\frac{\mu_f}{\sigma_f} = 1.3237$, and the normal, for which $\frac{\mu_f}{\sigma_f}$ is infinite. Approximate normality is attained if , for which $\frac{\mu_f}{\sigma_f} > 3$.

2.2.5 *Truncated Distributions*

Following Rohatgi and Saleh (2001), and Lawless (2004), we first present an overview of the truncated distributions.

2.2.5.1 Overview of Truncated Distributions

Suppose we have a probability distribution defined for a continuous random variable X. If some set of values in the range of X are excluded, then the probability distribution for the random variable X is said to be truncated. We defined the truncated distributions as follows.

Definition: Let X be a continuous random variable on a probability space (Ω, S, P), and let $T \in B$ such that $0 < P\{X \in T\} < 1$, where B is a σ- field on the set of real numbers $\Re$. Then the conditional distribution $P\{X \leq x \mid X \in T\}$, defined for any real x, is called the truncated distribution of X. Let $f_X(x)$ and $F_X(x)$ denote the probability density function (pdf) and the cumulative distribution function (cdf), respectively, of the parent random variable X. If the random variable with the truncated distribution function $P\{X \leq x \mid X \in T\}$ be denoted by Y, then Y has support T. Then the cumulative distribution function (cdf), say, $G(y)$, and the probability density function (pdf), say, $g(y)$, for the random variable Y are, respectively, given by

$$G_Y(y) = P\{Y \leq y \mid Y \in T\} = \frac{P\{Y \leq y,\ Y \in T\}}{P\{Y \in T\}} = \frac{\int_{(-\infty,\, y] \cap T} f_X(u)du}{\int_T f_X(u)du}, \tag{2.46}$$

and

$$g_Y(y) = \begin{cases} \dfrac{f_X(y)}{\int_T f_X(u)du}, & y \in T \\ 0, & y \notin T. \end{cases} \tag{2.47}$$

Clearly $g_Y(y)$ in (2.47) defines a pdf with support T, since $\int_T g_Y(y)dy = \frac{\int_T f_X(y)dy}{\int_T f_X(u)du} = 1$. Note that here T is not necessarily a bounded set of real numbers. In particular, if the values of Y below a specified value a are excluded from the distribution, then the remaining values of Y in the population have a distribution with the pdf given by $g_L(y;\ a) = \frac{f_X(y)}{1 - F_X(a)}$, $a \leq y < \infty$, and the distribution is said to be left truncated at a. Conversely, if the values of Y above a specified value a are excluded from the distribution, then the remaining values of Y in the population have a distribution with the pdf given by $g_R(y;\ a) = \frac{f_X(y)}{F_X(a)}$, $0 \leq y \leq a$, and the distribution is said to be right truncated at a. Further, if Y has a support $T = [a_1,\ a_2]$, where $-\infty < a_1 < a_2 < \infty$, then the conditional distribution of Y, given that $a_1 \leq y \leq a_2$, is called a doubly truncated distribution with the cdf, say, $G(y)$, and the pdf, say, $g(y)$, respectively, given by

$$G_Y(y) = \frac{F_X\{\max(\min(y,\ a_2),\ a_1)\} - F_X(a_1)}{F_X(a_2) - F_X(a_1)}, \tag{2.48}$$

and

$$g_Y(y) = \begin{cases} \dfrac{f_X(y)}{F_X(a_2) - F_X(a_1)}, & y \in [a_1,\ a_2] \\ 0, & y \notin [a_1,\ a_2]. \end{cases} \tag{2.49}$$

The truncated distribution for a continuous random variable is one of the important research topics both from the theoretical and applications point of view. It arises in many probabilistic modeling problems of biology, crystallography, economics, engineering, forecasting, genetics, hydrology, insurance, lifetime data analysis, management, medicine, order statistics, physics, production research, psychology, reliability, quality engineering, survival analysis, etc, when sampling is carried out from an incomplete population data. For details on the properties and estimation of parameters of truncated distributions, and their applications to the statistical analysis of truncated data, see, for example, Hald (1952), Chapman (1956), Hausman and Wise (1977), Thomopoulos (1980), Patel and Read (1982), Levy (1982), Sugiura and Gomi (1985), Schneider (1986), Kimber and Jeynes (1987), Kececioglu (1991), Cohen (1991), Andersen et al. (1993), Johnson et al. (1994), Klugman et al. (1998), Rohatgi and Saleh (2001), Balakrishnan and Nevzorov (2003), David and Nagaraja (2003), Lawless (2003), Jawitz (2004), Greene (2005), Nadarajah and Kotz (2006a), Maksay and Stoica (2006) and Nadarajah and Kotz (2007) and references therein.

The truncated distributions of a normally distributed random variable, their properties and applications have been extensively studied by many researchers, among them Bliss (1935 for the probit model which is used to model the choice probability of a binary outcome), Hald (1952), Tobin (1958) for the probit model which is used to model censored data), Shah and Jaiswal (1966), Hausman and Wise (1977), Thomopoulos (1980), Patel and Read (1982), Levy (1982), Sugiura and Gomi (1985), Schneider (1986), Kimber and Jeynes (1987), Cohen (1959, 1991), Johnson et al. (1994), Barr and Sherrill (1999), Johnson (2001), David and Nagaraja (2003), Jawitz (2004), Nadarajah and Kotz (2007), and Olive (2007), are notable. In what follows, we present the pdf, moment generating function (mgf), mean, variance and other properties of the truncated normal distribution most of which is discussed in Patel and Read (1982), Johnson et al. (1994), Rohatgi and Saleh (2001), and Olive (2007).

Definition: Let $X \sim N\left(\mu,\, \sigma^2\right)$ be a normally distributed random variable with the mean μ and the variance σ^2. Let us consider a random variable Y which represents the truncated distribution of X over a support T = [a, b], where $-\infty < a < b < \infty$. Then the conditional distribution of Y, given that $a \le y \le b$, is called a doubly truncated normal distribution with the pdf, say, $g_Y(y)$, given by

$$g_Y(y) = \begin{cases} \dfrac{\frac{1}{\sigma}\phi\left(\frac{y-\mu}{\sigma}\right)}{\left[\Phi\left(\frac{b-\mu}{\sigma}\right) - \Phi\left(\frac{a-\mu}{\sigma}\right)\right]}, & y \in [a,\, b] \\ 0, & y \notin [a,\, b] \end{cases}, \tag{2.50}$$

where $\phi(.)$ and $\Phi(.)$ are the pdf and cdf of the standard normal distribution, respectively. If $a = -\infty$, then the we have a (singly) truncated normal distribution from above, (that is, right truncated). On the other hand, if $b = \infty$, then the we have a (singly) truncated normal distribution from below, (that is, left truncated). The following are some examples of the truncated normal distributions.

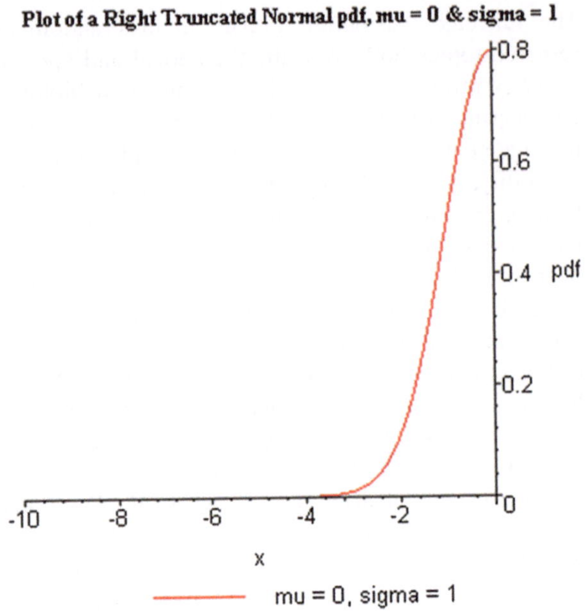

Fig. 2.7 Example of a right truncated normal distribution

(i) **Example of a Left Truncated Normal Distribution:** Taking $a = 0$, $\mathrm{b} = \infty$, and $\mu = 0$, the pdf $g_Y(y)$ in (2.50) reduces to that of the half normal distribution in (2.15), which is an example of the left truncated normal distribution.

(ii) **Example of a Right Truncated Normal Distribution:** Taking $a = -\infty, b = 0$, $\mu = 0$, and $\sigma = 1$ in (2.50), the pdf $g_Y(y)$ of the right truncated normal distribution is given by

$$g_Y(y) = \begin{cases} 2\phi(y), & -\infty < y \leq 0 \\ 0, & y > 0 \end{cases}, \tag{2.51}$$

where $\phi(.)$ is the pdf of the standard normal distribution. The shape of right truncated normal pdf $g_Y(y)$ in (2.51) is illustrated in the following Fig. (2.7).

2.2.5.2 MGF, Mean, and Variance of the Truncated Normal Distribution

These are given below.

(A) Moment Generating Function: The mgf of the doubly truncated normal distribution with the pdf $g_Y(y)$ in (2.50) is easily obtained as

$$
\begin{aligned}
M(t) &= E\left(e^{tY} | Y \in [a, b]\right) \\
&= e^{\mu t + \frac{\sigma^2 t^2}{2}} \left\{ \frac{\left[\Phi\left(\frac{b-\mu}{\sigma} - \sigma t\right) - \Phi\left(\frac{a-\mu}{\sigma} - \sigma t\right)\right]}{\left[\Phi\left(\frac{b-\mu}{\sigma}\right) - \Phi\left(\frac{a-\mu}{\sigma}\right)\right]} \right\}
\end{aligned} \quad (2.52)
$$

(B) Mean, Second Moment and Variance: Using the expression for the mgf (2.52), these are easily given by

(i)
$$
\begin{aligned}
Mean &= E(Y|Y \in [a, b]) = M'(t)\big|_{t=0} \\
&= \mu + \sigma \left[\frac{\phi\left(\frac{a-\mu}{\sigma}\right) - \phi\left(\frac{b-\mu}{\sigma}\right)}{\Phi\left(\frac{b-\mu}{\sigma}\right) - \Phi\left(\frac{a-\mu}{\sigma}\right)} \right]
\end{aligned} \quad (2.53)
$$

Particular Cases:

(I) If $b \to \infty$ in (2.52), then we have

$$E(Y|Y > a) = \mu + \sigma h,$$

where $h = \frac{\phi\left(\frac{a-\mu}{\sigma}\right)}{1 - \Phi\left(\frac{a-\mu}{\sigma}\right)}$ is called the Hazard Function (or the Hazard Rate, or the Inverse Mill's Ratio) of the normal distribution.

(II) If $a \to -\infty$ in (2.53), then we have

$$E(Y|Y < b) = \mu - \sigma \left[\frac{\phi\left(\frac{b-\mu}{\sigma}\right)}{\Phi\left(\frac{b-\mu}{\sigma}\right)} \right]. \quad (2.54)$$

(III) If $b \to \infty$ in (2.54), then Y is not truncated and we have

$$
\begin{aligned}
E(Y) &= \mu \\
V(Y) &= \sigma^2[1 + \alpha\phi]
\end{aligned}
$$

(ii)
$$
\begin{aligned}
\textit{Second Moment} &= E\left(Y^2 | Y \in [a, b]\right) = M''(t)\big|_{t=0} \\
&= 2\mu \{E(Y|Y \in [a, b])\} - \mu^2 \\
&= \mu^2 + 2\mu\sigma \left[\frac{\phi\left(\frac{a-\mu}{\sigma}\right) - \phi\left(\frac{b-\mu}{\sigma}\right)}{\Phi\left(\frac{b-\mu}{\sigma}\right) - \Phi\left(\frac{a-\mu}{\sigma}\right)} \right]
\end{aligned}
$$

$$+\sigma^2\left[1+\frac{\left(\frac{a-\mu}{\sigma}\right)\phi\left(\frac{a-\mu}{\sigma}\right)-\left(\frac{b-\mu}{\sigma}\right)\phi\left(\frac{b-\mu}{\sigma}\right)}{\Phi\left(\frac{b-\mu}{\sigma}\right)-\Phi\left(\frac{a-\mu}{\sigma}\right)}\right] \tag{2.55}$$

and

$$\textbf{(iii)} \quad Variance = Var\left(Y|Y\in[a,b]\right) = \left\{E\left(Y^2|Y\in[a,b]\right)\right\} - \left\{E\left(Y|Y\in[a,b]\right)\right\}^2$$
$$= \sigma^2\left\{1+\frac{\left(\frac{a-\mu}{\sigma}\right)\phi\left(\frac{a-\mu}{\sigma}\right)-\left(\frac{b-\mu}{\sigma}\right)\phi\left(\frac{b-\mu}{\sigma}\right)}{\Phi\left(\frac{b-\mu}{\sigma}\right)-\Phi\left(\frac{a-\mu}{\sigma}\right)} - \left[\frac{\phi\left(\frac{b-\mu}{\sigma}\right)-\phi\left(\frac{a-\mu}{\sigma}\right)}{\Phi\left(\frac{b-\mu}{\sigma}\right)-\Phi\left(\frac{a-\mu}{\sigma}\right)}\right]^2\right\} \tag{2.56}$$

Some Further Remarks on the Truncated Normal Distribution:

(i) Let $Y \sim TN\left(\mu,\sigma^2, a=\mu-k\sigma, b=\mu+k\sigma\right)$, for some real k, be the truncated version of a normal distribution with mean μ and variance σ^2. Then, from (2.53) and (2.56), it easily follows that $E(Y) = \mu$ and $Var(Y) = \sigma^2\left\{1-\frac{2k\phi(k)}{2k\Phi(k)-1}\right\}$, (see, for example, Olive, 2007).

(ii) The interested readers are also referred to Shah and Jaiswal (1966) for some nice discussion on the pdf $g_Y(y)$ of the truncated normal distribution and its moments, when the origin is shifted at a.

(iii) A table of the mean μ_t, standard deviation σ_t, and the ratio (mean deviation/σ_t) for selected values of $\Phi\left(\frac{a-\mu}{\sigma}\right)$ and $1-\Phi\left(\frac{b-\mu}{\sigma}\right)$ have been provided in Johnson and Kotz (1994).

2.2.6 Inverse Normal (Gaussian) Distribution (IGD)

The inverse Gaussian distribution (IGD) represents a class of distribution. The distribution was initially considered by Schrondinger (1915) and further studied by many authors, among them Tweedie (1957a, b) and Chhikara and Folks (1974) are notable. Several advantages and applications in different fields of IGD are given by Tweedie (1957), Johnson and Kotz (1994), Chhikara and Folks (1974, 1976,1977), and Folks and Chhikara (1978), among others. For the generalized inverse Gaussian distribution (GIG) and its statistical properties, the interested readers are referred to Good (1953), Sichel (1974, 1975), Barndorff-Nielsen (1977, 1978), Jorgensen

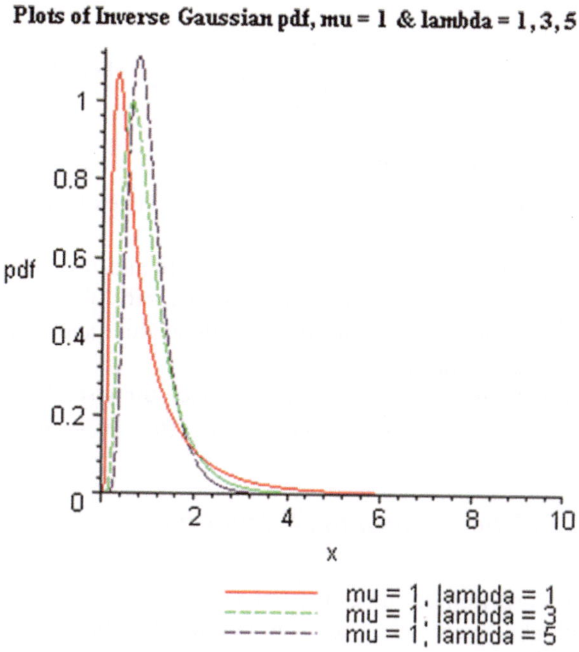

Fig. 2.8 Plots of the inverse Gaussian pdf

(1982), and Johnson and Kotz (1994), and references therein. In what follows, we present briefly the pdf, cdf, mean, variance and other properties of the inverse Gaussian distribution (IGD).

Definition: The pdf of the Inverse Gaussian distribution (IGD) with parameters μ and λ is given by

$$f(x,\ \mu,\ \lambda) = \left(\frac{\lambda}{2\pi x^3}\right)^{1/2} \exp\left\{-\frac{\lambda}{2\mu^2 x}(x - \mu)^2\right\} \quad x > 0, \mu > 0, \lambda > 0 \tag{2.57}$$

where μ is location parameter and λ is a shape parameter. The mean and variance of this distribution are μ and μ^3/λ respectively. To describe the shapes of the inverse Gaussian distribution, the plots of the pdf (2.57), for $\mu = 1$ and $\lambda = 1, 3, 5$ are provided in Fig. 2.8 by using Maple 10. The effects of the parameters can easily be seen from these graphs. Similar plots can be drawn for others values of the parameters.

Properties of IGD:

Let $x_1, x_2, \ldots, x_n$ be a random sample of size n from the inverse Gaussian distribution (1.1). The maximum likelihood estimators (MLE's) for μ and λ are respectively given by

$$\hat{\mu} = \bar{x} = \sum_{i=1}^{n} x_i/n,\ \tilde{\lambda} = \frac{n}{V},\ \text{where } V = \sum_{i=1}^{n}\left(\frac{1}{x_i} - \frac{1}{\bar{x}}\right).$$

It is well known that

(i) the sample mean $\bar{x}$ is unbiased estimate of μ where as $\tilde{\lambda}$ is a biased estimate of λ.
(ii) $\bar{x}$ follows IGD with parameters μ and $n\,\lambda$, whereas $\lambda\,V$ is distributed as chi –square distribution with (n-1) degrees of freedom
(iii) $\bar{x}$ and V are stochastically independent and jointly sufficient for (μ,λ) if both are unknown.
(iv) the uniformly minimum variance unbiased estimator (UMVUE) of λ is $\hat{\lambda} = (n-3)/V$ and $Var\left(\hat{\lambda}\right) = 2\,\lambda^2/(n-5) = MSE\left(\hat{\lambda}\right)$.

2.2.7 Skew Normal Distributions

This section discusses the univariate skew normal distribution (SND) and some of its characteristics. The skew normal distribution represents a parametric class of probability distributions, reflecting varying degrees of skewness, which includes the standard normal distribution as a special case. The skewness parameter involved in this class of distributions makes it possible for probabilistic modeling of the data obtained from skewed population. The skew normal distributions are also useful in the study of the robustness and as priors in Bayesian analysis of the data. It appears from the statistical literatures that the skew normal class of densities and its applications first appeared indirectly and independently in the work of Birnbaum (1950), Roberts (1966), O'Hagan and Leonard (1976), and Aigner et al. (1977). The term skew normal distribution (SND) was introduced by Azzalini (1985, 1986), which give a systematic treatment of this distribution, developed independently from earlier work. For further studies, developments, and applications, see, for example, Henze (1986), Mukhopadhyay and Vidakovic (1995), Chiogna (1998), Pewsey (2000), Azzalini (2001), Gupta et al. (2002), Monti (2003), Nadarajah and Kotz (2003), Arnold and Lin (2004), Dalla Valle (2004), Genton (2004), Arellano-Valle et al. (2004), Buccianti (2005), Azzalini (2005, 2006), Arellano-Valle and Azzalini (2006), Bagui and Bagui (2006), Nadarajah and Kotz (2006), Shkedy et al. (2006), Pewsey (2006), Fernandes et al. (2007), Mateu-Figueras et al. (2007), Chakraborty and Hazarika (2011), Eling (2011), Azzalini and Regoli (2012), among others. For generalized skew normal distribution, the interested readers are referred to Gupta and Gupta (2004), Jamalizadeh, et al. (2008), and Kazemi et al. (2011), among others. Multivariate versions of SND have also been proposed, among them Azzalini and Dalla Valle (1996), Azzalini and Capitanio (1999), Arellano-Valle et al. (2002), Gupta and Chen (2004), and Vernic (2006) are notable. Following Azzalini (1985, 1986, 2006), the definition and some properties, including some graphs, of the univariate skew normal distribution (SND) are presented below.

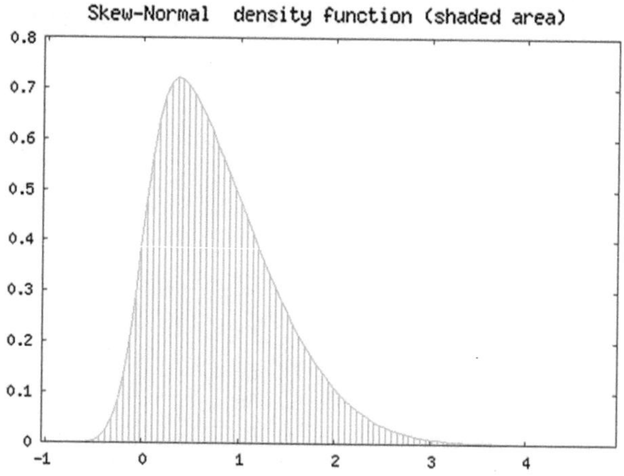

Fig. 2.9 Plot of the skew normal pdf: ($\mu = 0, \sigma = 1, \lambda = 5$)

Definition: For some real-valued parameter λ, a continuous random variable X_λ is said to have a skew normal distribution, denoted by $X_\lambda \sim SN(\lambda)$, if its probability density function is given by

$$f_X(x;\ \lambda) = 2\,\phi(x)\,\Phi(\lambda x),\ -\infty < x < \infty,, \tag{2.58}$$

where $\phi(x) = \left(\frac{1}{\sqrt{2\pi}}\right) e^{-\frac{1}{2}x^2}$ and $\Phi(\lambda x) = \int\limits_{-\infty}^{\lambda x} \phi(t)\,dt$ denote the probability density function and cumulative distribution function of the standard normal distribution respectively.

2.2.7.1 Shapes of the Skew Normal Distribution

The shape of the skew normal probability density function given by (2.58) depends on the values of the parameter λ. For some values of the parameters (μ, σ, λ), the shapes of the pdf (2.58) are provided in Figs. 2.9, 2.10 and 2.11. The effects of the parameter can easily be seen from these graphs. Similar plots can be drawn for others values of the parameters.

Remarks: The continuous random variable X_λ is said to have a skew normal distribution, denoted by $X_\lambda \sim SN(\lambda)$, because the family of distributions represented by it includes the standard $N(0,\ 1)$ distribution as a special case, but in general its members have a skewed density. This is also evident from the fact that $X_\lambda^2 \sim \chi^2$ for all values of the parameter λ. Also, it can be easily seen that the skew normal density function $f_X(x;\ \lambda)$ has the following characteristics:

1. when $\lambda = 0$, we obtain the standard normal density function $f_X(x; 0)$ with zero skewness;
2. as $|\lambda|$ increases, the skewness of the skew normal distribution also increases;

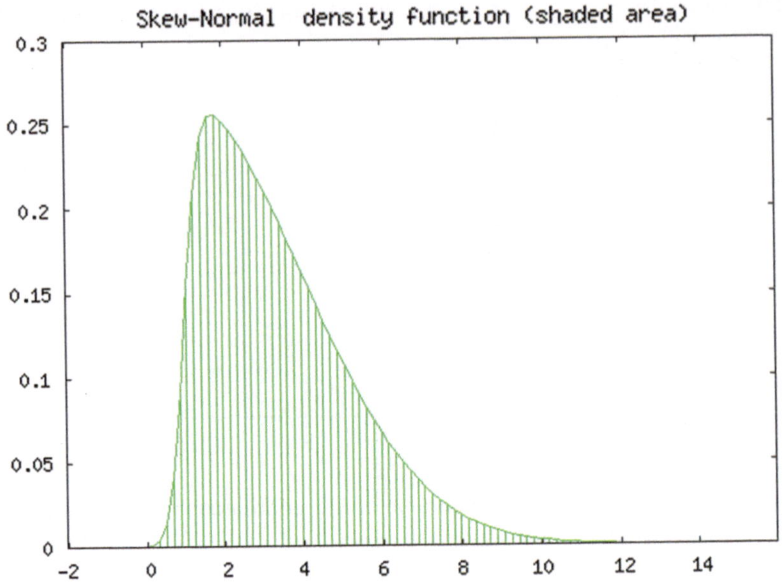

Fig. 2.10 Plot of the skew normal pdf: ($\mu = 1,\ \sigma = 3,\ \lambda = 10$)

3. when $|\lambda| \to \infty$, the skew normal density function $f_X(x;\ \lambda)$ converges to the half-normal (or folded normal) density function;
4. if the sign of λ changes, the skew normal density function $f_X(x;\ \lambda)$ is reflected on the opposite side of the vertical axis.

2.2.7.2 Some Properties of Skew Normal Distribution

This section discusses some important properties of the skew normal distribution, $X_\lambda \sim SN(\lambda)$.
Properties of $SN(\lambda)$:

(a) $SN(0) = N(0,\ 1)$.
(b) If $X_\lambda \sim SN(\lambda)$, then $-X_\lambda \sim SN(-\lambda)$.
(c) If $\lambda \to \pm\infty$, and $Z \sim N(0,\ 1)$, then $SN(\lambda) \to \pm|Z| \sim HN(0,\ 1)$, that is, $SN(\lambda)$ tends to the half-normal distribution.
(d) If $X_\lambda \sim SN(\lambda)$, then $X_\lambda^2 \sim \chi^2$.
(e) The MGF of X_λ is given by $M_\lambda(t) = 2e^{\frac{t^2}{2}}\Phi(\delta t),\ t \in \Re$, where $\delta = \frac{\lambda}{\sqrt{1+\lambda^2}}$.
(f) It is easy to see that $E(X_\lambda) = \delta\left(\sqrt{\frac{2}{\pi}}\right)$, and $Var(X_\lambda) = \frac{\pi - 2\delta^2}{\pi}$.
(g) The characteristic function of X_λ is given by $\psi_\lambda(t) = e^{\frac{-t^2}{2}}[1 + ih(\delta t)],\ t \in \Re$, where $h(x) = \left(\sqrt{\frac{2}{\pi}}\right)\int_0^x e^{\frac{y^2}{2}}dy$ and $h(-x) = -h(x)$ for $x \geq 0$.

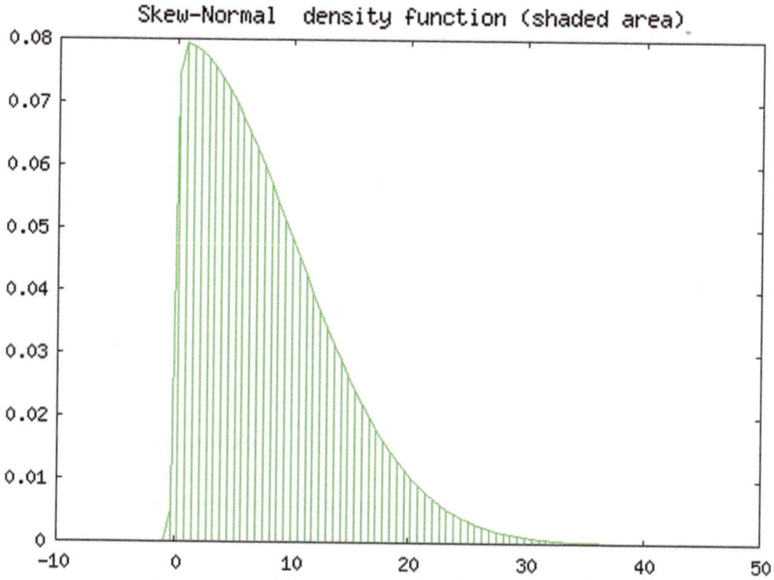

Fig. 2.11 Plots of the skew normal pdf: $(\mu = 0, \sigma = 10, \lambda = 50)$

(h) By introducing the following linear transformation $Y = \mu + \sigma X$, that is, $X = \frac{Y - \mu}{\sigma}$, where $\mu \geq 0$, $\sigma > 0$, we obtain a skew-normal distribution with parameters (μ, σ, λ), denoted by $Y \sim SN\left(\mu, \sigma^2, \lambda\right)$, if its probability density function is given by

$$f_Y(y; \mu, \sigma, \lambda) = 2\phi\left(\frac{y - \mu}{\sigma}\right)\Phi\left(\frac{\lambda(y - \mu)}{\sigma}\right), \ -\infty < y < \infty, \tag{2.59}$$

where $\phi(y)$ and $\Phi(\lambda y)$ denote the probability density function and cumulative distribution function of the normal distribution respectively, and $\mu \geq 0$, $\sigma > 0$ and $-\infty < \lambda < \infty$ are referred as the location, the scale and the shape parameters respectively. Some characteristic values of the random variable Y are as follows:

I. Mean: $E(Y) = \mu + \left(\sigma\delta\sqrt{\frac{2}{\pi}}\right)$

II. Variance: $Var(Y) = \frac{\sigma^2(\pi - 2\delta^2)}{\pi}$

III. Skewness: $\gamma_1 = \left(\frac{4-\pi}{2}\right)\frac{[E(X_\lambda)]^3}{[Var(X_\lambda)]^{\frac{3}{2}}}$

IV. Kurtosis: $\gamma_2 = 2(\pi - 3)\frac{[E(X_\lambda)]^4}{[Var(X_\lambda)]^2}$

2.2.7.3 Some Characteristics Properties of Skew Normal Distribution

Following Gupta et al. (2004), some characterizations of the skew normal distribution (SND) are stated below.

(i) Let X_1 and X_2 be *i.i.d.* F, an unspecified distribution which admits moments of all order. Then $X_1^2 \sim \chi_1^2$, $X_2^2 \sim \chi_1^2$, and $\frac{1}{2}(X_1 + X_2)^2 \sim H_0(\lambda)$ if and only if $F = SN(\lambda)$ or $F = SN(-\lambda)$ where $H_0(\lambda)$ is the distribution of $\frac{1}{2}(X + Y)^2$ when X and Y are *i.i.d.* $SN(\lambda)$.

(ii) Let $H_0(\lambda)$ be the distribution of $(Y + a)^2$ where $Y \sim SN(\lambda)$ and $a \neq 0$ is a given constant. Let X be a random variable with a distribution that admits moments of all order. Then $X^2 \sim \chi_1^2$, $(X + a)^2 \sim H_0(\lambda)$ if and only if $X \sim SN(\lambda)$ for some λ.

For detailed derivations of the above and more results on other characterizations of the skew normal distribution (SND), see Gupta et al. (2004) and references therein. The interested readers are also referred to Arnold and Lin (2004), where the authors have shown that the skew-normal distributions and their limits are exactly the distributions of order statistics of bivariate normally distributed variables. Further, using generalized skew-normal distributions, the authors have characterized the distributions of random variables whose squares obey the chi-square distribution with one degree of freedom.

2.3 Goodness-of-Fit Test (Test For Normality)

The goodness of fit (or GOF) tests are applied to test the suitability of a random sample with a theoretical probability distribution function. In other words, in the GOF test analysis, we test the hypothesis if the random sample drawn from a population follows a specific discrete or continuous distribution. The general approach for this is to first determine a test statistic which is defined as a function of the data measuring the distance between the hypothesis and the data. Then, assuming the hypothesis is true, a probability value of obtaining data which have a larger value of the test statistic than the value observed, is determined, which is known as the p-value. Smaller p-values (for example, less than 0.01) indicate a poor fit of the distribution. Higher values of p (close to one) correspond to a good fit of the distribution. We consider the following parametric and non-parametric goodness-of-fit tests

2.3.1 χ^2(Chi-Squared) Test

The χ^2 test, due to Karl Pearson, may be applied to test the fit of any specified continuous distribution to the given randomly selected continuous data. In χ^2 analysis, the data is first grouped into, say, k number of classes of equal probability. Each class should contain at least 5 or more data points. The χ^2 test statistic is given by

$$\chi^2 = \sum_{i=1}^{k} \frac{(O_i - E_i)^2}{E_i} \tag{2.60}$$

where O_i is the observed frequency in class $i, i = 1, \ldots, k$ and E_i is the expected frequency in class i, if the specified distribution were the correct one, and is given by

$$E_i = F(x_i) - F(x_{i-1}),$$

where $F(x)$ is the cumulative distribution function (CDF) of the probability distribution being tested, and x_i, x_{i-1} are the limits for the class i. The null and alternative hypotheses being tested are, respectively, given by:

H_0: The data follow the specified continuous distribution;
H_1: The data do not follow the specified continuous distribution.

The null hypothesis (H_0) is rejected at the chosen significance level, say, α, if the test statistic is greater than the critical value denoted by $\chi^2{}_{1-\alpha, k-1}$, with $k - 1$ degrees of freedom (df) and a significance level of α. If r parameters are estimated from the data, df are $k - r - 1$.

2.3.2 Kolmogorov-Smirnov (K-S) Test

This test may also be applied to test the goodness of fit between a hypothesized cumulative distribution function (CDF) $F(x)$ and an empirical CDF $F_n(x)$. Let $y_1 < y_2 < \ldots < y_n$ be the observed values of the order statistics of a random sample $x_1, x_2, \ldots, x_n$ of size n. When no two observations are equal, the empirical CDF $F_n(x)$ is given by, see Hogg and Tanis (2006),

$$F_n(x) = \begin{cases} 0, & x < y_1, \\ \frac{i}{n}, & y_i \le x < y_{i+1}, \quad i = 1, 2, \ldots, n-1, \\ 1, & y_n \le x. \end{cases} \tag{2.61}$$

Clearly,

$$F_n(x) = \frac{1}{n}\,[\text{Number of Observations} \le x].$$

Following Blischke and Murthy (2000), the Kolmogorov-Smirnov test statistic, D_n, is defined as the maximum distance between the hypothesized CDF $F(x)$ and the empirical CDF $F_n(x)$, and is given by

$$D_n = \max\left\{D_n^+, D_n^-\right\},$$

where

$$D_n^+ = \max_{i = 1, 2, \ldots, n}\left[\frac{i}{n} - F_n(y_i)\right]$$

and

$$D_n^- = \max_{i = 1, 2, \ldots, n}\left[F_n(y_i) - \frac{i-1}{n}\right].$$

For calculations of fractiles (percentiles) of the distribution of D_n, the interested readers are referred to Massey (1951). In Stephens (1974), one can find a close approximation of the fractiles of the distribution of D_n, based on a constant denoted by d_α which is a function of n only. The values of d_α can also be found in Table 11.2 on p. 400 of Blischke and Murthy (2000) for $\alpha = 0.15,\ 0.10,\ 0.05$, and 0.01. The critical value of D_n is calculated by the formula $d_\alpha / \left(\sqrt{n} + \frac{0.11}{\sqrt{n}} + 0.12\right)$. The null and alternative hypotheses being tested are, respectively, given by:

H_0: The data follow the specified continuous distribution;
H_1: The data do not follow the specified continuous distribution.

The null hypothesis (H_0) is rejected at the chosen significance level, say, α, if the Kolmogorov-Smirnov test statistic, D_n, is greater than the critical value calculated by the above formula.

2.3.3 Anderson-Darling (A-D) Test

The **Anderson-Darling test** is also based on the difference between the hypothesized CDF $F(x)$ and the empirical CDF $F_n(x)$. Let $y_1 < y_2 < \ldots < y_n$ be the observed values of the order statistics of a random sample $x_1, x_2, \ldots, x_n$ of size n. The A-D test statistic (A^2) is given by

$$A^2 = A_n^2 = \frac{-1}{n} \sum_{i=1}^{n} (2i - 1)\left\{\ln F_n(y_i) + \ln\left[1 - F_n(y_{n-i+1})\right]\right\} - n.$$

Fractiles of the distribution of A_n^2 for $\alpha = 0.15,\ 0.10,\ 0.05$, and 0.01, denoted by a_α, are given in Table 11.2 on p. 400 of Blischke and Murthy (2000). The null and alternative hypotheses being tested are, respectively, given by:

H_0 : The data follow the specified continuous distribution
H_1: The data do not follow the specified continuous distribution.

The null hypothesis (H_0) is rejected if the A-D test statistic, A_n^2, is greater than the above tabulated constant a_α (also known as the critical value for A-D test analysis) at one of the chosen significance levels, $\alpha = 0.15,\ 0.10,\ 0.05$, and 0.01. As pointed out in Blischke and Murthy (2000), "the critical value a_α does not depend on n, and has been found to be a very good approximation in samples as small as $n = 3$".

2.3.4 The Shapiro-Wilk Test for Normality

The **Shapiro-Wilk test** (also known as the W test) may be applied to test the goodness of fit between a hypothesized cumulative distribution function (CDF)$F(x)$ and an empirical CDF $F_n(x)$.

Let $y_1 < y_2 < \ldots < y_n$ be the observed values of the order statistics of a random sample $x_1, x_2, \ldots, x_n$ of size n with some unknown distribution function $F(x)$. Following Conover (1999), the Shapiro-Wilk test statistic, W, is defined as

$$W = \frac{\sum_{i=1}^{k} a_i (y_{n-i+1} - y_i)^2}{\sum_{i=1}^{n} (x_i - \overline{x})^2},$$

where $\overline{x}$ denotes the sample mean, and, for the observed sample size $n \leq 50$, the coefficients $a_i, i = 1, \ldots, k$, where k is approximately $\frac{n}{2}$, are available in Table A16, pp. 550–552, of Conover (1999). For the observed sample size $n > 50$, the interested readers are referred to D'Agostino (1971) and Shapiro and Francia (1972) (Fig. 2.12).

For the Shapiro-Wilk test, the null and alternative hypotheses being are, respectively, given by:

H_0: $F(x)$ is a normal distribution with unspecified mean and variance
H_1: $F(x)$ is non-normal.

The null hypothesis (H_0) is rejected at one of the chosen significance levels α if the Shapiro-Wilk test statistic, W, is less than the α quantile as given by Table A 17,

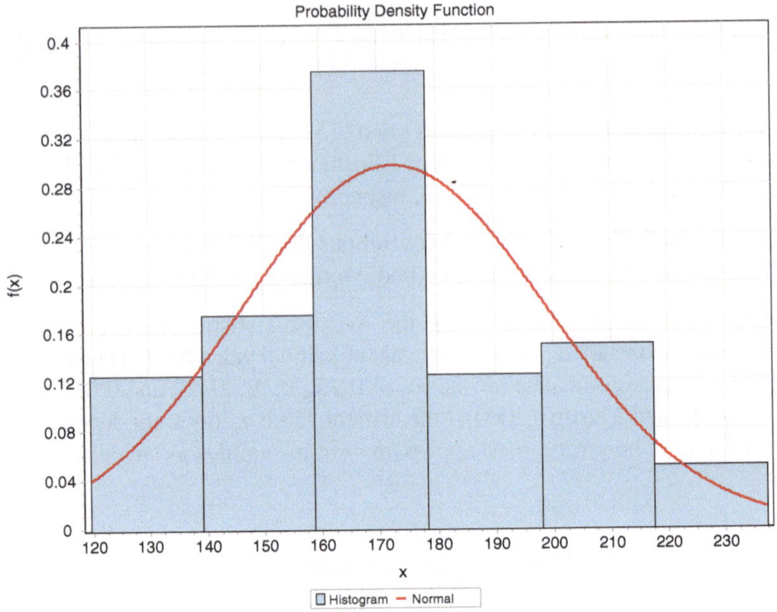

Fig. 2.12 Frequency histogram of the weights of 40 adult men

pp. 552–553, of Conover (1999). The *p*- value for the Shapiro-Wilk test may be calculated by following the procedure on p. 451 of Conover (1999).

Note: The Shapiro-Wilk test statistic may be calculated using the computer softwares such as R, Maple, Minitab, SAS, and StatXact, among others.

2.3.5 *Applications*

In order to examine the applications of the above tests of normality, we consider the following example of weights of a random sample of 40 adult men (Source: *Biostatistics for the Biological and Health Sciences*, Mario F Triola, Publisher: Pearson, 2005).

Example: We consider the weights of a random sample of 40 adult men as given below:

{169.1, 144.2, 179.3, 175.8, 152.6, 166.8, 135.0, 201.5, 175.2, 139.0, 156.3, 186.6, 191.1, 151.3, 209.4, 237.1, 176.7, 220.6, 166.1, 137.4, 164.2, 162.4, 151.8, 144.1, 204.6, 193.8, 172.9, 161.9, 174.8, 169.8, 213.3, 198.0, 173.3, 214.5, 137.1, 119.5, 189.1, 164.7, 170.1, 151.0}.

Table 2.2 Descriptive statistics

Statistic	Value	Percentile	Value
Sample size	40	Min	119.5
Range	117.6	5 %	135.11
Mean	172.55	10 %	137.56
Variance	693.12	25 % (Q1)	152.0
Standard deviation	26.327	50 % (Median)	169.95
Coefficient of variation	0.15258	75 % (Q3)	190.6
Standard error	4.1627	90 %	212.91
Skewness	0.37037	95 %	220.29
Excess Kurtosis	−0.16642	Max	237.1

Table 2.3 Normality for the Weights of 40 Adult Men

Test statistics	Value of the test statistics	P-value	Decision at 5 % level of significance
K-S test	0.112	0.652	Do not reject H_0
A-D test	0.306	0.552	Do not reject H_0
Chi-Squared test	2.712	0.844	Do not reject H_0
Shapiro-Wilk test	0.967	0.379	Do not reject H_0

Using the software EasyFit, the descriptive statistics are computed in the Table 2.2 below. The frequency histogram of the weights of 40 adult men is drawn in Fig. 2.12.

The goodness of fit (or GOF) tests, as discussed above, are applied to test the compatibility of our example of weights of the random sample of 40 adult men with our hypothesized theoretical probability distribution, that is, normal distribution, using various software such as EasyFit, Maple, and Minitab. The results are summarized in the Table 2.2 below. The chosen significance level is $\alpha = 0.05$. The null and alternative hypotheses being tested are, respectively, given by:

H_0: The data follow the normal distribution;
H_1: The data do not follow the normal distribution.

It is obvious from Table 2.3 is that the normal distribution seems to be an appropriate model for the weights of 40 adult men considered here. Since the sample size is large enough, all tests are valid for this example. In this section, we have discussed various tests of normality to test the suitability of a random sample with a theoretical probability distribution function. In particular, we have applied to test the applicability of normal distribution to a random sample of the weights of 40 adult men. It is hoped that this study may be helpful to apply these goodness of fit (or GOF) tests to other examples also.

2.4 Summary

The different forms of normal distributions and their various properties are discussed in this chapter. The entropy of a random variable having the normal distribution has been given. The expressions for the characteristic function of a normal distribution are provided. Some goodness of fit tests for testing the normality along with applications is given. By using Maple 10, various graphs have been plotted. As a motivation, different forms of normal distributions (folded and half normal etc.) and their properties have also been provided.

Chapter 3
Student's *t* Distribution

3.1 Student's *t* Distribution

The Student's *t* distribution or *t* distribution defines a family of continuous probability distributions. It has a wide range of applications in probability, statistics, and other fields of sciences. It was first developed by Willieam S. Gosset (1908) in his work on "the probable error of a mean," published by him under the *nom de plume* of *Student*. Further developments continued with the contributions of Fisher (1925) and others later. For detailed discussions on the development of the *t* distribution and its usages, see, example, Pearson (1967, 1970), Eisenhart (1979), Box (1981), Patel and Read (1982), Johnson et al. (1995), Wiper et al. (2005), Finner (2008), and Zabell (2008), and references therein. The graph of the probability density function of the Student's *t* distribution is a symmetric and bell-shaped curve, differing for different sample sizes. The Student's *t* distribution has mean $= 0$ and standard deviation is greater 1 for degrees of freedom greater than 2 and it does not exists for 1 and 2 degrees of freedom. As the sample size $n \to \infty$, the Student's *t* distribution approaches the standard normal distribution. If we compare the z-table and *t* table, we can see that the percentile points for of normal distribution and Student $-t$ distribution for large degrees of freedom are approximately equal.

The 95th and 99th percentile points of Standard Normal with pdf $\mathrm{f_Z(z)}$ and Student$-t$ distribution with n ($\geq$1) degrees of freedom with pdf $f_{t_n}(x)$, where

$$fz(z) = \frac{1}{\sqrt{2\pi}} e^{-\frac{z^2}{2}}, -\infty < z < \infty \quad \text{and}$$

$$f_{t_n}(x) = \frac{1}{\sqrt{n}B\left(\frac{n}{2}, \frac{1}{2}\right)} \left(1 + \frac{x^2}{n}\right)^{\frac{n+1}{2}}, -\infty < x < \infty.,$$

are given below.

The classical theory of statistical inference is mainly based on the assumption that errors are normally and independently distributed. Recently many researchers

M. Ahsanullah et al., *Normal and Student's t Distributions and Their Applications*, Atlantis Studies in Probability and Statistics 4, DOI: 10.2991/978-94-6239-061-4_3,

Percentile points	N(0,1)	t_{40}	t_{50}	t_{60}	t_{80}	t_{100}
95	1.645	1.684	1.676	1.671	1.664	1.660
99	2.33	2.423	2.403	2.390	2.374	2.364

have investigated that how inferences are affected if the population model departs from normality. In reality, many economic, finance and business data exhibit fat-tailed distributions. The suitability of independent t-distributions for stock return data was performed by Blattberg and Gonedes (1974). Zellner (1976) analyzed the stock prices data by a simple regression model under the assumption that errors have a multivariate t-distribution. However, errors in this model are uncorrelated but not independent. In a later date, Prucha and Kelejian (1984) discussed the inadequacy of normal distribution and suggested a correlated t-model for many real world problems as a better alternative of normal distribution. Kelejian and Prucha (1985) proved that the uncorrelated t-distributions are better to capture heavy-tailed behavior than independent t-distributions. For detailed on the multivariate t distribution and its applications in linear regression model, we refer our interested readers to Kelker (1970), Canmbanis et al. (1981), Fang and Anderson (1990), Kibria (1996), Kibria and Haq (1998, 1999), Kibria and Saleh (2003), Kotz and Nadarajah (2004), Joarder (1998), Joarder and Ali (1997), Joarder and Sing (1997), and the references therein.

In this chapter, we present some basic properties of student's t distribution, (for details, see, for example, Whittaker and Robinson (1967), Feller (1968, 1971), Patel et al. (1976), Patel and Read (1982), Johnson et al. (1994), Joarder and Ali (1997), Joarder and Singh (1997), Joarder (1998), Evans et al. (2000), Balakrishnan and Nevzorov (2003), and Kapadia et al. (2005), among others).

Definition 3.1.1 (General Form of Student's t Distribution): A continuous random variable X with location parameter μ, scale parameter $\sigma > 0$, and degrees of freedom $v > 0$ is said to have the general form of the Student's t distribution if its pdf $g_X(x)$ is given by, (for details, see Blattberg and Gonedes (1974)):

$$g_X(x) = \frac{1}{\sigma\sqrt{v}B\left(\frac{v}{2},\ \frac{1}{2}\right)}\left[1 + \frac{1}{v}\left(\frac{x-\mu}{\sigma}\right)^2\right]^{\frac{-(1+v)}{2}},$$

$$-\infty < x < \infty,\ v > 0,\ \sigma > 0, \tag{3.1}$$

where B (., .) denotes beta function.

The general Student's t distribution has the following properties, (for details, see Moix (2001)):

(i) $E(X) = \mu$ for $v > 1$ and $E(X)$ does not exist for $v = 1$;
(ii) $\text{var}(X) = = \frac{v\sigma^2}{v-2}$ for $v > 2$ and var (X) does not exist for $v \le 2$;
(iii) In general, all moments of order $r < v$ are finite;
(iv) When $v = 1$, the general Student's t distribution reduces to the Cauchy distribution;

(iv) When $v \to \infty$, the general Student's t distribution converges to the normal distribution;

(v) For $\mu = 0$ and $\sigma = 1$, the general Student's t distribution reduces to the standard Student's t distribution;

(vi) The probability density function of the standardized Student's t random variable exhibits fatter tails than the probability density function of the standardized normal random variable;

For a comparison between the stable and the Student's t distributions, see, for example, Embrechts et al. (1997), and Moix (2001), among others.

Definition 3.1.2 (Student's t as a Mixture of Normal and Inverted Gamma Distributions): It is well-known that the pdf of the Student's t distribution can be expressed as

$$g_X(x) = \int_0^\infty \frac{1}{\sqrt{2\pi\omega^2\sigma^2}} e^{-\frac{1}{2}\left(\frac{x-\mu}{\omega\sigma}\right)^2} h(\omega)\,d\omega, \quad -\infty < x < \infty, \omega > 0, \sigma > 0, \tag{3.2}$$

which is the mixture of the normal distribution $N\left(\mu, \omega^2\sigma^2\right)$ and the inverted gamma distribution with v degrees of freedom and pdf given by

$$h_\Omega(\omega) = \frac{2\left(\frac{v}{2}\right)^{\frac{v}{2}}}{\Gamma\left(\frac{v}{2}\right)} \omega^{-(v+1)} e^{-\frac{1}{2}\left(\frac{v}{\omega^2}\right)}, \quad \omega > 0, v > 0.$$

Definition 3.1.3 (Student's t as a Scale Mixture of Normal Distributions): Let T_v be a Student's t random variable with v degrees of freedom and pdf $f_{T_v}(t)$. The pdf of the Student's t distribution can be expressed as a scale mixture of normal distributions given by

$$f_{T_v}(t) = \int_0^\infty \frac{1}{\sqrt{2\pi}} e^{-\frac{t^2 x}{2v}} \frac{\sqrt{x}}{\sqrt{v}} \frac{1}{\Gamma\left(\frac{v}{2}\right) 2^{\frac{v}{2}}} x^{\frac{v}{2}-1} e^{-\frac{x}{2}} dx, \quad v > 0, \tag{3.3}$$

(for details, see Casella and Berger (2002)).

Definition 3.1.4 (Student's t as a Predictive Distribution, a Bayesian Approach): Let $X \sim N\left(0, \frac{1}{\theta}\right)$ be a normal random variable with pdf $f_X(x|\theta)$ given by

$$f_X(x|\theta) = \frac{\sqrt{\theta}}{\sqrt{2\pi}} e^{-\frac{\theta x^2}{2}}, \quad -\infty < x < \infty,$$

where $\theta = \frac{1}{\sigma^2}$, the inverse of the variance, is called the precision of X. Suppose that θ has the gamma distribution with pdf $h(\theta)$ given by

$$h(\theta) = \frac{1}{\Gamma(\alpha)\beta^{\alpha}} \theta^{\alpha-1} e^{-\frac{\theta}{\beta}}, \quad 0 < \theta < \infty.$$

Then the predictive pdf is given by

$$k_1(x) = \int_0^{\infty} \frac{\theta^{\alpha+\frac{1}{2}-1}}{\Gamma(\alpha)\beta^{\alpha}\sqrt{2\pi}} \; e^{-\left(\frac{x^2}{2}+\frac{1}{\beta}\right)\theta} d\theta$$

$$= \frac{\Gamma\left(\alpha+\frac{1}{2}\right)}{\Gamma(\alpha)\;\beta^{\alpha}\sqrt{2\pi}} \left(\frac{1}{\beta}+\frac{x^2}{2}\right)^{-\left(\alpha+\frac{1}{2}\right)}, \quad -\infty < x < \infty, \tag{3.4}$$

which, for $\alpha = \frac{r}{2}$ and $\beta = \frac{2}{r}$, reduces to a Student's t pdf with r degrees of freedom given by

$$k_1(x) \propto \left(1+\frac{x^2}{r}\right)^{-\frac{r+1}{2}}, \quad -\infty < x < \infty,$$

(for details, see Hogg et al. (2005), and Hogg and Tanis (2006), among others).

Definition 3.1.5 (Standard Student's t Distribution): A continuous random variable X is said to have the standard Student's t distribution with v degrees of freedom if, for some integer $v > 0$, its pdf $g_X(x)$ and cdf $G_X(x) = P(X \leq x)$ are, respectively, given by

$$g_X(x) = \frac{1}{\sqrt{v}B\left(\frac{v}{2},\,\frac{1}{2}\right)} \left(1+\frac{x^2}{v}\right)^{-(1+v)/2}, \quad -\infty < x < \infty, \quad v > 0 \quad \ldots \quad (3.5)$$

and

$$G_X(x) = \begin{cases} 1-\frac{1}{2}I_t\left(\frac{v}{2},\,\frac{1}{2}\right), & if\ x > 0, \\ \frac{1}{2}I_t\left(\frac{v}{2},\,\frac{1}{2}\right), & otherwise, \end{cases} \quad \ldots \quad (3.6)$$

with $t = \left(1+\frac{x^2}{v}\right)^{-1}$, where $B(.,.)$ and $I_t(.,.)$ denote beta and incomplete beta functions, respectively. For special cases, we have

for $v = 1$, $g_1(x) = \dfrac{1}{\pi(1+x^2)}$, $G_1(x) = \dfrac{1}{2} + \dfrac{1}{\pi}\arctan(x)$,

for $v = 2$, $g_2(x) = \dfrac{1}{(2+x^2)^{\frac{3}{2}}}$, $G_2(x) = \dfrac{1}{2} + \dfrac{x}{2\sqrt{2+x^2}}$.

for $\nu = 3,\ g_3(x) = \frac{6\sqrt{3}}{\pi(3+x^2)^2}$ and as $\nu \to \infty$, $G_\nu(x)$ converges in distribution to the cdf of N(0,1). Also,

$$G_1(x) = \frac{1}{2} + \frac{1}{\pi}\text{arc tan } (x)$$

and letting $\theta = \text{arc tan } (x/\nu)$, we have

$$G_{\nu(x)} = \frac{1}{2} + \frac{1}{\pi}\left[\theta + \left\{\cos\theta + \frac{2}{3}\cos^2\theta + \frac{2(4)\ldots(\nu-3)}{3(5)\ldots(\nu-2)}\cos^{\nu-2}\theta\right\}\sin\theta\right],$$

for $\nu = 2n+1, n = 1, 2, \ldots$ and

$$G_{\nu(x)} = \frac{1}{2} + \frac{1}{2}\left\{1 + \frac{1}{2}\cos^2\theta + \frac{1(3)}{2(4)}\cos^4\theta + \cdots + \frac{1(3)\ldots(\nu-3)}{3(5)\ldots(\nu-2)}\cos^{\nu-2}\theta\right\}\sin\theta,$$

for $\nu = 2n, n = 1, 2, \ldots$

Another simple form of cdf $G_X(x)$ of the Student's t distribution with ν degrees of freedom, which appears in the literature, is given by

$$G_X(x) = \frac{1}{2} + \frac{x\,\Gamma\left(\frac{\nu+1}{2}\right)\,{}_2F_1\left(\frac{1}{2}, \frac{\nu+1}{2}; \frac{3}{2}; -\frac{x^2}{\nu}\right)}{\sqrt{\pi\nu}\,\Gamma\left(\frac{\nu}{2}\right)}, \tag{3.7}$$

where ${}_2F_1$ (.) denote the generalized hypergeometric function of order (2.1), (see, for example, Wikipedia (2007), among others). To describe the shapes of the Student's t distribution, the plots of the pdf (3.5) and cdf (3.6), for different values of degrees of freedom, ν, are provided in Figs. 3.1 and 3.2, respectively, by using Maple 10. The effects of the parameter, ν, can easily be seen from these graphs. It is also clear that the graph of the pdf $g_X(x)$ of a Student's t distribution is symmetric about mean.

3.2 Some Properties of the Student's *t* Distribution

This section discusses the mode, moments, mean, variance, coefficients of skewness and kurtosis, and entropy of the Student's t distribution. For detailed derivations of these, see, for example, Lukacs (1972), Dudewicz and Mishra (1988), Johnson et al. (1995), Rohatgi and Saleh (2001), Balakrishnan and Nevzorov (2003), and Kapadia et al. (2005), among others.

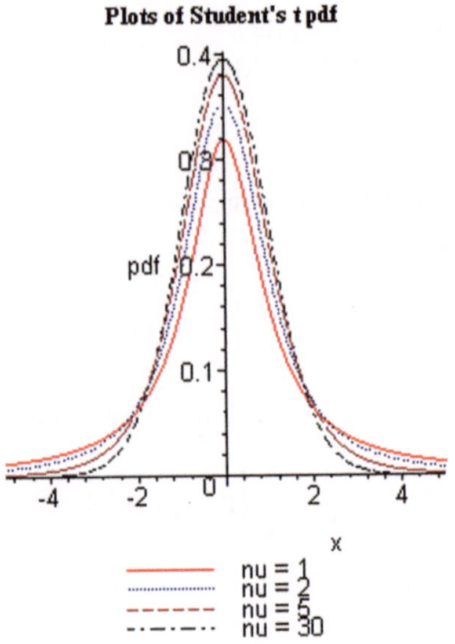

Fig. 3.1 Plots of the Student's t pdf, for different values of v

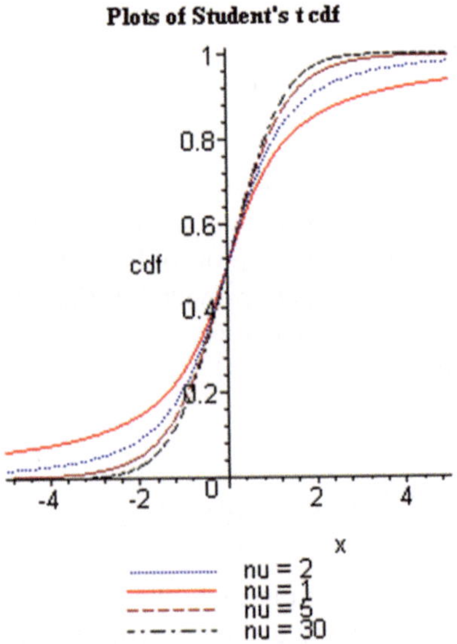

Fig. 3.2 Plots of the Student's t cdf, for different values of v

3.2.1 Mode

The Mode or modal value is that value of x for which the probability density function $g_x(x)$ defined by (3.5) is maximum. Now, differentiating Eq. (3.6), we have

$$g'_X(x) = -\frac{(v+1)\,x\left(1+\frac{x^2}{v}\right)^{-\frac{(v+3)}{2}}}{v\sqrt{v}B\left(\frac{v}{2},\frac{1}{2}\right)} \tag{3.8}$$

which, when equated to 0, gives the mode to be $x = 0$. It can be easily seen that $g''_X(0) < 0$. Thus, the maximum value of the probability density function $g_X(x)$ is easily obtained from (3.5) as $g_X(0) = \frac{1}{\sqrt{v}B\left(\frac{v}{2},\frac{1}{2}\right)}$. Since the equation $g'_v(x)$ has a unique root at x = 0 $g''_v(x)$ <0, the Student $-t$ distribution is unimodal.

3.2.2 *Moments*

For some degrees of freedom $v > 0$ and some integer $r > 0$, the rth moment about the mean of a random variable X having the pdf (3.5) is given by

$$E\left(X^r\right) = \begin{cases} \frac{v^{r/2}\Gamma\left(\frac{r+1}{2}\right)\Gamma\left(\frac{v-r}{2}\right)}{\Gamma\left(\frac{1}{2}\right)\Gamma\left(\frac{v}{2}\right)} & , \ when\ r\ is\ even,\ r < v; \\ 0, \ \ when\ r\ is\ odd \end{cases} \tag{3.9}$$

Using the properties of gamma function, Eq. (3.9) can be written as

$$E(X^r) = v^{r/2}\prod_{j=1}^{r/2}\frac{2i-1}{v-2i}, \text{ for even r and } 0 < \text{ r} < v \text{ and } v > 2.$$

3.2.3 *Mean, Variance, and Coefficients of Skewness and Kurtosis*

From the expression (3.9), the mean, variance, coefficients of skewness and kurtosis of a Student's t random variable X having the pdf (3.5) are easily obtained as follows:

(i) **Mean:** $\alpha_1 = E\ (X) = 0, \quad v > 0;$

(ii) **Variance:** $Var\ (X) = \beta_2 = \frac{v}{v-2}, \quad v > 2;$

(iii) **Coefficient of Skewness:** $\gamma_1\ (X) = \frac{\beta_3}{\beta_2^{3/2}} = 0;$

(iv) **Coefficient of Kurtosis:** $\gamma_2\ (X) = \frac{\beta_4}{\beta_2^2} = \frac{3\,(v-2)}{v-4}, \quad v > 4.$

Since the coefficient of kurtosis, $\gamma_2\ (X) > 4$ for $v > 4$, it follows that the Student's t distributions are leptokurtic distributions for $v > 4$.

3.2.4 Mean Deviation and Coefficient of Variation of a Student's t Random Variable

(i) **Mean Deviation:** $\left(\frac{2\sigma^2}{\pi}\right)^{\frac{1}{2}}$
(ii) **Coefficient of Variation:** Undefined

3.2.5 Moment Generating Function

Does not exist (for details, see, for example, Mood et al. (1974), among others).

3.2.6 Entropy

For some degrees of freedom $\nu > 0$, entropy of a random variable X having the pdf (3.5) is easily given by

$$H_X\left[g_x(x)\right] = E[-\ln(g_x(X)] = -\int_{-\infty}^{\infty} g_x(x)\ln\left[g_x(x)\right]dx$$
$$= \left(\frac{\nu+1}{2}\right)\left[\psi\left(\frac{\nu+1}{2}\right) - \psi\left(\frac{\nu}{2}\right)\right] + \ln\left(\sqrt{\nu}B\left(\frac{\nu}{2},\frac{1}{2}\right)\right), \tag{3.10}$$

where $\psi(.)$ and $B(.)$ denote digamma and beta functions, respectively, (see, for example, Lazo and Rathie (1978), and Kapur (1993), among others). The possible shape of the entropy for different values of the parameter ν is provided in Fig. 3.3, by using Maple 10. The effects of the parameter ν on entropy can easily be seen from the graph. Clearly, the entropy $H_X(.)$ of Student's t distribution is monotonic decreasing in ν. Moreover, as $\nu \to \infty$, the entropy $H_X(.)$ of Student's t distribution tends to $\frac{1}{2} + \frac{1}{2}\ln(2\pi)$, which is the entropy of standard normal distribution.

3.2.7 Characteristic Function

The characteristic function of the student's t distribution is a research topic of considerable importance and interest in statistics both from the theoretical and applications point of view. It has been studied by many authors, (among them, Ifram (1970), Mitra

Fig. 3.3 Plot of the entropy for $\nu = 1..40$

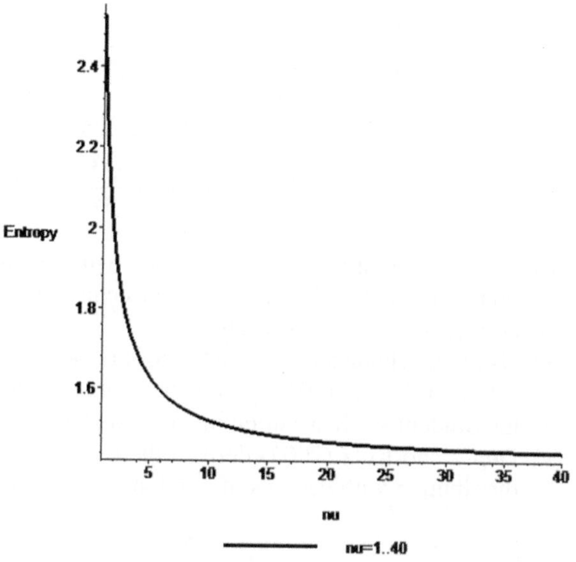

(1978), Pastena (1991), Hurst (1995), Dreier and Kotz (2002) for the characteristic function of the univariate t distributions, and Joarder and Ali (1996) for the characteristic function of the multivariate t distributions, are notable). The purpose of this section is to present briefly some of the expressions for the characteristic function of the Student's t distribution with degrees of freedom ν as developed by different authors which are provided below (for details, see Johnson et al. (1995), and Dreier and Kotz (2002), and references therein).

(i) Ifram (1970) derives the characteristic function of the Student's t distribution with degrees of freedom ν as given by

$$\psi_x(t) = E\left(e^{itX}\right) = \frac{1}{B\left(\frac{1}{2}, \frac{\nu}{2}\right)} \int_{-\infty}^{\infty} e^{it(\sqrt{\nu})x} \left(1 + x^2\right)^{-\left(\frac{1}{2}+\frac{\nu}{2}\right)} dx, \tag{3.11}$$

and discusses both the cases for odd and even degrees of freedom.

(ii) The following expressions for the characteristic function of the Student's t distribution are obtained by Mitra (1978):

$$\psi_X(t) = e^{-|\sqrt{\nu}t|} \sum_{j=0}^{m-1} c_{j,(m-1)} \left|\sqrt{\nu}t\right|^j,$$

where $m = \frac{\nu+1}{2}$ and the $c_{j,m}$'s satisfy the following recurrence relations

$$
\begin{aligned}
c_{0,m} &= 1, \\
c_{1,m} &= 1, \\
c_{(m-1),m} &= \frac{1}{1.3\ldots(2m-5)(2m-3)}, \\
c_{j,m} &= \frac{c_{(j-1),(m-1)} + (2m-3-j)c_{j,(m-1)}}{(2m-3)}, \quad 1 \le j \le m-1.
\end{aligned}
$$

(iii) Further development on the characteristic function of the Student's t distribution continued with the work of Pastena (1991) who provides comments and corrections to Ifram's results.

(iv) Using the characteristic function of the symmetric generalized hyperbolic distribution, Hurst (1995) derives the expression for the characteristic function of the Student's t distribution in terms of Bessel functions.

(v) Dreier and Kotz (2002) derived the characteristic function of the Student's t distribution with degrees of freedom v as given by

$$
\psi_X(t) = E\left(e^{itX}\right) = \frac{2^v v^{\frac{v}{2}}}{\Gamma(v)} \int_0^\infty e^{-(\sqrt{v})(2x+|t|)} \left[x(x+|t|)\right]^{\left(\frac{v}{2}-\frac{1}{2}\right)} dx, \quad t \in \Re. \tag{3.12}
$$

3.3 Percentiles

This section computes the percentiles of the Student's t distribution, by using Maple 10. For any $p(0 < p < 1)$, the $(100p)$th percentile (also called the quantile of order p) of the Student's t distribution with the pdf $g_X(x)$ is a number t_p such that the area under $g_X(x)$ to the left of t_p is p. That is, t_p is any root of the equation

$$
G(t_p) = \int_{-\infty}^{t_p} g_X(u)du = p. \tag{3.13}
$$

Using the following Maple program, the percentiles t_p of the Student's t distribution are computed for some selected values of p for the given values of v, which are provided in Table 3.1.

3.4 Different Forms of t Distribution

This section presents different forms of t distribution and some of their important properties, (for details, see, for example, Whittaker and Robinson (1967), Feller (1968, 1971), Patel et al. (1976), Patel and Read (1982), Johnson et al. (1994),

Table 3.1 Percentiles of Student's t distribution

ν	75 %	80 %	85 %	90 %	95 %	99 %
1	1.00000	1.37638	1.96261	3.07768	6.31375	31.82051
5	0.72668	0.91954	1.15576	1.47588	2.01504	3.36493
15	0.69119	0.86624	1.07353	1.34060	1.75305	2.60248
30	0.68275	0.86624	1.05466	1.31041	1.69726	2.45726

Evans et al. (2000), Balakrishnan and Nevzorov (2003), and Kapadia et al. (2005), among others).

3.4.1 Half t Distribution

A random variable X said to have a half-t distribution with parameters ξ, τ, and λ if its pdf can be written as

$$f(x|\xi,\tau,\lambda) = 2\frac{\Gamma(\lambda+1/2)\sqrt{\tau}}{\Gamma(\lambda/2)\sqrt{\lambda\pi}}\left[1+\frac{1}{\lambda}(\sqrt{\tau}(x-\xi)^2\right]^{-(\lambda+1/2)}$$
$$\text{for } x > \xi, -\infty < \xi < \infty, \tau > 0, \lambda > 0 \qquad (3.14)$$

Note that, as $\lambda \to \infty$ in Eq. (3.14), the half-t distribution approaches the half-normal distribution, which follows from the definition of the exponential function, $\lim_{t\to\infty} \left(1+\frac{x}{t}\right)^t = e^x$. Also, note that as $\lambda \to 0$ in Eq. (3.14), the right-hand tail of the half-t distribution becomes increasingly heavier relative to that of the limiting half-normal distribution, obtained as $\lambda \to \infty$.

3.4.2 Skew t Distribution

A random variable X is said to have the skew-t distribution if its pdf is $f(x) = 2g(x)G(\lambda x)$, where g(x) and G(x), respectively, denote the pdf and the cdf of the Student's t distribution with degrees of freedom ν. For different degrees of freedoms the skew t density are given in Fig. 3.4.

3.5 Summary

In this chapter, we first present the motivation and importance of studying the Student's t distribution. Then some basic ideas, definitions and properties of the

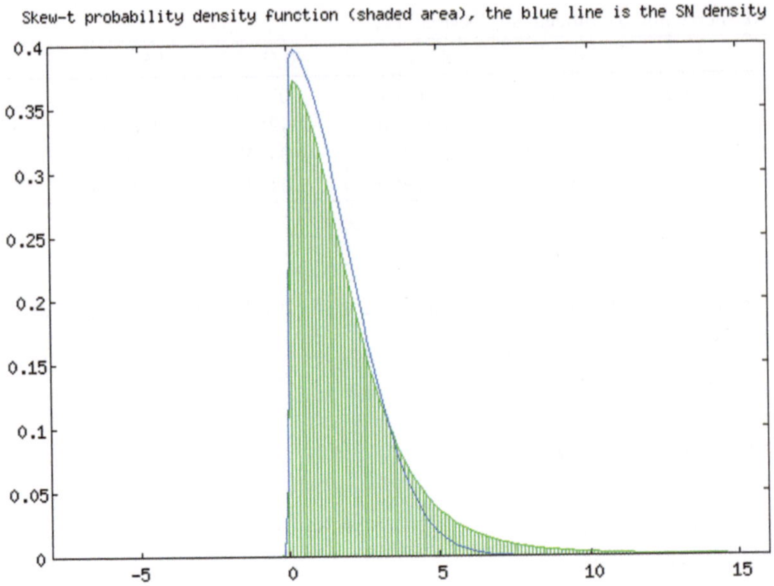

Fig. 3.4 Parameters of the Skew-t density: location = 0; scale = 2; $\alpha = 50$; $v = 4$ and 40

Student's t distributions have been reviewed. The entropy of a random variable having the Student's t distributions has been given. The expressions for different forms of the characteristic function of the Student's t distribution are provided. As a motivation, different forms of the Student's distribution, such as Half-t and Skew-t distributions, which are areas of current research, have been provided.

Chapter 4
Sum, Product and Ratio for the Normal Random Variables

4.1 Introduction

The distributions of the sum, product, and ratio of two independent random variables arise in many fields of research, for example, automation, biology, computer science, control theory, economics, engineering, fuzzy systems, genetics, hydrology, medicine, neuroscience, number theory, statistics, physics, psychology, reliability, risk management, etc. (for details, see Grubel (1968), Rokeach and Kliejunas (1972), Springer (1979), Kordonski and Gertsbakh (1995), Ladekarl et al. (1997), Amari and Misra (1997), Sornette (1998), Cigizoglu and Bayazit (2000), Brody et al. (2002), Galambos and Simonelli (2005), among others). The distributions of the sum $X + Y$, product XY, and ratio X/Y, when X and Y are independent random variables and belong to the same family, have been extensively studied by many researchers, among them, the following are notable:

(a) Ali (1982), Farebrother (1984), Moschopoulos (1985), Provost (1989a), Pham-Gia and Turkkan (1994), Kamgar-Parsi et al. (1995), Hitezenko (1998), Hu and Lin (2001), Witkovsky (2001), and Nadarajah (2006a) for the sum $X + Y$.
(b) Sakamoto (1943), Harter (1951) and Wallgren (1980), Springer and Thompson (1970), Stuart (1962) and Podolski (1972), Steece (1976), Bhargava and Khatri (1981), Abu-Salih (1983), Tang and Gupta (1984), Malik and Trudel (1986), Rathie and Rohrer (1987), Nadarajah (2005a, b), Nadarajah and Gupta (2005), Nadarajah and Kotz (2006a) for the product XY.
(c) Marsaglia (1965), and Korhonen and Narula (1989), Press (1969), Basu and Lochner (1971), Shcolnick (1985), Hawkins and Han (1986), Provost (1989b), Pham-Gia (2000), Nadarajah (2005c, 2006b), Nadarajah and Gupta (2005, 2006), and Nadarajah, and Kotz (2006b) for the ratio X/Y.

The algorithms for computing the probability density function of the sum and product of two independent random variables, along with an implementation of the algorithm in a computer algebra system, have also been developed by many authors, among them, Agrawal and Elmaghraby (2001), and Glen et al. (2004) are notable.

M. Ahsanullah et al., *Normal and Student's t Distributions and Their Applications*, Atlantis Studies in Probability and Statistics 4, DOI: 10.2991/978-94-6239-061-4_4,

This chapter presents the distributions of the sum $X+Y$, product XY, and ratio X/Y when X and Y are independent random variables and have the same normal distributions. For the sake of completeness, the definitions of sum, product, and ratio of two independent random variables are given below, (for details, see Lukacs (1972), Dudewicz and Mishra (1988), Rohatgi and Saleh (2001), Kapadia et al. (2005), and Larson and Marx (2006), among others).

Let X and Y be any two independent, absolutely continuous random variables with pdfs $f_X(x)$ and $f_Y(y)$, respectively. Note that the sum $X+Y$, product XY, and ratio X/Y of X and Y are also random variables, (for details, see Lukacs (1972), among others).

4.1.1 Definition (Distribution of a Sum)

Let $Z = X + Y$ for $-\infty < X, Y < +\infty$. Then

$$\text{(i)} \quad F_Z(z) = P(Z \le z) = P(X + Y \le z) = \int_{-\infty}^{\infty} f_X(x) F_Y(z - x)\, dx \tag{4.1}$$

$$\text{(ii)} \quad f_Z(z) = \int_{-\infty}^{\infty} f_X(x) f_Y(z - x)\, dx \tag{4.2}$$

(iii) **Convolution:** The probability density function $f_Z(z)$ in (ii) above is also called the convolution of the pdfs $f_X(x)$ and $f_Y(y)$, which is expressed as $\{f_Z(z)\} = \{f_X(z)\} * \{f_Y(z)\}$.

4.1.2 Definition (Distribution of a Product)

Let $W = XY$ for $-\infty < X, Y < +\infty$. Then

$$\text{(i)} \quad F_W(w) = P(W \le w) = P(XY \le w) = P\left(X \le \tfrac{w}{Y}\right)$$
$$= \begin{cases} \int_0^{\infty} F_X\left(\frac{w}{y}\right) f_Y(y)\, dy, & Y > 0 \\ F_Y(0) - \int_{-\infty}^{0} F_X\left(\frac{w}{y}\right) f_Y(y)\, dy, & Y < 0 \end{cases} \tag{4.3}$$

$$\text{(ii)}\quad f_W(w) = \begin{cases} \int_0^{\infty} \frac{1}{y} f_X\left(\frac{w}{y}\right) f_Y(y)\,dy, & Y > 0,\ -\infty < w < \infty \\ \int_{-\infty}^{\infty} \left|\frac{1}{y}\right| f_X\left(\frac{w}{y}\right) f_Y(y)\,dy, & Y < 0,\ -\infty < w < \infty \end{cases} \tag{4.4}$$

4.1.3 Definition (Distribution of a Ratio)

Let $U = \frac{X}{Y}$ for $-\infty < X, Y < +\infty$. Then

(i)

$$F_U(u) = P(U \le u) = P\left(\frac{X}{Y} \le u\right) = P(X \le uY)$$

$$= \begin{cases} \int_0^{\infty} F_X(uy) f_Y(y)\,dy, & Y > 0 \\ F_Y(0) - \int_{-\infty}^{0} F_X(uy) f_Y(y)\,dy, & Y < 0 \end{cases} \tag{4.5}$$

(ii)

$$f_U(u) = \begin{cases} \int_0^{\infty} y f_X(uy) f_Y(y)\,dy, & Y > 0,\ -\infty < u < \infty \\ \int_{-\infty}^{\infty} |y|\, f_X(uy) f_Y(y)\,dy, & Y < 0,\ -\infty < u < \infty \end{cases} \tag{4.6}$$

4.1.4 Mean and Variance of the Sum and Product of Independent Random Variables

Let X and Y be any two independent, absolutely continuous random variables with pdfs $f_X(x)$ and $f_Y(y)$, respectively. Suppose that the first and second moments of both X and Y exist. Then

(i) $\mathrm{E}(X + Y) = \mathrm{E}(X) + \mathrm{E}(Y)$
(ii) $\mathrm{E}(XY) = \mathrm{E}(X)\mathrm{E}(Y)$
(iii) $Var(X + Y) = Var(X) + Var(Y)$.

Note that the above results can be extended to any finite sums or products of independently distributed random variables, (for details, see Lukacs (1972), and Dudewicz and Mishra (1988), among others).

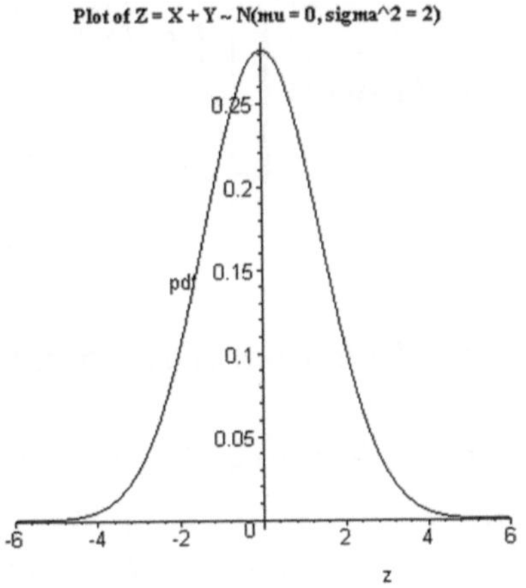

Fig. 4.1 Plot of the $Z = X + Y \sim N\left(\mu = 0, \sigma^2 = 2\right)$

4.2 Distribution of the Sum of Independent Normal Random Variables

This section presents the distributions of the sum of independent normal random variables as described below.

(i) Let X and Y be two independent $N\ (0, 1)$ random variables. Then $Z = X + Y \sim N\ (0, 2)$ with pdf given by

$f_Z(z) = \frac{1}{2\sqrt{\pi}} e^{-z^2/4}, -\infty < z < \infty,$
(for details, see Dudewicz and Mishra (1988), and Balakrishnan and Nevzorov (2003), among others). A Maple plot of this pdf is given in Fig. 4.1.

(ii) Let $X \sim N\left(\mu_X, \sigma_X^2\right)$ and $Y \sim N\left(\mu_Y, \sigma_y^2\right)$ be two independent random variables. Then $Z = X + Y \sim N\left(\mu_X + \mu_Y, \sigma_X^2 + \sigma_Y^2\right)$ with pdf given by

$$f_Z(z) = \frac{1}{\sqrt{2\pi\left(\sigma_X^2 + \sigma_Y^2\right)}} e^{-(z - \mu_X - \mu_Y)^2/2(\sigma_X^2 + \sigma_Y^2)}, \quad -\infty < z < \infty, \qquad (4.7)$$

(for details, see Lukacs (1972), and Balakrishnan and Nevzorov (2003), among others).

(iii) Let $X_1, X_2, \ldots, X_n$ be a set of independently distributed $N\left(\mu_i, \sigma_i^2\right)$ $(i = 1, 2, \ldots, n)$ random variables. Let $Z = \sum_{i=1}^{n} c_i X_i$ where c_i are some constants. Then $Z \sim N\left(\sum_{i=1}^{n} c_i \mu_i, \sum_{i=1}^{n} c_i^2 \sigma_i^2\right)$, (for details, see Lukacs (1972), Dudewicz and Mishra (1988), and Kapadia et al. (2005), among others).

Note that the mean and variance of the sum of independent normal random variables can be easily derived by following the above Sect. 4.1.4.

4.3 Distribution of the Product of Independent Normal Random Variables

This section presents the distributions of the product of independent normal random variables as described below, (for details, see Epstein (1948), Zolotarev (1957), Kotlarski (1960), Donahue (1964), Springer and Thompson (1966, 1970), Lomnicki (1967), and Glen et al. (2002), among others).

(i) Let $X_1, X_2, ..., X_n$ be a set of independently distributed $N\left(0, \sigma_i^2\right)$ $(i = 1, 2, ..., n)$ random variables. Let $W = \prod_{i=1}^{n} X_i$. Then the random variable W follows a distribution with pdf given, in terms of a Meijer G-function, by

$$g_W(w) = H\ G_{0\,n}^{n\,0}\left(w^2 \prod_{i=1}^{n} \frac{1}{2\,\sigma_i}\ |\ 0\right) \tag{4.8}$$

where H is a normalizing constant given by

$$H = \left[(2\pi)^{\frac{n}{2}} \prod_{i=1}^{n} \sigma_i\right]^{-1},$$

(for details, see Springer and Thompson (1966, 1970)), where $G_{0n}^{n0}\left(w^2 \prod_{i=1}^{n} \frac{1}{2\sigma_i}\ |0\right)$ denotes the Meijer G-Function. It is defined as follows

$$G_{p,q}^{m,n}\left(x\Big|_{b_1,...,b_q}^{a_1,...,a_p}\right) = \frac{1}{2\pi i}\int_L \frac{x^{-t}\Gamma(b_1+t)\cdots\Gamma(b_m+t)\Gamma(1-a_1-t)\cdots\Gamma(1-a_n-t)}{\Gamma(a_{n+1}+t)\cdots\Gamma(a_p+t)\Gamma(1-b_{m+1}-t)\cdots\Gamma(1-b_q-t)}dt,$$

where $(e)_k = e(e+1)\cdots(e+k-1)$ denotes the ascending factorial and L denotes an integration path (for details on Meijer G-Function, see, Gradshteyn and Ryzhik (2000), Sect. 9.3, p. 1068).

(ii) Let X and Y be independently distributed as N (0, 1), and Z = XY. Then the characteristic function of the product of two independent normal random variables is given by:

$$E(e^{itXY}) = E_Y(E(^{itXY}|Y)$$

Fig. 4.2 Plot of the PDF of $W = XY$, when $X \sim N\,(0, 1)$ and $Y \sim N\,(0, 1)$

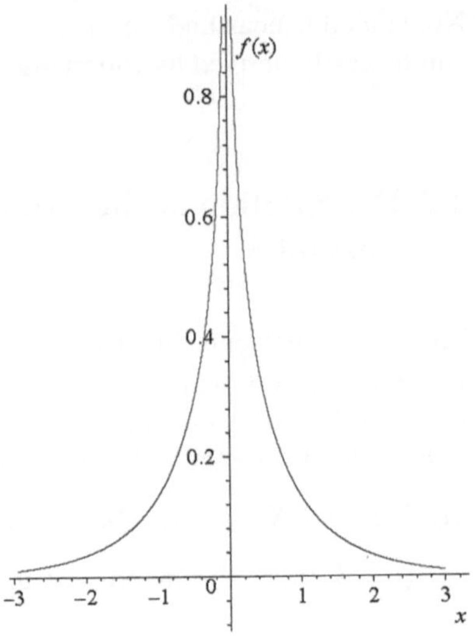

That is,

$$E(e^{itXY}|Y) = e^{-\frac{1}{2}t^2Y^2}$$

Thus we have,

$$E_Y(E(e^{itXY}|Y) = \frac{1}{\sqrt{2\pi}}\int_{-\infty}^{\infty} e^{-\frac{1}{2}t^2y^2}\,e^{-\frac{1}{2}y^2}dy$$
$$= \frac{1}{\sqrt{1+t^2}}$$

Inverting this characteristic function we get the pdf of Z as

$$f_Z(z) = \frac{1}{\pi}K_0(z),$$

where $K_0(z)$ is the Bessel function of the second kind.

(iii) Let $X \sim N\left(0, \sigma_X^2\right)$ and $Y \sim N\left(0, \sigma_y^2\right)$ be two independent random variables. Let $W = XY$. Then the random variable $W = XY$ follows a distribution with pdf given by

$$f_W(w) = \frac{1}{\pi\sigma_X\sigma_Y}\mathrm{K}_0\left(\frac{w}{\sigma_X\sigma_Y}\right), \tag{4.9}$$

where K_0 (.) denotes modified Bessel function of the second kind, (for details, see Lomnicki (1967), and Glen et al. (2002), among others). A Maple plot of the pdf of the random variable $W = XY$, when $X \sim N(0, 1)$ and $Y \sim N(0, 1)$, is given in Glen et al. (2002), which, for the sake of completeness, is reproduced and presented in Fig. 4.2.

(iv) **Alternative Derivation of the PDF (4.9) of the Product of Two Independently Distributed Normal Random Variables:**
Suppose X_1 and X_2 are two independently distributed $N(0, \sigma_i^2)$, $i = 1.2$, random variables, and $Y = X_1X_2$, then the pdf $f_Y(y)$ of Y is given by

$$f_Y(y) = \frac{1}{\pi\sigma_1\sigma_2} K_0\left(\frac{y}{\sigma_1\sigma_2}\right),$$

where $K_0(x)$ is the modified Bessel function of the second kind (see Abramowitz and Stegun (1970, p. 376).

Proof. Let $\phi_Y(t)$ be the characteristic of Y, then

$$\begin{aligned}
\phi_Y(t) &= \int_{-\infty}^{\infty} \frac{1}{\sigma_1\sqrt{2\pi}} e^{=\frac{x^2}{2\sigma_1^2}} \left(\int_{-\infty}^{\infty} e^{itxy} \frac{1}{\sigma_2\sqrt{2\pi}} e^{=\frac{y^2}{2\sigma_2^2}} dy \right) dx \\
&= \int_{-\infty}^{\infty} \frac{1}{\sigma_1\sqrt{2\pi}} e^{-\frac{x^2}{2\sigma_1^2}} e^{-\frac{t^2x^2\sigma_2^2}{2}} dx \\
&= \int_{-\infty}^{\infty} \frac{1}{\sigma_1\sqrt{2\pi}} e^{-\frac{x^2}{2\sigma_1^2}(1+t^2\sigma_1^2\sigma_2^2)} dx
\end{aligned}$$

Let $x\sqrt{(1 + t^2\sigma_1^2\sigma_2^2)} = u$, then

$$\begin{aligned}
\phi_Y(t) &= \frac{1}{\sqrt{1 + t^2\sigma_1^2\sigma_2^2}} \int_{-\infty}^{\infty} \frac{1}{\sigma_1\sqrt{2\pi}} e^{-\frac{u^2}{2\sigma_2^2}} du \\
&= \frac{1}{\sqrt{1 + t^2\sigma_1^2\sigma_2^2}}
\end{aligned}$$

Using the inverse of the characteristic function, the pdf $f_Y(y)$ of Y is given by

$$\begin{aligned}
f_Y(y) &= \frac{1}{2\pi} \int_{-\infty}^{\infty} e^{-ity} \frac{1}{\sqrt{1 + t^2\sigma_1^2\sigma_2^2}} dt \\
&= \frac{1}{\pi} \int_0^{\infty} \frac{\cos ty}{\sqrt{1 + t^2\sigma_1^2\sigma_2^2}} dt
\end{aligned}$$

Using the transform $t\sigma_1\sigma_2 = w$, we obtain

$$f_Y(y) = \frac{1}{\pi\sigma_1\sigma_2}\int_0^\infty \frac{\cos(\frac{y}{\sigma_1\sigma_2}w)}{\sqrt{1+w^2}}dw$$
$$= \frac{1}{\pi\sigma_1\sigma_2}K_0(\frac{y}{\sigma_1\sigma_2}),$$

where $K_0(.)$ is the Bessel function of the second kind. This completes the proof.

(v) **Distribution of the Product of Two Independently Distributed Standard Normal Random Variables:** As a special case of the Eq. (4.9) in (iii) above, for the sake of completion, using the definition of the characteristic function of a random variable, we derive below independently the distribution of the product of two independently distributed standard normal random variables.

Let X and Y be independently distributed random variables as N (0, 1), and Z = XY. Then the characteristic function of the product of two independent normal random variables is given by:

$$E(e^{itXY}) = E_Y(E(^{itXY}|Y).$$

That is,

$$E(^{itXY}|Y) = e^{-\frac{1}{2}t^2Y^2}.$$

Thus, we have

$$E_Y(E(^{itXY}|Y) = \frac{1}{\sqrt{2\pi}}\int_{-\infty}^{\infty} e^{-\frac{1}{2}t^2y^2}e^{-\frac{1}{2}y^2}dy$$
$$= \frac{1}{\sqrt{1+t^2}}$$

Inverting this characteristic function, we get the pdf of Z as given below

$$f_Z(z) = \frac{1}{\pi}K_0(z), \tag{4.10}$$

where $K_0(z)$ is the Bessel function of the second kind. Obviously, Eq. (4.10) is a special case of Eq. (4.9), when $\sigma_X = 1$ and $\sigma_Y = 1$.

(vi) Let $X_1, X_2, ..., X_n$ be a set of independently distributed N (0, 1) ($i = 1, 2, ..., n$) random variables. Let $K = \sum_{i=1}^{n} X_i^2$. Then the random variable $K = \sum_{i=1}^{n} X_i^2$ follows a chi-square distribution with n degrees of freedom whose pdf is given by

$$\begin{cases} p_K(k) &= \frac{1}{2^{\frac{n}{2}}\Gamma(\frac{n}{2})} w^{\frac{n}{2}-1} e^{\frac{-k}{2}} \quad \text{if } k > 0, \\ &= 0 \quad \text{if } w < 0. \end{cases}, \tag{4.11}$$

Fig. 4.3 Plots of $K = \sum_{i=1}^{n} X_i^2$, where each $X_i \sim N(0, 1)$

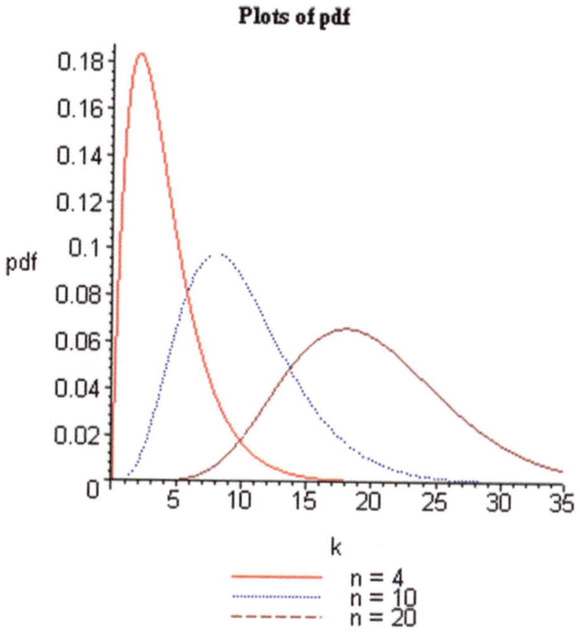

with the Mean $(K) = n$ and Variance $(K) = 2n$. Note that the random variable $K = \sum_{i=1}^{n} X_i^2$ also follows a *gamma* $\left(\frac{1}{2}, \frac{n}{2}\right)$ distribution, (for details, see Lukacs (1972), and Kapadia et al. (2005), among others). A Maple plot of the pdf (4.11) in (vi) is given in Fig. 4.3.

4.4 Distribution of the Ratio of Independent Normal Random Variables

In many statistical analysis problems, the ratio of two normally distributed random variables plays a very important role. The distributions of the ratio *X/Y*, when *X* and *Y* are normally random variables, have been extensively studied by many researchers, notable among them are Geary (1930), Fieller (1932), Curtiss (1941), Kendall (1952), Marsaglia (1965), Hinkley (1969), Hayya et al. (1975), Springer (1979), Cedilnik (2004), and Pham-Gia et al. (2006). This section presents the distributions of the ratio of independent normal random variables as described below.

(i) Let $X \sim N(0, 1)$ and $Y \sim N(0, 1)$ be two independent random variables. Let $U = X/Y$. Then the random variable $U = X/Y$ follows a standard Cauchy distribution with pdf and cdf respectively given by

Fig. 4.4 Plot of $U = X/Y$, where $X \sim N(0, 1)$ and $Y \sim N(0, 1)$

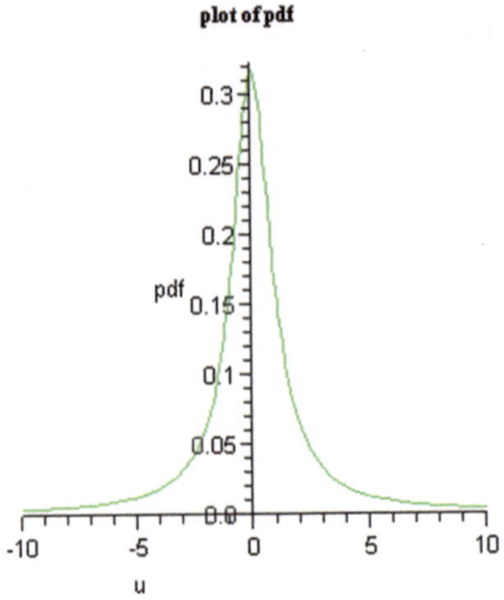

$$f_U(u) = \frac{1}{\pi\left(1 + u^2\right)}, \quad -\infty < u < \infty, \text{ and}$$

$$F_U(u) = \frac{1}{2} + \frac{1}{\pi} \arctan(u), \quad -\infty < u < \infty, \tag{4.12}$$

(for details, see Feller (1971), Casella and Berger (2002), and Severini (2005), among others). A Maple plot of the pdf (4.12) in (i) is given in Fig. 4.4.

(ii) Let $X \sim N\left(0, \sigma_X^2\right)$ and $Y \sim N\left(0, \sigma_y^2\right)$ be two independent random variables. Let $U = X/Y$. Then the random variable $U = X/Y$ follows a generalized form of Cauchy distribution with pdf given by

$$f_U(u) = \frac{\sigma_X \sigma_Y}{\pi\left(\sigma_X^2 + \sigma_Y^2 u^2\right)}, \quad -\infty < u < \infty \tag{4.13}$$

(for details, see Pham-Gia et al. (2006), among others).

(iii) Let $X \sim N\left(0, \sigma_X^2\right)$ and $Y \sim N\left(0, \sigma_y^2\right)$ be two independent random variables. Let $U = X/Y$. If $\sigma_X \neq 0, \sigma_Y \neq 0$ or $\rho \neq 0$, then the random variable $U = X/Y$ follows a more general Cauchy distribution with pdf given by

$$f_U(u) = \frac{1}{\pi}\left[\frac{\beta}{(u - \alpha)^2 + \beta^2}\right], \quad -\infty < u < \infty, \tag{4.14}$$

where ρ is the coefficient of correlation between X and Y, and $\alpha = \rho \frac{\sigma_X}{\sigma_Y}$ and $\beta = \frac{\sigma_X}{\sigma_Y}\sqrt{1 - \rho^2}$. Note that for $\rho = 0$, this pdf reduces to Cauchy distribution $C\left(0, \frac{\sigma_X}{\sigma_Y}\right)$ as given in Eq. (4.13) in (ii) above. Also for $\rho = 0, \sigma_X = \sigma_Y = 1$, we have a standard Cauchy distribution $C(0, 1)$ as given in Eq. (4.12) in (i) above.

(iv) Let $X \sim N\left(\mu_X, \sigma_X^2\right)$ and $Y \sim N\left(\mu_Y, \sigma_y^2\right)$ be two independent random variables, (that is, $cor\,(X, Y) = 0$). Let $U = X/Y$. Then, following Hinkley (1969), the pdf of the random variable $U = X/Y$ is given by

$$f_U(u) = \frac{b\,(u) \cdot c\,(u)}{a^3\,(u)} \frac{1}{\sqrt{2\pi}\;\sigma_X \sigma_Y} \left[2\,\Phi\left(\frac{b\,(u)}{a\,(u)}\right) - 1\right] + \frac{1}{a^2\,(u)\,.\,\pi \sigma_X \sigma_Y} \left[e^{-\frac{1}{2}\left\{\left(\frac{\mu_X}{\sigma_X}\right)^2 + \left(\frac{\mu_Y}{\sigma_Y}\right)^2\right\}}\right] \tag{4.15}$$

where

$$a(u) = \sqrt{\frac{1}{\sigma_x^2}u^2 + \frac{1}{\sigma_Y^2}}, \quad b(u) = \frac{\mu_X}{\sigma_x^2}u^2 + \frac{\mu_Y}{\sigma_Y^2},$$

$$c(u) = e^{\frac{1}{2}\frac{b^2(u)}{a^2(u)} - \frac{1}{2}\left\{\left(\frac{\mu_X}{\sigma_X}\right)^2 + \left(\frac{\mu_Y}{\sigma_Y}\right)^2\right\}},$$

and

$$\Phi\,(t) = \int_{-\infty}^{t} \frac{1}{\sqrt{2\pi}}\, e^{-\frac{1}{2}v^2} dv,$$

denote the standard normal cumulative distribution function. It is easy to see that the above pdf reduces to a standard Cauchy distribution $C\,(0, 1)$ if $\mu_X = \mu_Y = 0$, and $\sigma_X = \sigma_Y = 1$, that is, $b\,(u) = 0$.

(v) Recently, in a very detailed paper, Pham-Gia et al. (2006) have considered the density of the ratio X/Y of two normal random variables X and Y and applications, and have obtained some closed form expressions of the pdf of X/Y in terms of Hermite and Kummer's confluent hypergeometric functions, by considering all cases, that is, when X and Y are standardized and nonstandardized, independent or correlated, normal random variables. For the sake of brevity, only the case when X and Y are independent normal random variables is stated below. For other cases and applications, please visit Pham-Gia et al. (2006). Let $X \sim N\left(\mu_X, \sigma_X^2\right)$ and $Y \sim N\left(\mu_Y, \sigma_y^2\right)$ be independent normal random variables. Let $U = \frac{X}{Y}$. Then, following Pham-Gia et al. (2006), the random variable $U = \frac{X}{Y}$ has a distribution with pdf given by

$$f_U(u) = \frac{K_1.\left[{}_1F_1\left(1;\ \frac{1}{2};\ \theta_1\,(u)\right)\right]}{\sigma_X^2 + \sigma_Y^2 u^2}, \quad -\infty < u < \infty \tag{4.16}$$

where ${}_1F_1(.)$ denotes Kummer's confluent hypergeometric function, and

$$\theta_1(u) = \frac{1}{2\sigma_X^2\sigma_Y^2} \cdot \left[\frac{\left(\mu_Y\sigma_X^2 + u\mu_X\sigma_Y^2\right)^2}{\sigma_X^2 + \sigma_Y^2u^2}\right] \geq 0,$$

and $K_1 = \frac{\sigma_X\sigma_Y}{\pi} e^{-\frac{1}{2}\left\{\left(\frac{\mu_X}{\sigma_X}\right)^2 + \left(\frac{\mu_Y}{\sigma_Y}\right)^2\right\}}$.
By using the following relation

$$H_{-2}(s) + H_{-2}(-s) = {}_1F_1\left(1; \frac{1}{2}; s^2\right), \quad \forall s,$$

where $H_\nu(s)$ denotes Hermite function (for details, see Lebedev (1972, pp. 283–299, among others), it is easy to see that the random variable $U = \frac{X}{Y}$ follows a distribution with pdf given, in terms of Hermite function, as

$$f_U(u) = \frac{K_1.\left\{H_{-2}\left(\sqrt{\theta_1(u)}\right)\right\}}{\sigma_X^2 + \sigma_Y^2u^2}, \quad 0 \leq u < \infty, \tag{4.17}$$

where

$$\theta_1(u) = \frac{1}{2\sigma_X^2\sigma_Y^2} \cdot \left[\frac{\left(\mu_Y\sigma_X^2 + u\mu_X\sigma_Y^2\right)^2}{\sigma_X^2 + \sigma_Y^2u^2}\right] \geq 0,$$

$$K_1 = \left\{\frac{\sigma_X\sigma_Y}{\prod_{i=1}^2\left[1 - \Phi\left(\frac{-\mu_i}{\sigma_i}\right)\right]}\right\} e^{-\frac{1}{2}\left\{\left(\frac{\mu_X}{\sigma_X}\right)^2 + \left(\frac{\mu_Y}{\sigma_Y}\right)^2\right\}},$$

and

$$\Phi(t) = \int_{-\infty}^{t} \frac{1}{\sqrt{2\pi}} e^{-\frac{1}{2}v^2} dv$$

denote the standard normal cumulative distribution function.

The above results of Pham-Gia et al. are valid $\forall \sigma_X, \sigma_Y > 0$, and $\forall \mu_X, \mu_Y \in R$(*set of real numbers*). Since ${}_1F_1\left(1; \frac{1}{2}; 0\right) = 1$, it is easy to see that, when $\mu_X = \mu_Y = 0$, the random variable $U = X/Y$ follows a generalized form of Cauchy distribution with pdf given by

$$f_U(u) = \frac{\sigma_X\sigma_Y}{\pi\left(\sigma_X^2 + \sigma_Y^2u^2\right)}, \quad -\infty < u < \infty, \tag{4.18}$$

which reduces to the pdf of a standard Cauchy distribution $C(0, 1)$ when $\sigma_X = \sigma_Y$.

(vi) The following results about the ratio X/Y of two normal random variables X and Y found in the literature are also worth noting.

A. Kendall (1952), Kapadia et al. (2005, pp. 210–211): Let $X \sim N\left(\mu_X, \sigma_X^2\right)$ and $Y \sim N\left(\mu_Y, \sigma_y^2\right)$ be independent normal random variables. Assume that μ_Y is so large compared to σ_y that the range of Y is effectively positive. Then, as obtained by Kendall (1952), the random variable $U = X/Y$ has a distribution with pdf given by

$$f_U(u) = \frac{1}{\sqrt{2\pi}} \left[\frac{\left(\mu_Y \sigma_X^2 + \mu_X u \sigma_Y^2\right)}{\left(\sigma_X^2 + u^2 \sigma_Y^2\right)^{\frac{3}{2}}} \right] e^{\left[-\frac{1}{2} \frac{(u\mu_Y - \mu_X)^2}{\left(\sigma_X^2 + \sigma_Y^2 u^2\right)} \right]},$$

$$-\infty < u < \infty \tag{4.19}$$

B. Kamerud (1978) has obtained a particular Cauchy-like distribution by considering the ratio X/Y of two non-centered independent normal random variables X and Y.

C. Marsaglia (1965) has investigated the ratio X/Y of two arbitrary normal random variables X and Y, and has also obtained a Cauchy-like distribution.

D. Mood et al. (1974, p. 246), Patel et al. (1976, p. 209): Let $X \sim N(0, 1)$ and $Y \sim N(0, 1)$ be independent standard normal random variables. Let $U = (X/Y)^2$. Then the random variable $U = (X/Y)^2$ has an F distribution with pdf $f_U(u, 1, 1)$.

4.5 Distributions of the Sum, Product and Ratio of Dependent Normal Variables

Since the distribution of the sums, differences, products and ratios (quotients) of random variables arise in many fields of research such as automation, biology, computer science, control theory, economics, engineering, fuzzy systems, genetics, hydrology, life testing, medicine, neuroscience, number theory, statistics, physics, psychology, queuing processes, reliability and risk management, among others, the derivations of these distributions for dependent (correlated) random variables have also received the attention of many authors and researchers. For detailed discussions on the sum, difference, product and ratio of dependent (correlated) random variables, the interested readers are referred to Springer (1979), and references therein.

In what follows, we provide the distributions of the sums, differences, products and ratios of dependent (correlated) normal variables.

4.5.1 Some Basic Definitions

For the sake of completeness, some basic definitions are given below.
Let X and Y be any two absolutely continuous random variables with p.d.f.'s $f_X(x)$ and $f_Y(y)$ respectively. Let $f_{X,Y}(x, y)$ be the joint p.d.f. of X and Y.

Definition 4.5.1. The random variables X and Y are said to **dependent** if and only if

$$f_{X,Y}(x, y) \neq f_X(x) . f_Y(y), \quad \forall (x, y) \in \Re^2, \Re = \{\text{all real numbers}\};$$

otherwise X and Y are said to be **independent**.

Definition 4.5.2. The correlation coefficient between the random variables X and Y, denoted by r_{XY}, is defined as

$$r_{XY} = \frac{\sigma_{XY}}{\sigma_X \sigma_Y},$$

where σ_{XY} denotes the covariance of X and Y, σ_X is the standard deviation of X, and σ_Y is the standard deviation of Y. The random variables X and Y are said to correlated if the correlation coefficient between them, that is, $r_{XY} \neq 0$; otherwise X and Y are said to be **uncorrelated**.

Remark 4.5.1. It can easily be seen that two independent random variables are uncorrelated. But the converse is not true, that is, two uncorrelated random variables may not be independent. For example, if a random variable X has a standard normal distribution and $Y = X^2$, then it is easy to see that X and Y are uncorrelated but not independent. For details on correlation and dependence (independence), the interested readers are referred to Lukacs (1972), Tsokos (1972), Springer (1979), Dudewicz and Mishra (1988), Rohatgi and Saleh (2001), Mari and Kotz (2001), and Gupta and Kapoor (2002), among others.

Definition 4.5.3. Let

$$Z = X + Y, U = X - Y, V = XY, \text{and } W = X/Y$$

denote the sum, difference, product, and ratio of the random variables X and Y respectively. Then, following Theorem 3, p. 139, of Rohatgi and Saleh (2001), the p.d.f.'s of Z, U, V, and W are, respectively, given by

$$f_Z(z) = \int_{-\infty}^{+\infty} f(x, z - x) dx,$$

$$f_U(u) = \int_{-\infty}^{+\infty} f(u + y,\ y)dy,$$

$$f_V(v) = \int_{-\infty}^{\infty} f\left(x,\ \frac{v}{x}\right)\frac{1}{|x|}dx,$$

and

$$f_W(w) = \int_{-\infty}^{\infty} f(xw,\ x)\,|x|\,dx.$$

4.5.2 Distributions of the Sums and Differences of Dependent (Correlated) Normal Random Variables

Here, the distributions of the sums and differences of dependent (correlated) normal random variables are briefly provided. For details, one is referred to Springer (1979, pp. 67–75), among others.

Let X_1 and X_2 denote the two dependent (correlated) normal random variables (r.v.'s) with zero mean, correlation coefficient ρ, and variances σ_1^2 and σ_2^2, respectively. Let $Z_S^* = X_1 + X_2$ denote the sum of X_1 and X_2, with the p.d.f. $g(z^*)$. Then, following Springer (1979), the joint p.d.f. of X_1 and X_2 is given by

$$f(x_1,\ x_2) = \frac{1}{2\pi\sigma_1\sigma_2(1-\rho^2)^{1/2}}\exp\left[-\frac{1}{2(1-\rho^2)}\left(\frac{x_1^2}{\sigma_1^2} - \frac{2\rho x_1 x_2}{\sigma_1\sigma_2} + \frac{x_2^2}{\sigma_2^2}\right)\right],$$
$$|\rho| < 1,\quad -\infty < x_i < \infty,\quad \sigma_i > 0,\quad i = 1,\ 2, \tag{4.20}$$

whereas the p.d.f of the sum $Z_S^* = X_1 + X_2$ is given by

$$g(z^*) = \frac{1}{\sqrt{2\pi}\sqrt{\sigma_1^2 + 2\rho\sigma_1\sigma_2 + \sigma_2^2}}\exp\left[-\frac{z^2}{2(\sigma_1^2 + 2\rho\sigma_1\sigma_2 + \sigma_2^2)}\right],$$
$$|\rho| < 1,\quad -\infty < z < \infty. \tag{4.21}$$

Obviously $Z_S^* \sim N\left(0, \sigma_1^2 + 2\rho\sigma_1\sigma_2 + \sigma_2^2\right)$. It follows that if $X_1 \sim N\left(\mu_1, \sigma_1^2\right)$ and $X_2 \sim N\left(\mu_2, \sigma_2^2\right)$, and $Z_S = X_1 + X_2$, then $Z_S \sim N\left(\mu_1 + \mu_2, \sigma_1^2 + 2\rho\sigma_1\sigma_2 + \sigma_2^2\right)$.

Remark 4.5.2. Similar to the above, it can easily be shown that if $X_1 \sim N\left(\mu_1, \sigma_1^2\right)$ and $X_2 \sim N\left(\mu_2, \sigma_2^2\right)$, and $Z_D = X_1 - X_2$, then $Z_D \sim N\left(\mu_1 - \mu_2, \sigma_1^2 - 2\rho\sigma_1\sigma_2 + \sigma_2^2\right)$.

Remark 4.5.3. It is interesting to note that the following transformation

$$y_1 = \frac{1}{\left(1 - \rho^2\right)^{1/2}} \left(\frac{x_1}{\sigma_1} - \frac{\rho x_2}{\sigma_2}\right), \text{ and } y_2 = \frac{x_2}{\sigma_2},$$

to the above joint p.d.f. of the two dependent (correlated) normal random variables X_1 and X_2 with zero mean, correlation coefficient ρ, and variances σ_1^2 and σ_2^2, respectively, gives

$$g\left(y_1, y_2\right) = \left(\frac{1}{\sqrt{2\pi}} e^{-y_1^2/2}\right) \left(\frac{1}{\sqrt{2\pi}} e^{-y_2^2/2}\right),$$

that is, the transformed random variables Y_1 and Y_2 are independent and standard normally distributed, but the original random variables X_1 and X_2 are dependent (correlated) normal random variables. For details, see Springer (1979).

Remark 4.5.4. For the p.d.f. of the sum of n dependent (correlated) normal r.v.'s, which can be obtained in the same manner, see Springer (1979, Eq. 3.4.20, p. 72).

Remark 4.5.5. However, in general, normality of marginal random variables X_1 and X_2 does not imply normality of their joint distribution and thus does not imply normality of their sum. For example, it has been observed by Holton (2003) and Novosyolov (2006) that the sum of dependent normal variables may be not normal.

4.5.3 Distributions of the Products and Ratios of Dependent (Correlated) Normal

Random Variables: In what follows, the distributions of the product and ratio of dependent (correlated) normal random variables are briefly provided. For details, one is referred to Springer (1979, p. 151), and references therein.

Let $X_1 \sim N(0, 1)$ and $X_2 \sim N(0, 1)$ denote the two dependent (correlated) normal random variables (r.v.'s) with zero mean, variances 1, and correlation coefficient ρ. Let $Y = X_1 X_2$ denote the product of X_1 and X_2, with the p.d.f. $h(y)$. Also, let $W = \frac{X_1}{X_2}$ denote the quotient of X_1 and X_2, with the p.d.f. $g(w)$. Then, following Springer (1979), the p.d.f. of the product $Y = X_1 X_2$ is given by

$$h(y) = \frac{1}{\pi\left(1 - \rho^2\right)^{1/2}} \exp\left[\frac{\rho\, y}{1 - \rho^2}\right] K_0\left[\frac{y}{1 - \rho^2}\right], \quad |\rho| < 1,$$
$$-\infty < y < \infty, \tag{4.22}$$

where $K_0(x)$ is the modified Bessel function of the second kind of order zero. The p.d.f of the quotient $W = \frac{X_1}{X_2}$ is given by

$$g(w) = \frac{\left(1 - \rho^2\right)^{1/2}}{\pi} \frac{1}{w^2 - 2\rho w + 1}, \quad |\rho| < 1, \quad -\infty < w < \infty.$$

Remark 4.5.6. For the p.d.f. of the quotient $W = X_1/X_2$, where $X_1 \sim N\left(\mu_1, \sigma_1^2\right)$ and $X_2 \sim N\left(\mu_2, \sigma_2^2\right)$ denote the two dependent (correlated) normal random variables (r.v.'s), with means μ_i, variances $\sigma_i^2 \quad (i = 1, 2)$, and correlation coefficient ρ, see Hinkley (1969).

4.6 Summary

First, this chapter presents some basic definitions and ideas on the sum, product, and ratio of two independent random variables. The distributions of the sum $X + Y$, product XY, and ratio, X/Y, when X and Y are independent random variables and have the normal distributions, have been reviewed in details. A short discussion, when X and Y are dependent or correlated, is also provided. The expressions for the pdfs as proposed by different authors are presented. By using Maple programs, various graphs have been plotted.

Chapter 5
Sum, Product and Ratio for the Student's *t* Random Variables

This chapter presents the distributions of the sum $X + Y$, product XY, and ratio X/Y when X and Y are independent random variables and have the Student's t distributions with appropriate degrees of freedoms.

5.1 Distribution of the Sum of Independent Student's *t* Random Variables

The distributions of the sum of independent Student's t random variables play an important role in many areas of research. The general theories of the distributions of linear combinations of independent Student's t random variables began with the work of Behrens (1929) and later by Fisher (1935). Further developments continued with the contributions of Bose and Roy (1938), Sukhatme (1938), Fisher (1939, 1941), Chapman (1950), Fisher and Healy (1956), Fisher and Yates (1957), James (1959), Ruben (1960), Patil (1965), Scheffe (1970), Ghosh (1975), Walker and Saw (1978), and recently by Fotowicz (2006) and Nadarajah (2006a). The distributions of the weighted sum of independent Student's t random variables in some cases are described below.

(i) Distribution of the weighted sum of two Student's t random variables: Fisher (1935), in his studies of the fiducial argument in statistical inference, developed a statistic, called the Behrens-Fisher statistic, in the form of a weighted sum $Z = \alpha X_m + \beta Y_n$ for the given Student's t random variables X_m and Y_n of degrees of freedom m and n, respectively. By approximating the integral of the fiducial density of Z through a Riemann sum, Sukhatme (1938) computed the percentage points of the distribution of Z. On the other hand, Chapman (1950) considered the case when $\alpha = \beta = 1$ and $m = n = \nu$ (say), and derived an expression for the density of Z for odd values of ν and computed the values of the cdf of Z through numerical integration.

M. Ahsanullah et al., *Normal and Student's t Distributions and Their Applications*, Atlantis Studies in Probability and Statistics 4, DOI: 10.2991/978-94-6239-061-4_5,

(ii) Distribution of the difference (or sum) of two Student's t random variables: Let X_m and Y_n be two independent Student's t random variables with pdfs $f_X(x)$ and $f_Y(y)$, and degrees of freedom $m > 0$ and $n > 0$, respectively. Then, for $m = n > 0$, Ghosh (1975) derived the pdf of the random variable $Z = X_m - Y_n$ given by

$$f_z(z) = \int_{-\infty}^{+\infty} f_Y(y) f_Y(z + y)\, dy = \frac{\Gamma\left(\frac{1}{2}n+\frac{1}{2}\right)\Gamma\left(n+\frac{1}{2}\right)}{2^n \sqrt{n}\Gamma^2\left(\frac{1}{2}n\right)\Gamma\left(\frac{1}{2}n+1\right)} \left(\frac{4n}{4n+z^2}\right)^{n+1} \times$$
$$\times {}_2F_1\left(\frac{1}{2}, n+\frac{1}{2}; \frac{1}{2}n+1; \frac{z^2}{4n+z^2}\right), \quad -\infty < z < \infty \tag{5.1}$$

where ${}_2F_1(.)$ denotes Gauss hypergeometric function. Further, when $m \neq n$, Ghosh (1975) also derived an approximation of the distribution of $Z = X_m - Y_n$ by using a series expansion. The results of Ghosh (1975) are applicable to the sum $Z = X_m + Y_n$ as well, since $X_m - Y_n$ and $X_m + Y_n$ are identically distributed.

(iii) **Distribution of the linear combinations of Student's t random variables:** Walker and Saw (1978) have considered the distribution of a linear combination of independent Student's t random variables $X_1, X_2, \ldots, X_n$, with degrees of freedom $v_1, v_2, \ldots, v_n$, respectively, given by

$$Z^* = c_1 X_1 + c_2 X_2 + \cdots + c_n X_n.$$

where $c_1, c_2, \ldots, c_n$ denote the real arbitrary coefficients. For odd degrees of freedom, they derived the cdf of Z^* as a mixture of Student's t distributions, which enables one to calculate the percentage points using only tables of the t distribution. If $\nu_i = 1$ for $i = 1, 2, \ldots, n$, the random variable Z^* will have the Cauchy distribution. Further, Z^* will have the normal distribution if $\nu_i \to \infty \;\; \forall i$. The distribution of Z^* does not have a closed form when $1 < \nu_i < \infty$, for some i, (for details, see Walker and Saw (1978), among others).

(iv) Special cases:
(a) Let X and Y be two independent standard Student's t random variables each with 1 degree of freedom, and pdfs $f_X(x)$ and $f_Y(y)$, respectively. Let $Z = X + Y$. Then, the pdf of the random variable Z given by

$$f_z(z) = \int_{-\infty}^{+\infty} f_X(x) f_Y(z - x)\, dx = \frac{2}{\pi(4 + z^2)}, \quad -\infty < z < \infty, \tag{5.2}$$

which is also the pdf of the sum of two standard Cauchy random variables $X \sim C(0, 1)$ and $Y \sim C(0, 1)$, (for details, see Kapadia et al. (2005), among others). The above expression is also easily obtained by substituting $n = 1$ in the Eq. (5.1) above, and noting that ${}_2F_1(\alpha, \beta; \beta; t) = (1 - t)^{-\alpha}$, (for details, see Abramowitz and Stegun (1970), Lebedev (1972), Prudnikov et al. (1986), and Gradshteyn and Ryzhik (2000), among others). A Maple plot of the above pdf is given in Fig. 5.1.

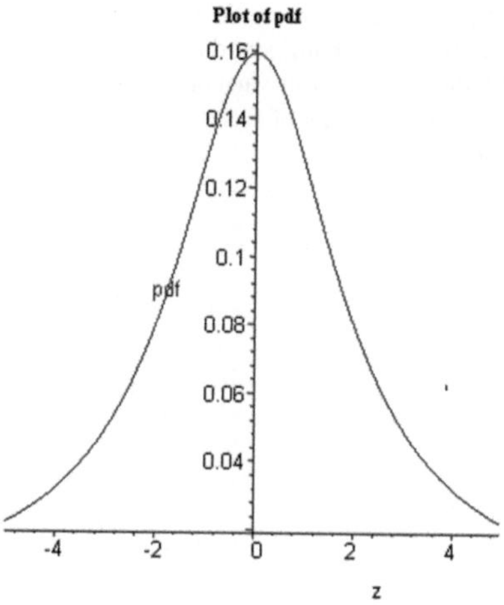

Fig. 5.1 Plot of $Z = X + Y$, where X and Y are standard Student's t random variables having 1 degree of freedom each

(b) The general form of the pdf of the sum of n independent Student's t random variables $X_1, X_2, \ldots, X_n$, each having 1 degree of freedom, is given by

$$f_z(z) = \frac{n}{\pi(n^2 + z^2)}, \quad -\infty < z < \infty, \tag{5.3}$$

which is also the pdf of the sum of n standard Cauchy random variables, (for details, see Kapadia et al. (2005), among others).

(c) Let X and Y be two independent standard Student's t random variables with 2 degrees of freedom, and pdfs $f_X(x)$ and $f_Y(y)$, respectively. Let $Z = X + Y$. Then, by substituting $n = 2$ in the expression for pdf in Eq. (5.1), and using the equation (7.3.2.106/p 474) from Prudnikov et al. volume 3 (1986), the pdf of the random variable Z is obtained as follows

$$f_z(z) = 2\left[\left(8 + z^2\right) \mathrm{K}\left(\frac{z}{\sqrt{8 + z^2}}\right) + \left(8 - z^2\right) D\left(\frac{z}{\sqrt{8 + z^2}}\right)\right], -\infty < z < \infty, \tag{5.4}$$

where $D(.)$ and $\mathrm{K}(.)$ denote the complete elliptic integrals.

(d) Let X and Y be two independent standard Student's t random variables with 3 degrees of freedom, and pdfs $f_X(x)$ and $f_Y(y)$, respectively. Let $Z = X + Y$. Then, by substituting $n = 3$ in the expression for pdf in (i) above as derived by Ghosh (1975), and using the equation (7.3.2.121/p 475) from Prudnikov et al. volume 3 (1986), the pdf of the random variable Z is obtained as follows

$$f_z(z) = \frac{12\sqrt{3}\left(60 + z^2\right)}{\pi\left(12 + z^2\right)^3}, \quad -\infty < z < \infty \tag{5.5}$$

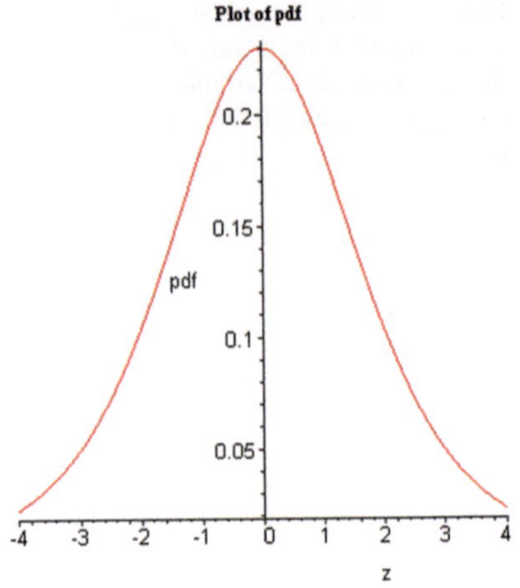

Fig. 5.2 Plot of $Z = X + Y$, where X and Y are standard Student's t random variables having 3 degrees of freedom each

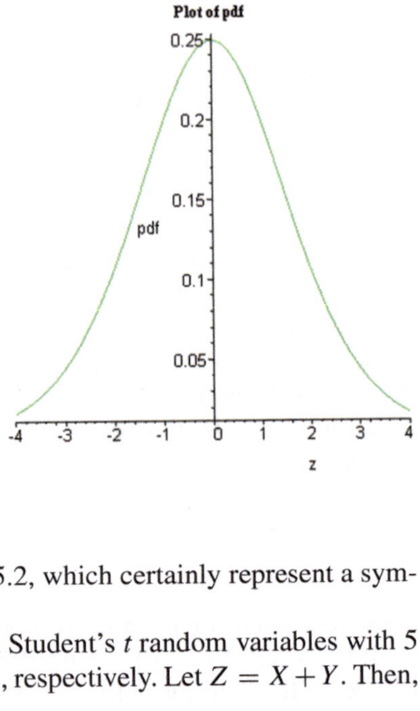

Fig. 5.3 Plot of $Z = X + Y$, where X and Y are standard Student's t random variables with 5 degrees of freedom

A Maple plot of the above pdf is given in Fig. 5.2, which certainly represent a symmetric distribution.

(e) Let X and Y be two independent standard Student's t random variables with 5 degrees of freedom, and pdfs $f_X(x)$ and $f_Y(y)$, respectively. Let $Z = X + Y$. Then, the pdf of the random variable Z is given by

$$f_Z(z) = \frac{400\sqrt{5}\left(8400 + 120z^2 + z^4\right)}{3\pi\left(20 + z^2\right)^5}, \quad -\infty < z < \infty,$$

(for details, see Casellla and Berger (2002)). A Maple plot of the above pdf is given in Fig. 5.3, which is a form of symmetric distribution.

(v) Computation of the Coverage Interval:
Recently, the computation of the coverage interval for the convolution of two independent Student's *t* random variables has been investigated by Fotowicz (2006), and Nadarajah (2006a), which is defined as follows.

Let X_m and Y_n be two independent Student's *t* random variables with pdfs $f_X(x)$ and $f_Y(y)$, and degrees of freedom $m > 0$ and $n > 0$, respectively. Let Z denote the convolution $Z = X_m + Y_n$. Obviously, Z is a symmetric random variable. Let the pdf and cdf of Z be denoted by $g(\cdot)$ and $G(\cdot)$ respectively. Then, the coverage interval of $Z = X_m + Y_n$ corresponding to coverage probability $p = 1 - \alpha$ is defined by

$$[z_{low}, z_{high}],$$

where $z_{low} = G^{-1}\left(\frac{\alpha}{2}\right)$ and $z_{high} = G^{-1}\left(1 - \frac{\alpha}{2}\right)$. For details of the methods for the computation of the coverage interval for the convolution of two independent Student's *t* random variables, see Fotowicz (2006), and Nadarajah (2006a).

5.2 Distribution of the Product of Independent Student's *t* Random Variables

This section discusses the distribution of the product of independent Student's *t* random variables. The distribution of the product independent random variables is one of the important research topics both from theoretical and applications point of view. It arises in many applied problems of biology, economics, engineering, genetics, hydrology, medicine, number theory, order statistics, physics, psychology, etc, (see, for example, Cigizoglu and Bayazit (2000), Frisch and Sornette (1997), Galambos and Simonelli (2005), Grubel (1968), Ladekarl et al. (1997), Rathie and Rohrer (1987), Rokeach and Kliejunas (1972), Springer (1979), and Sornette (1998, 2004), among others). The distributions of the product $Z = XY$, when X and Y are independent random variables and belong to the same family, have been studied by many authors, (see, for example, Sakamoto (1943) for the uniform family, Springer and Thompson (1970) for the normal family, Stuart (1962) and Podolski (1972) for the gamma family, Steece (1976), Bhargava and Khatri (1981) and Tang and Gupta (1984) for the beta family, AbuSalih (1983) for the power function family, and Malik and Trudel (1986) for the exponential family, among others). The distribution of the product of two correlated *t* variates has been considered by Wallgren (1980). Recently, Nadarajah and Dey (2006) have studied the distribution of the product $Z = XY$ of two independent Student's *t* random variables X and Y. It appears from the literature that not much study has been done for the distributions of the product of independent Student's *t* random variables and applications, which need further investigation. Following Nadarajah and Dey (2006), the distributions of the product of independent Student's *t* random variables are summarized below. For details, see Nadarajah and Dey (2006).

(i) Distribution of the product of two independent Student's t random variables:
(a) Let X_m and Y_n be two independent Student's t random variables with pdfs $f_X(x)$ and $f_Y(y)$, and degrees of freedom $m > 0$ and $n > 0$, respectively. Then, as derived by Nadarajah and Dey (2006), the cdf of the random variable $Z = X_m Y_n$, for m odd, is given by

$$F(z) = \left(\frac{4}{\pi\sqrt{n}B\left(\frac{n}{2},\frac{1}{2}\right)}\right)\int_0^\infty \tan^{-1}\left(\frac{z}{\sqrt{m}y}\right)\left(1+\frac{y^2}{n}\right)^{-\frac{1}{2}-\frac{n}{2}} dy + \left(\frac{z}{\pi\sqrt{mn}B\left(\frac{n}{2},\frac{1}{2}\right)}\right)$$
$$\times \sum_{k=1}^{(m-1)/2}\left[B\left(\tfrac{1+n}{2},k\right)B\left(k,\tfrac{1}{2}\right){}_2F_1\left(k,\tfrac{1+n}{2};k+\tfrac{1+n}{2};1-\tfrac{z^2}{mn}\right)\right], \tag{5.6}$$

where $B(\cdot)$ and ${}_2F_1(\cdot)$ denotes the beta and Gauss hypergeometric functions respectively.

(b) Let X_m and Y_n be two independent Student's t random variables with pdfs $f_X(x)$ and $f_Y(y)$, and degrees of freedom $m > 0$ and $n > 0$, respectively. Then, as derived by Nadarajah and Dey (2006), the cdf of the random variable $Z = X_m Y_n$, for m even, is given by

$$F(z) = \left(\frac{z}{\pi\sqrt{mn}B\left(\frac{n}{2},\frac{1}{2}\right)}\right)$$
$$\times\sum_{k=1}^{m/2}\left[B\left(\frac{1+n}{2},k-\frac{1}{2}\right)B\left(k-\frac{1}{2},\frac{1}{2}\right){}_2F_1\left(k-\frac{1}{2},\frac{1+n}{2};k+\frac{n}{2};1-\frac{z^2}{mn}\right)\right], \tag{5.7}$$

where $B(\cdot)$ and ${}_2F_1(\cdot)$ denotes the beta and Gauss hypergeometric functions respectively.

(c) For the cdf of the random variable $Z = X_m Y_n$ for particular cases of the degrees of freedom $m = 2, 3, 4, 5$ and $n = 1, 2, 3, 4, 5$, see Nadarajah and Dey (2006).

(d) The possible shapes of the pdf of $Z = |X_m Y_n|$ for a range of values of degrees of freedom $m > 0$ and $n > 0$ are illustrated in Fig. 5.4a, b. We observe that the shapes of the pdf of $Z = X_m Y_n$ are unimodal. The effects of the parameters can easily be seen from these graphs.

(ii) Special Cases:
(a) Let X and Y be two independent Student's t random variables with 1 degree of freedom, and pdf $f_X(x)$ and pdf $f_Y(y)$, respectively. Then, the pdf of the random variable $Z = XY$ is given by

$$g_z(z) = \frac{2\,\mathrm{In}|z|}{\pi^2(z^2-1)} \quad -\infty < z < \infty,$$

(for details, see Rider (1965), Springer and Thompson (1966), and Springer (1979)). A Maple plot of the above pdf is given in Fig. 5.5.

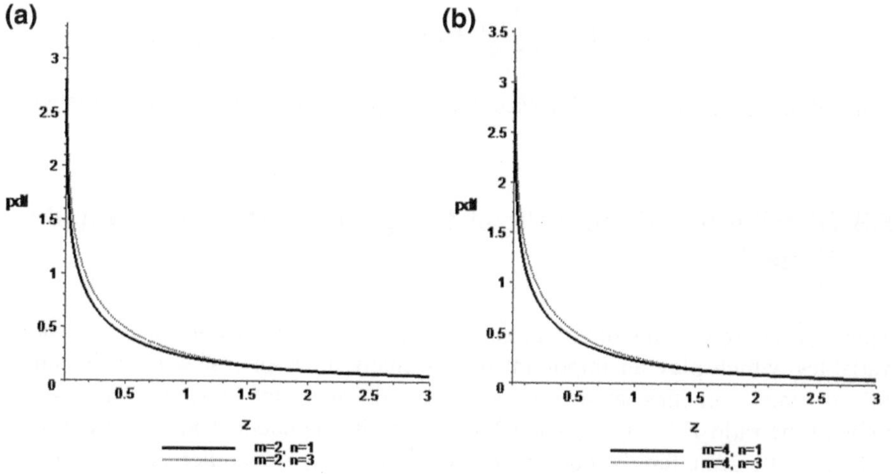

Fig. 5.4 Plots of the pdf of $Z = |X_m Y_n|$ for (**a**) $m = 2$ and $n = 1, 3$ and (**b**) $m = 4$ and $n = 1, 3$

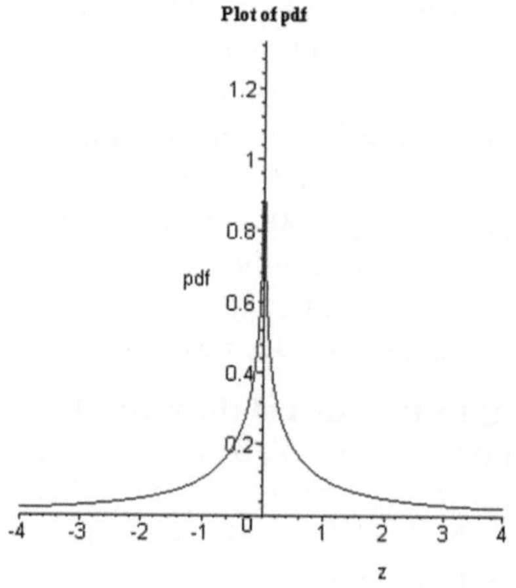

Fig. 5.5 Plot of the pdf of $Z = XY$, where and are standard Student's t random variables with 1 degree of freedom

(b) Let X be a Student's t random variables with pdf $f_X(x)$ and degrees of freedom $\nu > 0$. Then, the pdf of the random variable $Z = X^2$ is given by

$$g_Z(z) = \frac{\Gamma\left(\frac{\nu+1}{2}\right)}{\sqrt{\nu\pi}\,\Gamma\left(\frac{\nu}{2}\right)} \left(1 + \frac{z}{2}\right)^{-\left(\frac{1+\nu}{2}\right)} z^{-\frac{1}{2}}, z > 0, \nu > 0, \tag{5.8}$$

which defines a $F_{1,\nu}$ distribution (see, for example, Lukacs (1972) and Patel and Read (1982), among others).

(c) Let X be a Student's t random variables with pdf $f_X(x)$ and 1 degree of freedom. Then, the pdf of the random variable $Z = X^2$ is given by

$$g_Z(z) = \frac{1}{\pi}(1+z)^{-1} z^{-\frac{1}{2}}, z > 0$$

which defines a $F_{1,1}$ distribution (see, for example, Lukacs (1972), among others).

5.3 Distribution of the Ratio of Independent Student's *t* Random Variables

This section presents the distributions of the ratio of independent Student's random variables, which play an important role in many areas of research, for example, in economics, genetics, meteorology, nuclear physics, statistics, etc. The distributions of the ratio $Z = X/Y$, when X and Y are independent random variables and belong to the same family, have been studied by many researchers, (see, for example, Marsaglia (1965) and Korhonen and Narula (1989) for the normal family, Basu and Lochner (1971) for the Weibull family, Shcolnick (1985) for the stable family, Hawkins and Han (1986) for the non-central chi-squared family, Provost (1989b) for the gamma family, and Pham-Gia (2000) for the beta family, among others). Recently, Nadarajah and Dey (2006) have studied the distributions of the ratio $Z = X/Y$ of two independent Student's t random variables X and Y. It appears from the literature that not much attention has been paid to the study of the distributions of the ratio of independent Student's t random variables and applications, except Nadarajah and Dey (2006), which need further investigation. For the sake of completeness of this project, following Nadarajah and Dey (2006), the distributions of the ratio of independent Student's t random variables are summarized below. For details, see Nadarajah and Dey (2006).

(i) Distribution of the ratio of two independent Student's *t* random variables:
(a) Let X_m and Y_n be two independent Student's t random variables with pdfs $f_X(x)$ and $f_Y(y)$, and degrees of freedom $m > 0$ and $n > 0$, respectively. Then, as derived by Nadarajah and Dey (2006), the cdf of the random variable $Z = X_m/Y_n$, for m odd, is given by

$$F(z) = \left(\frac{4}{\pi\sqrt{n}B\left(\frac{n}{2},\frac{1}{2}\right)}\right)\int_0^{\infty}\tan^{-1}\left(\frac{zy}{\sqrt{m}}\right)\left(1+\frac{y^2}{n}\right)^{-\frac{1}{2}-\frac{n}{2}}dy + \left(\frac{2\sqrt{n}z}{\pi\sqrt{m}B\left(\frac{n}{2},\frac{1}{2}\right)}\right) \times \sum_{k=1}^{(m-1)/2}\left[\frac{m^k B(k,\frac{1}{2})}{z^{2k}n^k(2k+n-1)}\,{}_2F_1\left(k, k+\frac{n-1}{2}; k+\frac{1+n}{2}; 1-\frac{m}{nz^2}\right)\right], \tag{5.9}$$

where $B(\cdot)$ and ${}_2F_1(\cdot)$ denotes the beta and Gauss hypergeometric functions respectively.

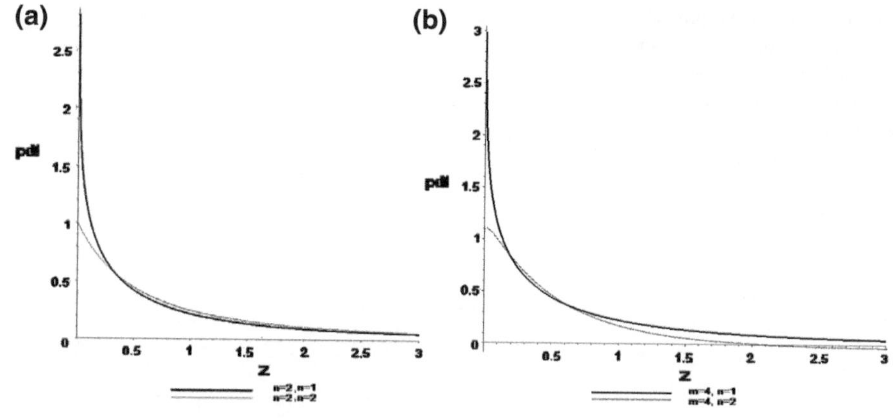

Fig. 5.6 Plots of the pdf of $Z = \left|\frac{X_m}{Y_n}\right|$ for (**a**) $m = 2$ and $n = 1, 2$ and (**b**) $m = 4$, and $n = 1, 2$

(b) Let X_m and Y_n be two independent Student's t random variables with pdfs $f_X(x)$ and $f_Y(y)$, and degrees of freedom $m > 0$ and $n > 0$, respectively. Then, as derived by Nadarajah and Dey (2006), the cdf of the random variable $Z = X_m/Y_n$, for m even, is given by

$$F(z) = \left(\frac{2\sqrt{n}z}{\pi\sqrt{m}B\left(\frac{n}{2}, \frac{1}{2}\right)}\right) \times \sum_{k=1}^{(m/2)} \left[\frac{m^{k-\frac{1}{2}}B\left(k-\frac{1}{2}, \frac{1}{2}\right)}{z^{2k-1}n^{k-\frac{1}{2}}(2k-2+n)} {}_2F_1\left(k-\frac{1}{2}, k+\frac{n}{2}-1; k+\frac{n}{2}; 1-\frac{m}{nz^2}\right)\right], \tag{5.10}$$

where $B(\cdot)$ and ${}_2F_1(\cdot)$ denotes the beta and Gauss hypergeometric functions respectively.

(c) For the cdf of the random variable $Z = X_m/Y_n$ for particular cases of the degrees of freedom $m = 2, 3, 4, 5$ and $n = 1, 2, 3, 4, 5$ see Nadarajah and Dey (2006).

(d) The possible shapes of the pdf of $Z = \left|\frac{X_m}{Y_n}\right|$ for a range of values of degrees of freedom and are illustrated in Fig. 5.6a, b. We observe that the shapes of $Z = \left|\frac{X_m}{Y_n}\right|$ are unimodal. The effects of the parameter can easily be seen from these graphs.

(ii) Special Case: Let X and Y be two independent Student's t random variables with 1 degree of freedom, and pdfs $f_X(x)$ and $f_Y(y)$, respectively. Then, the pdf of the random variable $Z = |X/Y|$ is given by

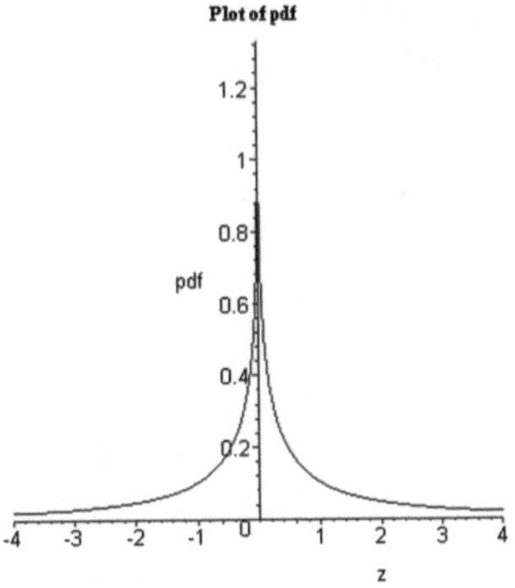

Fig. 5.7 Plot of the pdf of $Z = X/Y$, where X and Y are standard Student's t random variables with 1 degree of freedom

$$g_Z(z) = \frac{2\ln|z|}{\pi^2(z^2-1)} \quad -\infty < z < \infty,$$

which is identical to the pdf of the product random variable $Z = |XY|$ of two independent Student's t random variables X and Y as discussed above, (for details, see Rider (1965), Springer and Thompson (1966), and Springer (1979)). A Maple plot of the above pdf is given in Fig. 5.7.

Since the Cauchy distribution with median zero (standard) the pdf of X and 1/X are identical, the pdfs of product and the ratio of two standard Cauchy distribution are identical.

5.4 Distributions of the Sums, Differences, Products and Ratios of Dependent (Correlated) Student *t* Random Variables

It appears from literature that not much attention has been paid to the distributions of the sums, differences, products and ratios of dependent (correlated) Student's t random variables, and therefore needs further research investigation.

5.5 Summary

This chapter has reviewed the distributions of the sum $X + Y$, product XY, and ratio X/Y, when X and Y are independent random variables and have the Student's t distributions. The expressions for the pdfs as proposed by different authors are presented. Some special cases of the sum, product and ratio distributions are given. By using Maple programs, various graphs have been plotted.

Chapter 6
Sum, Product and Ratio for the Normal and Student's *t* Random Variables

The distributions of the sum $X + Y$, product XY, and ratio X/Y, when X and Y are independent random variables and belong to different families, are of considerable importance and current interest. These have been recently studied by many researchers, (among them, Nadarajah (2005b, c, d) for the linear combination, product and ratio of normal and logistic random variables, Nadarajah and Kotz (2005c) for the linear combination of exponential and gamma random variables, Nadarajah and Kotz (2006d) for the linear combination of logistic and Gumbel random variables, Nadarajah and Kibria (2006) for the linear combination of exponential and Rayleigh random variables, Nadarajah and Ali (2004) for the distributions of the product XY when X and Y are independent Laplace and Bessel random variables respectively, Ali and Nadarajah (2004) for the product and the ratio of t and logistic random variables, Ali and Nadarajah (2005) for the product and ratio of t and Laplace random variables, Nadarajah and Kotz (2005b) for the ratio of Pearson type VII and Bessel random variables, Nadarajah (2005c) for the product and ratio of Laplace and Bessel random variables, Nadarajah and Ali (2005) for the distributions of XY and X/Y, when X and Y are independent Student's and Laplace random variables respectively, Nadarajah and Kotz (2005a) for the product and ratio of Pearson type VII and Laplace random variables, Nadarajah and Kotz (2006a) for the product and ratio of gamma and Weibull random variables, Shakil, Kibria and Singh (2006) for the ratio of Maxwell and Rice random variables, Shakil and Kibria (2007) for the ratio of Gamma and Rayleigh random variables, Shakil, Kibria and Chang (2007) for the product and ratio of Maxwell and Rayleigh random variables, and Shakil and Kibria (2007) for the product of Maxwell and Rice random variables, are notable). This chapter studies the distributions of the sum $X + Y$, product XY, and ratio X/Y, when X and Y are independent normal and Student's t random variables respectively.

M. Ahsanullah et al., *Normal and Student's t Distributions and Their Applications*,
Atlantis Studies in Probability and Statistics 4, DOI: 10.2991/978-94-6239-061-4_6,

6.1 Distribution of the Sum of Normal and Student's *t* Random Variables

This section discusses the distributions of the sum of the independent normal and Student's *t* random variables. The distributions of the sum of the normal and Student's *t* random variables were first studied by Kendall (1938) who formulated the pdf of the random variable $Z = X + Y$, for 1 degree of freedom. Recently, Nason (2006) has studied the distributions of the sum $X+Y$, when X and Y are independent normal and sphered Student's *t* random variables respectively. It appears from the literature that not much attention has been paid to the distributions of the sum of the normal and Student's *t* random variables. This chapter introduces and develops some new results on the pdfs for the sum of the normal and Student's *t* random variables, which have been independently derived here.

6.1.1 Kendall's Pdf for the Sum X + Y

Let $X \sim N(\mu, \sigma^2)$ be a normal random variable with pdf $f_X(x)$. Let Y be a Student's *t* random variable with degrees of freedom $\nu = 1$, and pdf $f_Y(y)$. Let $Z = X + Y$ Then, the pdf of the random variable Z, when $\nu = 1$, as derived by Kendall (1938), is given by

$$f_z(z) = \frac{\sqrt{2}}{\pi\, \sigma} \operatorname{Re}\left\{e^{z^2} erfc(z)\right\}, \tag{6.1}$$

where $z = d - \frac{ip}{2}, d = \frac{1}{\sqrt{2}\sigma}, p = \frac{\sqrt{2}\mu}{\sigma}$, $erfc\,(.)$, denotes complementary error function, and means 'take the real part'.

6.1.2 Nason's Pdfs for the Sum X + Y, Based on Sphered Student's t Density

Recently, Nason (2006) has studied the distributions of the sum $X + Y$, when X and Y are independent normal and sphered student's *t* random variables respectively. The pdf of a sphered student's *t* random variable is defined as.

Definition (Sphered Student's *t* density): The sphered Student's *t*-density on $\nu \geq 3$ degrees of freedom is defined by $t_\nu : \Re \to (0, \infty)$ such that

$$t_\nu(x) = \frac{\Gamma\left(\frac{\nu}{2} + \frac{1}{2}\right)}{\sqrt{\pi(\nu - 2)}\Gamma\left(\frac{\nu}{2}\right)} \left[1 + \frac{x^2}{\nu - 2}\right]^{\frac{-(\nu+1)}{2}}, \quad -\infty < x < \infty, \nu > 0 \tag{6.2}$$

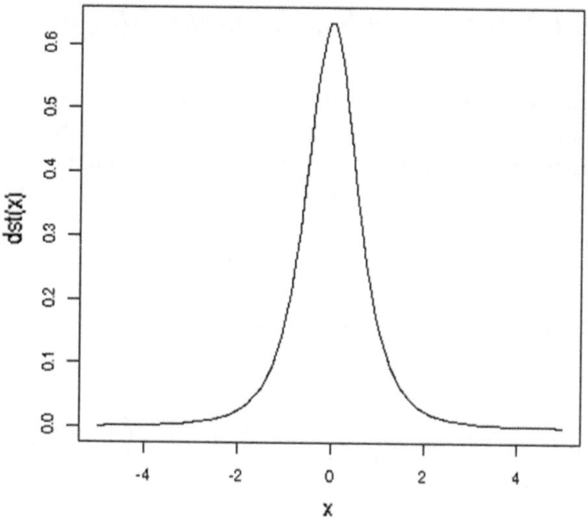

Fig. 6.1 Plot of the pdf of sphered student's t for degrees of freedom $\nu = 3$

Note that the sphered Student's t density is the standard Student's t distribution rescaled to have unit variance. For a multivariate version of the sphered Student's t distribution, see, for example, Krzanowski and Marriott (1994), and Nason (2001), among others. The possible shape of the above sphered Student's t density for degrees of freedom $\nu = 3$ is illustrated in Fig. 6.1, which is reproduced from Nason (2006).

Nason (2006) has studied the distributions of the sum of normal and sphered student's t random variables. Let $X \sim N(\mu, \sigma^2)$ be a normal random variable with pdf $\varphi_{\mu,\,\sigma}(\mathrm{x})$. Let T_ν be a random variable distributed according to sphered Student's t-distribution on degrees of freedom, with pdf $h_\nu(t)$. Let $\mathrm{Y} = \mathrm{X} + \mathrm{T}_\nu$. Then, the pdf of the random variable Y can be represented as the convolution of the density functions of X and T_ν , as follows:

$$f_Y(y) = \int_{-\infty}^{+\infty} \varphi_{\mu,\sigma}(y-t)h_\nu(t)dt = \int_{-\infty}^{+\infty} \varphi_{\mu+y,\sigma}(t)h_\nu(t)dt = \left\langle \varphi_{\mu+y,\,\sigma},\, h_\nu \right\rangle, \quad (6.3)$$

where $\langle , \rangle$ is the usual inner product $\langle f, g\rangle = \int_{-\infty}^{+\infty} f(u)g(u)du$. Then, Nason has derived some nice results for the inner products of $h_\nu(t)$ with $\varphi_{\mu,\sigma}(x)$ for both cases when $\nu = 3$ and $\nu > 3$. For details of these results, see Nason (2006).

As pointed out by Nason (2006), some of the applications and importance of these results are in Bayesian wavelet shrinkage, Bayesian posterior density derivations, calculations in the theoretical analysis of projection indices, and computation of certain moments.

6.1.3 Some New Results on the Pdfs for the Sum X + Y

In what follows, we introduce and develop some new results on the pdfs for the sum of the normal and Student's t random variables. Let $X \sim N\left(0, \sigma^2\right)$ be a normal random variable with pdf $f_X(x)$. Let Y be a Student's t random variable with ν degrees of freedom, and pdf $f_Y(y)$. Let $Z = X + Y$. Then, the pdf of the random variable Z is given by

$$
\begin{aligned}
f_Z(z) &= \int_{-\infty}^{+\infty} f_X(z-t) f_Y(t) dt \\
&= \frac{1}{\sqrt{2\pi\nu}\,\sigma B\left(\frac{\nu}{2}, \frac{1}{2}\right)} \int_{-\infty}^{\infty} e^{-\frac{(z-t)^2}{2\sigma^2}} \left(1 + \frac{t^2}{\nu}\right)^{\frac{-(1+\nu)}{2}} dt \\
&= \frac{2\,e^{-\frac{z^2}{2\sigma^2}}}{\sqrt{2\pi\nu}\,\sigma B\left(\frac{\nu}{2}, \frac{1}{2}\right)} \int_{0}^{\infty} e^{-\frac{t^2}{2\sigma^2}} \cosh\left(\frac{zt}{\sigma^2}\right) \left(1 + \frac{t^2}{\nu}\right)^{\frac{-(1+\nu)}{2}} dt \\
&= \frac{2\,\nu^{\frac{\nu}{2}} e^{-\frac{z^2}{2\sigma^2}}}{\sqrt{2\pi\nu}\,\sigma B\left(\frac{\nu}{2}, \frac{1}{2}\right)} \sum_{n=0}^{\infty} \frac{1}{n!} \left(\frac{z^2}{2\sigma^4}\right)^n \int_{0}^{\infty} t^{2n} e^{-\frac{t^2}{2\sigma^2}} \left(t^2 + \nu\right)^{\frac{-(1+\nu)}{2}} dt \qquad (6.4)
\end{aligned}
$$

Substituting $t^2 = u$ in the above integral, and using the Eq. (2.3.6.9), We obtain,

$$
\int_{0}^{\infty} w^{\alpha-1} e^{-pw} (w+\kappa)^{-\rho}\, dw = \Gamma(\alpha) \kappa^{\alpha-\rho} \psi(\alpha, \alpha+1-\rho; p\kappa)
$$

from Prudnikov et al. (1986, volume 1), the above expression for the pdf reduces to:

$$
f_Z(z) = \frac{\nu^{\frac{\nu}{2}} e^{-\frac{z^2}{2\sigma^2}}}{\sqrt{2\pi}\,\sigma B\left(\frac{\nu}{2}, \frac{1}{2}\right)} \sum_{n=0}^{\infty} \frac{1}{n!} \left(\frac{z^2}{2\sigma^4}\right)^n \Gamma\left(n + \frac{1}{2}\right) \nu^{n-\frac{\nu}{2}} \psi\left(n + \frac{1}{2}, n + 1 - \frac{\nu}{2}; \frac{\nu}{2\sigma^2}\right),
$$

where $|z| < \infty$, $\nu > 0$, $\sigma > 0$ and $\psi(.)$ denotes Kummer's hypergeometric function. In view of the fact that $|z| < \infty$, ignoring all the terms after the first term in the above series, an approximate pdf of the random variable $Z = X + Y$ is easily obtained as

$$
f_Z(z) = \frac{1}{\sqrt{2}\,\sigma B\left(\frac{\nu}{2}, \frac{1}{2}\right)} \psi\left(\frac{1}{2}, 1 - \frac{\nu}{2}; \frac{\nu}{2\sigma^2}\right) e^{-\frac{z^2}{2\sigma^2}} \qquad (6.5)
$$

Substituting $\nu = 1$, and using the Eq. (13.6.39) from Abramowitz and Stegun (1972), the above expression for the pdf easily reduces to:

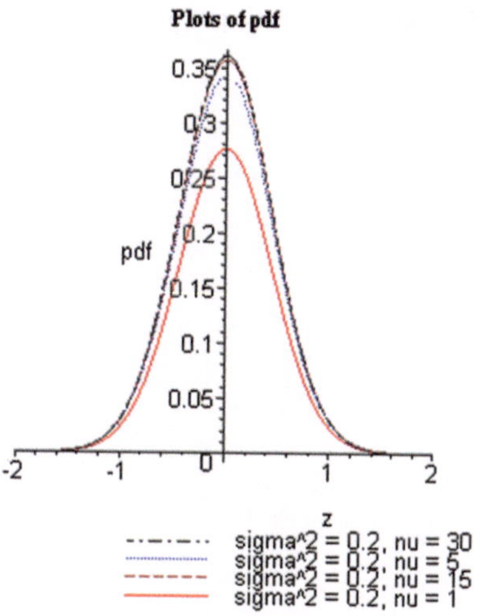

Fig. 6.2 Plots of the pdf of $Z = X + Y$

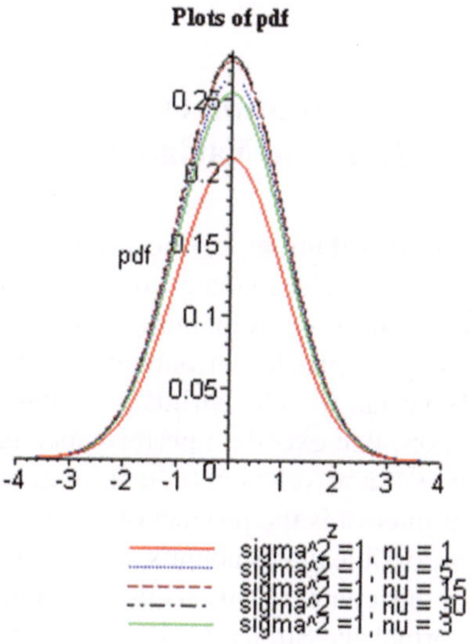

Fig. 6.3 Plots of the pdf of $Z = X + Y$

$$f_Z(z) = \frac{1}{\sqrt{2\pi}\,\sigma} e^{\frac{1}{2\sigma^2}} erfc\left(\frac{1}{\sqrt{2}\,\sigma}\right) e^{-\frac{z^2}{2\sigma^2}},$$

where *erfc* (.) denotes complementary error function. Using Maple, the possible shapes of the above approximate pdf of $Z = X + Y$ for a range of values of $\sigma > 0$ and degrees of freedom $\nu > 0$ are illustrated in Figs. 6.2, 6.3, and 6.4.

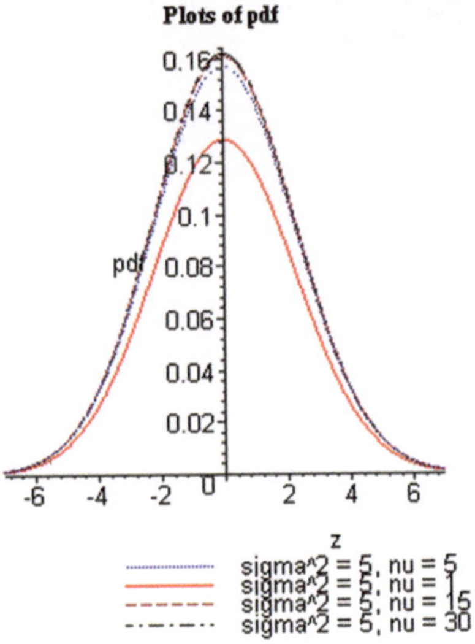

Fig. 6.4 Plots of the pdf of $Z = X + Y$

6.2 Distribution of the Product of Normal and Student's *t* Random Variables

The distributions of the product of independent normal and Student's *t* random variables arise in many areas of engineering, medicine, science, and statistics. These distributions play an important role in many areas of statistics, for example, Bayesian analysis, projection pursuit, and wavelet shrinkage, to mention a few. In case of Bayesian wavelet shrinkage, it has been shown by Johnstone and Silverman (2004, 2005) that excellent performance is obtained by using heavy-tailed distributions as part of a wavelet coefficient mixture prior instead of the standard normal. A quantity of interest is the product of the heavy-tailed distribution with the standard normal. It is possible that Student's *t* distribution might also be an interesting distribution to use in this context. This section presents the distributions of the product of independent normal and Student's *t* random variables, independently derived by the author. These results are believed to be new.

6.2.1 Derivation of the Pdf for the Product of Normal and Student's t Random Variables

Let $X \sim N\left(0, \sigma^2\right)$ be a normal random variable with pdf $f_X(x)$. Let Y be a Student's t random variable with ν degrees of freedom, and pdf $f_Y(y)$. Let $Z = XY$. Then, the pdf of the random variable Z is given by

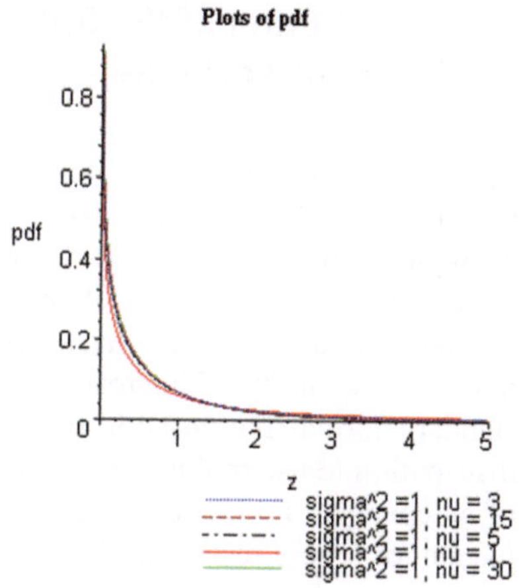

Fig. 6.5 Plots of the pdf of $Z = |XY|$

$$f_Z(z) = \sqrt{\frac{2}{\pi \nu}} \frac{1}{B\left(\frac{\nu}{2}, \frac{1}{2}\right)} \int_0^{\infty} \frac{1}{y} e^{-\frac{z^2 y^2}{2\sigma^2}} \left(1 + \frac{y^2}{\nu}\right)^{-\frac{\nu+1}{2}} dy \tag{6.6}$$

Substituting $\frac{1}{y^2} = t$, and using following result

$$\int_0^{\infty} w^{\alpha-1} e^{-pw} (w + \kappa)^{-\rho} dw = \Gamma(\alpha) \kappa^{\alpha-\rho} \psi(\alpha, \alpha + 1 - \rho; p\kappa),$$

in the Eq. (6.6), (see the Eq. (2.3.6.9), p. 324, Prudnikov et al. (1986, volume 1)), the pdf of the random variable $Z = XY$ reduces to

$$f_Z(z) = \sqrt{\frac{1}{2\nu}} \frac{\Gamma\left(\frac{\nu}{2}, \frac{1}{2}\right)}{B\left(\frac{\nu}{2}, \frac{1}{2}\right)} \psi\left(\frac{\nu}{2} + \frac{1}{2}, 1, \frac{z^2}{2\sigma^2 \nu}\right), \ -\infty < z < \infty, \ \nu > 0.\sigma > 0. \tag{6.7}$$

where $\psi(.)$ denotes Kummer's hypergeometric function. Substituting $\nu = 1, \sigma = 1$, and using the Eq. (13.6.39), p. 510, from Abramowitz and Stegun (1972), the above expression for the pdf easily reduces to:

$$f_Z(z) = \frac{1}{\sqrt{2\pi}} e^{-\frac{z^2}{2}} erfc\left(\frac{z}{\sqrt{2}}\right), \tag{6.8}$$

where $erfc\,(.)$ denotes complementary error function. Using Maple, the possible shapes of the pdf of the random variable $Z = |XY|$ for a range of values $\sigma > 0$ of and degrees of freedom $\nu > 0$ are illustrated in Fig. 6.5.

6.3 Distribution of the Ratio of Normal and Student's *t* Random Variables

The distributions of the ratio of independent normal and Student's *t* random variables are of interest in many areas of engineering, medicine science, and statistics. For example, in the analysis of Bayesian wavelet shrinkage, Johnstone and Silverman (2004, 2005) show that excellent performance is obtained by using heavy-tailed distributions as part of a wavelet coefficient mixture prior instead of the standard normal. A quantity of interest is the ratio of the heavy-tailed distribution with the standard normal. It is possible that Student's *t* distribution might also be an interesting distribution to use in this context. This section presents the distributions of the ratio of independent normal and Student's *t* random variables, independently derived by the author. These results are believed to be new.

6.3.1 Derivation of the Pdf for the Ratio of Normal and Student's t Random Variables

Let $X \sim N\left(0, \sigma^2\right)$ be a normal random variable with pdf $f_X(x)$. Let Y be a Student's t random variable with ν degrees of freedom, and pdf $f_Y(y)$. Let $Z = X/Y$. Then, the pdf of the random variable Z is given by

$$f_Z(z) = \int_{-\infty}^{\infty} |y| f_X(zy) f_Y(y) dy$$

$$= \frac{\sqrt{2}}{\sqrt{\pi \nu} \sigma \, \mathrm{B}\left(\frac{\nu}{2}, \frac{1}{2}\right)} \int_0^{\infty} y e^{-\frac{z^2 y^2}{2\sigma^2}} \left(1 + \frac{y^2}{\nu}\right)^{-\frac{1+\nu}{2}} dy \tag{6.9}$$

Substituting $y^2 = t$, and using following result

$$\int_0^{\infty} w^{\alpha-1} e^{-pw} (w + \kappa)^{-\rho} \, dw = \Gamma(\alpha) \kappa^{\alpha - \rho} \psi(\alpha, \alpha + 1 - \rho; p\kappa),$$

in the Eq. (6.9), (see the Eq. (2.3.6.9), p. 324, Prudnikov et al. (1986, volume 1)), the pdf of the random variable $Z = X/Y$ reduces to

$$f_Z(z) = \frac{\sqrt{\nu}}{\sqrt{2\pi} \sigma \, B\left(\frac{\nu}{2}, \frac{1}{2}\right)} \psi\left(1, \frac{3-\nu}{2}; \frac{\nu z^2}{2\sigma^2}\right), -\infty < z < \infty, \ \nu > 0, \ \sigma > 0. \tag{6.10}$$

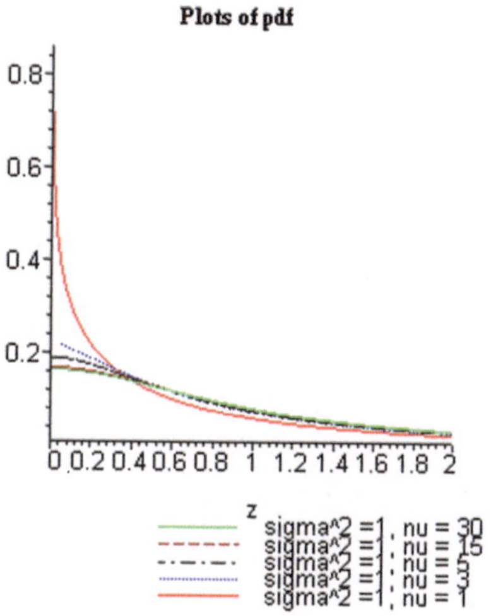

Fig. 6.6 Plots of the Pdf of $Z = |X/Y|$

where ψ (.) denotes Kummer's hypergeometric function. Substituting $\nu = 1, \sigma = 1$, and using the Eq. (13.6.39), p. 510, from Abramowitz and Stegun (1972), the above expression for the pdf easily reduces to:

$$f_Z(z) = \frac{1}{\sqrt{2\pi}} e^{-\frac{z^2}{2}} erfc\left(\frac{z}{\sqrt{2}}\right),$$

where *erfc* (.) denotes complementary function. Note that, for $\nu = 1, \sigma = 1$, the expressions for the pdfs of $Z = XY$ and $Z = X/Y$ are identical. Using Maple 10, the possible shapes of the pdf of the random variable $Z = X/Y$ for a range of values of $\sigma > 0$ and degrees of freedom $\nu > 0$ are illustrated in Fig. 6.6.

6.4 Summary

This chapter has studied the distributions of the sum $X+Y$, product XY, and ratio X/Y, when X and Y are independent normal and Student's t random variables respectively. The distributions of the sum, product, and ratio of the independent normal and Student's t random variables as proposed by different researchers have been reviewed. Some new results have been included. To describe the possible shapes of these pdfs, several graphs have been plotted using Maple programs.

Chapter 7
Product of the Normal and Student's *t* Densities

7.1 Introduction

The distributions of the product of two random variables have a great importance in many areas of research both from the theoretical and applications point of view. Srivastava and Nadarajah (2006) have studied some families of Bessel distributions and their applications by taking products of a Bessel function pdf of the first kind and a Bessel function pdf of the second kind. It appears from the literature that not much attention has been paid to this kind of study. The normal and Student's t distributions arise in many fields and have been extensively studied by many researchers in different times. This chapter introduces and develops a new symmetric type distribution with its probability density function (pdf) taken to be of the form $p_{\mathrm{X}}(x) = C.f_{\mathrm{X}}(x).g_{\mathrm{X}}(x)$, where C is the normalizing constant, and $f_{\mathrm{X}}(x)$ and $g_{\mathrm{X}}(x)$ denote the pdfs of normal and Student's distributions, respectively. More on this topic the readers are referred to Frisch and Sornette (1997), Sornette (1998, 2004), Galambos and Simonelli (2005), Shakil and Kibria (2007), and recently Shakil and Kibria (2009), among others.

7.1.1 Some Useful Lemmas

The following Lemmas will be used to complete the derivations.

Lemma 7.1.1 (Prudnikov et al. 1986, volume 1, Equation 2.3.6.9, page 324).
For $|\arg z| < \pi$, $\rho > 0$, $z > 0$, $\mathrm{Re}(\alpha) > 0$, and $p > 0$,

$$\int_0^\infty \frac{x^{\alpha-1}}{(x+z)^\rho} e^{-px} dx = \Gamma(\alpha)\, z^{\alpha-\rho}\, \psi\,(\alpha,\ \alpha+1-\rho;\ pz)$$

M. Ahsanullah et al., *Normal and Student's t Distributions and Their Applications*,
Atlantis Studies in Probability and Statistics 4, DOI: 10.2991/978-94-6239-061-4_7,

where $\psi(\cdot)$ denotes Kummer's function.

Proof: The proof of Lemma 7.1.1 easily follows by replacing z by $z.\ p$, and substituting $\gamma = \alpha + 1 - \rho$ and $t = \frac{x}{z}$ in the Eq. (1.1) of Chap. 1.

Lemma 7.1.2 (Prudnikov et al. 1986, volume 1, equation 2.3.8.1, page 328). For $|\arg\left(1+\frac{a}{z}\right)| < \pi, a > 0, \rho > 0, z > 0, \mathrm{Re}(\alpha) > 0, \mathrm{Re}(\beta) > 0$, and $p > 0$,

$$\int_0^a \frac{x^{\alpha-1}(a-x)^{\beta-1}}{(x+z)^\rho} e^{-px} dx = B(\alpha, \beta)\, z^{-\rho} a^{\alpha+\beta-1} \Phi_1\left(\alpha, \rho; \alpha+\beta; \frac{-a}{z}, ap\right),$$

where $B(\cdot)$ and $\Phi_1(\cdot)$ denote the beta and generalized hypergeometric functions respectively.

Proof: The proof of Lemma 7.1.2 easily follows by using the substitution $x = a \cdot u$, and applying the Eq. (1.2) of Chap. 1 and the definition of beta function.

7.2 Product of the Densities of Normal and Student's t Random Variables

This section develops a new symmetric Student's t-type distribution with its pdf taken to be the product of the densities of normal and Student's t random variables.

7.2.1 Expressions for the Normalizing Constant and PDF

Consider the product function $p_X(x) = C \cdot f_X(x) \cdot g_X(x), -\infty < x < \infty, C > 0$, where C denotes a normalizing constant, and $f_X(x)$ and $g_X(x)$ denote the pdf of the standard normal and Student's t distributions respectively. The pdf $f_X(x)$ of $X \sim N(0, 1)$ is given by

$$f_X(x) = \frac{1}{\sqrt{2\pi}} e^{-x^2/2}, -\infty < x < \infty, \tag{7.1}$$

whereas the pdf $g_X(x)$ of a Student's t distribution with ν degrees of freedom (for some integer $\nu > 0$), in terms of beta function, is, given by

$$g_X(x) = \frac{1}{\sqrt{\nu} B\left(\frac{\nu}{2}, \frac{1}{2}\right)} \left(1 + \frac{x^2}{\nu}\right)^{-(1+\nu)/2}, -\infty < x < \infty, \nu > 0 \tag{7.2}$$

Theorem 7.2.1 The product function $p_X(x) = C.f_X(x).g_X(x)$, $-\infty < x < \infty$, $C > 0$, defines a pdf if the normalizing constant is given by

$$C = \frac{\sqrt{2}B\left(\frac{\nu}{2}, \frac{1}{2}\right)}{\psi\left(\frac{1}{2}, 1 - \frac{\nu}{2}; \frac{\nu}{2}\right)}, \qquad \nu > 0, \tag{7.3}$$

where $B(\cdot)$ and $\psi(\cdot)$ denote beta and Kummer's functions respectively.

Proof: Clearly, $p_X(x) \geq 0$, $\forall x \in (-\infty, +\infty)$ and $C > 0$. Hence for $p_X(x)$ to be a pdf, we must have $\int_{-\infty}^{+\infty} p_X(x)dx = 1$, where $p_X(x) = C.f_X(x).g_X(x)$. Clearly, in view of the even properties of the pdfs of normal and Student's t distributions, we have $p_X(-x) = p_X(x)$ that is, $p_X(x)$ is an even function. Thus, we have

$$\int_{-\infty}^{+\infty} p_X(x)dx = 2\int_0^{\infty} p_X(x)dx = 2\int_0^{\infty} C.f_X(x).g_X(x)dx = 1,$$

which, on substituting the expressions (7.1) and (7.2) for the pdfs of standard normal and Student's t distributions respectively, gives

$$2\int_0^{\infty} C.\frac{1}{\sqrt{2\pi}}e^{-x^2/2}.\frac{1}{\sqrt{\nu}B\left(\frac{\nu}{2}, \frac{1}{2}\right)}\left(1 + \frac{x^2}{\nu}\right)^{-(1+\nu)/2} dx = 1,$$

that is,

$$\int_0^{\infty} e^{-x^2/2}.\left(1 + \frac{x^2}{\nu}\right)^{-(1+\nu)/2} dx = \frac{\sqrt{2\pi\nu}B\left(\frac{\nu}{2}, \frac{1}{2}\right)}{2C}. \tag{7.4}$$

Now, by substituting $\frac{x^2}{2} = t$ and using Lemma 7.1.1 in the above integral (7.4), the proof of Theorem 7.2.1 easily follows.

Theorem 7.2.2 If $f_X(x)$ and $g_X(x)$ are the pdfs of standard normal and Student's t distributions, as defined in (7.1) and (7.2), respectively, for some continuous random variable X, and C denotes the normalizing constant given by (7.3), the product function given by

$$\begin{aligned} p_X(x) &= C.f_X(x) \cdot g_X(x) \\ &= \frac{e^{-x^2/2}.\left(1 + \frac{x^2}{\nu}\right)^{\frac{-(1+\nu)}{2}}}{(\sqrt{\pi\nu})\psi\left(\frac{1}{2}, 1 - \frac{\nu}{2}; \frac{\nu}{2}\right)}, \qquad -\infty < x < \infty, \nu > 0, \end{aligned} \tag{7.5}$$

defines a pdf of the random variable X, where $\psi(\cdot)$ denotes Kummer's function. It appears that $p_X(x)$ is symmetric and t-type distribution with ν degrees of freedom.

Proof: The proof easily follows from Theorem 7.2.1.
Special Case: For $\nu = 1$, the pdf in (7.5) reduces to the product of densities of normal and Cauchy, with the normalizing constant given by

$$C = \frac{(\sqrt{2})\pi}{\psi\left(\frac{1}{2}, \frac{1}{2}; \frac{1}{2}\right)}.$$

7.2.2 Derivation of the CDF

This section derives the associated cdf of the random variable X, when the normalizing constant $C(> 0)$ satisfies the requirements for the product function $p_X(x)$ to be a density function, as shown in Sect. 7.2.1.

Theorem 7.2.3 Let $f_X(x)$ and $g_X(x)$ be the pdfs of standard normal and Student's t distributions defined by (7.1) and (7.2).Then, the cdf of the random variable X is given by

$$F_X(x) = \frac{1}{2} + \frac{1}{\sqrt{\pi\nu}\,\psi\left(\frac{1}{2},\ 1-\frac{\nu}{2}; \frac{\nu}{2}\right)}\left[x\,\Phi_1\left(\frac{1}{2}, \frac{1}{2}+\frac{\nu}{2}; \frac{3}{2}; \frac{-x^2}{\nu}, \frac{x^2}{2}\right)\right], \tag{7.6}$$

where $|\arg\left(1+\frac{x^2}{\nu}\right)| < \pi$, $\nu > 0$, and $\psi(\cdot)$ and $\Phi_1(\cdot)$ denote Kummer's and generalized hypergeometric functions respectively.

Proof: Using the expression for the pdf (7.5) as derived in Theorem 7.2.2 above, we have

$$\begin{aligned}
F_X(x) &= \Pr(X \le x) \\
&= \frac{1}{\sqrt{\pi\nu}\psi\left(1/2, 1-\frac{\nu}{2}; \frac{\nu}{2}\right)} \int_{-\infty}^{x} \left\{\left(1+\frac{t^2}{\nu}\right)^{-(1+\nu)/2} e^{-t^2/2}\right\} dt \\
&= 1 - \frac{1}{\sqrt{\pi\nu}\psi\left(1/2, 1-\frac{\nu}{2}; \frac{\nu}{2}\right)} \left[\int_{0}^{\infty} \left\{\left(1+\frac{t^2}{\nu}\right)^{-(1+\nu)/2} e^{-t^2/2}\right\} dt \right.\\
&\quad \left. - \int_{0}^{x} \left\{\left(1+\frac{t^2}{\nu}\right)^{-(1+\nu)/2} e^{-t^2/2}\right\} dt\right]
\end{aligned} \tag{7.7}$$

Now, by substituting $\frac{t^2}{2} = u$ and using Lemma 7.1.2 in the above integral (7.7), the proof of Theorem 7.2.3 easily follows.

7.3 Some Properties of the Symmetric Distribution

This section discusses some characteristics of the proposed new symmetric distribution.

7.3.1 Mode

Mode is the value of x for which the product probability density function $p_{\mathrm{X}}(x)$ defined by (7.5) is maximum. Now, differentiating equation (2.5), we have

$$p'_{\mathrm{X}}(x) = -\frac{xe^{-x^2/2}\left(1+\frac{x^2}{v}\right)^{-(1+v)/2}}{\sqrt{\pi v}\psi\left(1/2, 1-\frac{v}{2}; \frac{v}{2}\right)}\left[1+\left(\frac{1+v}{v}\right)\left(1+\frac{x^2}{v}\right)^{-1}\right],$$

which, when equated to 0, gives the mode to be $x = 0$. It can be easily seen that $p''(x) < 0$. Thus, the maximum value of the product probability density function $p_{\mathrm{X}}(x)$ is easily obtained from (7.5) as $p_{\mathrm{X}}(0) = \frac{1}{\sqrt{\pi v}\psi(1/2,1-\frac{v}{2};\frac{v}{2})}$. Clearly, the product probability density function (7.5) is unimodal.

7.3.2 Moments

Theorem 7.3.4 For some degrees of freedom $v > 0$ and some integer $k > 0$, kth moment of a random variable x having the pdf (7.5) is given by

$$E\left(X^k\right) = \begin{cases} \frac{\Gamma\left(\frac{k+1}{2}\right)v^{\frac{k}{2}}\psi\left(\frac{k+1}{2}, \frac{k-v}{2}, +1; \frac{v}{2}\right)}{\sqrt{\pi}\,\psi\left(\frac{1}{2}, 1-\frac{v}{2}; \frac{v}{2}\right)}, & when\, k\, is\, even; \\ 0, & when\, k\, is\, odd; \end{cases} \tag{7.8}$$

where $k > 0$ is integer, and $\psi(\cdot)$ denotes Kummer's function.

Proof: Using the expression for the pdf (7.5), we have

$$E\left(X^k\right) = \frac{1}{\sqrt{\pi v}\,\psi\left(\frac{1}{2}, 1-\frac{v}{2}; \frac{v}{2}\right)}\int_{-\infty}^{+\infty}\left\{x^k\left(1+\frac{x^2}{v}\right)^{-\frac{(1+v)}{2}} e^{\frac{-x^2}{2}}\right\}dx \tag{7.9}$$

Case I: When k is even.
Let $k = 2n$, where $n > 0$ is an integer. Then, since, clearly, the integrand in (7.9) is an even function, we have

$$E\left(X^k\right) = \frac{2v^{\frac{v}{2}}}{\sqrt{\pi}\,\psi\left(\frac{1}{2},\,1-\frac{v}{2};\,\frac{v}{2}\right)} \int_0^{+\infty} \left\{ x^{2n}\left(x^2+v\right)^{-\frac{(1+v)}{2}} e^{\frac{-x^2}{2}} \right\} dx \tag{7.10}$$

Now, by substituting $\frac{x^2}{2} = u$ and using Lemma 7.1.1 in the above integral (7.10), the proof of Theorem 7.3.4 (when k is an even integer) easily follows.

Case II: When k is odd.
Let $k = 2n - 1$, where $n > 0$ is an integer. Then, since, clearly, the integrand in (7.10) is an odd function, the proof of Theorem 7.3.4 (when k is an odd integer) easily follows.

Special Cases: By taking $k = 2$ and $k = 4$ respectively in (7.10), for some degrees of freedom $v > 0$, the second and fourth moments are easily obtained as follows

$$E\left(X^2\right) = \frac{1}{2}\frac{v\psi\left(\frac{3}{2},2-\frac{v}{2};\frac{v}{2}\right)}{\psi\left(\frac{1}{2},1-\frac{v}{2};\frac{v}{2}\right)}, \tag{7.11}$$

and

$$E\left(X^4\right) = \frac{3}{4}\frac{v^2\psi\left(\frac{5}{2},3-\frac{v}{2};\frac{v}{2}\right)}{\psi\left(\frac{1}{2},1-\frac{v}{2};\frac{v}{2}\right)}, \tag{7.12}$$

where $\psi(\cdot)$ denotes Kummer's function.

7.3.3 *Mean, Variance, and Coefficients of Skewness, and Kurtosis*

From (7.12), the mean, variance, and coefficients of skewness and kurtosis of probability density function (7.5) are easily obtained as follows:

(i) Mean: $\mu = E(x) = 0$;

(ii) Variance: $Var\,(X) = \sigma^2 = \frac{v\psi\left(\frac{3}{2},2-\frac{v}{2};\frac{v}{2}\right)}{2\psi\left(\frac{1}{2},1-\frac{v}{2};\frac{v}{2}\right)}, \quad v > 0$

(iii) Coefficient of Skewness: $\gamma_1 = \frac{\beta_3}{\beta_2^{3/2}} = 0$

(iv) Coefficient of Kurtosis: $\gamma_2 = \frac{\beta_4}{\beta_2^2} = \frac{3\psi\left(\frac{5}{2},3-\frac{v}{2};\frac{v}{2}\right)\psi\left(\frac{1}{2},1-\frac{v}{2};\frac{v}{2}\right)}{\left[\psi\left(\frac{3}{2},2-\frac{v}{2};\frac{v}{2}\right)\right]^2}, \quad v > 0$

7.4 Entropy

The Shannon (1948) entropy of an absolutely continuous random variable x having the probability density function $\phi_X(x)$ is defined as

$$H[X] = \mathrm{E}[-\ln(\phi_X(x)] = -\int_S \phi_X(x) \ln[\phi_X(x)]dx \tag{7.13}$$

where $S = \{x : \phi_X(x) > 0\}$.

Theorem 7.4.1 For $v > 0$, the entropy of a random variable x having the probability density function $p_X(x)$ in (7.5) , is given by

$$H[X] = \ln\left\{\sqrt{\pi v}\psi\left(\frac{1}{2}, 1-\frac{v}{2}; \frac{v}{2}\right)\right\} + \frac{1}{4}\frac{v\psi\left(\frac{3}{2}, 2-\frac{v}{2}; \frac{v}{2}\right)}{\psi\left(\frac{1}{2}, 1-\frac{v}{2}; \frac{v}{2}\right)}$$
$$+\left[\frac{(1+v)}{2\sqrt{\pi}}\right]\sum_{j=1}^{\infty}(-1)^{j-1}\frac{\Gamma\left(j+\frac{1}{2}\right)}{j}\frac{\psi\left(j+\frac{1}{2}, j-\frac{v}{2}+1; \frac{v}{2}\right)}{\psi\left(\frac{1}{2}, 1-\frac{v}{2}; \frac{v}{2}\right)},$$

where $\psi(\cdot)$ denotes Kummer's function.

Proof: From (7.13), we have

$$H[X] = \mathrm{E}[-\ln(p_X(x))] = -\int_{-\infty}^{+\infty} p_X(x) \ln[p_X(x)]dx$$

$$= -\int_{-\infty}^{+\infty}\frac{\left(1+\frac{x^2}{v}\right)^{-\frac{(1+v)}{2}} e^{\frac{-x^2}{2}}}{\sqrt{\pi v}\,\psi\left(\frac{1}{2}, 1-\frac{v}{2}; \frac{v}{2}\right)} \mathrm{x} \ln\left\{\frac{\left(1+\frac{x^2}{v}\right)^{-\frac{(1+v)}{2}} e^{\frac{-x^2}{2}}}{\sqrt{\pi v}\,\psi\left(\frac{1}{2}, 1-\frac{v}{2}; \frac{v}{2}\right)}\right\} dx \tag{7.14}$$

In view of the definitions of the pdf (7.5) (Theorem 7.2.2) and moments, the above expression (7.14) of entropy easily reduces as

$$H[X] = \ln\left\{\sqrt{\pi v}\,\psi\left(\frac{1}{2}, 1-\frac{v}{2}; \frac{v}{2}\right)\right\} + \frac{1}{2}E\left(X^2\right)$$
$$+\left(\frac{1+v}{2}\right)E\left[\ln\left(1+\frac{X^2}{v}\right)\right]$$
$$= \ln\left\{\sqrt{\pi v}\,\psi\left(\frac{1}{2}, 1-\frac{v}{2}; \frac{v}{2}\right)\right\} + \frac{1}{2}E\left(X^2\right)$$

Table 7.1 Percentiles of the new symmetric distribution

v	50 %	60 %	70 %	75 %	80 %	85 %	90 %	95 %	99 %
3	≈ 0	0.1750	0.3640	0.4692	0.5881	0.7287	0.9095	1.1865	1.7385
4	≈ 0	0.1761	0.3656	0.4714	0.5902	0.7304	0.9098	1.1831	1.7230
5	≈ 0	0.1767	0.3667	0.4727	0.5914	0.7313	0.9097	1.1804	1.7118
10	≈ 0	0.1780	0.3689	0.4750	0.5940	0.7325	0.9090	1.1734	1.6840
15	≈ 0	0.1784	0.3696	0.4757	0.5942	0.7327	0.9081	1.1704	1.6721
20	≈ 0	0.1786	0.3699	0.4760	0.5944	0.7328	0.9077	1.1687	1.6659
25	≈ 0	0.1787	0.3701	0.4762	0.5946	0.7329	0.9074	1.1677	1.6620
30	≈ 0	0.1789	0.3702	0.4763	0.5947	0.7329	0.9073	1.1670	1.6593
35	≈ 0	0.1788	0.3703	0.4764	0.5947	0.7329	0.9071	1.1664	1.6573
40	≈ 0	0.1788	0.3703	0.4765	0.5948	0.7329	0.9070	1.1660	1.6560
50	≈ 0	0.1789	0.3704	0.4766	0.5949	0.7329	0.9069	1.1655	1.6537
60	≈ 0	0.1790	0.3705	0.4767	0.5914	0.7329	0.9068	1.1651	1.6524
70	≈ 0	0.1792	0.3708	0.4770	0.5953	0.7333	0.9073	1.1660	1.6560
75	≈ 0	0.1790	0.3706	0.4767	0.5950	0.7329	0.9067	1.1648	1.6513

$$+\left(\frac{1+v}{2}\right)\sum_{j=1}^{\infty}\frac{(-1)^{j-1}}{jv^{j}}E\left(X^{2j}\right) \tag{7.15}$$

By using the moment expressions (7.10) and (7.11) in (7.15), the proof of Theorem 7.4.1 easily follows.

7.5 Percentage Points

This section computes the percentage points of the new symmetric distribution. For any α, where $0 < \alpha < 1$, the (100α)th percentile or the quantile of order α of the new symmetric distribution with the pdf $p_X(x)$ is a number t_α such that the area under $p_X(x)$ to the left of t_α is α. That is, t_α is any root of the equation

$$F\left(t_\eta\right) = \int_{-\infty}^{t_\alpha} p_X(u)\,du = \alpha\ .$$

Using Maple, the percentiles t_α of the new symmetric distribution are computed for some selected values of α for the given values of degrees of freedom v, which are provided in Table 7.1.

7.6 Summary

This chapter has derived a new symmetric type distribution and its properties, with its probability density function (pdf) taken to be the product of a normal pdf $f_X(x)$and a Student's t distribution pdf $g_X(x)$ for some continuous random variable X. The expressions for the associated pdf, cdf, *kth* moment, mean, variance, skewness, kurtosis, and entropy have been derived in terms of Kummer's functions. It is shown that the pdf of the proposed distribution is unimodal. It is noted that for $v = 1$, we obtain a new symmetric distribution which is the product of the standard normal and Cauchy distributions, and for large v we will obtain a distribution which is product of two standard normal densities. The percentage points have also been provided. We hope the findings of this chapter paper will be useful for the practitioners in various fields of sciences.

Chapter 8
Characterizations of Normal Distribution

8.1 Introduction

Before a particular probability distribution model is applied to fit the real world data, it is necessary to confirm whether the given probability distribution satisfies the underlying requirements by its characterization. Thus, characterization of a probability distribution plays an important role in probability and statistics. A probability distribution can be characterized through various methods. In recent years, many authors have studied characterizations of various distributions. This chapter discusses characterizations of normal distribution.

8.2 Characterizations of Normal Distribution

In this section, we will consider several characterizations of normal distribution. Let Z be a standard normal distribution (N(0, 1)) with pdf f(z), then

$$f(z) = \frac{1}{\sqrt{2\pi}} e^{-\left(\frac{1}{2}\right)z^2}, -\infty < z < \infty \tag{8.1}$$

Suppose the random variable X has a normal distribution with mean μ and standard deviation σ (N(μ, σ)) with the pdf f(x) as

$$f(x) = \frac{1}{\sqrt{2\pi}\sigma} e^{-\frac{1}{2}\left(\frac{x-\mu}{\sigma}\right)^2} \tag{8.2}$$

$$-\infty < x < \infty, -\infty < \mu < \infty, \sigma > 0.$$

The characteristic functions $\phi(t)$ and $\phi_1(t)$ of Z and X are respectively

M. Ahsanullah et al., *Normal and Student's t Distributions and Their Applications*, Atlantis Studies in Probability and Statistics 4, DOI: 10.2991/978-94-6239-061-4_8,

$$\phi(t) = e^{-\frac{1}{2}t^2}. \tag{8.3}$$

and

$$\phi_1(t) = e^{i\mu t - \frac{1}{2}\sigma^2 t^2}, \quad i = \sqrt{-1} \quad \text{for all t.} \tag{8.4}$$

Here we present some basic properties of the characteristic function $\phi(t)$ which we need for our characterization theorems.

1. $\phi(t)$ always exists.
2. $\phi(t)$ is uniformly continuous on entire space.
3. $\phi(t)$ is non vanishing in a region around zero, $\phi(0) = 1$.
4. $\phi(t)$ is bounded. $|\phi(t)| \le 1$.
5. $\phi(t)$ is Hermitian, $\phi(-t) = \overline{\phi}(t)$, where $\overline{\phi}(t)$ is the complex conjugate of $\phi(t)$.
6. $\phi(t)$ for a symmetric (around zero) random variable is real and even function.
7. There is a one to one correspondence between $\phi(t)$ and the cdf

Polya (1923) gave the following characterization theorem.

Theorem 8.2.1. Suppose X_1 and X_2 are independent and identically distributed random variables. Then X_1 and $\frac{(X_1+X_2)}{\sqrt{2}}$ are identically distributed if and only if X_1 and X_2 are normally distributed.

Proof. It is easy to show that $E(X_1) = E(X_2) = 0$. Let $\phi(t)$ and $\phi_1(t)$ be the characteristic functions of X_1 and $\frac{(X_1+X_2)}{\sqrt{2}}$ respectively.

If X_1 and X_2 are $N(0, \sigma^2)$, then

$$\phi_1(t) = [\phi(t/\sqrt{2})]^2 = \left(e^{-(1/2)\sigma^2(t/\sqrt{2})^2}\right) = e^{-(1/2)\sigma^2 t^2} \tag{8.5}$$

Thus $\frac{(X_1+X_2)}{\sqrt{2}}$ is distributed as $N(0, \sigma^2)$.
Suppose X_1 and $\frac{(X_1+X_2)}{\sqrt{2}}$ are identically distributed.
Then

$$\phi_1(t) = [\phi(t/\sqrt{2})]^2$$

i.e.

$$\phi_1(\sqrt{2}t) = [\phi(t)]^2 \quad \text{for all t.}$$

Therefore recurrently

$$\phi_1\left(t2^{\frac{k}{2}}\right) = (\phi(t))^{2^k} \quad \text{for all t.} \tag{8.6}$$

Let us take a t_0 such that $\phi(t_0) \neq 0$, such a t_0 can be found since $\phi(t)$ is continuous and $\phi(0) = 1$. Let $\sigma^2 > 0$ be such that $\phi(t_0) = e^{-\sigma^2}$, then we have

$$\phi_1(t_0 2^{-\frac{k}{2}}) = e^{-\sigma^2 2^{-k}} \quad \text{for } k = 0, 1, 2, \ldots\ldots \tag{8.7}$$

Thus $\phi_1(t) = e^{-t^2\sigma^2}$ for all t and the theorem is proved.
The following theorem is due to Cramer (1936).

Theorem 8.2.2. Let X_1 and X_2 be two independent but not necessarily identically distributed random variables and $\mathrm{Z} = \mathrm{X}_1 + \mathrm{X}_2$. If Z is normally distributed, then X_1 and X_2 are normally distributed.

To prove the theorem, we need the following two Lemmas.

Lemma 8.2.1. Hadamard's factorial Theorem.
If g(t) is a integral function of finite order ρ which has zeros $\beta_1, \beta_2, \ldots$and does not vanish at the origin, then g(t) can be written as

$$g(t) = m(t)e^{n(t)},$$

where m(t) is the canonical product formed with the zeros of $\beta_1, \beta_2, \ldots$ and n(t) is a polynomial of degree not exceeding ρ.

Lemma 8.2.2. If $e^{n(t)}$, where n(t) is a polynomial, is a characteristic function, then the degree of n(t) cannot exceed 2.

Proof of Theorem 8.2.2. The necessary condition is easy to prove. We will prove here the sufficiency. We will prove the Theorem under the assumption that Z has mean zero and standard deviation $= \sigma$. The characteristic function ϕ(t) of Z can be written as $\phi(t) = \phi_1(t)\phi_2(t)$, where $\phi_1(t)$ and $\phi_2(t)$ are the characteristic functions of X_1 and X_2 respectively. Now $\phi_1(t) = e^{-\frac{1}{2}\sigma^2 t^2}$ is an entire function without zero. Thus $\phi_1(t) = e^{p(t)}$, where p(t) is a polynomial of degree not exceeding 2. Hence we can write

$$\phi_1(t) = e^{-a_0 + a_1 t + a_2 t^2} \quad \text{for some real } a_0,\ a_1 \text{ and } \mathrm{a}_2.$$

For any characteristic function $\phi(t)$, $|\phi(t)\} \le 1$, hence a_2 must be negative. Assuming mean of $X_1 - = \mu$ and standard deviation $\mathrm{X}_1 = \sigma$. We obtain

$$\phi_1(t) = e^{i\mu t - (\frac{1}{2})\sigma^2 t^2} \quad \text{for all t.}$$

Thus X_1 is normally distributed. Similarly it can be proved that X_2 is also normally distributed.

Remark 8.2.1. If Z is normally distributed, then we can write $\mathrm{Z} = \mathrm{X}_1 + \mathrm{X}_2 + \cdots + \mathrm{X}_\mathrm{n}$, where $X_i's$, $i = 1, 2, \ldots, n$ are independent and normally distributed.

Remark 8.2.2. If $\mathrm{X}_1, \mathrm{X}_2, \ldots, \mathrm{X}_n$ are n independent and identically distributed random variable with mean zero and variance 1, then by central limit theorem

$$S_n = \frac{X_1}{\sqrt{n}} + \frac{X_2}{\sqrt{n}} + \cdots + \frac{X_n}{\sqrt{n}} \rightarrow N(0, 1).$$

But by Cramer's theorem if S_n is N(0, 1), then each $\frac{X_i}{\sqrt{n}}, i = 1, 2, \ldots n$ must be normal.
Gani and Shanbag (1975) proved that if $Z = X_1 + X_2$ is a sum of two independent random variables is normal, then Z can be decomposed as such that the conditional distribution of one given the other is normal.

Basu and Ahsanullah (1983) gave a generalization of Cramer's decomposition theorem in the case of sum of dependent random variables based on Gani and Shanbag's decomposition in the following theorem.

Theorem 8.2.3. Let Z be a normal random variable with zero mean and variance σ^2 which has Gani and Shanbag decomposition as sum of two random variables $(X_1 + X_2)$ with additional property that X_1 and X_2 are identically distributed with finite strictly positive second moment and correlation coefficient $\rho, 0 <\mid \rho \mid< 1$ Then X_1 and X_2 both follow the normal distribution.

Proof. See Basu and Ahsanullah (1983).

The following characterization theorem is due to Darmois (1951) and Basu (1951).

Theorem 8.2.4. Suppose $X_1, X_2, \ldots, X_n$ be a set of independent but need not be identically distributed random variables and let

$$L_1 = a_1X_1 + a_2X_2 + \cdots + a_nX_n \tag{8.8}$$

and

$$L_2 = b_1X_1 + b_2X_2 + \cdots + b_nX_n \tag{8.9}$$

where a's and b's are constant. If L_1 and L_2 are independently distributed, then X_i for which $a_ib_i \neq 0$ is normally distributed.

For an interesting proof of the theorem see Linnik (1964 p. 97).

Kagan et al. (1965) showed that if $n(\geq 3)$ independent and identically distributed random variable with $E(X_i) = 0$ and $E(\bar{X} \mid X_1 - \bar{X}, X_2 - \bar{X}, \ldots, X_n - \bar{X}_n) = 0$, where $n\bar{X} = \sum_{i=1}^{n} X_i$, then$X_i's(i = 1, 2, \ldots, n)$ are normally distributed.

Rao (1967) proved that for $n(n \geq 3)$ i.i.d.rv's $X_1, \ldots, X_n$ if $E(X_i) = 0$ and $E(X_i^2) < \infty$, $i = 1, 2, \ldots, n$. Then if $E(\bar{X} \mid X_i - \bar{X}) = 0$ for a fixed i, then X's are normal.

The following example shows that the result need not be true for $n = 2$.

Example 8.2.1. Let the random variable X_1 and X_2 have the following joint pdf,

$$f(x_1, x_2) = \frac{1}{4}, -1 \leq x_1, x_2 \leq 1, \tag{8.10}$$
$$= 0, \text{otherwise.}$$

Let $Y_1 = X_1 + X_2$ and $Y_2 = X_1 - X_2$, the conditional pdf of $Y_1 \mid Y_2 = Y_2$ is given by

$$f(y_1 \mid Y_2 = y_2) = \frac{1}{2(2 - y_2)}, \quad 0 < y_2 < 2, \ -2 < y_1 < 2 - y_2$$

Hence the result.

Kagan and Zinger (1971) proved the normality of the X's under the following conditions.

$$E(\mid X_i \mid^2) < \infty, \text{i} = 1, 2, \ldots, n$$
$$E(L_1^{k-1} \mid L_2) = 0, k = 1, 2, \ldots, n$$

where $L_1 = a_1X_1 + a_2X_2 + \cdots + a_nX_n$ and $L_2 = b_1X_1 + b_2X_2 + \cdots + b_nX_n$.

Theorem 8.2.5. Suppose X_1 and X_2 independent and identically distributed random variables with mean zero and variance 1. If $X_1 + X_2$ and $X_1 - X_2$ are independent, then the common distribution of X_1 and X_2 is normal.

Proof. Let $\phi(t)$ be the common characteristic function of X_1 and X_2. Since $X_1 + X_2$ and $X_1 - X_2$ are independent, we must have

$$\phi(2t) = (\phi(t))^3 \phi(-t).$$

The function $\phi(t)$ never vanishes. Writing $\rho(t) = \frac{\phi(t)}{\phi(-t)}$, we have $\rho(2t) = (\rho(t))^2$. By induction, we obtain

$$\rho(t) = \left(\rho(\frac{t}{2^k})\right)^k - \left(1 + o(\frac{t}{2^k})\right)^{2^k} \to 1.$$

Thus $\rho(t) - 1$ and $\phi(t) = \phi(-t)$ for all t.

We have $\phi(t) = [\phi(t/2^k)]^{2^k} = e^{-(1/2)t^2}$
The theorem is proved.

The following lemma is due to Roberts (1971).

Lemma 8.2.3. If is a characteristic function of the random variable Z. Suppose Z^2 is distributed as chi-square with one degree of freedom, then

$$\phi(t) + \phi(-t) = 2e^{-(\frac{1}{2})t^2} \tag{8.11}$$

Proof. Let g(.) be the probability density function of Z^2 and h(.) be the pdf of Z. Then

$$g(u) = \frac{u^{-(\frac{1}{2})}}{\sqrt{2\pi}} e^{-(\frac{1}{2})u} = \frac{1}{2\sqrt{u}} h(u) + \frac{1}{2\sqrt{u}} h(-u).$$

Thus

$$h(u) + h(-u) = \sqrt{\frac{2}{\pi}} e^{-\frac{1}{2}u}$$

Now

$$\phi(t) + \phi(-t) + \int_{-\infty}^{\infty} e^{itx} h(x)dx + \int_{-\infty}^{\infty} e^{-itx} h(x)dx$$
$$= \int_{-\infty}^{\infty} e^{itx} h(x)dx + \int_{-\infty}^{\infty} e^{itx} h(-x)dx - \int_{-\infty}^{\infty} e^{itx} \sqrt{\frac{2}{\pi}} e^{-\frac{1}{2}x^2} dx$$
$$= 2e^{-\frac{1}{2}t^2} \tag{8.12}$$

Ahsanullah (1989) gave the following characterization theorem using the above lemma.

Theorem 8.2.6. Suppose $X_1, X_2, \ldots X_n (n \geq 2)$ are n independent and identically distributed random variables. Suppose $L_1 = a_1 X_1 + \cdots + a_n X_n$, where $a_1, \ldots, a_n$ are constants, not all them are zero and X's are symmetric around zero. Then if L_1^2 is distributed as a chi-square with one degree of freedom, then X's are normal.

Let $\varnothing_1^*(t)$ be the characteristic function of L_1, then by Lemma 8.1,

$$2e^{-\frac{1}{2}t^2} = \varnothing_1^*(t) + \varnothing_1^*(-t)$$
$$= \prod_{j=1}^{n} \phi(a_j t) + \prod_{j=1}^{n} \phi(-a_j t)$$

where $\varnothing(t)$ is the characteristic function of $X_i's, i = 1, \ldots, n$

$= 2 \prod_{j=1}^{n} \phi(a_j t)$ by the symmetry of the X's.

It is known (Linnik and Zinger (1955) that if $\varnothing_1(t), \varnothing_2(t) \ldots, \varnothing_n(t)$ are characteristic functions and $a_1, a_2, \ldots a_n$ are positive constants, then if

$$(\phi_1(t))^{a_1} (\phi_2(t))^{a_2} \ldots\ldots (\phi_n(t))^{a_n} = e^{-[i\mu t - \frac{\sigma^2 t^2}{2}] \sum_{i=1}^{n} a_n},$$

$-\infty < \mu < œ, \sigma > 0$ for $|t| < \delta, \delta > 0$. holds if then $\varnothing_1(t), \ldots \varnothing_n(t)$ are the characteristic function of normal distribution. Then it follows that $\varnothing(t)$ is the characteristic function of a normal distribution.

Remark 8.2.3. Taking $a_1 = a_2 = \cdots = a_n = \frac{1}{\sqrt{n}}$, it follows from the above theorem that $n\bar{X}^2$ is distributed as chi-square with one degree of freedom and X's are symmetric around zero imply the normality of the X's.

The following Theorem (Ahsanullah 1990) gives a characterization of normal distribution using chi-square distribution.

Theorem 8.2.7. Let X_1 and X_2 be two independent and identically distributed random variables. Suppose $L_1 = aX_1 + \sqrt{1-a^2}X_2,\ 0 < |a| < 1$, and assume L_1^2 and X_1^2 are each distributed as chi-square with one degree of freedom, this X_1 and X_2 are normally distributed.

Proof. Let $\phi(t)$ be the characteristic function of X_1, then $2e^{-\frac{1}{2}t^2} = \emptyset(at)\phi(\sqrt{1-a^2}t) + \phi(-at) + \phi(-\sqrt{1-a^2}t)$ and $2e^{-\frac{1}{2}t^2} = \emptyset(t) + \emptyset(-t)$

Then $\emptyset(at)\emptyset\sqrt{1-a^2}t) = e^{-\frac{1}{2}t^2}$

It is known (Linnik and Zinger (1955) that if $\emptyset_1(t), \emptyset_2(t) \cdots, \emptyset_n(t)$ are characteristic functions and $a_1, a_2, \ldots a_n$ are positive constants, then if

$$(\phi_1(t))^{a_1}(\phi_2(t))^{a_2}\ldots(\phi_n(t))^{a_n} = e^{-\left(i\mu t - \frac{\sigma^2 t^2}{2}\right)\sum_{k=1}^{n} a_k} \tag{8.13}$$

where $-\infty < \mu < \infty, \sigma < 0$ and $i = \sqrt{-1}$, holds for all $|t| < \delta$, for all $\delta > 0$, then $\phi_1(t), \ldots, \phi_n(t)$ are characteristic functions of normal distributions. Thus $\phi(t)$ is the characteristic function of a normal distribution and X's are normally distributed.

Ahsanullah and Hamedani (1988) proved the following theorem.

Theorem 8.2.8. If X_1 and X_1 be independent and identically distributed symmetric (about zero) random variables with pdf $f(\cdot)$ and let $Z = \min(X_1, X_2)$. If Z^2 is distributed as chi-square with one degree of freedom. then X_1 and X_2 are normally distributed.

Proof. Let $\emptyset(t)$ be the characteristic function of Z, then $\emptyset(t) = 2\int_{-\infty}^{\infty} e^{itx}\bar{F}(x) f(x)dx$, where

$$\bar{F}(x) = 1 - F(x) \text{ and } F(x) = \int_{-\infty}^{x} f(u)du.$$

Hence

$$\begin{aligned}
\emptyset(t) + \emptyset(-t) &= 4\int_{-\infty}^{\infty} \cos(tx)\bar{F}(x)f(x)dx \\
&= 4\int_{0}^{\infty} \cos(tx)(1 - F(x))f(x)dx \\
&\quad + 4\int_{0}^{\infty} \cos(tx)F(x)f(x)dx \\
&= 4\int_{0}^{\infty} \cos(tx)f(x)dx
\end{aligned} \tag{8.14}$$

$= 2\int_{-\infty}^{\infty} \cos(tx) f(x)dx$ (by the symmetric property of $f(x)$)
$= 2\phi_X(t)$, where $\phi_X(t)$ is the characteristic function of X.
By Lemma 2.3, $\emptyset(t) + \emptyset(-t) = 2e^{-\frac{t^2}{2}}$, and hence $\phi_X(t) = e^{-\frac{t^2}{2}}$.
Thus the common distribution of X_1 and X_2 is normal.

Remark 8.2.4. It is easy to see that we can replace in Z the min by the max.

The following two theorems were proved by Ahsanullah (1989).

Theorem 8.2.9. Let X_1 and X_2 be independent and identically distributed random variables and

Suppose

$$Z_1 = a_1 X_1 + a_2 X_2$$

$$Z_2 = b_1 X_1 + b_2 X_2$$

Such that

$$-1 < a_1, a_2 < 1, -1 < b_1, b_2 < 1, 1 = a_1^2 + a_2^2 = b_1^2 + b_2^2$$

and $a_1 b_2 + a_2 b_1 = 0$.

If z_1^2 and z_2^2 are each distributed as chi-square with one degree of freedom, then X_1 and X_2 are normally distributed.

Proof. Let $\emptyset_3$(t) and $\emptyset_4$(t) be the characteristic functions of Z_1 and Z_2 respectively. Then we have

$$\phi_1(t) + \phi_1(-t) = 2e^{-\frac{1}{2}t^2} = \phi_2(t) + \phi_2(-t)$$

Now if $\emptyset(t)$ is the characteristic function of X, then

$$\emptyset_3(t) = \emptyset(a_1 t)\emptyset(a_2 t), \text{ while } \emptyset_4(t) = \emptyset(b_1 t)\emptyset(b_2 t). \tag{8.15}$$

Substituting $b_1 = -a_1 b_2 \mid a_2$ and using the relation $a_1^2 + a_2^2 = 1$, we must have $a_2^2 = b_2^2$, Taking $a_2 = b_2$ or $a_2 = -b_2$ and writing $\emptyset_3(t), \emptyset_4(t)$ in terms of $\emptyset(t)$, we get on simplification

$$\begin{aligned} &\emptyset(a_1 t)\emptyset(a_2 t) + \emptyset(-a_1 t)\emptyset(-a_2 t) \\ &\quad = \emptyset(-a_1 t)\emptyset(-a_2 t) + \emptyset(a_1 t)\emptyset(-a_2 t) \\ &\quad = 2e^{-(\frac{1}{2})t^2}, \text{ for all } t, -\infty < t < \infty. \end{aligned} \tag{8.16}$$

From (8.16) we obtain directly

$$(\emptyset(a_1 t) + \emptyset(-a_1 t))(\emptyset(a_2 t) + \emptyset(-a_2 t)) = 4e^{-\frac{t}{2}^2} \tag{8.17}$$

$$(\emptyset(a_1t) - \emptyset(-a_1t))(\emptyset(a_2t) - \emptyset(-a_2t)) = 0 \tag{8.18}$$

for all t, $-\infty < t < \infty$. From (8.18) it follows that we must have $\emptyset(t) = \emptyset(-t)$ for all t. Thus

$$\emptyset(a_i t)\emptyset(a_2 t) = e^{-\frac{t^2}{2}}. \tag{8.19}$$

Hence by Cramer's theorem X_1 and X_2 are normally distributed.

Theorem 8.2.10. Let X_1 and X_2 be independent and identically distributed random variables and suppose $U = aX_1 + bX_2$ such that $0 < a, b < 1$ and $a^2 + b^2 = 1$. If U^2 and X_1^2 are each distributed as chi-square with one degree of freedom, then X_1 and X_2 are both distributed as normal.

Proof. Let $\phi_1(t)$ and $\phi(t)$ be the characteristic function of U and X_1 respectively. Then by lemma 8.1, we have

$$\begin{aligned} 2e^{-\frac{1}{2}t^2} &= \phi_1(t) + \phi_1(-t) \\ &= \phi(at)\phi(bt) + \phi(-at)\phi(-bt) \\ &= \phi(t) + \phi(-t), \end{aligned} \tag{8.20}$$

for all t.

From (8.20), we have

$$\phi(at) + \phi(-at) = 2e^{-(1/2)a^2t^2},$$
$$\phi(bt) + \phi(-bt) = 2e^{-(1/2)b^2t^2}$$

and hence

$$(\phi(at) + \phi(-at))(\phi(bt) + \phi(-bt)) = 4e^{-(1/2)t^2} \tag{8.21}$$

We have also

$$(\phi(at) - \phi(-at))(\phi(bt) - \phi(-bt)) = 0 \tag{8.22}$$

Since (8.22) is true for all t, $-\infty < t < \infty$, we must have
$\phi(t) = \phi(-t)$ for all t, $-\infty < t < \infty$.
Hence we obtain

$$\phi(at)\phi(bt) = e^{-(1.2)t^2}, \quad \text{for all t, } -\infty < t < \infty. \tag{8.23}$$

By Cramer's theorem it follows that X_1 and X_2 are normally distributed.

The following theorem is due to Ahsanullah et al. (1991).

Theorem 8.2.11. Suppose $X_1, X_2, \ldots, X_n$ be n independent and identically distributed random variables for some fixed n, $(n \geq 2)$. Let $T_1 = \sum_{i=1}^{n} X_i^2$ and $T_2 = n\bar{X}^2$.

If T_1 and T_2 are distributed as chi-squares with n and 1 degrees of freedom, then X_1's, $i = 1, 2, \ldots$ are $N(0, 1)$.

Proof. Let $\varphi_2(t)$ be the characteristic function of $\sqrt{n}\bar{X}$, then by Lemma 8.3.

$$\varphi_2(t)\,\varphi_2(-t) = 2e^{-(1/2)t^2}.$$

Now $\varphi_2(t) = \left(\varphi\left(\frac{t}{\sqrt{n}}\right)\right)^n$, where $\varphi(t)$ is the characteristic function of t. Thus we have

$$2e^{-(1/2)t^2} = \varphi_2(t) + \varphi_2(-t) = \left(\varphi\left(\frac{t}{\sqrt{n}}\right)\right)^n + \left(\varphi\left(\frac{-t}{\sqrt{n}}\right)\right)^n$$

$$\text{i.e., } 2e^{-\left(\frac{1}{2}\right)nt^2} = \varphi_2\left(\sqrt{nt}\right) + \varphi_2\left(\sqrt{-nt}\right) = (\varphi(t))^n + (\varphi(-t))^n. \tag{8.24}$$

Since $X_i^2 (i = 1, 2, \ldots, n)$ is distributed as a chi-square with one degree of freedom, we have

$$2e^{-\left(\frac{1}{2}\right)t^2} = \varphi(t) + \varphi(-t)$$

and hence

$$2e^{-(1/2)nt^2} = \varphi\left(\sqrt{nt}\right) + \varphi\left(-\sqrt{nt}\right). \tag{8.25}$$

Now $e^{-(1/2)nt^2} = \left(\frac{\varphi(t)+\varphi(t)}{2}\right)^n$, and therefore

$$2 = \frac{\varphi\left(\sqrt{nt}\right) + \varphi\left(-\sqrt{nt}\right)}{\left(\frac{\varphi(t)+\varphi(-t)}{2}\right)^n} = \left(\frac{2\varphi(t)}{\varphi(t) + \varphi(-T)}\right)^n + \left(\frac{2\varphi(-t)}{\varphi(t) + \varphi(-t)}\right)^n \quad \text{for all } t. \tag{8.26}$$

Thus we have

$$\frac{\varphi(\sqrt{nt}) + \varphi(-\sqrt{nt})}{2} = \left(\frac{\varphi(t) + \varphi(-t)}{2}\right)^n = \frac{(\varphi(t))^n + (\varphi(-t))^n}{2} \quad \text{for } n \geq 2.$$

Since $\left(\frac{\varphi(t)+\varphi(-t)}{2}\right)^n = \frac{(\varphi(t))^n+(\varphi(-t))^n}{2}$ for some fixed $n \geq 2$ and all t and further $\varphi(t) \neq 0$ and $\varphi(t) \neq 1$, hence $\varphi(t) = \varphi(-t)$ and x's are symmetric around zero. Thus

$$\varphi(t) = \left(\varphi\left(\sqrt{nt}\right)\right)^{1/n}, \text{ for all } t. \tag{8.27}$$

$$= (\varphi(nt))^{1/n^2} = \left(\varphi\left(n^k t\right)\right)^{1/n^{2k}}$$
$$= \left(\varphi\left(t\sqrt{N}\right)\right)^{1/N}, \quad N = (n)^{2k},\ k \geq 1. \tag{8.28}$$

Since the X's have zero mean and unit variance, writing $\varphi(t) = 1 - \frac{t^2}{2} + \eta(t)$, where $\eta(t)/t^2 \to 0$ as $t \to 0$, we have

$$\varphi(t) = \lim_{N\to\infty}\left(1 - \frac{t^2}{2N} + \eta\left(\frac{t}{\sqrt{N}}\right)\right)^N = e^{-\frac{1}{2}t^2}$$

Thus X's are normal. The proof is complete.

The result of the following theorem has lots of application in statistical inferences.

Theorem 8.2.12. Let $X_1, X_2, \ldots, X_n$ be a simple random sample from a normal population with pdf *f(x)*. Then the sample mean $\overline{X}\left(= \frac{1}{n}\sum_{k=1}^{n} X_k\right)$ and sample variance $S^2\left(= \frac{1}{n}\sum_{k=1}^{n}\left(X_k - \overline{X}\right)^2\right)$ are independent if and only if the distribution of the X's is normal.

Proof. The proof of necessity.

Suppose X_i, i = 1, 2, ..., *n*, is distributed as normal with mean = μ and variance = σ^2. The joint pdf of X_i, i = 1, 2, , ..., *n* is

$$f_X(x_1, x_2, \ldots, x_n) = \frac{1}{(2\pi)^{n/2}\sigma^n} e^{-\frac{1}{2}\sum_{I=1}^{n}\left(\frac{x_i-\mu}{\sigma}\right)^2}$$

Let us make the following transformation

$$Y_1 = \overline{X}$$
$$Y_2 = X_2 - \overline{X}$$
$$Y_3 = X_3 - \overline{X}$$
$$\ldots\ldots\ldots\ldots$$
$$Y_n = X_n - \overline{X}$$

The jacobian of the transformation is n. We have

$$\sum_{i=1}^{n}\left(\frac{X_i-\mu}{\sigma^2}\right)^2 = \frac{1}{\sigma^2}[(X_1-\overline{X})^2+\sum_{i=2}^{n}(X_i-\overline{X})^2+n(\overline{X}-\mu)^2]$$

$$= \frac{1}{\sigma^2}[(\sum_{i=2}^{n}(X_i-\overline{X})^2+\sum_{i=2}^{n}(X_i-\overline{X})^2+n(\overline{X}-\mu)^2]$$

$$= \frac{1}{\sigma^2}\left[\left(\sum_{i=2}^{n}Y_i\right)^2+\sum_{i=2}^{n}Y_i^2+nY_1^2\right]$$

The joint pdf of $Y_1 \ldots, Y_n$ is

$$f_Y(y_1, y_2, \ldots, y_n) = \frac{n}{(2\pi)^{n/2}\sigma^n}e^{-\frac{1}{2\sigma^2}[\sum_{i=2}^{n}y_i^2+(\sum_{i=2}^{n}y_i)^2+nY_1^2]},$$
$$-\infty < y_i < \infty, \ \mathrm{i} = 1, 2, \ldots, \mathrm{n}$$

Thus $Y_1(=\overline{X})$ is independent of $Y_2, \ldots, Y_n$.

Now

$$nS^2 = \sum_{k=1}^{n}(X_k-\overline{X})^2 = \sum_{k=2}^{n}Y_k^2+(\sum_{k=2}^{n}Y)^2.$$

Thus $\overline{X}$ and S^2 are independent.

The proof of the sufficiency.

Let the characteristic function, $\phi(t)$, of the distribution is given by $\phi(t) = \int e^{itx} f(x)dx$.

The joint characteristic function of the statistics $\overline{X}$ and S^2 is given by

$$\phi(t_1, t_2) = \int \cdots \int, e^{it_1\overline{x}+it_2s^2} f(x_1)\ldots f(x_n)dx_1\ldots dx_n$$

The characteristic function of the mean $\overline{X}$ is

$$\phi_1(t_1) = \phi(t_1, 0) = \int \cdots \int, e^{it_1\overline{x}} f(x_1)\ldots f(x_n)dx_1\ldots dx_n$$

and the characteristic function of the variance S^2 is given by

$$\phi_2(t_2) = \phi(0, t_2) = \int \cdots \int, e^{it_2s^2} f(x_1)\ldots f(x_n).$$

The independence of the distribution of $\overline{X}$ and S^2 means in terms of the characteristic function

$$\phi(t_1, t_2) = \phi_1(t_1)\phi_2(t_2).$$

Using the fact $\overline{X} = \frac{1}{n}\sum_{j=1}^{n} X_j$, we can write

$$\phi_1(t_1) = \prod_{k=1}^{n} \int e^{it_1 x_k/n} f(x_k)dx_k = (\phi(t_1/n))^n$$

We have

$$\begin{aligned}
&\frac{d}{dt_2}\phi(t_1, t_2) \mid_{t_2=0} \\
&= \int \cdots \int is^2 e^{it_1x_1/n} \ldots e^{it_1x_n/n} f(x_1)\cdots f(x_n)dx_1\cdots dx_n \\
&= i(\phi(t_1/n))^{n-1}E(s^2) \\
&= \frac{n-1}{n} i\sigma^2(\phi(t_1/n))^{n-1},
\end{aligned}$$

where σ^2 is the variance of the X's.

Using the relation $\frac{d}{dt_2}\phi(t_1, t_2)\big|_{t_2=0} = \phi_1(t_1)\frac{d}{dt_2}\phi_2(t_2)\big|_{t_2=0,}$ we obtain

$$\begin{aligned}
&\textstyle\int\cdots\int is^2 e^{it\bar{x}} f(x_1)\ldots f(x_n)\,dx_1\ldots dx_n \\
&= \tfrac{n-1}{n} i\sigma^2 \textstyle\int\cdots\int e^{it\bar{x}} f(x_1)\ldots f(x_n)\,dx_1\ldots dx_n
\end{aligned}$$

Now $s^2 = \frac{1}{n}\sum\limits_{i=1}^{n}(x_1 - \bar{x})^2 = \frac{1}{n}[\frac{n-1}{n}\sum\limits_{i=1}^{n} x_i^2 - \frac{1}{n}\sum\limits_{i,j=1,i\neq j}^{n} x_i x_j.$

We have

$$\frac{d}{dt}\phi(t) = \phi'(t) = i\int xe^{itx} f(x)\,dx,$$

and

$$\frac{d^2}{dt^2}\phi(t) = \phi''(t) = -\int x^2 e^{itx} f(x)\,dx,$$

Thus we can write the following differential equation

$$\phi''(t)(\phi(t))^{n-1} - \left(\phi'(t)\right)^2(\phi(t))^{n-2} = \sigma^2\phi(t)^n$$

Since $\phi(0) = 1$ and ϕ(t) is continuous, there exist a neighborhood around zero where ϕ(t) is not zero. Now restricting t in that region, we can write

$$\phi''(t) - (\phi'(t))^2(\phi(t))^{-1} = \sigma^2\phi(t)$$

i.e

$$\frac{\phi''(t)}{\phi(t)} - \frac{(\phi'(t))^2}{(\phi(t))^2} = -\sigma^2$$

We can write the above equation as

$$\frac{d^2}{dt^2}\ln\phi(t) = -\sigma^2.$$

The solution of the above equation is $\phi(t) = e^{at+bt^2}$, where a and b are constants.
Using the condition $E(X) = \mu$ and Var(X) $= \sigma^2$, we have

$$\phi(t) = e^{i\mu t - \frac{1}{2}\sigma^2 t^2}.$$

Thus X is normally distributed.

We have presented here the proof of the sufficiency given by Lukacs (1942). Geary (1934) was the first to prove the sufficiency of the theorem.

We know that if X_1 and X_2 are distributed as N(0, 1), then the ratio X_1/X_2 is distributed as Cauchy with median zero (C(0)). The following example shows that the converse is not true.

Example 8.2.2. Consider the following density function.

$$f(x) = \frac{2^{1/2}}{\pi(1+x^4)}, -\infty < x < \infty.$$

Then X_1/X_2 is C(0).

It is natural to ask that if X_1 and X_2 are *i.i.d* and X_1/X_2 is distributed as C(0), what additional condition will guarantee the normality of X_1 and X_2. The following theorem (Ahsanullah and Hamedani (1988) gives an answer to the question.

Theorem 8.2.13. Let X and Y be independent and identically distributed random variables with absolutely continuous (with respect to Lebesgue measure) distribution function and let Z = min (X,Y). If Z^2 and X/Y are distributed as chi-square with one degree of freedom and C(0) respectively, then X and Y are distributed as standard normal.

Proof. Let f(x) be the pdf of X. Since X/Y is distributed as C(0), we have

$$\int_{-\infty}^{\infty} f(uv)f(v)|v|dv = \frac{1}{\pi(1+u^2)}, -\infty < u < \infty.$$

Or

$$\int_0^{\infty} [f(uv)f(v) + f(-uv)f(-v)]vdu = \frac{1}{\pi(1+u^2)}, -\infty < u < \infty..$$

Now letting $u \to 1$ and $u \to -1$, we obtain respectively

$$\int_0^\infty [(f(v))^2 + (f(-v))^2]vdu = \frac{1}{2\pi} \tag{8.29}$$

and

$$\int_0^\infty 2f(v)f(-v)vdu = \frac{1}{2\pi} \tag{8.30}$$

From (8.29) and (8.30) it follows that

$$\int_0^\infty [(f(v) - f(-v))^2]vdu = 0 \tag{8.31}$$

Hence f(v) = f(−v) for almost all v, v< ∞. Thus f(x) is symmetric and hence the conclusion follows from Theorem 8.2.8.

8.3 Summary

A probability distribution can be characterized through various methods. The purpose of this chapter was to discuss various characterizations of normal distribution. It is hoped that the findings of the chapter will be useful for researchers in different fields of applied sciences.

Chapter 9
Characterizations of Student's *t* Distribution

9.1 Introduction

As pointed out in Chap. 8, characterization of a probability distribution plays an important role in probability and statistics. In this chapter, we will present some characterizations of Student's t distribution.

9.2 Characterizations of Student's *t* Distribution

Student's *t* distributions have been widely used in both the theoretical and applied work in statistics. In this section, we will consider some characterizations of Student's t distribution.

The probability density function of the t-distribution with v degrees of freedom is given by

$$f_v(x) = \frac{\Gamma\left(\frac{v+1}{2}\right)}{\sqrt{v\pi}\,\Gamma\left(\frac{v}{2}\right)} \left(1 + \frac{x^2}{v}\right)^{-(v+1)/2}, -\infty < x < \infty. \tag{9.1}$$

We will denote Student's t distribution with v degrees of freedom as t_v distribution. In this chapter We will present some characterizations of Student's t-distribution.

For $v = 1$, the pdf of t_1 is given by

$$f_1(x) = \frac{1}{\pi(1 + x^2)}, -\infty < x < \infty.$$

This is a the standard Cauchy (C(0)) distribution. The following are some of the characteristic properties of standard Cauchy distribution.
If X is distributed as C(0), then

M. Ahsanullah et al., *Normal and Student's t Distributions and Their Applications*, Atlantis Studies in Probability and Statistics 4, DOI: 10.2991/978-94-6239-061-4_9,

(i) 1/X is also distributed as C(0).
(ii) $2X/(1-X^2)$ is also distributed as C(0),

However identical distribution of X and 1/X does not characterize the Cauchy distribution (C(0)). The identical distribution of X and $2X/(1-X^2)$ Characterizes the distribution C(0). If X_1 and X_2 are two independent and identically distributed continuous random variables, then any one of following two conditions (see Arnold, (1979)) characterize the standard Cauchy distribution.

(i) X_1 and $(X_1+X_2)/(1-X_1X_2)$ are independent,
(ii) X_1 and $(X_1+X_2)/(1-X_1X_2)$ are identically distributed,

The pdf of t_2 distribution is as given below:

$$f_2(x) = \frac{1}{(2+x^2)^{\frac{3}{2}}}, \tag{9.2}$$

and the corresponding cdf is

$$F_2(x) = \frac{1}{2}\left(1 + \frac{x}{\sqrt{2+x^2}}\right) \tag{9.3}$$

Several characterizations of t_2 distribution based on regression properties of order statistics were obtained in Akundov et al. (2004); Balakrishnan and Akundov (2003); Nevzorov et al. (2003).

It is interesting to note that t_2 distribution is a member of a general family of distribution satisfying the relation

$$(F(x))(1-F(x))^{\alpha} = cf(x), \tag{9.4}$$

where c is a constant. The logistic distribution satisfies the relation with $\alpha = 1$, the uniform distribution satisfies the relation with $\alpha = 0$ and the squared sign distribution with cdf $F(x) = \sin^2(x)$, $0 \le x \le \pi/2$ satisfies the relation with $\alpha = 1/2$. Here we will present a characterization of t_2 distribution as given by Nevzorov et al. (2003) in Theorem 9.1 satisfying the relation (9.4) with $\alpha = 3/2$.

Theorem 9.2.1. Suppose that X_1, X_2 and X_3 are independent and identically distributed with cdf F(x) and pdf f(x). Let $X_{1,2} \le X_{2,3} \le X_{3,3}$ be the corresponding order statistics. Further let $W_3 = (X_{1,3}+X_{3,3})/2$ and $M_3 = X_{2,3}$. Then $E(W_3|M_3 = x) = x$, where $\gamma < x < \delta$, $\gamma = \inf\{x|F(x) > 0\}$, $\delta = \sup\{x\}F(x) < 1\}$ if and only if

$$F(x) + F_2\left[\frac{x-\mu}{\sigma}\right], -\infty < x < \infty, \sigma < 0,$$

where $F_2(.)$ is the cdf as given in (9.3).

Proof.

$$E(W_3|M_3 = x) = x$$

$$\text{i.e. } E\left(\frac{X_{1,3} + X_{2,3}}{2} + \frac{X_{2,3}}{2}|X_{2,3} = x\right) = \frac{3x}{2}$$

and

$$E(\overline{X}|X_{2,3} = x) = x \qquad \text{or} \qquad E(X_1|X_{2,3} = x) = x$$

For t_2 distribution,

$$E(X_1|X_{2,3}) = \frac{1}{3}\left\{x + \frac{1}{F_2(x)}\int_{-\infty}^{0} uf(u)du + \frac{1}{1 - F_2(x)}\int_{0}^{\infty} uf(u)du\right\},$$

where $F_2(x) = \frac{1}{2}\left(1 + \frac{x}{\sqrt{2+x^2}}\right)$
We have

$$\frac{1}{F_2(x)}\int_{-\infty}^{x} uf_2(u)du = x - \frac{1}{F_2(x)}\int_{x}^{\infty} F_2(u)du,$$

$$\frac{1}{1 - F_2(x)}\int_{-\infty}^{x} uf_2(u)du = x + \frac{1}{1 - F_2(x)}\int_{x}^{\infty} F_2(u)du,$$

Thus

$$E(X_1|X_{2,3}) = x - \frac{1}{3}\left\{\frac{1}{F_2(x)}\int_{-\infty}^{0} F_2(u)du - \frac{1}{1 - F_2(x)}\int_{0}^{\infty}(1 - F_2(u))du\right\}.$$

But

$$-\frac{1}{F_2(x)}\int_{-\infty}^{x} F_2(u)du + \frac{1}{1 - F_2(x)}\int_{x}^{\infty}(1 - F_2(u))dy = 0$$

Hence for t_2 distribution

$$E(W_3|M_3 = x) = x.$$

We now prove the sufficiency condition.

$$E(W_3|M_3 = x) = x \text{ implies}$$

$$E(X_1|X_{2,3} = x) = x,$$

$$E(W_3|M_3 = x) = x \text{ implies}$$

$$\frac{1}{F(x)}\int_{-\infty}^{x} F(u)du - \frac{1}{1 - F(x)}\int_{x}^{\infty}(1 - F(u))du = 0 \tag{9.5}$$

The Eq. (9.5) is equivalent to

$$(1 - F(x)) \int_{-\infty}^{x} F(u)du - F(x) \int_{x}^{\infty} (1 - F(u))du = 0$$

or

$$\frac{d}{dx}\left[\int_{x}^{\infty} (1 - F(u))du \int_{-\infty}^{x} F(u)du\right] = 0.$$

Thus we have

$$\int_{-\infty}^{x} F(u)du \int_{x}^{\infty} (1 - F(u))du = c \tag{9.6}$$

where c is a constant. We can write the Eq. (9.6) as

$$\int_{x}^{\infty} (1 - F(u))du = \frac{c}{\int_{-\infty}^{x} c, F(u)du}$$

Differentiating the above equation with respect to x, we obtain

$$1 - F(u) = \frac{cF(x)}{\left(\int_{-\infty}^{x} F(u)du\right)^2},$$

which is equivalent to

$$\int_{-\infty}^{x} F(u)du = \left(\frac{cF(x)}{(1 - F(x))}\right)^{1/2.}$$

Differentiating the above equation with respect to x, we obtain

$$\{F(x)(1 - F(x))\}^{3/2} = cf(x), c > 0. \tag{9.7}$$

Let G(x) be the inverse function of F(x) such that F(G(x)) = x, 0 < x<1, we obtain from (9.7)

$$(u(1 - u))^{3/2} = cf(G(x)), 0 > x < 1. \tag{9.8}$$

Since $\frac{d}{dx}G(x) = \frac{1}{f(G(x))}$, the general solution of (9.8) is

$$G(x) = d + \int_{1/2}^{x} \frac{du}{(u(1 - u))^{3/2}}, \text{c > 0 and, } -\infty < \text{d} < \infty.$$

We have $G(0) = \gamma = -\infty$ and $G(1) = \delta = \infty$, hence

$$G(x) = \frac{2x-1}{2\sqrt{x(1-x)}}, 0 < x < 1 \tag{9.9}$$

The inverse of G(x) is

$$F(x) = \frac{1}{2}\left(1 + \frac{x}{\sqrt{1+x^2}}\right), -\infty < x < \infty. \tag{9.10}$$

So we have

$$\begin{aligned} F(x) &= F_0\left(\frac{x-\mu}{\sigma}\right) \\ &= \frac{1}{2}\left\{1 + \frac{x-\mu}{\sqrt{\{\sigma^2 + (x-\mu)^2\}}}\right\}, -\infty < x < \infty, \quad -\infty < \mu < \infty, \sigma > 0. \end{aligned}$$

We now consider t_3 distribution.
The pdf of t_3 distribution is given by

$$f_3(x) = \frac{2}{\pi\sqrt{3}\left(1+\frac{x^2}{3}\right)^2}, -\infty < x < \infty. \tag{9.11}$$

The following characterization Theorem (Theorem 9.2) is due to Akundov and Nevzorov (2010).

Theorem 9.2.2. Suppose that X_1, X_2 and X_3 are independent and identically distributed with cdf F(x) and pdf f(x). We assume that $E(X_1^2) < \infty$. We assume without any loss of generality E(X) = 0 and $E(X_1^2) = 1$. Let $X_{1,2} \le X_{2,3} \le X_{3,3}$ be the corresponding order statistics. Then the following conditions are equivalent.

$$\begin{aligned} &(a)\ E((X_{2,3} - X_{1,3})^2 | X_{2,3} = x) \\ &\qquad = E((X_{3,3} - X_{2,3})^2 | X_{2,3} = x) \quad a.s. \\ &(b)\ F(x) = F_3\left(\frac{x-\mu}{\varsigma}\right), -\infty < x < \infty, \sigma > 0, \end{aligned} \tag{9.12}$$

where $F_3(.)$ is the t_3 distribution with the pdf given in (9.11).

Proof. We can rewrite (9.12) as

$$\begin{aligned} &2xE(X_{3,3}|X_{2,3} = x) - 2xE(X_{1,3,}|X_{2,3} = x) \\ &\qquad = E(X_{3,3}^2|X_{2,3} = x) - E(X_{1,3}^2|X_{2,3} = x) \end{aligned} \tag{9.13}$$

We know that

$$E(X_{1:3}|X_{2:3} = x) = \frac{1}{F(x)}\int_{-\infty}^{x} t\,dF(t).$$

and

$$E(X_{3:3}|X_{2:3} = \mathrm{x}) = \frac{1}{1 - F(x)} \int_x^{\infty} t dF(t).$$

Using these results we obtain from (9.13) that

$$\frac{2x}{1 - F(x)} \int_x^{\infty} t dF(t) - \frac{2x}{F(x)} \int_{\infty}^{x} t dF(t)$$
$$= \frac{1}{1 - F(x)} \int_x^{\infty} t^2 dF(t) - \frac{1}{F(x)} \int_{\infty}^{x} t^2 dF(t)$$

If we define

$$I(x) = \int_{-\infty}^{x} t dF(t) \text{ and } R(x) = \frac{1}{1 - F(x)} \int_{-\infty}^{x} t^2 dF(t).$$

Then we have

$$\int_x^{\infty} t dF(t) = -I(x) \text{ and } \int_x^{\infty} t^2 dF(t) = 1 - R(x)$$

It follows immediately that

$$\int_t^{\infty} t dF(t) = I(x) \text{ and } \int_t^{\infty} t^2 dF(t) = 1 - R(x) \tag{9.14}$$

since E(X) = 0 and $E(\mathrm{X}^2) = 1$. Now we can write

$$2xI(x)\left(\frac{1}{1 - F(x)} + \frac{1}{F(x)}\right)$$
$$= \frac{1}{1 - F(x)} - R(x)\left(\frac{1}{1 - F(x)} + \frac{1}{F(x)}\right)$$

or

$$R(\mathrm{x}) = \mathrm{F(x)} + 2xl(x). \tag{9.15}$$

Differentiating (9.15) with respect to x, we obtain

$$x^2 f(x) = f(x) + 2I(x) + 2x^2 f(x)$$

or

$$-2l(x) = f(x) + x^2 f(x). \tag{9.16}$$

Since the left-hand side of (9.16) is differentiable it follows that f(x) is also differentiable and then differentiating (9.16) with respect to x, we obtain

$$-2xf(x) = f'(x) + 2xf(x) + x^2 f(x)$$

Or

$$\frac{f'(x)}{f(x)} = \frac{-4x}{1+x^2} \tag{9.17}$$

Upon solving the differential equation in (9.17), we arrive at

$$f(x) = \frac{2}{\pi\left(1+x^2\right)^2} \tag{9.18}$$

Note that if X has the probability density function given by (9.18), then the random variable
Y = $\sqrt{3}$ X has the Student's t_3 distribution as given by

$$f_3(x) = \frac{1}{\pi\sqrt{3}}\left(1+\frac{x^2}{3}\right)^{-2}$$

Since we restricted ourselves so far to E(x) = 0 and E(x^2) = 1, it is clear that considering now arbitrarily expected value (μ) and variance (σ^2) we arrive on the result of theorem that (a) $\rightarrow$ (b)

Next, looking at the steps above, it can be readily checked that (b) $\rightarrow$ (a). The proof of the theorem is complete.

Remark 9.2.1. The characterization result established in this theorem can equivalently be stated in terms of variances of the left-truncated and right-truncated random variables, i.e. the theorem holds if we replace its condition (a) by the condition

$$\begin{aligned}\text{Vac}(X_{1:3}X|_{2:3} = \text{x}) &- \text{Var}(X_{3:3}|X_{2:3} = x)\\ &= E(X_{3:3}X_{1:3}|X_{2:3} = \text{x})[E(X_{3:3} + X_{1:3}|X_{2:3} = \text{x}) - 2x].\end{aligned}$$

Let Q(x) be the quantile function of a random variable X with cdf F(x), i.e. F(Q(x) = x, $0 < x < 1$. Akundov et al. (2004) proved that for $0 < x < 1$, the relation

$$E(\lambda X_{2,3} + (1-\lambda)X_{3,3}|X_{2,3} = x) = x. \tag{9.19}$$

Characterizes a family of distributions with quantile function

$$Q_\lambda(x) = \frac{c(x-\lambda)}{\lambda(1-\lambda)(1-x)^\lambda x^{1-\lambda}} + d, 0 < x < 1, \tag{9.20}$$

where $0 < \text{c} < \infty$, $-\infty < \text{d} < \infty$. Let us call this family of distribution as Q family.

Theorem 9.2.3. (Q family) Assume that E|X| < ∞ and n ≥ 3 is a positive integer. The random variable X belongs to the Q family if for some k, 2 ≤ k ≤ n − 1 and λ, 0 < λ < 1,

$$\lambda E\left[\frac{1}{k-1}\sum_{i=1}^{k-1}(X_{k,n}-X_{i,n}|X_{k,n}=x\right]$$
$$=(1-\lambda)E\left[\frac{1}{n-k}\sum_{j=k+1}^{n}(X_{j,n}-X_{k,n}|X_{k,n}=x\right] \tag{9.21}$$

Proof. The Eq. (9.21) can be written as

$$\lambda E\left[\frac{1}{k-1}\sum_{i=1}^{k-1}X_{i,n}|X_{k,n}=x\right]$$
$$+(1-\lambda)E\left[\frac{1}{n-k}\sum_{j=k+1}^{n}(X_{j,n}|X_{k,n}=x\right]=x \tag{9.22}$$

Clearly for n = 3 and k = 2 Eq. (9.22) reduces to (9.20).
Q family with different values of λ approximates well member of common distribution including Tukey's Lambda, Cauchy and Gumbel (for maximum). t_2 distribution belongs to the Q family having quantile function (9.20) with

$$Q_{1/2}(x)=\frac{2^{1/2}(x-1/2)}{(x(1-x))^{1/2}},\ 0<x<1.$$

A generalization of Theorem 9.1 can easily be established by considering 2n+1 samples and using the condition $E(\overline{X}/M_{2n+1}=x)=x$, where $\overline{X}=\frac{1}{2n+1}\sum_{j=1}^{2n+1}X_j$ and $M_{2n+1}=[X_{1,2n+1}+X_{2n+1,2n+1}]/2$.
The following Theorem 9.2.4 (See Yanev and Ahsanullah (2012)) is a generalization of Theorem 9.2.3.

Theorem 9.2.4. Assume that the random variable X has cdf F(x) with $E(X^2)<\infty$. Let $v\geq 3$ and n ≥ 3 positive integers. Then

$$F(x)=Fv\left(\frac{x-\mu}{\sigma}\right).\text{ for }-\infty<\mu<\infty,\sigma>0. \tag{9.23}$$

where $F_v(.)$ is the t distribution with v degrees of freedom, if and only if

$$E\left[\frac{1}{k-1}\sum_{i=1}^{k}\left(\frac{v-1}{2}X_{k,n}-(v-2)X_{i,n}\right)^2|X_{k,n}=x\right]$$
$$=E\left[\frac{1}{n-k}\sum_{j=1}^{k}\left((v-2)X_{j,n})-\frac{v-1}{2}X_{k,n}\right)^2|X_{k,n}=x\right] \tag{9.24}$$

To prove the theorem, we need the following two lemmas.

Lemma 9.2.1. The cdf F(x) of a random variable X with quantile function (9.20) is the only continuous cdf solution of the equation

$$(F(x))^{\lambda-1}(1-F(x))^{1+\lambda}=cF'(x), c>0. \tag{9.25}$$

Lemma 9.2.2. Let $v \geq 1, n \geq 2$, integers. Then

$$\frac{1}{\mathrm{k}-1}\sum_{\mathrm{i}=1}^{\mathrm{k}-1}\mathrm{E}([\mathrm{X}_{\mathrm{i,n}}^{\mathrm{r}}|\mathrm{X}_{\mathrm{k,n}}=\mathrm{x}]=\frac{1}{\mathrm{F(x)}}\int_{-\infty}^{x}t^r dF(t), \quad 2\leq \mathrm{k}\leq \mathrm{n}, \tag{9.26}$$

and

$$\frac{1}{n-k}\sum_{j=1}^{k-1}E([X_{j,n}^{r}|X_{k,n}=x)=\frac{1}{1-F(x)}\int_{x}^{\infty}t^r dF(t), 1\leq k\leq n-1. \tag{9.27}$$

Proof. Using the formulas of the conditional density of $X_{j,n}$ given $X_{k,n}=x (j<k)$ (see Ahsanullah and Hamedani (2010), p.13, Ahsanullah and Nevzorov (2001), p.3 and Ahsanullah et al. (2013)), we obtain for r > 1

$$\frac{1}{k-1}\sum_{j=1}^{k-1}E(X_{j,n}^{r}|X_{k,n}=x$$
$$=\frac{1}{k-1}\frac{k-1}{(F(x))^{k-1}}\sum_{j=1}^{k-1}\binom{k-2}{j-1}\int_{-\infty}^{x}(F(t))^{j-1}(F(x)-F(t))^{k-1-j}t^r dF(t)$$
$$=\frac{1}{[F(s)]^{k-1}}\sum_{i=0}^{k-2}\binom{k-2}{i}\int_{-\infty}^{x}[F(t)]^{i}[F(x)-F(t)]^{k-2-i}t^r dF(t)$$
$$=\frac{1}{F(x)}\int_{-\infty}^{x}t^r dF(t)$$

This verifies the relation (9.26). The relation (9.27) can be proved similarly.

Proof of Theorem 9.2.4. First we prove (9.21) implies (9.20). Applying Lemma (9.2) for the left hand side of Eq. (9.22), we obtain

$$\lambda \sum_{j=1}^{k-1} E(X_{j,n}|X_{k,n} = x) + \frac{1-\lambda}{n-k} \sum_{j=k+1}^{n} E([X_{j,n}|X_{k,n} = x].$$

$$= \frac{\lambda}{F(x)} \int_{-\infty}^{x} tF(t) + \frac{1-\lambda}{1-F(x)} \int_{x}^{\infty} tF(t) \tag{9.28}$$

Since that $E(|X|) < \infty$, we have

$$\lim_{x \to -\infty} xF(x) = 0 \text{ and } \lim_{x \to \infty} x[1 - F(x)] = 0. \tag{9.29}$$

Therefore integrating by parts, we obtain from (9.28)

$$= \frac{\lambda}{F(x)} \int_{-\infty}^{x} tF(t) + \frac{1-\lambda}{1-F(x)} \int_{x}^{\infty} tF(t)$$

$$= x - \frac{\lambda}{F(x)} \int_{-\infty}^{x} F(t)dt + \frac{1-\lambda}{1-F(x)} \int_{x}^{\infty} (-F(t))dt \tag{9.30}$$

Thus, from Eqs. (9.28) and (9.30), it follows from Eq. (9.21) is equivalent to

$$\lambda(1 - F(x)) \int_{-\infty}^{x} F(t)dt = (1-\lambda)F(x) \int_{x}^{\infty} (1 - F(t))dt$$

The above equation can be written as

$$-\frac{\lambda}{1-\lambda} \int_{-\infty}^{x} F(t)dt \frac{d}{dx} \int_{x}^{\infty} (1 - F(t))dt = \int_{x}^{\infty} (1 - F(t))dt \frac{d}{dx} \int_{-\infty}^{x} F(t)dt$$

which leads to

$$\int_{-\infty}^{x} F(t)dt = c\left(\int_{x}^{\infty} (1 - F(t))dt\right)^{-\lambda/(1-\lambda)}, x > 0$$

Differentiating both sides of the above equation with respect to x, we obtain

$$\int_{x}^{\infty} (1 - F(t))dt) = c_1 \left(\frac{1}{F(x)} - 1\right)^{1-\lambda}, x > 0$$

Differentiating one more time, we have

$$(F(x))^{2-\lambda}(1 - F(x))^{1+\lambda} = c_2 F'(x), c_2 > 0, \tag{9.31}$$

which is Eq. (9.25). Referring to Lemma 9.1, we see that (9.21) implies (9.20).

To complete the proof of the theorem it remains to verify that F(x) with quantile function (9.20) satisfies Eq. (9.21). Differentiating Eq. (9.20) with respect to x, we obtain

$$Q'_\lambda(x) = c(1-x)^{-(1+\lambda)}x^{-(2+\lambda)}, c > 0,$$

On the other hand, since F(Q(x)) = x, we have $Q'(x) = F'(Q(x))^{-1}$.
Note that the left hand side is differentiable so is the right hand side. Therefore

$$(1-x)^{1-\lambda}x^{2-\lambda} = cF'(Q_\lambda(x))$$

which is equivalent to Eq. (9.30) and then to Eq. (9.21).

Proof of Theorem 9.2.4. The Eq. (9.21) can be written as

$$(v-1)x\left\{\frac{1}{n-k}\sum_{j=k+1}^{n} E(X_{j,n}|X_{k,n}-x) - \frac{1}{k-1}\sum_{j=1}^{k-1} E(X_{j.n}|X_{k,n}=x)\right\}$$
$$= (v-2)\left\{\frac{1}{n-k}\sum_{j=k+1}^{n} E(X_{j,n}^2|X_{k.n}-x) - \sum_{j=n}^{k-1} E(X_{j,n}^2|X_{k,n}=x)\right\}$$

Referring to Lemma 9.2, with r = 1 and r = 2, we see that this equivalent to

$$(v-1)x\left\{\frac{1}{F(x)}\int_x^\infty tdF(t) - \frac{1}{F(x)}\int_{-\infty}^x tdF(t)\right\}$$
$$= (v-2)\left\{\frac{1}{1-F(x)}\int_x^\infty t^2dF(t) - \frac{1}{F(x)}\int_{-\infty}^x t^2dF(t)\right\} \tag{9.32}$$

If E(X) = 0 and $E(X^2) = 1$. Then

$$\int_x^\infty dF(t) = -\int_{-x}^\infty tdF(t) \text{ and } \int_x^\infty t^2dF(t) = 1 - \int_{-\infty}^x t^2dF(t)$$

and the Eq. (9.32) is equivalent to

$$-(v-1)x\left\{\frac{1}{1-F(x)} + \frac{1}{F(x)}\right\}\int_{-\infty}^x tdF(t)$$
$$= \frac{v-2}{1-F(x)} - (v-2)\left\{\frac{1}{1-F(x)} + \frac{1}{F(x)}\right\}\int_{-\infty}^x t^2dF(t)$$

Multiplying both sides of the above equation by F(x)(1−F(x)), we obtain

$$-(v-1)x\int_{-\infty}^x tdF(t) = (v-2)[F(x)-1]\int_{-\infty}^x t^2dF(t)] \tag{9.33}$$

Differentiating both sides of the above equation, we have

$$-(v-1)\int_{-\infty}^{x} t dF(t) = f(x)(x^2+v-2)$$

Since the left hand side of the above equation is differentiable, we have $f'(x)$ exists. Differentiating with respect to x, we obtain

$$\frac{f'(x)}{f(x)} = -\frac{v+1}{v-2}\frac{x}{1+x^2/(v-2)}$$

Integrating both sides of the equation and making use of the fact that f(x) is a pdf, we have

$$f(x) = c\left(1+\frac{x^2}{v-2}\right)^{-(v-1)/2}, \text{ where } c = \frac{\Gamma((v+1)/2)}{\Gamma(v/2)\sqrt{(v-2)\pi}} \tag{9.34}$$

It is not difficult to see that if Z has the pdf as given in (9.34), then

$$X = Z\sqrt{\frac{v}{v-2}}$$

follows t_v distribution with pdf as given in (9.1).
The following theorem (Theorem 9.2.5) gives a characterization of folded t_3 distribution based on truncated first moment.

Theorem 9.2.5. Suppose that the random variable X has an absolutely continuous (with respect to Lebesgue measure) cdf F(x), with $f(x) = \frac{2}{\pi\sqrt{3}}\left(1+\frac{x^2}{3}\right)^{-2}$ (folded t_3 distribution) if and only if $E(X|X \le x) = g(x)\tau(x)$, where $g(x) = \frac{x^2(x^2+3)}{6}$ and $\tau(x) = \frac{f(x)}{F(x)}$ is the reversed hazard rate.

Proof. Suppose that $f(x) = \frac{2}{\pi\sqrt{3}}\left(1+\frac{x^2}{3}\right)^{-2}$, x ≥ 0, then

$$g(x) = \frac{\int_0^x x\left(1+\frac{t^2}{3}\right)dt}{\left(1+\frac{x^2}{3}\right)^{-2}} = \frac{1}{6}x^2(x^2+3)$$

Suppose $g(x) = \frac{1}{6}x^2(x^2+3)$,
Then $E(X|X \le x) = g(x)\tau(x)$ implies that

$$\int_0^x tf(t)dt = g(x)f(x) \tag{9.35}$$

Differentiating (9.35) and using $g(x) = \frac{1}{6}x^2(x^2+3)$, we obtain

$$xf(x) = \frac{1}{6}x^2(x^2+3)f'(x) + \frac{1}{6}x(4x^2+6)f(x) \tag{9.36}$$

On simplification we obtain from (9.36)

$$\frac{f'(x)}{f(x)} = -\frac{4x}{x^2+3} \tag{9.37}$$

On integrating (9.37) and using the fact f(x) is a pdf, we have

$$f(x) = \frac{2}{\pi\sqrt{3}}(1+\frac{x^2}{3})^{-2}, x \geq 0,$$

9.3 Summary

In this chapter, some characterizations of Student's t distribution have been discussed. It is hoped that the findings of this chapter will be useful for the practitioners in various fields of studies and further enhancement of research in the field of distribution theory and its applications.

Chapter 10
Concluding Remarks and Some Future Research

The normal and Student's t distributions are two of the most important distributions in statistics. This book has reviewed the normal and Student's t distributions, and their applications. The sum, product and ratio for the normal distributions, and the sum, product and ratio for the Student's t distributions have been discussed extensively. Their properties and possible applications are discussed. Some special cases for each of the chapters are given. The distributions of the sum, product, and ratio of independent random variables belonging to different families are also of considerable importance and one of the current areas of research interest. This book introduces and develops some new results on the distributions of the sum of the normal and Student's t random variables. Some properties are discussed. Further, a new symmetric distribution has been derived by taking the product of the probability density functions of the normal and Student's t distributions. It is observed that the new distribution is symmetric and carries most of the properties of symmetric distributions. Some characteristics of the new distributions are presented. The entropy expression has been given. The percentage points have also been provided. It is shown that the pdf of the proposed distribution is unimodal. It is noted that for the degrees of freedom $\nu = 1$, we obtain a new symmetric distribution which is the product of the standard normal and Cauchy distributions and for large ν we will obtain a distribution which is the product of two standard normal densities. The percentage points of the new distributions have also been provided. The characterizations of normal and Student's t are given. We hope the findings of the book will be useful for the practitioners in various fields of sciences. Finally, the given references of the book will be a valuable asset for those researchers who want to do research in these areas.

The purpose of this book was to provide the distribution of the sums, differences, products and ratios of independent (uncorrelated) normal and student t random variables. Most of the recent works are reviewed. It appears that not much attention has been paid to the distribution of the sums, differences, products and ratios of dependent (correlated) student t random variables, and therefore needs further research investigation. It is also evident that very little attention has been paid to the estimates of parameters, inferential and prediction properties based the distribution of the sums, differences, products and ratios of dependent (correlated) normal and

M. Ahsanullah et al., *Normal and Student's t Distributions and Their Applications*, Atlantis Studies in Probability and Statistics 4, DOI: 10.2991/978-94-6239-061-4_10,

student's t random variables. Also, one can investigate the distribution of order statistics and record values based on the distribution of the sums, differences, products and ratios of dependent (correlated) normal and student's t random variables. The inferential properties and prediction of future order statistics and record values based on existing ones from the distribution of order statistics and record values based on the distribution of the sums, differences, products and ratios of dependent (correlated) normal and student t random variables are also open problems. We hope that the materials of this book will be useful for the practitioners in various fields of studies and further enhancement of research on the distribution of the sums, differences, products and ratios (quotients) of random variables and their applications.

Finally, as stated above, this book primarily provided the theoretical contributions of normal and student's t distributions and their possible applications. However, it appears that not much attention has been paid to the estimates of parameters and inferential properties based on the distribution of the sums, differences, products and ratios of dependent (correlated) normal and student's t random variables. Therefore, we hope that, using a real world data, one can pursue further research based on the results provided in this book, specially, parameter estimates, inferences about the parameters, goodness-of-fit, and prediction of the future observations for these distributions are open problems.

References

Abramowitz, M., & Stegun, I. A. (1970). *Handbook of mathematical functions, with formulas, graphs, and mathematical tables*. New York: Dover.

Abu-Salih, M. S. (1983). Distributions of the product and the quotient of power-function random variables. *Arab Journal of Mathematics*, *4*, 77–90.

Agrawal, M. K., & Elmaghraby, S. K. (2001). On computing the distribution function of the sum of independent random variables. *Computers and Operations Research*, *28*, 473–483.

Ahsanullah, M. (1987). A note on the characterization of the normal distribution. *Biometrical Journal*, *29*(7), 885–888.

Ahsanullah, M. (1989a). Characterizations of the normal distribution by properties of linear functions and chi-squared. *Pakistan Journal of Statistics*, *5*(3), 13, 267–276.

Ahsanullah, M. (1989b). On characterizations of the normal law. *Bulletin of the Institute of Mathematics Academia Sinica*, *17*(2), 229–232.

Ahsanullah, M. (1990). Some characterizations properties of normal distribution. *Computational Statistics and Data Analysis*, *10*, 117–120.

Ahsanullah, M. (2011). *Handbook of univariate statistical distributions*. MI, USA: Trafford Publishers.

Ahsanullah, M., Bansal, N., & Hamedani, G. G. (1991). On characterization of normal distribution. *Calcutta Statistical Assciation Bulletin*, *41*(161–164), 157–162.

Ahsanullah, M., & Hamedani, G. G. (1988). Some characterizations of normal distribution. *Calcutta Statistical Association Bulletin*, *37*, 145–146.

Ahsanullah, M., & Hamedani, G. G. (2010). *Exponential distribution*. New York: Nova Science Publishers.

Ahsanulllah, M., & Kirmani, S. N. U. A. (2008). *Topics in extreme value distributions*. New York, USA: Nova Science Publishers.

Ahsanullah, M., & Nevzorov, M. (2001). *Ordered random variables*. New York: Nova Science Publishers.

Ahsanullah, M., & Nevzorov, V. B. (2005). *Order statistics :Examples and exercises*. New York: Nova Publishers.

Ahsanullah, M., Nevzorov, V. B., & Shakil, M. (2013). *An introduction to order statistics*. Paris, France: Atlantis Press.

Aigner, D. J., Lovell, C. A. K., & Schmidt, P. (1977). Formulation and estimation of stochastic frontier production function models. *Journal of Econometrics*, *6*, 21–37.

Airy, G. B. (1861). *On the algebraical and numerical theory of errors of observations and the combination of observations*. London: Macmillan.

M. Ahsanullah et al., *Normal and Student's t Distributions and Their Applications*, Atlantis Studies in Probability and Statistics 4, DOI: 10.2991/978-94-6239-061-4,

Akundov, I. S., Balakrishnan, N., & Nevzorov, V. B. (2004). New characterizations by properties of midrange and related statistics. *Communications in Statistics-Theory and Methods*, *33*(12), 3133–3143.

Akundov, I. S., & Nevzorov, V. B. (2010). A simple characterization of Student t_2 distribution. *Statistics and Probability Letters*, *80*(5–6), 293–295.

Ali, M. M. (1982). Distribution of linear combinations of exponential variates. *Communications in Statistics-Theory and Methods*, *11*, 1453–1463.

Ali, M. M., & Nadarajah, S. (2004). On the product and the ratio of *t* and logistic random variables. *Calcutta Statistical Association Bulletin*, *55*, 1–14.

Ali, M. M., & Nadarajah, S. (2005). On the product and ratio of *t* and Laplace random variables. *Pakistan Journal of Statistics*, *21*(1), 1–14.

Altman, D. G. (1993). Construction of age-related reference centiles using absolute residuals. *Statistics in Medicine*, *12*(10), 917–924.

Amari, S. V., & Misra, R. B. (1997). Closed-form expressions for distribution of sum of exponential random variables. *IEEE Transactions on Reliability*, *46*, 519–522.

Amoroso, L. (1925). Richerche intorno alla curve die redditi. *Annali di Matematica Pura ed Applicata*, *21*, 123–159.

Andersen, P. K., Borgan, O., Gill, R. D., & Keiding, N. (1993). *Statistical models based on counting processes*. New York: Springer.

Arellano-Valle, R. B., & Azzalini, A. (2006). On the unification of families of skew-normal distributions. *Scandinavian Journal of Statistics*, *33*, 561–574.

Arellano-Valle, R. B., Del Pino, G., & San Martin, E. (2002). Definition and probabilistic properties of skew-distributions. *Statistics and Probability Letters*, *58*(2), 111–121.

Arellano-Valle, R. B., G6mez, H. W., & Quintana, F. A. (2004). A new class of skew-normal distributions. *Communications in Statistics-Theory and Methods*, *33*(7), 1465–1480.

Arnold, B. C. (1979). Some characterizations of the Cauchy distribution. *Australian Journal of Statistics*, *21*, 166–169.

Arnold, B. C., & Lin, G. D. (2004). Characterizations of the skew-normal and generalized chi distributions. *Sankhyā: The Indian Journal of Statistics*, *66*(4), 593–606.

Azzalini, A. (1985). A class of distributions which includes the normal ones. *Scandinavian Journal of Statistics*, *12*, 171–178.

Azzalini, A. (1986). Further results on a class of distributions which includes the normal ones. *Statistica*, *46*, 199–208.

Azzalini, A. (2001). A note on regions of given probability of the skew-normal distribution. *Metron*, *LIX*, 27–34.

Azzalini, A. (2005). The skew-normal distribution and related multivariate families. *Scandinavian Journal of Statistics*, *32*, 159–188.

Azzalini, A. (2006). Some recent developments in the theory of distributions and their applications. Atti della XLIII Riunione della SocietÃ Italiana di Statistica, volume Riunioni plenarie e specializzate, pp. 51–64.

Azzalini, A., & Capitanio, A. (1999). Statistical applications of the multivariate skew-normal distribution. *Journal of Royal Statistical Society B*, *61*, 579–602.

Azzalini, A., & Dalla_Valle, A. (1996). The multivariate skew-normal distribution. *Biometrika*, *83*, 715–726.

Azzalini, A., & Regoli, G. (2012). Some properties of skew-symmetric distributions. *Annals of the Institute of Statistical Mathematics*, *64*, 857–879.

Babbitt, G. A., Kiltie, R., & Bolker, B. (2006). Are fluctuating asymmetry studies adequately sampled? Implications of a new model for size distribution. *The American Naturalist*, *167*(2), 230–245.

Bagui, S., & Bagui, S. (2006). Computing percentiles of Skew normal distributions. *Journal of Modern Applied Statistical Methods*, *5*(2), 283–595.

Balakrishnan, N., & Nevzorov, V. B. (2003). *A primer on statistical distributions*. New Jersey: Wiley.

Barndorff-Nielsen, O. (1977). Exponentially decreasing log-size distributions. *Proceedings of the Royal Society of London Series A, 353*, 401–419.

Barndorff-Nielsen, O. (1978). Hyperbolic distributions and distributions on hyperbolae. *Scandinavian Journal of Statistics, 5*, 151–159.

Barranco-Chamorro, I., Moreno-Rebollo, J. L., Pascual-Acosta, A., & Enguix-Gonzalez, A. (2007). An overview of asymptotic properties of estimators in truncated distributions. *Communications in Statistics-Theory and Methods, 36*(13), 2351–2366.

Basu, D. (1951). On the independence of linear functions of independent chance variables. *Bulletin of the International Statistics Institute, 39*, 83–96.

Basu, A. K., & Ahsanullah, M. (1983). On decomposition of normal and Poisson law. *Calcutta Statistical Association Bulletin, 32*, 103–109.

Basu, A. P., & Lochner, R. H. (1971). On the distribution of the having generalized life distributions. *Technometrics, 13*, 281–287.

Behrerens, W. V. (1929). Ein Beitrag zur Fehlerberechnung bei wenigen Beobachtungen. *Landw Jb, 68*, 807–837.

Bessel, F. W. (1818). Astronomiae pro anno MDCCLV deducta ex observationibus viri incomparabilis James Bradley specula Grenovicensi per annos 1750–1762.

Bessel, F. W. (1838). Untersuchungen uber der Wahrscheinlichkeit der Beobachtungsfehler. *Astronomische Nachrichten, 15*, 368–404.

Bhargava, R. P., & Khatri, C. G. (1981). The distribution of product of independent beta random variables with application to multivariate analysis. *Annals of the Institute of Statistical Mathematics, 33*, 287–296.

Birnbaum, Z. W. (1950). Effect of linear tuncation on a multinormal population. *The Annals of Mathematical Statistics, 21*, 272–279.

Bland, J. M. (2005). The Half-Normal distribution method for measurement error: two case studies. Unpublished talk available on http://www-users.york.ac.uk/~mb55/talks/halfnor.pdf

Bland, J. M., & Altman, D. G. (1999). Measuring agreement in method comparison studies. *Statistical Methods in Medical Research, 8*, 135–160.

Blatberg, R. C., & Gonedes, N. J. (1974). A comparison of the stable and student distributions as statistical models for stock prices. *Journal of Business, 47*, 224–280.

Blischke, W. R., & Murthy, D. N. P. (2000). *Reliability, modeling, prediction, and optimization.* New York: Wiley.

Bose, R. C., & Roy, S. N. (1938). The distribution of the D^2 studentized-statistic. *Sankhya, 4*, 19–38.

Box, J. F. (1981). Gosset, Fisher and t distribution. *The American Statistician, 35*, 61–67.

Bravais, A. (1846). *Analyse mathematique sur les probabilities des erreurs de situation d'un point, Memores des Savans Etrangers* (Vol. 9, pp. 255–332). Paris: Academie (Royale) des Sciences del'Institut de France.

Brody, J. P., Williams, B. A., Wold, B. J., & Quake, S. R. (2002). Significance and statistical errors in the analysis of DNA microarray data. *Proceedings of the National Academy of Sciences USA, 99*(20), 12975–12978.

Buccianti, A. (2005). Meaning of the parameter of skew-normal and log-skew normal distributions in fluid geochemistry. *CODAWORK'05, Girona, Spain* October 19–21, 2005.

Buckland, S. T., Anderson, D. R., Burnham, K. P., & Laake, J. L. (1993). *Distance sampling: Estimating abundance of biological populations.* London: Chapman and Hall.

Cambanis, S., Huang, S., & Simons, G. (1981). On the theory of elliptically contoured distributions. *Journal of Multivariate Analysis, 11*, 368–385.

Casella, G., & Berger, R. L. (2002). *Statistical inference*. California: Thomson Learning.

Cedilnik, A., Kosmelj, K., & Blejee, A. (2004). The distribution of the ratio of jointly normal variables. *Metodoloskizvezki, 1*(1), 99–108.

Chakraborty, S., & Hazarika, P. J. (2011). A survey on the theoretical developments in univariate skew normal distributions. *Assam Statistical Review, 25*(1), 41–63.

Chapman, D. G. (1950). Some Two Sample Tests. *Annals of Mathematical Statistics, 21*, 601–606.

Chapman, D. G. (1956). Estimating the parameters of a truncated Gamma distribution. *The Annals of Mathematical Statistics*, *27*(2), 498–506.

Charlier, C. V. L. (1905). Uber die Darstellung willkurlicher Funktionen. *Arkiv for Matematik, Astronomi och Fysik*, *2*(20), 1–35.

Chen, Y. Y., & Wang, H. J. (2004). A method of moments estimator for a stochastic frontier model with errors in variables. *Economics Letters*, *85*(2), 221–228.

Chiogna, M. (1998). Some results on the scalar skew-normal distribution. *Journal of the Italian Statistical Society*, *7*(1), 1–13.

Chou, C. Y., & Liu, H. R. (1998). Properties of the half-normal distribution and its application to quality control. *Journal of Industrial Technology*, *14*(3), 4–7.

Cigizoglu, H. K., & Bayazit, M. (2000). A generalized seasonal model for flow duration curve. *Hydrological Processes*, *14*, 1053–1067.

Coffey, C. S., Kairalla, J. A., & Muller, K. E. (2007). Practical methods for bounding type I error rate with an internal pilot design. *Communications in Statistics-Theory and Methods*, *36*(11), 2143–2157.

Conover, W. J. (1999). *Practical nonparametric statistics*. New York: Wiley.

Cohen, A. C. (1959). Simplified estimators for the normal distribution when samples are singly censored or truncated. *Technometrics*, *1*, 217–237.

Cohen, A. (1991). *Truncated and censored samples*. New York: Marcel Dekker.

Cooray, K., & Ananda, M. M. (2008). A generalization of the half-normal distribution with applications to lifetime data. *Communications in Statistics-Theory and Methods*, *37*(9), 1323–1337.

Copson, E. T. (1957). *An introduction to the theory of functions of a complex variable*. Oxford, U.K.: Oxford University Press.

Cramer, H. (1936). Uber cine eigensehaft der mornalen verteilungs function. *Mathematische Zeitschriff Chribt*, *11*, 405–411.

Curtiss, J. H. (1941). On the distribution of the quotient of two chance variables. *The Annals of Mathematical Statistics*, *12*(4), 409–421.

D'Agostino, R. B. (1971). An omnibus test of normality for moderate and large size samples. *Biometrika*, *58*, 341–348.

Dalla Valle, A. (2004). "The skew-normal distribution". In M. G. Genton (Ed.), *Skew-elliptical distributions and their applications: A journey beyond normality* (pp. 3–24). Boca Raton, FL: Chapman & Hall.

Daniel, C. (1959). Use of half-normal plots in interpreting factorial two-level experiments. *Technometrics*, *1*(4), 311–341.

Darmois D. (1951). Sur diverse propriete caracteristique de la loi de probabilite de Laplace- Gauss. *Bulletin of the Institute of Economics and Statistics Institute*, *13*, 79–82.

David, H. A., & Nagaraja, H. N. (2003). *Order statistics* (3rd ed.). Hoboken: Wiley.

Dobzhansky, T., & Wright, S. (1947). Genetics of natural populations. XV. Rate of diffusion of a mutant gene through a population of *Drosophila pseudoobscura*. *Genetics*, *32*(3), 303.

Donahue, J. D. (1964). *Products and quotients of random variables and their applications*. Denver, Colorado: The Martin Company.

Dreiera, I., & Kotz S. (2002). A note on the characteristic function of the *t*-distribution. *Statistics and Probability Letters*, *57*, 221–224.

Dudewicz, E. J., & Mishra, S. N. (1988). *Modern mathematical statistics*. New York: Wiley.

Edgeworth, F. Y. (1892). Correlated averages. *Philosophical Magazine Series*, *5*(34), 190–204.

Edgeworth, F. Y. (1883). The law of error. *Philosophical Magazine*, *16*, 300–309.

Edgeworth, F. Y. (1905). The law of error. *Proceedings of the Cambridge Philosophical Society*, *20*, 36–65.

Eisenhart, C. (1979). On the transition from "Student's" z to "Student's" t. *The American Statistician*, *33*, 6–11.

Elandt, R. C. (1961). The folded normal distribution: Two methods of estimating parameters from moments. *Technometrics*, *3*(4), 551–562.

Eling, M. (2011). Fitting insurance claims to skewed distributions: Are the Skew-normal and Skew-student good models? Working papers on risk management and insurance, No. 98-November, 2011, Institute of Insurance Economics, University of St. Gallen, St. Gallen, Switzerland.

Embrechts, P., Klüppelberg, C., & Mikosch, T. (1997). *Modelling extremal events: For insurance and finance*. Berlin: Springer.

Epstein, B. (1948). Some applications of the Mellin transform in statistics. *Annals of Mathematics Statistics, 19*, 370–379.

Evans, M., Hastings, N., & Peacock, B. (2000). *Statistical distributions* (3rd ed.). New York: Wiley.

Fang, K. T., & Anderson, T. W. (1990). *Statistical inference in elliptically contoured and related distributions*. New York: Allerton Press.

Farebrother, R. W. (1984). Algorithm AS 204: The distribution of a positive linear combination of chi-squared random variables. *Applied Statistics, 33*, 332–339.

Feller, W. (1968). *An introduction to probability theory and applications* (Vol. 1). New York: Wiley.

Feller, W. (1971). *An introduction to probability theory and applications* (Vol. 2). New York: Wiley.

Fernandes, E., Pacheco, A., & Penha-Gonçalves, C. (2007). Mapping of quantitative trait loci using the skew-normal distribution. *Journal of Zhejiang University Science B, 8*(11), 792–801.

Fieller, E. C. (1932). The distribution of the index in a normal bivariate population. *Biometrika, 24*(3/4), 428–440.

Fisher, R. A. (1925). Applications of student's *t* distribution. *Metron, 5*, 90–104.

Fisher, R. A. (1930). The moments of the distribution for normal samples of measures of departure from normality. *Proceedings of the Royal Society of London, A130*, 16–28.

Fisher, R. A. (1931). The truncated normal distribution. British Association for the Advancement of Science. *Mathematical Tables, 5*, xxxiii–xxxiv.

Fisher, R. A. (1935). The fiducial argument in statistical inference. *Annals of Eugenics, 6*, 391–398.

Fisher, R. A. (1939). The comparison of samples with possibly unequal variances. *Annals of Eugenics, 9*, 174–180.

Fisher, R. A. (1941). The asymptotic approach to Behrens' integral, with further tables for the d test of significance. *Annals of Eugenics, 11*, 141–173.

Fisher, R. A., & Healy, M. J. R. (1956). New tables of Behrens' test of significance. *Journal of the Royal Statistical Society Series B, 18*(2), 212–216.

Fisher, R. A., & Yates, F. (1957). *Statistical tables for biological agricultural and medical research*. Edinburgh: Oliver and Boyd.

Folks, J. L., & Chhikara, R. S. (1978). The inverse Gaussian distribution and its statistical application-a review. *Journal of the Royal Statistical Society Series B (Methodological), 40*, 263–289.

Fotowicz, P. (2006). An analytical method for calculating a coverage interval. *Metrologia, 43*, 42–45.

Frisch, U., & Sornette, D. (1997). Extreme deviations and applications. *Journal of Physics I France, 7*, 1155–1171.

Galambos, J., & Simonelli, I. (2005). *Products of random variables—applications to problems of physics and to arithmetical functions*. Boca Raton, Atlanta: CRC Press.

Galton, F. (1875). Statistics by inter comparison, with remarks on the law of frequency of error. *Philosophical Magazine, 49*, 33–46.

Galton, F. (1889). *Natural inheritance*. London: Macmillan.

Gauss, C. F. (1809). Theoria Motus Corporum Coelestium, Lib. 2, Sec. III, 205–224. Hamburg.: Perthes u. Besser.

Geary, R. C. (1930). The frequency distribution of the quotient of two normal variates. *Journal of the Royal Statistical Society, 93*(3), 442–446.

Geary, R. C. (1934). Distribution of student's ratio for non normal distribution. *Journal of Royal Statistitcs Society Series B, 3*, 178–184.

Genton, M. G. (Ed.). (2004). *Skew-elliptical distributions and their applications: A journey beyond normality*. Boca Raton, FL: Chapman & Hall.

Ghosh, B. K. (1975). On the distribution of the difference of two t-Variables. *Journal of the American Statistical Association*, *70*(350), 463–467.
Glen, A. G., Leemis, L. M., & Drew, J. H. (2004). Computing the distribution of the product of two continuous random variables. *Computational Statistics and Data Analysis*, *44*(3), 451–464.
Goldar, B., & Misra, S. (2001). Valuation of environmental goods: Correcting for bias in contingent valuation studies based on willingness-to-accept. *American Journal of Agricultural Economics*, *83*(1), 150–156.
Good, I. J. (1953). The population frequencies of species and the estimation of population parameters. *Biometrika*, *40*, 237–260.
Gosset, W. S. ("Student") (1908). The probable error of a mean. *Biometrika*, *6*, 1–25.
Gradshteyn, I. S., & Ryzhik, I. M. (2000). *Table of integrals, series, and products* (6th ed.). San Diego: Academic Press.
Greene, W. H. (2005). Censored data and truncated distributions. NYU Working Paper No. 2451/26101, Available at SSRN (Social Science Research Network) 825845.
Grubel, H. G. (1968). Internationally diversified portfolios: Welfare gains capital flows. *American Economic Review*, *58*, 1299–1314.
Gupta, R. C., & Gupta, R. D. (2004). Generalized skew normal model. *Test*, *13*(2), 501–524.
Gupta, A. K., Chang, F. C., & Huang, W. J. (2002). Some skew-symmetric models. *Random Operators Stochastic Equations*, *10*, 133–140.
Gupta, A. K., & Chen, J. T. (2004). A class of multivariate skew-normal models. *Annals of the Institute of Statistical Mathematics*, *56*(2), 305–315.
Gupta, S. C., & Kapoor, V. K. (2002). *Fundamentals of mathematical statistics*. New Delhi: Sultan Chand & Sons.
Gupta, A. K., Nguyen, T. T., & Sanqui, J. A. T. (2004). Characterization of the skew-normal distribution. *Annals of the Institute of Statistical Mathematics*, *56*(2), 351–360.
Haberle, J. G. (1991). Strength and failure mechanisms of unidirectional carbon fibre-reinforced plastics under axial compression, Unpublished Ph.D. thesis. London, U.K.: Imperial College.
Hald, A. (1952). *Statistical theory with engineering applications*. New York: Wiley.
Harter, H. L. (1951). On the distribution of Wald's classification statistic. *Annals of Mathematical Statistics*, *22*, 58–67.
Hausman, J. A., & Wise, D. A. (1977). Social experimentation, truncated distributions, and efficient estimation. *Econometrica: Journal of the Econometric Society*, *45*(4), 919–938.
Hawkins, D. I., & Han, C. P. (1986). Bivariate distributions noncentral chi-square random variables. *Communications in Statistics-Theory and Methods*, *15*, 261–277.
Hayya, J., Armstrong, D., & Gressis, N. (1975). A note on the ratio of two normally distributed variables. *Management Science*, *21*(11), 1338–1341.
Helmert, F. R. (1876). Die Genauigkeit der Formel von Peters zur Berechnung des wahrscheinlichen Beobachtungsfehlers director Beobachtungen tungen gleicher Genauigkeit. *Astronomische Nachrichten*, *88*, 113–120.
Henze, N. (1986). A probabilistic representation of the 'skew-normal' distribution. *Scandinavian Journal of Statistics*, *13*, 271–275.
Hinkley, D. V. (1969). On the ratio of two correlated normal random variables. *Biometrika*, *56*(3), 635–639.
Hitezenko, P. (1998). A note on a distribution of weighted sums of iid Rayleigh random variables. *Sankhya, A*, *60*, 171–175.
Hogg, R. V., McKean, J., & Craig, A. T. (2005). *Introduction to mathematical statistics*. New Jersey: Pearson Prentice Hall.
Hogg, R. V., & Tanis, E. A. (2006). *Probability and statistical inference*. New Jersey: Pearson Prentice Hall.
Holton, G. A. (2003). *Value-at-risk: Theory and practice*. San Diego, CA: Academic Press.
Hu, C.-Y., & Lin, G. D. (2001). An inequality for the weighted sums of pairwise i.i.d. generalized Rayleigh random variables. *Journal of Statistical Planning and Inference*, *92*, 1–5.

Hurst, S. (1995). The characteristic function of the Student t-distribution. Financial Mathematics Research Report 006–95. Australia: Australian National University, Canberra ACT 0200.

Ifram, A. F. (1970). On the characteristic functions of the F and t-distributions. *Sankhya, A32*, 350–352.

Jamalizadeh, A., Behboodian, J., & Balakrishnan, N. (2008). A two-parameter generalized skew-normal distribution. *Statistics and Probability Letters*, *78*(13), 1722–1726.

James, G. S. (1959). The Behrens-Fisher distribution and weighted means. *Journal of the Royal Statistics Society B*, *21*, 73–90.

Jawitz, J. W. (2004). Moments of truncated continuous univariate distributions. *Advances in Water Resources*, *27*, 269–281.

Joarder, A. H. (1998). Some useful Wishart expectations based on the multivariate t-model. *Statistical Papers*, *39*(2), 223–229.

Joarder, A. H., & Ali, M. M. (1996). On the characteristic function of the multivariate t distribution. *Pakistan Journal of Statistics*, *12*, 55–62.

Joarder, A. H., & Ali, M. M. (1997). Estimation of the scale matrix of a multivariate t-model under entropy loss. *Metrika*, *46*(1), 21–32.

Joarder, A. H., & Singh, S. (1997). Estimation of the trace of the scale matrix of a multivariate t-model using regression type estimator. *Statistics*, *29*, 161–168.

Johnson, A. C. (2001). *On the truncated normal distribution: characteristics of singly- and doubly-truncated populations of application in management science, PhD. Dissertation*. Illinois.: Stuart Graduate School of Business, Illinois Institute of Technology.

Johnson, N. L., Kotz, L., & Balakrishnian, N. (1994). *Continuous univariate distributions* (2nd ed., Vol. 1). New York: Wiley.

Johnson, N. L., Kotz, L., & Balakrishnan, N. (1995). *Continuous univariate distributions* (2nd ed., Vol. 2). New York: Wiley.

Johnstone, I., & Silverman, B. (2004). Needles and hay in haystacks: Empirical Bayes estimates of possibly sparse sequences. *Annals of Statistics*, *32*, 1594–1649.

Johnstone, I., & Silverman, B. (2005). Empirical Bayes selection of wavelet thresholds. *Annals of Statistics*, *33*, 1700–1752.

Jones, D. S. (1979). *Elementary information theory*. Oxford: Clarendon Press.

Jørgensen, B. (1982). *Statistical properties of the generalized inverse gaussian distribution*. New York: Springer.

Kagan, A. M., Linnik, Yu V, & Rao, C. R. (1973). *Characterization problems in mathematical statistics*. New York: Wiley.

Kagan, A. M., & Zinger, A. R. (1971). Sample mean as an alternative of location parameter: Case of non quadratic loss function. *Sankhya Series A*, *33*, 351–358.

Kamgar-Parsi, B., & Brosh, M. (1995). Distribution and moments of weighted sum of uniform random variables with applications in reducing Monte Carlo simulations. *Journal of Statistical Computation and Simulation*, *52*, 399–414.

Kamerud, D. (1978). The random variable X/Y, X, Y normal. *The American Monthly*, *85*(3), 206–207.

Kapadia, A. S., Chan, W., & Moyé, L. A. (2005). *Mathematical statistics with applications* (Vol. 176). New York.: Chapman & Hall.

Kapur, J. N. (1993). *Maximum-entropy models in science and engineering*. New Delhi: Wiley-Eastern.

Kappenman, R. F. (1971). A note on the multivariate t ratio distribution. *Annals of Mathematical Statistics*, *42*, 349–351.

Kazemi, M. R., Haghbin, H., & Behboodian, J. (2011). Another generalization of the Skew normal distribution. *World Applied Sciences Journal*, *12*(7), 1034–1039.

Keceioglu, D. (1991). *Reliability engineering handbook* (Vol. 1). Englewood Cliffs, NJ:Prentice Hall.

Kelejian, H. H., & Prucha, I. R. (1985). Independent or uncorrelated disturbances in linear regression. *Economic Letters*, *19*, 35–38.

Kelker, D. (1970). Distribution theory of spherical distributions and locations scale parameters. *Sankhya, A*, *32*, 419–430.

Kendall, D. (1938). Effect of radiation damping and doppler broadening on the atomic absorption coefficient. *Z. Astrophysics*, *16*, 308–317.

Kendall, M. G. (1952). *The advanced theory of statistics* (5th ed., Vol. 1). London: Charles Griffin and Company.

Kendall, M. G., & Stuart, A. (1958). *The advanced theory of statistics, 1*. New York: Hafner Pub. Co.

Kibria, B. M. G. (1996). On shrinkage ridge regression estimators for restricted linear models with multivariate t disturbances. *Students*, *1*(3), 177–188.

Kibria, B. M. G., & Haq, M. S. (1998). Marginal likelihood and prediction for the multivariate ARMA(1,1) linear models with multivariate t error distribution. *Journal of Statistical Research.*, *32*(1), 71–80.

Kibria, B. M. G., & Haq, M. S. (1999). The multivariate linear model with multivariate t and intra-class covariance structure. *Statistical Paper*, *40*, 263–276.

Kibria, B. M. G., & Nadarajah, S. (2007). Reliability modeling: Linear combination and ratio of exponential and Rayleigh. *IEEE Transactions on Reliability*, *56*(1), 102–105.

Kibria, B. M. G., Saleh, A. K. Md. E. (2003). Preliminary test ridge regression estimators with student's t errors and conflicting test-statistics. *Metrika*, *59*, 105–124.

Kim, H. J. (2006). On the ratio of two folded normal distributions. *Communications in Statistics-Theory and Methods*, *35*(6), 965–977.

Kimber, A. C., & Jeynes, C. (1987). An application of the truncated two-piece normal distribution to the measurement of depths of arsenic implants in silicon. *Applied statistics*, 352–357.

Klugman, S., Panjer, H., & Willmot, G. (1998). *Loss models: From data to decisions*. New York: Wiley.

Korhonen, P. J., & Narula, S. C. (1989). The probability distribution of the ratio of the absolute values of two normal variables. *Journal of Statistical Computation and Simulation*, *33*, 173–182.

Kotlarski, I. (1960). On random variables whose quotient follows the Cauchy law. *Colloquium Mathematicum*, *7*, 277–284.

Kotz, S., & Nadarajah, S. (2004). *Multivariate t Distributions and their Applications*. New York: Cambridge University Press.

Krzanowski, W. J., & Marriot, F. H. C. (1994). *Multivariate Analysis*. London: Arnold.

Ladekarl, M., Jensen, V., & Nielsen, B. (1997). Total number of cancer cell nuclei and mitoses in breast tumors estimated by the optical disector. *Analytical and Quantitative Cytology and Histology*, *19*, 329–337.

Laha, R. G. (1957). On a characterization of the normal distribution from properties of suitable linear statistics. *Ann. Math. Statistics*, *28*, 126–139.

Laplace, P. S. (1812). *Theorie Analytique des Probabilites* (Vol. 7). Paris: Oeuvres Completes.

Larson, R. J., & Marx, M. L. (2006). *An introduction to mathematical statistics and its applications*. New Jersey: Pearson Prentice Hall.

Lawless, J. F. (2003). *Statistical models and methods for lifetime data* (2nd ed.). New York: Wiley.

Lazo, A. C. G. V., & Rathie, P. N. (1978). On the entropy of continuous probability distributions. *IEEE Transactions on Information Theory*, *IT–24*, 120–122.

Lebedev, N. N. (1972). *Special functions and their applications*. New York: Dover.

Lee, R. Y., Holland, B. S., & Flueck, J. A. (1979). Distribution of a ratio of correlated gamma random variables. *SIAM Journal on Applied Mathematics*, *36*, 304–320.

Legendre, A. M. (1805). Nouvelles Methodes pour la Determination des Orbites des Cometes, Paris.

Leone, F. C., Nelson, L. S., & Nottingham, R. B. (1961). The folded normal distribution. *Technometrics*, *3*(4), 543–550.

Levy, H. (1982). Stochastic dominance rules for truncated normal distributions: A note. *The Journal of Finance*, *37*(5), 1299–1303.

Linnik, Yu V. (1964). *Decomposition of probability distributions*. London: Oliver & Boyd.

Lomnicki, Z. A. (1967). On the distribution of products of random variables. *Journal of the Royal Statistical Society. Series B*, *29*(3), 513–524.

Lukas, E. (1942). A characterization of the normal distribution. *Annals of Mathematical Statistics*, *13*, 91–93.

Lukacs, E. (1972). *Probability and mathematical statistical statistics—an introduction*. New York: Academic Press.

Lukacs, E., & Laha, R. G. (1964). *Applications of characteristic functions*. New York: Hefner Publishing Company.

Lyapunov, A. M. (1901). Nouvelle formed u theoreme sur la limite de probabilite, Memores de l'Academie Imperiale des Sciences de St. *Petersbourg*, *12*, 1–24.

Maksay, S., & Stoica, D. M. (2006). Considerations on modeling some distribution laws. *Applied Mathematics and Computation*, *175*(1), 238–246.

Malik, H. J., & Trudel, R. (1986). Probability density function of the product and quotient of two correlated exponential random variables. *Canadian Mathematical Bulletin*, *29*, 413–418.

Mari, D. D., & Kotz, S. (2001). *Correlation and dependence*. London: Imperial College Press.

Markov, A. A. (1899–1900). The law of large numbers and the method of least squares. *Izvestia Physiko-mathematicheskago Obschvestva pri Imperatorskom Kazanskom Universitet*, *8*, 110–128.

Marsaglia, G. (1965). Ratios of normal variables and ratios of sums of uniform variables. *Journal of the American Statistical Association*, *60*, 193–204.

Mateu-Figueras, G., Pedro, P., & Pewsey, A. (2007). Goodness-of-fit tests for the Skew-normal distribution when the parameters are estimated from the data. *Communications in Statistics: Theory and Methods*, *36*(9), 1735–1755.

McLaughlin, M. P. (1999). *A compendium of common probability distributions*. http://www.ub.edu/stat/docencia/Diplomatura/Compendium.pdf

Meeusen, W., & van den Broeck, J. (1977). Efficiency estimation from Cobb-Douglas production functions with composed error. *International Economic Review*, *8*, 435–444.

Mitra, S. S. (1978). *Recursive formula for the characteristic function of Student t- distribution for odd degrees of freedom, Manuscript*. University Park, PA: Pennsylvania State University.

Mood, A. M., Graybill, F. A., & Boes, D. C. (1974). *Introduction to the theory of statistics*. New York: McGraw-Hill Book Co.

Moivre, A. de (1733). Approximatio ad Summam Ferminorum Binomii in Seriem expansi, Supplementum II to Miscellanae. *Analytica*, 1–7.

Moivre, A. de (1738). *The doctrine of chances*. London: Fank Cass & Co (Reprint 1967).

Moix, P.-Y. (2001). *The measurement of market risk: Modelling of risk factors, asset pricing, and approximation of portfolio distributions*. New York: Springer.

Monti, A. C. (2003). A note on the estimation of the skew normal and the skew exponential power distributions. *Metron*, *61*(2), 205–219.

Moschopoulos, P. G. (1985). The distribution of the sum of independent gamma random variables. *Annals of the Institute of Statistical Mathematics*, *37*, 541–544.

Mukhopadhyay, N. (2006). *Introductory statistical inference*. London: Chapman & Hall.

Mukhopadhyay, S., & Vidakovic, B. (1995). Efficiency of linear Bayes rules for a normal meanskewed priors class. *The Statistician*, *44*, 389–397.

Nadarajah, S. (2005a). A generalized normal distribution. *Journal of Applied Statistics*, *32*(7), 685–694.

Nadarajah, S. (2005b). Products, and ratios for a bivariate gamma distribution. *Applied Mathematics and Computation*, *171*(1), 581–595.

Nadarajah, S. (2005c). On the product XY for some elliptically symmetric distributions. *Statistics and Probability Letters*, *72*, 67–75.

Nadarajah, S. (2005d). On the product and ratio of laplace and bessel random variables. *Journal of Applied Mathematics*, *4*, 393–402.

Nadarajah, S. (2005e). Linear combination, product and ratio of normal and logistic random variables. *Kybernetika*, *41*(6), 787–798.

Nadarajah, S. (2006a). Exact calculation of the coverage interval for the convolution of two student's t distributions. *Metrologia, 43*, L21–L22.

Nadarajah, S. (2006b). On the ratio X/Y for some elliptically symmetric distributions. *Journal of Multivariate Analysis, 97*(2), 342–358.

Nadarajah, S., & Ali, M. M. (2004). On the product of Laplace and Bessel random variables. *Journal of the Korean Data & Information Science Society, 15*(4), 1011–1017.

Nadarajah, S., & Ali, M. M. (2005). On the product and ratio of t and laplace random variables. *Pakistan Journal of Statistics, 21*, 1–1.

Nadarajah, S., & Dey, D. K. (2006). On the product and ratio of t random variables. *Applied Mathematics Letters, 19*, 45–55.

Nadarajah, S., & Gupta, A. K. (2005). On the product and ratio of bessel random variables. *International Journal of Mathematics and Mathematical Sciences, 18*, 2977–2989.

Nadarajah, S., & Gupta, A. K. (2006). On the ratio of logistic random variables. *Computational Statistics & Data Analysis, 50*(5), 1206–1219.

Nadarajah, S., & Kotz, S. (2003). Skewed distributions generated by the normal kernel. *Statistics and Probability Letters, 65*, 269–277.

Nadarajah, S., & Kotz, S. (2005a). On the product and ratio of Pearson type VII and laplace random variables. *Austrian Journal of Statistics, 34*(11), 11–23.

Nadarajah, S., & Kotz, S. (2005b). On the ratio of Pearson type VII and Bessel random variables. *Journal of Applied Mathematics and Decision Sciences, 9*(4), 191–199.

Nadarajah, S., & Kotz, S. (2005c). On the linear combination of exponential and gamma random variables. *Entropy, 7*(2), 161–171.

Nadarajah, S., & Kotz, S. (2006a). Skew distributions generated from different families. *Acta Applicandae Mathematica, 91*(1), 1–37.

Nadarajah, S., & Kotz, S. (2006b). On the product and ratio of gamma and Weibull random variables. *Econometric Theory, 22*, 338–344.

Nadarajah, S., & Kotz, S. (2006c). The linear combination of logistic and gumbel random variables. *Fundamenta Informaticae, 74*(2–3), 341–350.

Nadarajah, S., & Kotz, S. (2006d). On the product XY for the elliptically symmetric Pearson type VII distribution. *Mathematical Proceedings of the Royal Irish Academy, 106A*(2), 149–162.

Nadarajah, S., & Kotz, S. (2006e). The Bessel ratio distribution. *Comptes rendus de l'Académie des Sciences, Series I, 343*, 531–534.

Nadarajah, S., & Kotz, S. (2007). Programs in R for computing truncated t distributions. *Quality and Reliability Engineering International, 23*(2), 273–278.

Nason, G. (2001). Robust projection indices. *Journal Royal Statistical Society B, 63*, 551–567.

Nason, G. (2006). On the sum of t and Gaussian random variables. *Statistics & Probability Letters, 76*, 1280–1286.

Nevzorov, V. B. (2002). On a property of student t distribution with 2 degrees of freedom. Zap Nauchn Sem. POMI294, 148–157 (Russian).

Nevzorov, V. B., Balakrishnan, N, and Ahsanullah, M. (2003). Simple Characterizations. of Students $t_2$distribution. *Statistician, 52*(3), 395–400.

Nevzorov, L., Nevzorov, V.B. and Akundov, I. (2007). A simple characterization of Student's t_2 distribution. *Metron, LXV*(1), 53–57.

Novosyolov, A. (2006). The sum of dependent normal variables may be not normal. Krasnoyarsk, Russia: Institute of Computational Modelling, Siberian Branch of the Russian Academy of Sciences.

O'Hagan, A., & Leonard, T. (1976). Bayes estimation subject to uncertainty about parameter constraints. *Biometrika, 63*, 201–203.

Pastena, D. (1991). Note on a paper by Ifram. *Sankhya, A39*, 396–397.

Patel, J. K., Kapadia, C. H., & Owen, D. B. (1976). *Handbook of statistical distributions*. New York: Marcel Dekker Inc.

Patel, J. K., & Read, C. B. (1982). *Handbook of the normal distribution*. New York: Marcel Dekker Inc.

Patil, V. H. (1965). Approximation to the Behrens-Fisher distributions. *Biometrika, 52*, 267–271.

Pearson, E. S. (1967). Some reflexions on continuity in the development of mathematical statistics, 1885–1920. *Biometrika, 54*, 341–355.

Pearson, E. S. (1970). William Sealy Gosset, 1876–1937: "Student" as a statistician. In E. S. Pearson, & M. G. Kendall, (Eds.), *Studies in the history of statistics and probability* (pp. 360–403). New York: Hafner.

Pearson, K. (1896). Mathematical contributions to the theory of evolution, III: Regression, heredity, and panmixia. *Philosophical Transactions of the Royal Society of London, A187*, 253–318.

Pewsey, A. (2000). Problems of inference for Azzalini's skewnormal distribution. *Journal of Applied Statistics, 27*(7), 859–870.

Pewsey, A. (2002). Large-sample inference for the general half-normal distribution. *Communications in Statistics-Theory and Methods, 31*, 1045–1054.

Pewsey, A. (2004). Improved likelihood based inference for the general half-normal distribution. *Communications in Statistics-Theory and Methods, 33*, 197–204.

Pewsey, A. (2006). Modelling asymmetrically distributed circular data using the wrapped skew-normal distribution. *Environmental and Ecological Statistics, 13*(3), 257–269.

Pham-Gia, T. (2000). Distributions of the ratios of independent beta variables and applications. *Communications in Statistics-Theory and Methods, 29*, 2693–2715.

Pham-Gia, T. G., & Turkkan, N. (1994). Reliability of a standby system with beta-distributed component lives. *IEEE Transactions on Reliability, 43*, 71–75.

Pham-Gia, T., Turkkan, N., & Marchand, E. (2006). Density of the ratio of two normal random variables and applications. *Communications in Statistics-Theory and Methods, 35*(9), 1569–1591.

Podolski, H. (1972). The distribution of a product of n independent random variables with generalized gamma distribution. *Demonstratio Mathematica, 4*, 119–123.

Polya, G. (1923). Herleitung der Gausschen fehler genertzes aux einer functional gieichung. *Math-Zeit, 18*, 96–108.

Press, S. J. (1969). The t ratio distribution. *Journal of the American Statistical Association, 64*, 242–252.

Provost, S. B. (1989a). On sums of independent gamma random variables. *Statistics, 20*, 583–591.

Provost, S. B. (1989b). On the distribution of the ratio of powers of sums of gamma random variables. *Pakistan Journal Statistics, 5*, 157–174.

Prucha, I. R., & Kelejian, H. H. (1984). The structure of simultaneous equation estimators: A generalization towards nonnormal disturbances. *Econometrica, 52*, 721–736.

Prudnikov, A. P., Brychkov, Y. A., & Marichev, O. I. (1986). *Integrals and series (Vols. 1, 2, and 3)*. Amsterdam: Gordon and Breach Science Publishers.

Rathie, P. N., & Rohrer, H. G. (1987). The exact distribution of products of independent random variables. *Metron, 45*, 235–245.

Rényi, A. (1961). On measures of information and entropy. *Proceedings of the fourth Berkeley Symposium on Mathematics, Statistics and Probability, 1960*, 547–561.

Rider, P. R. (1965). Distributions of product and quotient of Cauchy variables. *The American Mathematical Monthly, 72*(3), 303–305.

Roberts, C. (1966). A correlation model useful in the study of twins. *Journal of the American Statistical Association, 61*, 1184–1190.

Rohatgi, V. K., & Saleh, A. K. M. E. (2001). *An introduction to probability and statistics*. New York: Wiley.

Rokeach, M., & Kliejunas, P. (1972). Behavior as a function of attitude-toward-object and attitude-toward-situation. *Journal of Personality and Social Psychology, 22*, 194–201.

Ruben, H. (1960). On the distribution of the weighted difference of two independent student variables. *Journal of the Royal Statistical Society, Series B, 22*(1), 188–194.

Sakamoto, H. (1943). On the distributions of the product and the quotient of the independent and uniformly distributed random variables. *Tohoku Mathematical Journal, 49*, 243–260.

Savchuk, V., & Tsokos, C. P. (2011). *Bayesian theory and methods with applications*. Paris, France: Atlantis Press.

Scheffe, H. (1970). Practical solutions of the Behrens-Fisher problem. *Journal of the American Statistical Association, 65*, 1501–1508.

Schneider, H. (1986). *Truncated and censored samples from normal populations*. New York: Marcel Dekker Inc.

Schrodinger, E. (1915). Zur Theorie der Fall-und Steigversuche an Teilchen mit Brownscher Bewegung. *Physikalische Zeitschrift, 16*, 289–295.

Severini, T. A. (2005). *Elements of distribution theory*. New York: Cambridge University Press.

Shah, S. M., & Jaiswal, M. C. (1966). Estimation of parameters of doubly truncated normal distribution from first four sample moments. *Annals of the Institute of Statistical Mathematics, 18*(1), 107–111.

Shakil, M., Kibria, B. M. G., & Singh, J. N. (2006). Distribution of the ratio of Maxwell and Rice random variables. *International Journal of Contemporary Mathematical Sciences, 1*(13), 623–637.

Shakil, M., & Kibria, B. M. G. (2006). Exact distribution of the ratio of gamma and rayleigh random variables. *Pakistan Journal of Statistics and Operations Research, 2*(2), 87–98.

Shakil, M., & Kibria, B. M. G. (2007). On the product of Maxwell and Rice random variables. *Journal of Modern Applied Statistical Methods, 6*(1), 212–218.

Shakil, M., & Kibria, B. M. G. (2009). A new student's t-type distribution and its properties. *Journal of Applied Statistical Science., 16*(3), 355–364.

Shakil, M., Kibria, B. M. G., & Chang, K.-C. (2008). Distributions of the product and ratio of Maxwell and Rayleigh random variables. *Statistical Papers, 49*, 729–747.

Shannon, C. E. (1948). A mathematical theory of communication. *Bell System Technical Journal, 27*, 379–432.

Shapiro, S. S., & Francia, R. S. (1972). An approximate analysis of variance test for normality. *Journal of the American Statistical Association, 67*, 215–216.

Shcolnick, S. M. (1985). On the ratio of independent stable random variables. *Lecture Notes in Mathematics* (Vol. 1155, pp. 349–354). Berlin, Germany: Springer.

Shkedy, Z., Aerts, M., & Callaert, H. (2006). The weight of Eurocoins: Its distribution might not be as normal as you would expect. *Journal of Statistics Education, 14*(2), 102–129.

Sichel, H. S. (1974). On a distribution representing sentence-length in written prose. *Journal of the Royal Statistical Society Series A, 137*, 25–34.

Sichel, H. S. (1975). On a distribution law for word frequencies. *Journal of the American Statistical Association, 70*, 542–547.

Sinha, S. K. (1983). Folded normal distribution: A Bayesian approach. *Journal of the Indian Statistical Association, 21*(1083), 31–34.

Song, K.-S. (2001). Rényi information, loglikelihood and an intrinsic distribution measure. *Journal of Statistical Planning and Inference, 93*(1–2), 51–69.

Sornette, D. (1998). Multiplicative processes and power laws. *Physical Review, E, 57*, 4811–4813.

Sornette, D. (2004). *Critical phenomena in natural sciences, chaos, fractals, self-organization and disorder: Concepts and tools* (2nd ed.). Heidelberg, Germany: Springer Series in Synergetics.

Springer, M. D. (1979). *The algebra of random variables*. New York: Wiley.

Springer, M. D., & Thompson, W. E. (1966). The distribution of product of independent random variables. *SIAM Journal of Applied Mathematics, 14*(3), 511–526.

Springer, M. D., & Thompson, W. E. (1970). The distribution of products of beta, gamma and Gaussian random variables. *SIAM Journal on Applied Mathematics, 18*, 721–737.

Srivastava, H. M., & Nadarajah, S. (2006). Some families of Bessel distributions and their applications. *Integral Transforms and Special Functions, 17*(1), 65–73.

Steece, B. M. (1976). On the exact distribution for the product of two independent beta-distributed random variables. *Metron, 34*, 187–190.

Stigler, S. M. (1999). *Statistics on the table*. Cambridge, MA: Harvard University Press.

Stuart, A. (1962). Gamma-distributed products of independent random variables. *Biometrika, 49*, 564–565.

Sugiura, N., & Gomi, A. (1985). Pearson diagrams for truncated normal and truncated Weibull distributions. *Biometrika, 72*(1), 219–222.

Suhir, E. (1997). *Applied probability for engineers and scientists.* New York: McGraw-Hill.

Sukhatme, P. V. (1938). On Fisher and Behrens test of significance for the difference in means of two normal samples. *Sankhya, 4*, 39–48.

Tang, J., & Gupta, A. K. (1984). On the distribution of the product of independent beta random variables. *Statistics & Probability Letters, 2*, 165–168.

Tchebyshev, P. L. (1890). Sur deux theorems relatifs aux probabilites. *Acta Mathematica, 14*, 305–315; Reprinted (1962) in Oeuvres, Vol. 2, Chelesa, New York.

Thomopoulos, N. T. (1980). *Applied forecasting methods.* Englewood Cliffs, New Jersey: Prentice-Hall.

Triola, M. F. (2005). *Biostatistics for the biological and health science.* New Jersey: Pearson.

Tsokos, C. P. (1972). *Probability distributions: An introduction to probability theory with applications.* California: Duxbury Press.

Tweedie, M. C. K. (1957a). Statistical properties of inverse Gaussian distribution. I. *Annals of Mathematical Statistics, 28*, 362–377.

Tweedie, M. C. K. (1957b). Statistical properties of inverse Gaussian distribution. II. *Annals of Mathematical Statistics, 28*, 696–705.

Vernic, R. (2006). Multivariate skew-normal distributions with applications in insurance. *Insurance: Mathematics and Economics, 38*, 413–426.

Walker, G. A., & Saw, J. G. (1978). The distribution of linear combinations of t-variables. *Journal of the American Statistical Association, 73*(364), 876–878.

Wallgren, C. M. (1980). The distribution of the product of two correlated variates. *Journal of the American Statistical Association, 75*, 996–1000.

Watson, G. N. (1962). *A treatise on the theory of Bessel functions* (2nd ed.). London: Cambridge University Press.

Weisstein, E. W. (2007). "Half-normal distribution". http://mathworld.wolfram.com

Weisstein, E. W. (2007). "Inverse Erf". From MathWorld-A Wolfram Web Resource. http://mathworld.wolfram.com/InverseErf.html

Whittaker, E. T., & Robinson, G. (1967). *The calculus of observations: A treatise on numerical mathematics* (4th ed.). New York: Dover.

Wiper, M. P., Giron, F. J., and Pewsey, A. (2005). Bayesian inference for the half-normal and half-t distributions. *Statistics and Econometrics Series 9*, Working Papers 05–47.

Witkovsky, V. (2001). Computing the distribution of a linear combination of inverted gamma variables. *Kybernetika, 37*, 79–90.

Yanev, G. P., & Ahsanullah, M. (2012). Characterizations of student's t-distribution via regression of order statistics. *Statistics, 46*(4), 425–429.

Zabell, S. L. (2008). On student's t 1908 article "The probable error of a mean". *Journal of the American Statistical Association, 103*(481), 1–7.

Zellner, A. (1976). Bayesian and non-Bayesian analysis of the regression model with multivariate student error terms. *Journal of the American Statistical Association, 71*, 400–405.

Zolotarev, V. M. (1957). Mellin-Stieltjes transforms in probability theory. *Teoriya Veroyat. Primen, 2*, 444–469.

